GEOENVIRONMENTAL ENGINEERING

GEOENVIRONMENTAL ENGINEERING

Principles and Applications

Lakshmi N. Reddi
Kansas State University
Manhattan, Kansas

Hilary I. Inyang
University of Massachusetts Lowell
Lowell, Massachusetts

Marcel Dekker, Inc. New York • Basel

ISBN: 0-8247-0045-7

This book is printed on acid-free paper.

Headquarters
Marcel Dekker, Inc.
270 Madison Avenue, New York, NY 10016
tel: 212-696-9000; fax: 212-685-4540

Eastern Hemisphere Distribution
Marcel Dekker AG
Hutgasse 4, Postfach 812, CH-4001 Basel, Switzerland
tel: 41-61-261-8482; fax: 41-61-261-8896

World Wide Web
http://www.dekker.com

The publisher offers discounts on this book when ordered in bulk quantities. For more information, write to Special Sales/Professional Marketing at the headquarters address above.

Current printing (last digit):
10 9 8 7 6 5 4 3 2 1

PRINTED IN THE UNITED STATES OF AMERICA

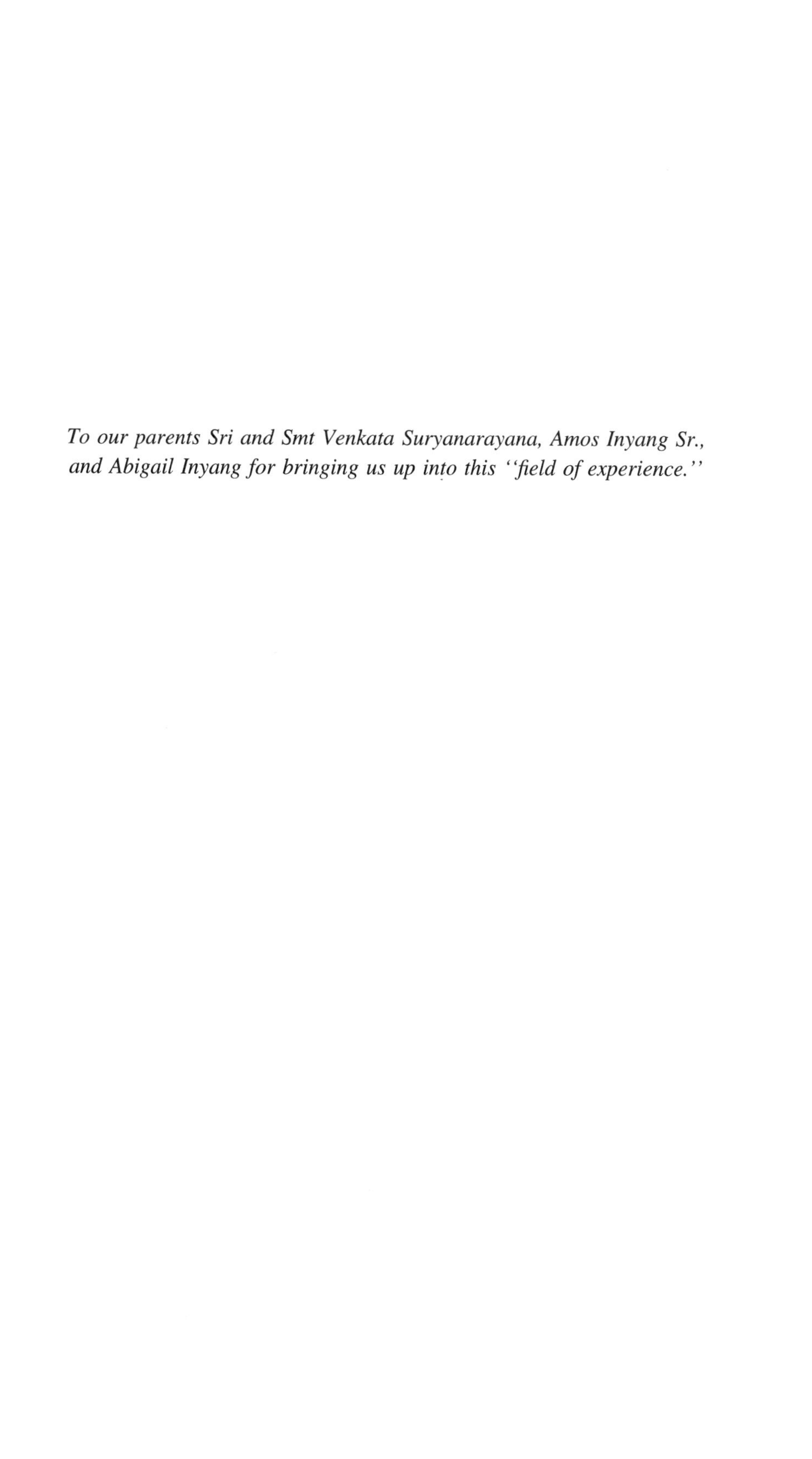

To our parents Sri and Smt Venkata Suryanarayana, Amos Inyang Sr., and Abigail Inyang for bringing us up into this "field of experience."

Preface

In the past few decades there has been immense growth in global population, industrial development, energy resource use, and civil infrastructure development. Growth in one sector often generates problems in other areas. Consequently, debates have intensified on industrialization and associated environmental issues such as waste generation, ecosystem and human health risk assessment, and waste management systems. An increasing number of research efforts are focusing on technologies for contaminated-site characterization, subsurface barriers for contaminants (waste containment), clean-up systems for contaminated ground, and assessment of the fate and transport of contaminants. As public agencies, private firms, global organizations, and academe embarked on projects aimed at seeking solutions to waste management and subsurface contamination problems, it became clear that the scientific and engineering issues involved are very diverse and require the adoption of cross-disciplinary and multidisciplinary approaches. Primarily, this need stemmed from the fact that none of the existing traditional disciplines, such as geotechnical engineering, geology, environmental engineering, water resource engineering, chemical engineering, biological sciences, and water resource management, adequately covers the principles that are essential to the assessment of contaminant generation, subsurface contamination, and development of relevant control systems. This was the setting for the birth of geoenvironmental engineering as an amalgam of principles drawn from a variety of engineering and applied science fields.

In a broad and most useful sense, geoenvironmental engineering must include all elements related to environmental problems of the ground surface and subsurface as well as relevant engineering control measures. We define it as a field that encompasses the application of science and engineering principles to the analysis of the fate of contaminants on and in the ground; transport of moisture, contaminant, and energy through geomedia; and design and implementation of schemes for treating, modifying, reusing, or containing wastes on and in the ground.

Based on current practice, the following general areas of professional activities are recognized. (The reader should note that these areas are categories and not listings of specific project titles or issues.)

Characterization of geomedia (soils, rocks, pore water, and pore gas) with respect to stability, contamination level, and fluid flow properties
Assessment of the response of terrains that host waste containment systems to natural and/or manmade hazards such as earthquakes, global warming, subsidence, and floods
Analysis of contaminant generation and migration through porous and fractured geomaterials and fabricated materials
Physicochemical, chemical, thermal, and biological treatment of wastes and contaminated geomaterials to reduce or eliminate pollutants
Design and analysis of surficial waste containment systems, such as landfills, monofills, slurry walls, grout curtains, and dewatering schemes, and deep disposal systems such as radioactive waste disposal chambers in rock

The need to address issues that pertain to the subdisciplines mentioned above has brought together scientists and engineers from diverse disciplines. In order to cover these subdisciplines, it is not uncommon to find geologists, agronomists, physicists, and chemists working side by side with geotechnical and environmental engineers in the consulting industry. Personnel from each of the major fields often bring different perspectives to the definition of issues and scope of projects in geoenvironmental engineering. For example, geotechnical engineers have focused mostly on site characterization and waste containment systems within a disciplinary framework that they often refer to as ''environmental geotechnics.'' This can be considered a subset of geoenvironmental engineering. Environmental engineers are active in site clean-up projects where they adapt their knowledge of water resource engineering and waste treatment technologies to treatment system design for contaminated soils. This has been a necessary expansion of the scope of environmental engineering beyond traditional coverage of wastewater and drinking water treatments, air pollution control, and surface/groundwater hydrology. Due to their deep coverage of stoichiometry, thermodynamics and kinetics of chemical reactions, and mass balances of reactants and products, chemical engineers and chemists have established a niche in geoenvironmental engineering in the area of contaminated soil and water treatment, although the basic principles they apply are much more suited to reactions among purer chemical substances than to soils, which are usually made up of grains of various sizes, mineralogies, and, often, uncertain chemical composition.

For decades, mining engineers, geochemists, and petroleum engineers have used chemical principles and mathematics to address the fate and transport of heavy metals in ores and tailings, petroleum in reservoir rock, and sands in aboveground heap-leaching projects. Some of these principles are being applied to the geoenvironmental engineering subfield of ''contaminant generation, fate, and transport modeling.'' Another major sector is bioremediation, including natural attenuation and phytoremediation, where the expertise of soil scientists and biologists, including microbiologists and plant physiologists, is truly needed.

On characterization of geomedia, the activities of geotechnical engineers at micro- and meso-scales are complemented by larger-scale investigations by geologists. Indeed, both professions intersect at the meso spatial scale. An example is site characterization and screening for siting of waste disposal facilities. Both geophysical techniques and laboratory-based sample characterization tests could be used. Advances in penetrometer and in-situ sensing technologies for contaminants have been gained through increased application of the principles of optics (a subfield of physics) and reagent chemistry at the micro-scale, coupled with mathematical techniques such as neural networks, fractals, geostatistics, and data inversion at the meso-scale. Indeed, for very long time scales (geologic time scales) that are considered in the design of deep disposal systems for high-level radioactive wastes, the potential impacts of relevant geologic and climatic processes such as seismic activity, subsidence, and global warming are appropriately treated by geologists and atmospheric scientists.

The need for interdisciplinarity in assessing and solving current and future geoenvironmental problems requires that students, program officers, researchers, and engineering project personnel understand and apply essential principles from the diverse set of disciplines discussed above. This is the central premise of this book, which is a synthesis of the most critical principles and their practical applications in geoenvironmental engineering.

The book is organized into three parts. Part I deals with the fundamental principles and processes related to soil, water, and chemical interactions. This is the prerequisite material to understand Parts II and III, which are devoted to applications in site and risk assessment, clean-up techniques, and waste containment, including the physico-chemical and biological principles on which they are based. We have taken a comprehensive approach in presenting relevant subject matter, and we have included the necessary science and engineering principles. The intent behind the sequence of topics in Part I is to provide the reader with a progressive understanding of the nature of soil as a porous media—its formation, composition, and structure, and its behavior in the presence of fluids. In dealing with the interactions between soils and fluids, we start with water in Chapter 3, take up the dissolved contaminants in Chapter 4, and discuss immiscible contaminants in Chapter 5. The fate and transport of the fluids make up an important theme dealt with in these three chapters.

Site remediation is the central theme of Part II. This is an evolving issue at present; however, all of the remediation "technologies" are, in one form or another, linked to the basic physico-chemical and/or biological processes. Chapters 6 and 7 lay the necessary foundations for site remediation choices by presenting clean-up criteria and the pathways for contaminant exposure. Chapter 8 outlines the basics of various remediation technologies.

Waste containment is the focus of Part III. The types of containment and considerations in containment site selection are introduced in Chapter 9, and the various containment system configurations are described in Chapter 10. Chapter 11 deals with the essentials of containment system design, and Chapter 12 concludes with a treatment of barrier composition and performance issues.

This book is intended to serve as a classroom instruction textbook at the graduate level and as a reference text for practicing engineers and scientists. It is assumed that students and other users have a background in undergraduate level soil mechanics and/or engineering geology; mathematics up to first-level calculus; and chemistry, waste management, and hydrogeology or groundwater hydrology. Most of these prerequisite courses are usually covered in undergraduate degree programs in civil engineering, environmental engineering, geology, soil science, and mining/petroleum engineering.

Obviously, the material covered in this book is too wide to be covered in a single geoenvironmental engineering course. For teachers at technical institutes and universities who wish to use this book as a class text, we propose the following graduate course titles as options and recommend the use of the chapters indicated for each course.

Course 1 Fluid Flow and Contaminant Interactions in Soils (Chapters 1, 2, 3, 4, and 5)
Course 2 Fundamentals of Contaminated Site Treatment Techniques (Chapters 1, 6, 7, and 8)
Course 3 Design and Analysis of Waste Containment Systems (Chapters 1, 4, 9, 10, 11, and 12)

This book incorporates the results of several research and analysis projects that have been implemented in this evolving field over several decades. We are grateful to scientists, engineers, and agencies that made contributions to advances in the issues discussed in this book. References have been made herein to techniques, diagrams, and tables developed by the U.S. Environmental Protection Agency, the U.S. Department of Agriculture, and several other agencies.

Although we have taken great pains to check the accuracy of the equations, charts, and tables included in this text, there may be some minor errors, either typographical or structural. We look forward to receiving feedback from readers in terms of critique, comments, identified errors, etc. We express our sincere gratitude to several individuals who have helped, in one form or another, to develop this book. Primary among them are Ming Xiao, Mohan Bonala, Jeremy Lin, Hui Wu, S. Lye, Abhijna Shukla, John Daniels, Diane Sparrow, Manav Shah, Jaydeep Parikh, and Vincent Ogunro.

Lakshmi N. Reddi
Hilary I. Inyang

Contents

GEOENVIRONMENTAL ENGINEERING

1
Soil Formation and Composition

1.1 INTRODUCTION

The origin and formation of mineral matter in nature continues to puzzle scientists. How and why the worlds were formed with the underlying universal laws of attraction and repulsion has always lured the intellect of mankind. The physical evolution of matter, which apparently begins in the mineral kingdom, proceeds with an unending impulse to plant, animal, and human kingdoms. The generalities in the formation of matter in the mineral kingdom may perhaps be extrapolated to the formation of subsequent kingdoms during the course of evolution. And perhaps the orbital movement of electrons around a neutron is a mere imitation of the orbital movement of planets around the sun, in accordance with the adage, ''As above, so below.'' Such generalities concerning cosmogenesis have never failed to inspire the intellect of mankind; they have caused various perspectives and scales of investigation.

Narrowing down to the formation and composition of our habitat—the earth—the scales of investigation are several. They range from studies of macroscopic geological processes such as plate tectonics, which are concerned with the earth's morphology, to microscopic mineralogical studies, which focus on the composition at a particle scale. An integrated knowledge of these rather vast scales of studies is neither practical nor needed. However, an appreciation of the various perspectives is essential to an engineer interested in interdisciplinary studies. The engineering behavior of a soil mass may often govern and be governed by the mineralogical processes occurring at the particle scale. This will be our guiding philosophy in approaching the subject of soil composition.

In our study of soil composition, we will start from general composition of soils in terms of soil profiles and gradually approach the particulars of mineral composition. We will therefore conveniently subdivide the problem, in sequential order of scale from macroscopic to microscopic, into the following categories:

1. Soil formation and macroscopic composition in terms of soil profiles
2. Phase composition in terms of solid, liquid, organic, and gas phases present in the soil mass
3. Solids composition and classification in terms of the relative sizes of particles
4. Mineral composition in terms of basic elements

1.2 SOIL FORMATION

Soils are formed by the disintegration (or more precisely, evolution) of rock material of the earth's relatively deeper crust, which itself is formed by the cooling of volcanic magma. The stability of crystalline structure governs the rock formation. As the temperature falls, new and often more stable minerals are formed. For instance, one of the most abundant minerals in soils known as quartz acquires a stable crystalline structure when the temperature drops below 573°C. The intermediate and less stable minerals (from which quartz has evolved) lend themselves to easy disintegration during the formation of soils.

The disintegration process of rocks leading to the formation of soils is called *weathering*. It is caused by natural agents, primarily wind and water (note that these are the same agents that aid the evolution and life in other kingdoms). The specific processes responsible for weathering of rocks are:

1. Erosion by the forces of wind, water, or glaciers, and alternate freezing and thawing of the rock material
2. Chemical processes, often triggered by the presence of water. These include: (a) hydrolysis (reaction between H^+ and OH^- ions of water and the ions of the rock minerals), (b) chelation (complexation and removal of metal ions), (c) cation exchange between the rock mineral surface and the surrounding medium, (d) oxidation and reduction reactions, and (e) carbonation of the mineral surface because of the presence of atmospheric CO_2.
3. Biological processes which, through the presence of organic compounds, affect the weathering process either directly or indirectly.

Once the rock material is weathered, the resultant soil may either remain in place or may be transported by the natural agencies of water, air, and glaciers. In the former case, the soils are called *residual* soils. Depending on the natural agent involved, the transported soils are called *alluvial* or *fluvial* (water-laid), *aeolian* (wind-laid), or *glacial* (ice-transported) soils. Several subdivisions are often made based on the transportation and deposition conditions. Prominent among these are:

1. *Braided stream deposits* formed along stream channels as a result of sedimentation from overloaded streams. The braided pattern consists of a stream being split into large number of small channels separated by islands.
2. *Meander belt deposits* formed due to the meandering of the stream, causing soil erosion on the concave side of the stream and deposition on the convex side of the stream.
3. *Lacustrine deposits* formed in lakes by sedimentation due to gravity.
4. *Marine deposits* formed by particle deposition in the seas.
5. *Glacial till* formed by the deposition of particles as glaciers melt, and *glacial-fluvial soils* formed by the stream channels (similar to braided stream deposits) created by the melting of the ice.
6. *Glacial lake deposits* formed when the glacial meltwater carrying small-sized particles form lakes and the particles settle in the lakes.
7. *Wind-blown sand* formed as dunes; and *loess* deposits formed when finer particles are wind-blown and occur in thick layers.
8. *Colluvial soils* formed at depressions at hillsides because of their movement downslope by gravity and the instability of the slope.

The properties of the soil deposits formed depend on the soil-forming factors. In general, five independent variables may be viewed as governing soil formation: (1) climate, (2) organisms present, (3) topography, (4) the nature of the parent material, and (5) time. It is generally established in the soil sciences literature that any property of soil is invariably linked to these five fundamental soil-forming factors (Jenny, 1941).

Soil formation due to weathering of rocks and subsequent transportation and deposition yields us the first scheme for soil composition. The deposition of soils occurs in layers and each layer possesses unique properties reflecting the parent material from which it arose. A host of environmental factors which were responsible for its formation, including climate, ground slope, and the presence of organic matter, are reflected in the properties of each of these layers. The composition of soil in terms of the various layers is usually illustrated in what is known as a *soil profile*. Each of the layers is called a *horizon*. The horizons are designated by the capital letters O, C, A, B, and E.

The sequence of formation of the layers is shown schematically in Figure 1.1. It starts with the horizon C indicating the exposed rock, which may be slightly altered from its parent material due to atmospheric factors. The atmospheric organisms and plants begin to colonize the surface of horizon C, resulting in the appearance of a layer consisting predominantly of organic matter. When the atmospheric conditions are not conducive to decomposition (lack of oxygen, for instance), the organic matter may be present at the top as sediments of peat and muck. This forms the O horizon, which is marked by undecomposed or partly

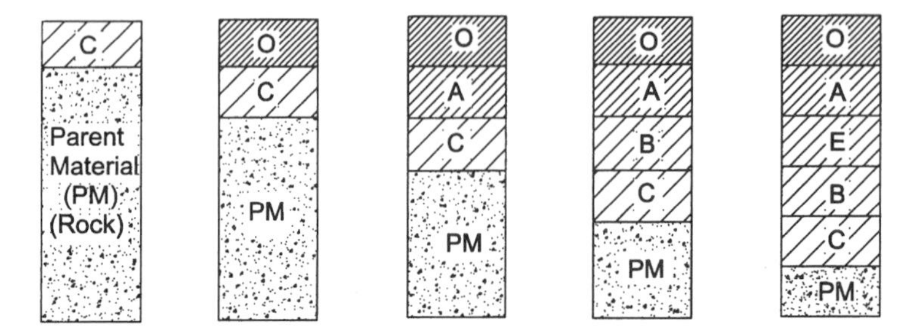

Figure 1.1 Sequence of layer formation in a soil profile.

decomposed organic matter. In the course of time, the organic matter is digested and decomposed by animals and bacteria feeding on them, resulting in *humus*, which is relatively resistant to further alteration. The layer consisting of humified organic matter forms the A horizon. Continued weathering processes produce fine clay-type particles which are transported downward through the A horizon with the help of percolating water. These fine particles are trapped eventually at intermediate depths between the A and C horizons, forming the B horizon. The processes of washing out of particles from the A horizon and washing of the same into the B horizon are sometimes refered as *eluviation* and *illuviation*. The formation of the B horizon is a long process occurring over thousands of years. Figure 1.2 shows this long process in the case of granitic materials weathered in

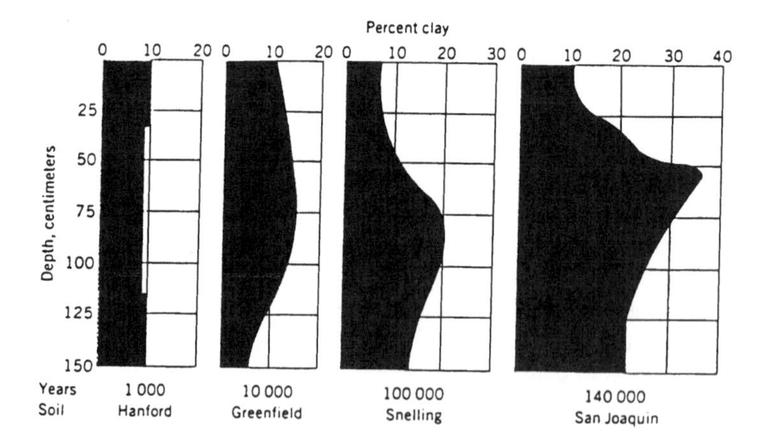

Figure 1.2 Gradual formation of the B horizon consisting of clay from granitic materials in the Central Valley of California. (*Source*: Arkley, 1964)

the Central Valley of California. It took 140,000 years for the formation of the B horizon, with a clay content of about 3.4 times that of the A horizon. In some cases, after a long period of time the eluviation process may be drastic enough to leave the bottom portion of the A horizon entirely cleaned of the fine particles, resulting in the E (totally *E*luviated) horizon.

This scheme of soil composition, commonly studied under *pedology*, is a useful one to study the characteristics of soil near the ground surface in terms of formational factors. A more detailed discussion of pedology may be found in textbooks such as those by McRae (1988), Foth (1990), and Lyon et al. (1952). However, pedological studies are qualitative in nature, and they have limited use in the study of mechanical and engineering characteristics of soils.

1.3 PHASE COMPOSITION

As a result of the interactions among the parent rocks, atmospheric agencies (primarily water and air), and organisms, during their formation, soils consist primarily of four components or phases: mineral matter, organic matter, water, and air. Nature introduces ''fluidity'' to the inert weathered rock mass through the water and air phases, and this is where the challenges of predicting the engineering behavior of soils arise. The myriad factors responsible for the coexistence of the four components during soil formation impart a great variety of properties to soils. The need to know the relative proportions of these components in our study of soil properties leads us to the soil composition at a scale finer than that of soil profiles.

Typical volumetric composition of the four phases of soils is shown in Figure 1.3. Of the four components, organic matter occupies the least amount of volume, and its quantity decreases with depth below the ground surface. In general, the quantity of organic matter varies widely from one region to the other. It may also exhibit considerable variation within a given region. However, as Figure 1.4 illustrates for four different soil profiles, it decreases rapidly with depth below the ground surface. In general, poor drainage results in higher organic matter contents in the surface horizons. The organic matter is typically ignored in phase composition studies in geotechnical engineering, primarily because its effects are limited to surficial layers only. However, it may play an important role in geoenvironmental engineering.

The relative proportions of water and air are usually of primary importance in engineering practice. These phases are both spatially and temporally variable in a given soil mass. The relative proportions of these fluid phases control to a great extent the contaminant transporting capacities of soils. The two phases interact continuously with the solid phase and participate in the mutual exchange of contaminants. We will define below the terms that are used to express quantita-

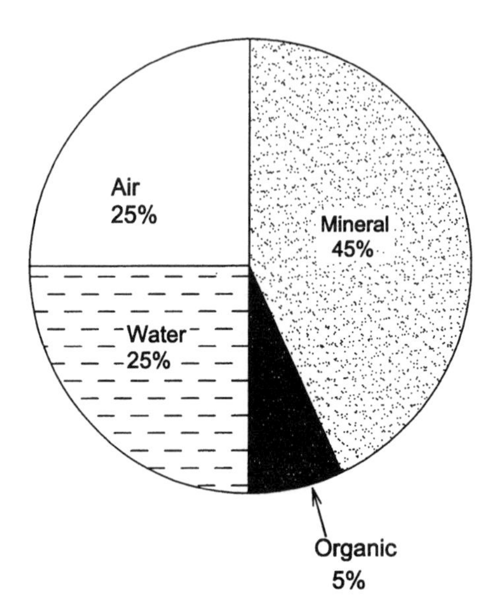

Figure 1.3 Typical volumetric composition of the four components of soils (proportions of water and air are highly variable).

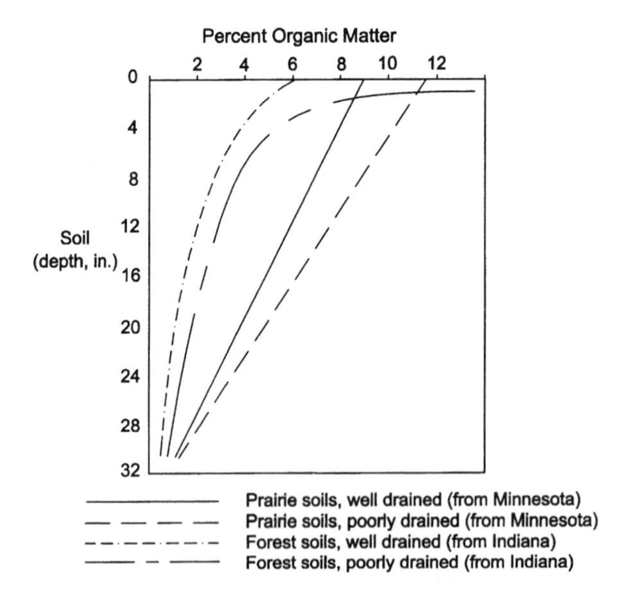

Figure 1.4 Typical variation of organic matter with depth.

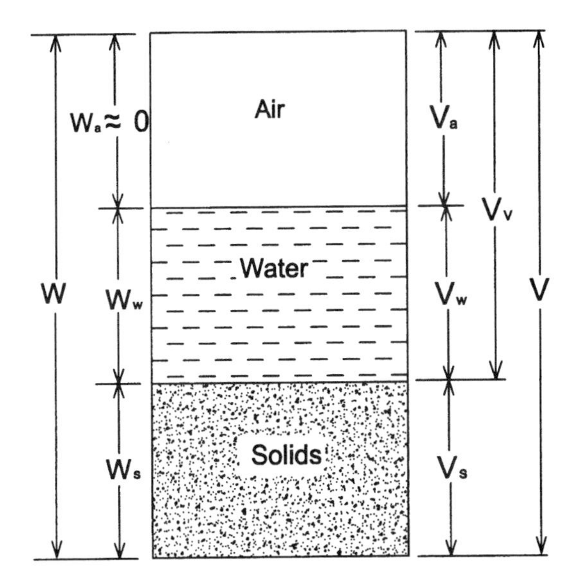

Figure 1.5 Schematic representation of solid, water, and air phases in soil.

tively the relative proportions of the solid, water, and air phases. In accordance with common practice, we will not make a distinction between the mineral matter and the organic matter in the solid phase. Figure 1.5 shows a schematic representation of the three phases with their weights indicated on the left-hand side and the corresponding volumes indicated on the right-hand side.

The volume-based indicators of phase composition are:

Void ratio, e, defined as the ratio of volume of void space (occupied by both water and air) to volume of solids, or

$$e = \frac{V_v}{V_s} \tag{1.1}$$

Porosity, n, defined as the ratio of volume of void space to total volume of soil mass, or

$$n = \frac{V_v}{V} \tag{1.2}$$

Degree of saturation, S, defined as the ratio of volume of water to volume of void space, or

$$S = \frac{V_w}{V_v} \tag{1.3}$$

Volumetric water content, θ, defined as the ratio of volume of water to total volume of soil mass, or

$$\theta = \frac{V_w}{V} \tag{1.4}$$

Of the four, the void ratio and porosity describe the volumetric extent of the void space, and the degree of saturation and volumetric water content indicate the relative wetness of the soil. The weight-based indicators of phase composition are:

Water content, w, defined as the ratio of weight of water to the weight of solids, or

$$w = \frac{W_w}{W_s} \tag{1.5}$$

Unit weight, γ, defined as the weight per unit volume of soil mass, or

$$\gamma = \frac{W}{V} \tag{1.6}$$

Note that water content is defined both in terms of volumes [Eq. (1.4)] and in terms of weights [Eq. (1.5)]. The former is used in hydrology and soil science, whereas the latter is more commonly found in geotechnical engineering literature. Geotechnical engineers often find it useful to exclude the water phase, and define the unit weight in terms of the weight of solid phase alone. One area where this is commonly done is soil compaction, discussed later in this chapter. The unit weight is then called *dry unit weight*, and is given as

$$\gamma_d = \frac{W_s}{V} \tag{1.7}$$

Using specific gravity, G_s, of solids (defined as the ratio of unit weight of solids to unit weight of water), and the unit weights of soil mass (γ or γ_d) and water (γ_w), it is possible to relate the volume-based and weight-based indicators of phase composition. With the aid of Figure 1.6, the following expressions can be derived:

$$n = \frac{e}{1 + e} \tag{1.8}$$

$$\theta = \frac{W\gamma_d}{\gamma_w} \tag{1.9}$$

$$Se = wG_s \tag{1.10}$$

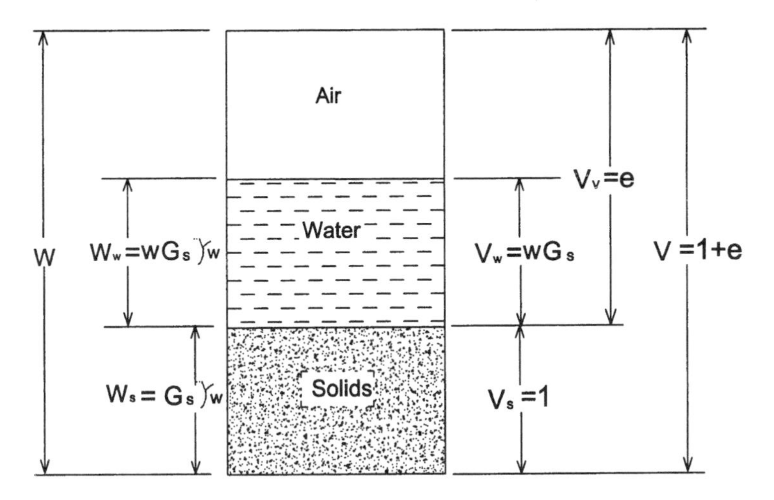

Figure 1.6 Phase relationships.

$$S = \frac{\theta}{n} \tag{1.11}$$

$$\gamma_d = \frac{\gamma}{1 + w} \tag{1.12}$$

$$\gamma = \frac{(1 + w)G_s\gamma_w}{1 + e} \tag{1.13}$$

$$\gamma_d = \frac{G_s\gamma_w}{1 + e} \tag{1.14}$$

Expressions such as the above are useful in assessing the relative quantities of the solid, water, and air phases in soils.

1.3.1 Geotechnical Processes Controlling Phase Composition

The phase composition of soils is highly variable and is controlled by both natural and man-made processes. Two such processes well known to geotechnical engineers are consolidation and compaction. Consolidation is a process involving reduction of void ratio as a result of load application on a soil mass, which is completely saturated with water. This is an important process that governs settlements of structures. Compaction involves reduction of void ratio as a result of expulsion of air phase, generally aimed at increasing the dry unit weight of a soil mass. The latter is an important process in the construction of embankments and other earthworks, and landfill liners. Although consolidation and compaction are used synonymously in some soil science and geology disciplines, we follow this important distinction throughout this book. Figure 1.7 shows schematic phase diagrams that illustrate the distinction between the two processes. The two processes are described briefly in the following subsections. For an in-depth description of these processes, the reader is referred to standard sources in soil mechanics or geotechnical engineering.

Soil Compaction

In the construction of facilities such as roadway embankments, dams, and lagoon or landfill liners, it is necessary to be able to manipulate the relative proportions of the fluid phases in soil. The water content at which a soil is molded and the energy with which it is molded govern to a large extent the density of the soil mass. The role of the water phase is to lubricate the solid particles and bring them close together. This was explained by Hogentogler (1936) in terms of four stages of wetting, namely, hydration, lubrication, swelling, and saturation. During the initial hydration stage, water is absorbed by the soil particles as cohesive films. Increased addition of water will lead to the lubrication stage, wherein the water will act as a lubricant, yielding a closer rearrangement of solid particles. This results in an increase in the dry unit weight of soil mass. Subsequent addition of water will result in swelling of the soil with some quantities of air still remaining. Although it is not practically achievable, the fourth stage of saturation is conceived to be possible when the air is completely displaced by water.

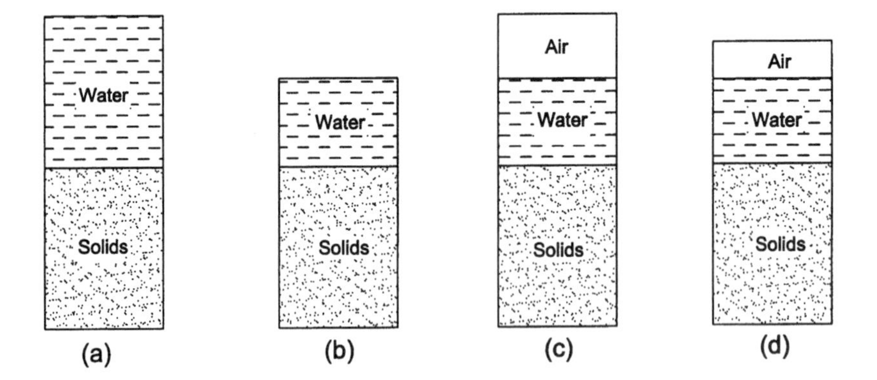

Figure 1.7 Illustration of changes during consolidation process (a, b) and compaction process (c, d).

The variation of dry unit weight with respect to initial molding water content is represented by a *compaction curve*. Two such curves are shown schematically in Figure 1.8, corresponding to two different compaction energies commonly employed in soil testing laboratories, one using a Standard Proctor Hammer and the other using a Modified Proctor Hammer. Note that for a given soil mass, dry unit weight has an inverse relationship with void ratio [Eq. (1.14)], as shown on the *y* axes in Figure 1.8. Complete expulsion of air from voids at a given molding water content would yield a theoretically possible maximum dry unit weight. Noting that $e = wG_s$ at complete saturation of soil by the water phase [Eq. (1.10)], one can obtain the theoretical maximum dry densities γ_{zav} corresponding to *z*ero *a*ir *v*oids using Eq. (1.14) as

$$\gamma_{zav} = \frac{G_s \gamma_w}{1 + wG_s} \tag{1.15}$$

The line representing Eq. (1.15) is usually shown along with the compaction curves as the zero-air-voids line (Fig. 1.8). An increase in compaction energy

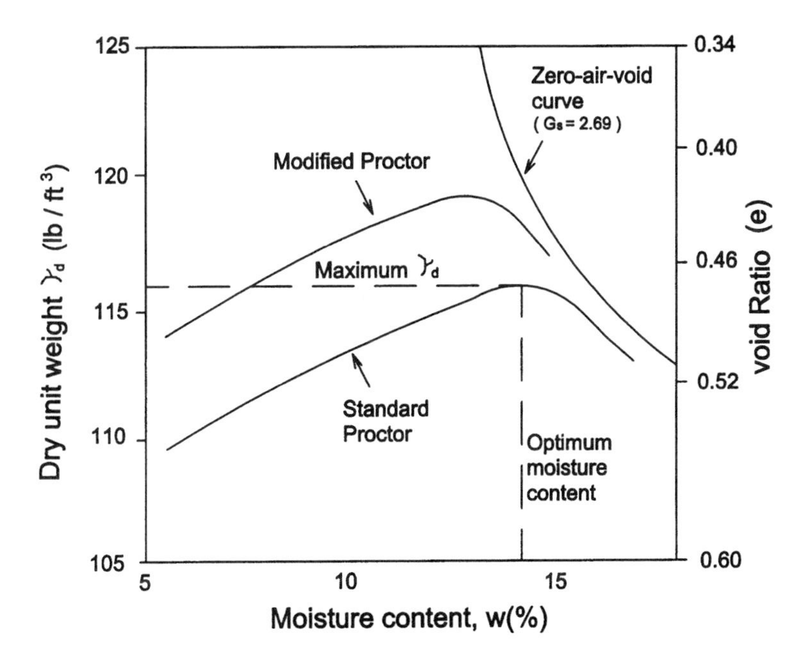

Figure 1.8 Typical variation of dry unit weight and void ratio due to compaction.

will result in higher dry unit weights, as one would expect; however, it is not practically possible to remove all the air voids and achieve γ_{zav} in the field. Increase in dry unit weight (or decrease in void ratio) and the associated changes in the skeleton of the soil mass play a crucial role in the fluid-transporting characteristics of compacted soils. This makes the soil compaction process very important in the context of waste-containment structures. The change in structure of soil as a result of compaction is discussed in detail in Chapter 2, and the change in fluid-transporting characteristics of compacted soils as a function of the compaction variables (molding water content, compaction effort, etc.) is discussed in Chapter 3.

Soil Consolidation

In contrast to compaction process, consolidation involves expulsion of water with a consequent reduction in void ratio. Since the process involves travel of water from a point of higher pressure to a point of lower pressure in soil, it is a time-dependent process. Depending on the type of soil, the thickness of the soil layer, and the magnitude of pressures developed in the pore water, the consolidation process may last several years or decades.

The magnitude of total void ratio reduction as a result of load application is often estimated by subjecting an in-situ soil sample to incremental loading in the laboratory and obtaining a stress–strain relationship. The relationship is obtained in terms of effective stresses applied on the soil sample and void ratios, and is shown schematically in Figure 1.9. The effective stress is an indicator of the intergranular stress borne by the soil skeleton. It is obtained using Terzaghi's effective stress principle, which states that the effective stress is the difference between total stress and pore water pressure. Once the slope of the consolidation curve is known, the magnitude of void ratio reduction due to a given load increase can be estimated. In general, two distinct slopes are exhibited in a consolidation curve. Region I (Fig. 1.9) is marked by low void ratio reductions, indicating that the soil sample was compressed in the past under similar stresses. Region II is marked by steep void ratio reductions, indicating that the soil sample is experiencing the stresses for the first time. The two regions meet at what is known as *preconsolidation pressure*, p_c, which is an indicator of the maximum pressure to which the soil sample has ever been subjected in the past. A knowledge of this parameter lets one track the history of the overburden stress at the site from where the laboratory soil sample was obtained. The site is said to be *normally consolidated* if the current overburden stress p_o is equal to p_c, or *overconsolidated* if p_o is less than p_c.

The change in void ratio Δe for a given load increment Δp may be obtained as follows in terms of the slopes of consolidation curve C_c (compression index) and C_s (swell index).

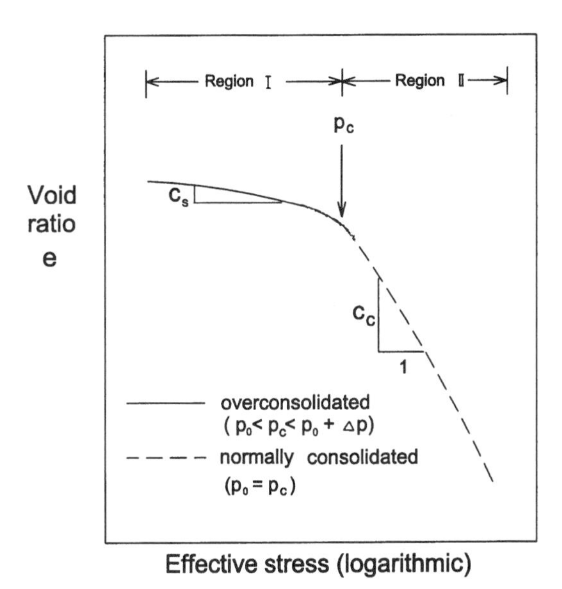

Figure 1.9 Void ratio reduction due to consolidation process.

$$\Delta e = C_c \log\left(\frac{p_o + \Delta p}{p_o}\right) \quad \text{for normally consolidated clays} \tag{1.16}$$

$$\Delta e = C_s \log\left(\frac{p_o + \Delta p}{p_o}\right)$$

$$\text{for overconsolidated clays when } p_o + \Delta p \leq p_c \tag{1.17}$$

$$\Delta e = C_s \log\left(\frac{p_c}{p_o}\right) + C_c \log\left(\frac{p_o + \Delta p}{p_c}\right)$$

$$\text{for overconsolidated clays when } p_o + \Delta p > p_c \tag{1.18}$$

The time rate at which void ratio is reduced during consolidation process is governed by a heat-conduction-type equation. This is dealt with in Chapter 3 as a special case of groundwater flow.

1.4 SOLIDS COMPOSITION AND CHARACTERIZATION

We now narrow down our focus to the solid phase and study its composition in terms of the nature and size of solid particles. Specifying the size of particles is one of the common ways of characterizing the solid phase. However, because of the nature of weathering processes, the sizes of solid particles encountered in the soil mass exhibit tremendous variation. Figure 1.10 illustrates the range of sizes of particles grouped under the broad categories of sand, silt, and clay. There is at least a 100-fold difference between the sizes of sand and clay particles. The coexistence of particles of this vast size range obviously creates a variety of pore space architectures in soils, and hence the importance of a study of pore structures in Chapter 2.

The size of particles was the original basis for grouping solids as sand, silt, and clay. A number of groupings have been developed over the years by various organizations. Figure 1.11 shows the groupings made by some of the major organizations. The particle sizes imply equivalent diameters of spheres. It is also common to characterize the solids under several subgroupings using the textural triangle (Fig. 1.12).

The experimental determination of particle sizes involves *mechanical sieving* for particles larger than 0.075 mm, and *hydrometer analysis* for particles smaller than 0.075 mm. The mechanical sieving involves using sieves of known mesh sizes to determine the fraction of solid particles passing through each sieve. The hydrometer analysis is based on the principle of sedimentation of soil particles in water. Stokes's law is used to relate the diameter D of particles (assumed to be spherical) to sedimentation velocity v.

$$v = \frac{\gamma_s - \gamma_w}{18\eta} D^2 \tag{1.19}$$

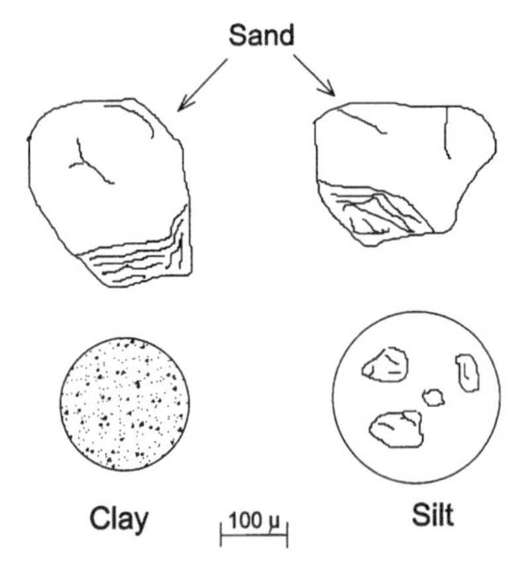

Figure 1.10 Relative sizes of particles grouped under sand, silt, and clay.

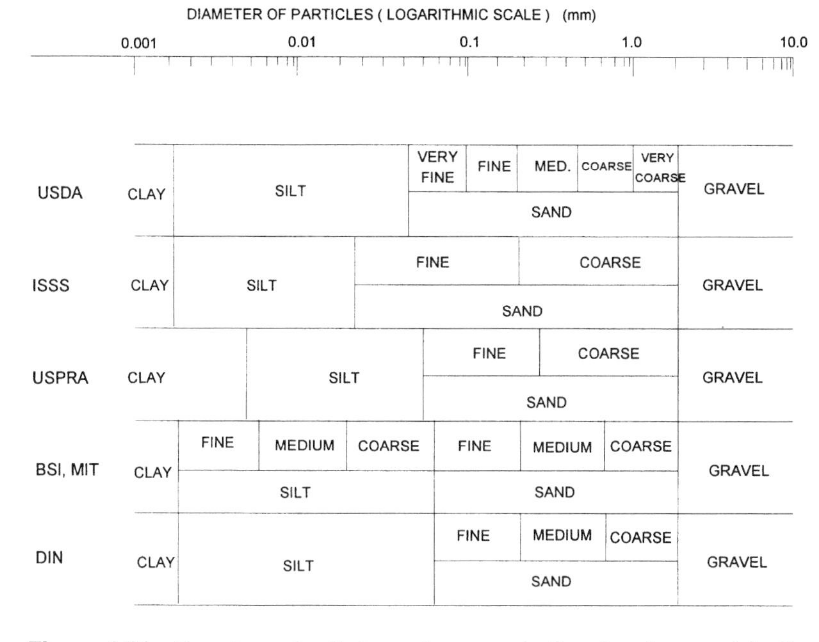

Figure 1.11 Groupings of soils by various organizations based on particle diameters (USDA, U.S. Department of Agriculture; ISSS, International Soil Science Society; USPRA, U.S. Public Roads Administration; DIN, German Standards; BSI, British Standards Institute; and MIT, Massachusetts Institute of Technology).

where γ_s = unit weight of solid particles, γ_w = unit weight of water, and η = viscosity of water. In the laboratory, a hydrometer is used to monitor the amounts of soil particles in suspension at various times. The amounts in suspension correspond to those particles whose diameter D could be predicted using Eq. (1.19).

Using mechanical sieving and hydrometer analyses, the cumulative fraction F_i of particles finer than a specific equivalent diameter, D_i, could be determined as

$$F_i = \frac{1}{W_s}\left(W_s - \sum_{1}^{i} W_i\right) \tag{1.20}$$

where ΣW_i is the cumulative weight of particles finer than the ith size, and W_s is the total weight of solids. One can then construct the size distribution of solid particles by plotting F_i against particle size, as shown in Figure 1.13. To enable

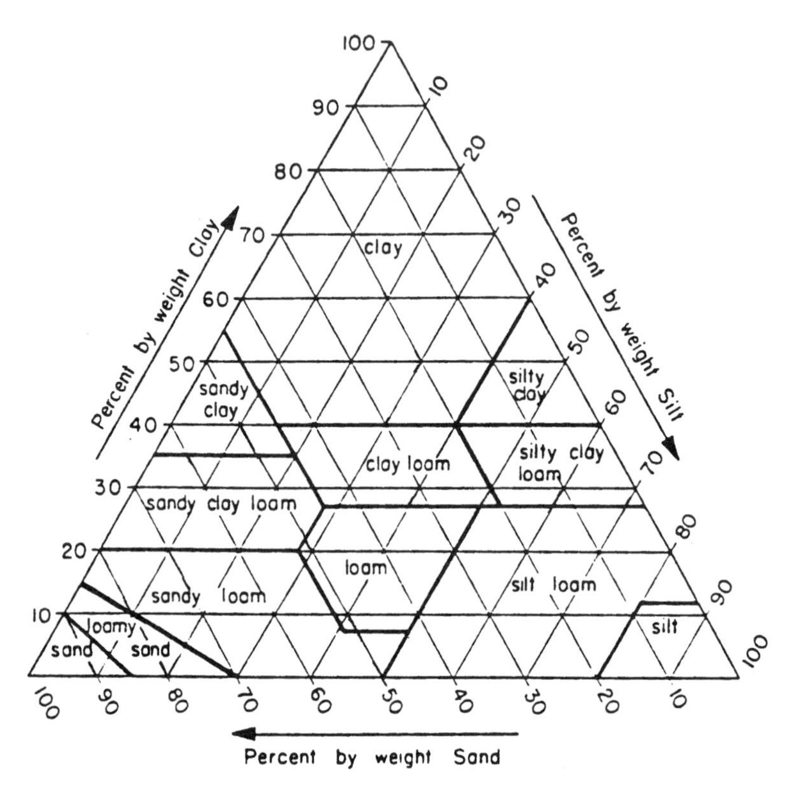

Figure 1.12 Textural classification of soils.

a comparison of different soils based on their particle size distributions, two parameters are often used. These are the *uniformity coefficient*, C_u, defined as

$$C_u = \frac{D_{60}}{D_{10}} \tag{1.21}$$

and the *coefficient of gradation*, C_g, defined as

$$C_g = \frac{D_{30}^2}{D_{60} \times D_{10}} \tag{1.22}$$

where D_{10}, D_{30}, and D_{60} are particle diameters corresponding to 10%, 30%, and 60% finer, respectively. D_{10} is also called *effective size* of solids.

Very often, the characterization scheme based on particle sizes is insufficient in the case of clayey soils. Clays, unlike sands and silts, are chemically active and undergo reactions with water. Two different types of clays may contain

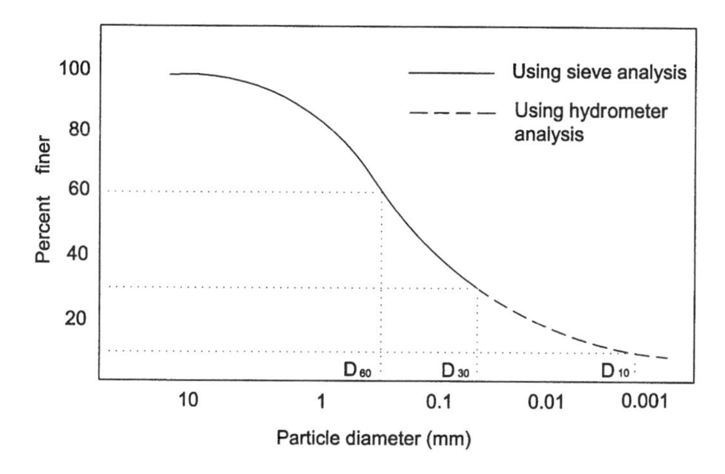

Figure 1.13 Typical particle size distribution of soils.

particles of the same size range but react quite differently with water. They may require different quantities of water to achieve the same consistency. At small water contents, the clayey soils behave as solids. As the water content is increased, they achieve semisolid and plastic stages, until a stage is reached where the water phase makes the solids behave like liquids. These stages were defined formally by Atterberg in the early 1900s in terms of what are now known as Atterberg limits. Simple experiments are routinely conducted in soil mechanics laboratories to determine the water contents at which clayey soils pass through the various stages. Three limits are in general used to characterize the clayey soils:

1. *Shrinkage limit*, which is the water content at which the soil passes from solid to semisolid state
2. *Plastic limit*, which is the water content at which transition from semisolid to plastic state takes place
3. *Liquid limit*, which indicates the water content required in order for the clayey soil to begin exhibiting flow characteristics like liquids

For simple procedures to determine these consistency limits in the laboratory, the reader is referred to standard textbooks or laboratory manuals in geotechnical engineering. It is in general accepted that the Atterberg limits are useful indicators of the engineering behavior of soils. Several engineering properties in geotechnical engineering are correlated to these limits. It is believed that the water contents at which clayey soils transit from one state to the other represent their entire physicochemical nature. The liquid limit, in particular, has been recently established to be a state at which many of the properties are the same for a variety of clay minerals. As we will see in the following section, different minerals need different quantities of water to bring them to the liquid limit state. However, once they are at the liquid limit state, all the widely differing soils seem to possess a relatively constant set of engineering properties such as pore water suction, shear strength, and hydraulic conductivity (Nagaraj et al., 1994).

1.5 MINERAL COMPOSITION

The characterization of solids based on particle sizes and consistency limits may not be sufficient to explain the variety of engineering behavior exhibited by natural soils. This is especially true with clay fraction, which shows tremendous differences in behavior. The mineral composition of the soils provides some important clues to the characteristics of clays. To introduce mineral composition, we note that the formation of soils is based on gradual weathering of rocks over eons of time periods. Sands and silts evolve first out of rocks and they in turn weather into clay fractions subsequently. It is therefore reasonable to expect that sands and silts are closer to the parent material in terms of their mineral composition than are clays. The sands and silts contain some *original or primary* minerals that are more or less unchanged from the parent material. The clays contain *secondary* minerals formed by the weathering of less resistant primary minerals. The relative proportion of secondary minerals present in sands, silts, and clays is shown schematically in Figure 1.14.

Some examples of primary and secondary minerals are listed in Table 1.1. An important feature that characterizes all the minerals is the bond between Si and O. It is generally established that the molar ratio of these two elements is an indicator of the resistance of the mineral. Minerals with a high molar ratio of 0.50 (quartz and feldspar) are more stable and resistant to weathering than those with a low molar ratio, such as olivine, which has a molar ratio of 0.25. Quartz grains come directly from the parent rock material and persist even in the secondary minerals along with the weathered products of less resistant primary minerals.

Most clays contain mixtures of the ideal minerals in varying composition. Each mineral type imparts to the soil its share of engineering properties; thus, knowledge of the behavior of ideal minerals is useful to understand the aggregate behavior of clays. We therefore study the composition of ideal minerals individually and formulate some general characteristics that the ideal minerals impart to the aggregate behavior of clays.

Minerals are formed in nature in accordance with a probabilistic minimum energy state. This is a law of economics that is consistently obeyed in creation. In order to form with the least energy possible, three conditions must be fulfilled:

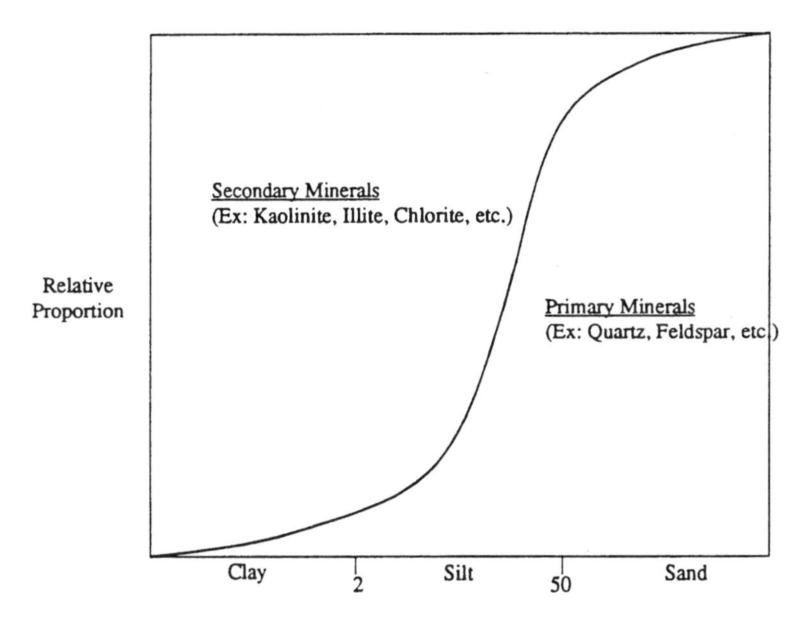

Figure 1.14 Relative proportions of primary and secondary minerals in soils.

1. The formation proceeds incorporating all the available atoms; obviously, the atoms that are most available in nature will occur frequently in the mineral structure.
2. The atoms forming the mineral should follow a geometric pattern which repeats itself in space.
3. The formation always strives for a neutrality of electronic charge, thus resulting in the alternation of anions and cations.

Table 1.1 Examples of Primary and Secondary Minerals

Primary minerals	Secondary minerals
Quartz (SiO_2)	Kaolinite $[Si_4]Al_4O_{10}(OH)_8$
Feldspar $(Na,K)AlO_2[SiO_2]_3$	Illite $(K,H_2O)_2(Si)_8(Al,Mg,Fe)_{4,6}O_{20}(OH)_4$
Mica $K_2Al_2O_5[Si_2O_5]_3Al_4(OH)_4$	Chlorite $(OH)_4(SiAl)_8(Mg,Fe)_6O_{20}$
Pyroxene $(Ca,Mg,Fe,Ti,Al)(Si,Al)O_3$	Montmorillonite $Si_8Al_4O_{20}(OH)_4.nH_2O$
Olivine $(Mg,Fe)_2SiO_4$	Gypsum $CaSO_4.2H_2O$

To see how condition 1 is fulfilled, we note the relative abundance of various elements in the earth's crust. As shown in Table 1.2, oxygen, silicon, and aluminum are the three most abundant elements, comprising 46.6%, 27.7%, and 8.1% in terms of weights, and 93.8%, 0.9%, and 0.5% in terms of volumes, respectively. It is therefore natural to expect the presence of these three elements in abundance in all the minerals (see Table 1.1). Condition 1 is fulfilled through accommodation of these elements in the mineral composition. Condition 2 enters in the packing arrangement of the elements. The elements pack themselves together in the most space-efficient manner. Thus, one would expect that the structure of minerals is such that small elements occupy the holes created by the bigger elements. However, the packing arrangement must simultaneously obey the third condition, that is, charge neutrality. As seen in Table 1.2, oxygen is the most prevalent anion in the earth's crust, and is the only anion present in the earth at more than 1% by weight. Therefore, the charge neutrality condition dictates oxygen to be present abundantly in the packing arrangement with the cations. In addition to oxygen, F^- and Cl^- are also observed as anions in the packing arrangement of the elements, although they are rare. In contrast to the anions, a variety of cations of different ionic radii are available. This makes it possible to balance the negative charge of O^- in a number of different ways. The cations compete to participate in the mineral structure. Which cation manages to participate in the structure will depend on the individual radii of ions and their capacity to fulfill charge neutrality. Table 1.2 shows the valence and ionic radii of the predominant elements in the earth.

Space-efficient packing of the different elements depends on *radius ratio*, defined as the ratio of the radius of cation to the radius of anion. The coordination number, i.e., the number of anions that can be situated around a cation, is directly

Table 1.2 Percent Weight and Volume in the Earth's Crust, Valence, Ionic Radii, and Radius Ratios (with Respect to Oxygen) of the Most Abundant Elements in the Earth's Crust

Element	Weight (%)	Volume (%)	Valence	Ionic radius (Å)	Radius ratio
O	46.6	93.8	−2	1.32	—
Si	27.7	0.9	+4	0.31–0.39	0.23–0.30
Al	8.1	0.5	+3	0.45–0.79	0.34–0.60
Fe	5.0	0.4	+2	0.67–0.82	0.51–0.62
Mg	2.1	0.3	+2	0.78–0.89	0.59–0.68
Na	2.8	1.3	+1	0.98	0.74
Ca	3.6	1.0	+2	1.06–0.17	0.80–0.89
K	2.6	1.8	+1	1.33	1.0

proportional to this radius ratio. The radius ratios given in Table 1.2 with respect to oxygen are conducive to two types of arrangements. One is a tetrahedral type of arrangement with a coordination number of 4, requiring a radius ratio of 0.22. The other is an octahedral arrangement with a coordination number of 6, requiring a radius ratio of 0.41. These radius ratios are limiting values; in other words, these are the lowest radius ratios that are required in order to achieve tetrahedral and octahedral arrangements. We therefore note that it is highly probable for Si to enter into a tetrahedral arrangement, and for Al, Fe, and others to enter into an octahedral arrangement with O^-. These two are the most frequently occurring arrangements in minerals and are shown schematically in Figure 1.15. Note that the popular schematic (shown in Fig. 1.15b) is only for a clarified view of the structure; in reality, the elements are close to each other (Fig. 1.15a), with the anions hiding the cation.

A single tetrahedron has a net negative charge of -4 and a single octahedron has a net negative charge of -10. The condition of charge neutrality forces these basic units to align themselves with other similar units. Thus, layers consisting of these units develop. For instance, a tetrahedral layer is formed when the net negative charge tends to be neutralized by a sharing of the oxygens with the surrounding tetrahedra. Similarly, an octahedral layer is formed due to a sharing of the anions by the adjacent octahedra. These layers are shown in Figure 1.16 along with the symbols commonly used to represent them as building blocks in minerals. They are the most fundamental layers existing in clay minerals.

The condition of charge neutrality requires these layers to associate with each other in the vertical dimension as well, with chemical bonding in between the layers. Ionic bonds, involving sharing of ions, are by far the strongest bonds and are facilitated between a tetrahedral and an octahedral layer. The sharing of

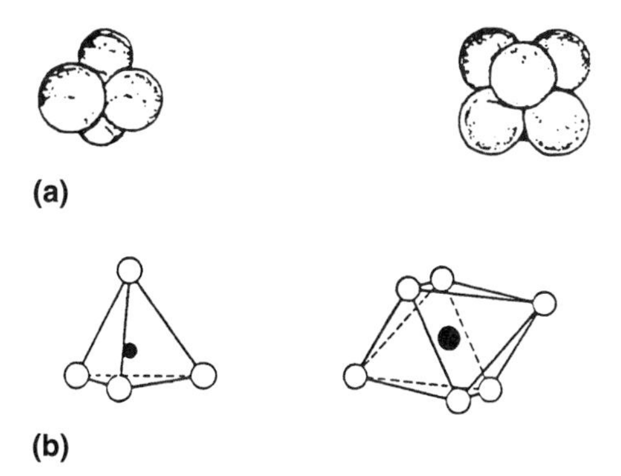

Figure 1.15 Tetrahedral and octahedral packing arrangements: (a) actual coordination, and (b) idealization.

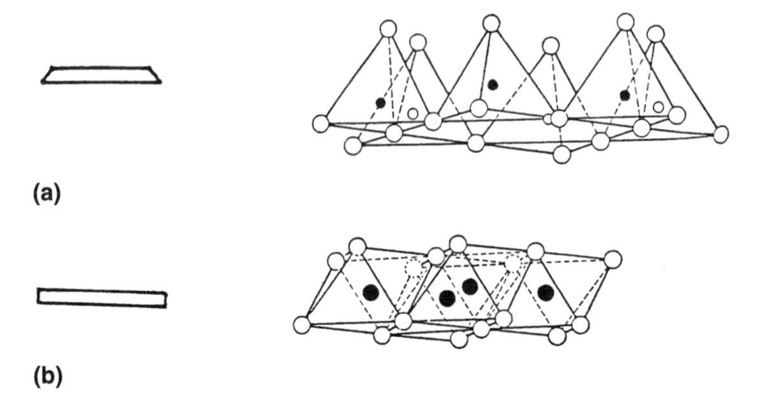

Figure 1.16 (a) Tetrahedral and (b) octahedral layers with symbolic notations.

anions at the interface of the two layers contributes to the charge neutrality. Such stacking of layers results in the so-called subbasic or semibasic units. The nomenclature used for the unit formed between one octahedral layer and one tetrahedral layer is 1:1. Another important unit is formed when an octahedral layer is sandwiched between two tetrahedral layers, with ionic bonds on either side of the octahedral layer, and is represented as 2:1. These two subbasic units, shown in Figure 1.17, together constitute a majority of the minerals encountered in clays.

It is important to emphasize that because of the variety of cations available in the earth, a number of different cations may be present in a given semibasic unit. The ionic radii of a number of cations (Table 1.2) are compatible enough in size to form either a tetrahedral or an octahedral arrangement with anions. For instance, the tetrahedral layer may contain Al^{3+} or Fe^{3+} ions, and the octahedral layer may contain Al^{3+}, Fe^{2+}, Zn, or Ni ions. Of course, such substitution takes place with due regard to the condition of charge neutrality. When a trivalent ion such as Al^{3+} occurs in an octahedral layer, no cation is observed in every third octahedron, because of this condition of charge neutrality. Such substitution is termed *isomorphous substitution* and is common in a variety of minerals. The large number of possibilities due to isomorphous substitution gives clays wide latitude of engineering behavior. Also, a number of possibilities arise for stacking of the two subbasic units either individually or in combination, thereby exhibiting widely different engineering properties.

Before we outline the common mineral types, it is important to note that the condition of charge neutrality may also be fulfilled in part by the liquid medium surrounding the subbasic units. An exchange of cations may result between the cation in the solution and the cation on the surface of the unit, thus altering the charge balance. The total positive charge that a given mineral is capable of ad-

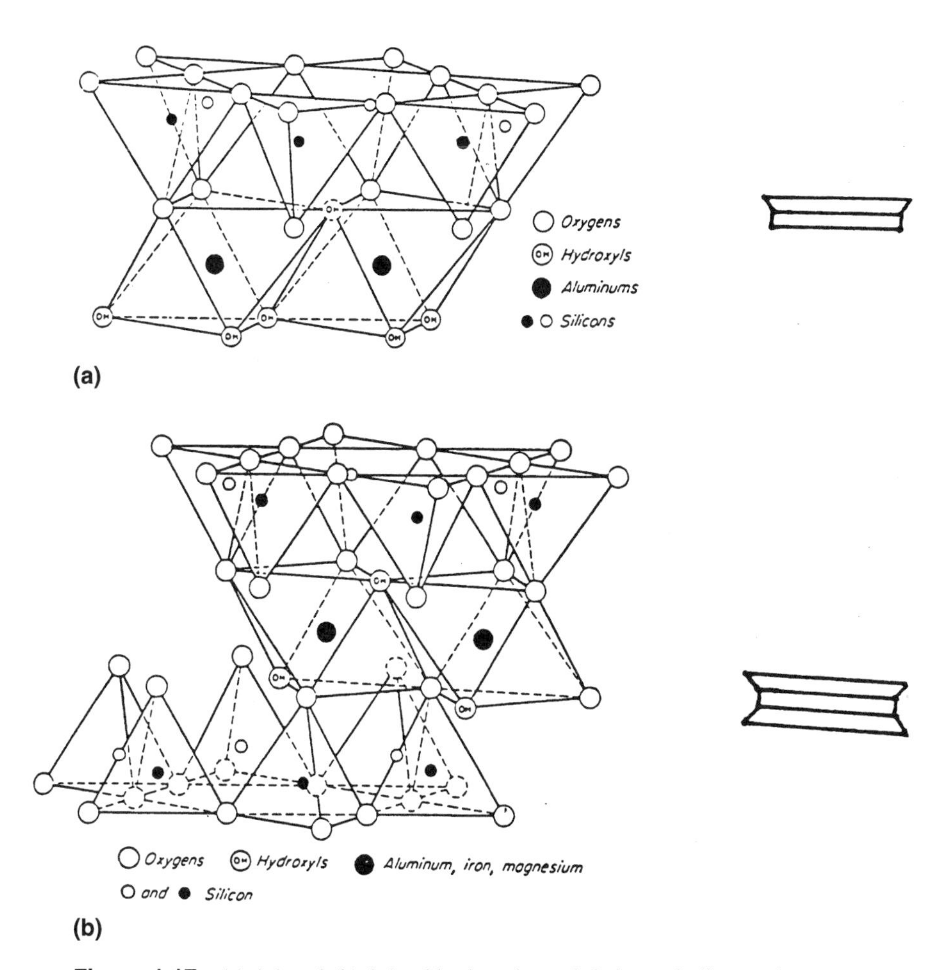

Figure 1.17 (a) 1:1 and (b) 2:1 subbasic units and their symbolic notations.

sorbing is equal to its net negative charge. This capacity of the mineral to adsorb cations is termed the *cation-exchange capacity* (CEC) of soils. It is expressed as milliequivalents of cations adsorbed per 100 g of dry soil (mEq/100 g) which is equivalent to centimoles of positive charge per kilogram of dry soil. CEC is an important property of the mineral that governs the interactions between solids and the pore fluid.

The common minerals formed by the stacking of the subbasic units are shown in Figure 1.18. Some important properties of these minerals are listed in Table 1.3, and their characteristics are outlined briefly below.

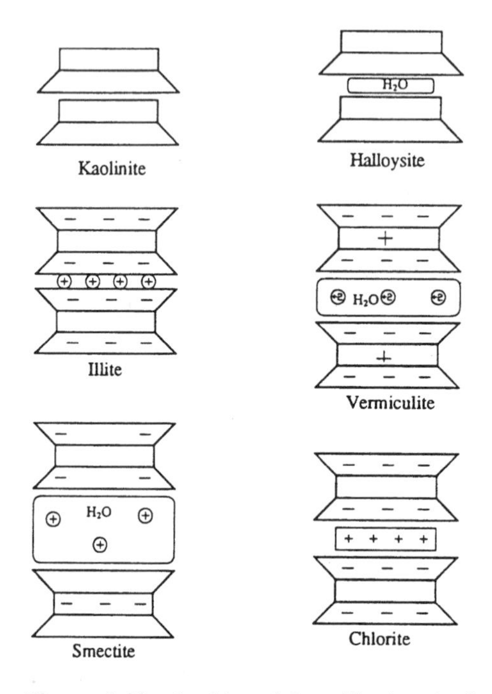

Figure 1.18 Stacking of the subbasic units for some common minerals.

Kaolinite is formed by the stacking of 1:1 units with strong hydrogen bonding between O^{2-} and OH^- of the tetrahedra and octahedra, respectively. Because of the strong bonding, kaolinite does not exhibit swelling in water. It is characterized by a low surface area and low CEC.

Halloysite is similar to kaolinite in that it has 1 : 1 layer structure; however,

Table 1.3 Important Properties of Clay Minerals

Mineral	Specific surface (m^2/g)	Specific gravity	CEC (mEq/100 g)	Basal spacing (Å)
Kaolinite	10–20	2.60–2.68	3–15	7.2
Halloysite	35–70	2.00–2.20	5–40	10.1
Illite	65–100	2.6–3.00	10–40	10
Vermiculite	40–80		100–150	10.5–14
Montmorillonite	700–840	2.35–2.70	80–150	9.6–infinity
Chlorite	80	2.6–2.96	10–40	14

it has a sheet of water molecules between the layers and therefore the thickness of stacked units is greater. Upon dehydration, halloysite collapses irreversibly to kaolinite structure.

Illite is formed by the stacking of 2 : 1 units bonded together by potassium ions. The negative charge inviting potassium ions is due to the substitution of aluminum for some silicon in the tetrahedral sheets. Illite's structure is similar to that of the primary mineral mica, except that it is less crystalline and contains less potassium. The potassium ions, fixed between the layers, prevent swelling of the mineral and provide little interlayer surface area for cation exchange.

Vermiculite has a 2 : 1 layer structure similar to that of illite, but the subbasic units in this case are separated by a couple of sheets of water molecules. Substitution of aluminium for some silicon is found in this mineral also, resulting in a high net negative charge.

Montmorillonite is again a 2 : 1 layer structure, but with extensive isomorphous substitution of magnesium and iron for aluminum in the octahedral sheets. Unlike illite, no potassium ions are found between the layers, and unlike vermiculite, the layers are separated by several sheets of water molecules. A distinguishing feature of this mineral is that it swells extensively when placed in water because of the penetration of water molecules into the interlayer spacing. It is due to this that the liquid limit of montmorillonites is very high.

Chlorite is often termed a 2 : 1 : 1 layer silicate because of the presence of a metal hydroxide sheet sandwiched between 2 : 1 units. The interlayer sheet is often known as a brucite layer, named after the mineral brucite, which has magnesium as the cation in an octahedral arrangement. Some isomorphous substitution of Mg^{2+} by Al^{3+} exists in this intermediate layer. Chlorite has low surface area and CEC because of the presence of this layer, and it does not swell in water.

It must be emphasized that the composition of any single ideal mineral outlined above is rarely exhibited in its entirety in a clay fraction. A number of minerals are present in a given soil and one or more of these minerals may dominate the composition, but rarely does the soil exhibit the extreme character of an ideal mineral. The successive layers may belong to different minerals. They may not be bonded together in a way that can be generalized. These are called mixed-layer minerals and may have a predictable structure (such as every other layer being an illite or montmorillonite) or a totally random structure.

1.6 ROLE OF COMPOSITION IN ENGINEERING BEHAVIOR OF SOILS

We observe in this section how the compositional aspects at various scales influence the engineering properties of soils. The fundamental character or signature of soils often lies in the mineral composition. For a given percentage of solids in soil, the mineral composition may dictate the overall consistency of the soil. Table 1.4 shows the ranges of Atterberg limits for the minerals discussed in the previous section.

Perhaps one of the important effects of mineral composition concerns with the interaction between solid and pore fluid phases. The vast differences in surface areas of minerals create differences in mass transfer of contaminants between the pore fluid and the solid matrix. The differences in CEC also cause differences in the conduction phenomena of soils. The negative charge imbalance of the minerals (as reflected by CEC) provides a need for the minerals to interact with the cations present in the pore fluid. This exchange of ions between the solid and the liquid phases alters the charge balance of the particles. As we will see in Chapter 2, the electrochemistry at the surface of clay particles plays an important role in how the particles are aligned and in how the pores form. An end result of this interaction is the change in conduction phenomena of soils. As will be evident when we discuss double-layer theory in Chapter 2, the structure acquired by the soil mass depends on the interactions between solid and pore fluid phases.

The variation in the relative proportions of air and water phases in soils also plays an important role in the transporting capacity of soils. The ease with which water and air can be transported in soils is related directly to the volumetric proportions of these phases. Coexistence of these two phases in the pore space creates some unique challenges in modeling the mass transfer and contaminant transport phenomena discussed in Chapters 3 and 4.

The presence of organic matter in soils is often ignored, although it influences some important engineering properties. Organic matter, although relatively small in volumetric proportion, significantly affects the water-absorbing capacity of the soil. The importance of organic matter is best illustrated in Figure 1.19, which shows the increase in liquid limit as the percentage carbon and montmorillonite contents are increased. It is clear that a mere 5% of organic carbon increased the liquid limit of soil much more than when montmorillonite proportion

Table 1.4 Atterberg Limits and Activities of Ideal Clay Minerals

Mineral	Liquid limit (%)	Plastic limit (%)	Shrinkage limit (%)	Activity[a]
Kaolinite	30–100	25–40	25–29	0.5
Halloysite	50–70	47–60		0.1
Illite	60–120	35–60	15–17	0.5–1
Montmorillonite	100–900	50–100	8.5–15	1–7
Chlorite	44–47	36–40		

[a] Activity of a soil is defined as the ratio of plasticity index (liquid limit − plastic limit) to percentage fraction less than 2μm in size. [Data from Mitchell (1993).]

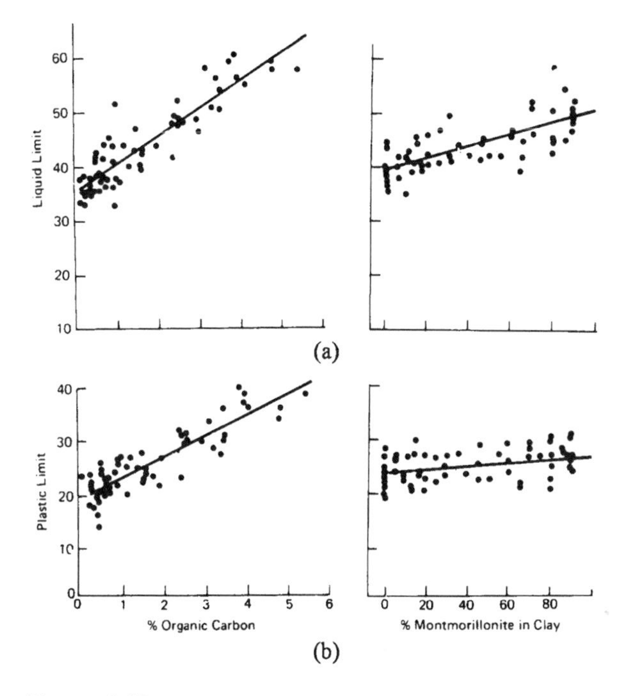

Figure 1.19 Effect of organic carbon and montmorillonite contents on (a) liquid limit and (b) plastic limit. (*Source*: Odell et al., 1960.)

Table 1.5 Engineering Considerations Pertinent to the Various Scales of Soil Composition

Scale	Descriptors of composition	Engineering issues governed by the composition
Macroscopic	Soil profile	Type of material: rock (fractured?) or soil Depth of study?
Phase composition	Void ratio, porosity, degree of saturation, water content, unit weight	Saturated or unsaturated flow; consolidation, compaction
Solids composition	Particle size distribution, Atterberg limits	Solid–liquid interaction: physical or physicochemical?
Mineral composition	CEC, surface area	Soil structure, mass transfer and transport

is increased up to 80%. The same effect is shown in plastic limit. In terms of mechanical properties of soils, the organic matter is known to reduce the maximum dry unit weight and cohesion of soils (Franklin et al., 1973). The cation-exchange capacity may increase significantly when organic matter is present in soils. Often, the organic matter accounts for 30–90% of the adsorbing capacity of minerals. On average, the cation-exchange capacity of the soil increases by 2 mEq/100 g for each 1% of well-humified organic matter (Lyon et al., 1952). Organic matter is also a very good source of nutrients, an important consideration in bioremediation of contaminated sites.

Table 1.5 summarizes the important engineering issues governed by each scale of soil composition. As we will see in the remaining chapters of the book, each scale is equally important in our consideration of practical geoenvironmental issues.

REFERENCES

Arkley, R. J. (1964). Soil survey of the Eastern Stanislaus Area, California. U.S. Dept. Agr. and Cal. Agr. Exp. Sta.

Foth, H. D. (1990). *Fundamentals of Soil Science*, 8th ed. Wiley, New York.

Franklin, A. F., Orozco, L. F., and Semrau, R. (1973). Compaction of slightly organic soils. *ASCE J. Soil Mech. Found. Div.* 99 (SM7):541–557.

Hogentogler, C. A. (1936). Essentials of soil compaction. *Proc. Highway Research Board*, 16:309.

Jenny, H. (1941). *Factors of Soil Formation; A System of Quantitative Pedology*. McGraw-Hill, New York.

Lyon, T. L., Buckman, H. O., and Brady, N. C. (1952). *The Nature and Properties of Soils; A College Text of Edaphology*. Macmillan, New York.

McRae, S. G. (1988). *Practical Pedology; Studying Soils in the Field*. Ellis Horwood, West Sussex, England.

Mitchell, J. K. (1993). *Fundamentals of Soil Behavior*, 2nd ed. Wiley, New York.

Nagaraj, T. S., Srinivasa Murthy, B. R., and Vatsala, A. (1994). *Analysis and Prediction of Soil Behavior*. Wiley Eastern, New Delhi.

Odell, R. T., Thornburn, T. H., and McKenzie, L. (1960). Relationships of Atterberg limits to some other properties of Illinois soils. *Proc. Soil Sci. Soc. Am.*, 24(5):297–300.

2
Soil Structure

2.1 INTRODUCTION

As we have seen in Chapter 1, nature creates a large number of variations in soil composition during the stages of soil formation. When man-made variations due to engineered processes (such as compaction) are added to these, the task of defining the physical appearance of soils becomes very difficult. The term *fabric* is commonly used to denote the physical arrangement of individual particles in soils. We will see shortly that identification of a ''particle'' is itself a matter of judgment in clays. Groups of single clay platelets aggregate to give an appearance of what one may loosely call a *particle*. The term *structure* is used to denote not only the physical arrangement within soils, as implied by ''fabric,'' but also its stability or integrity. Yong and Sheeran (1973) define fabric as the ''physical arrangement of soil particles and includes the particle spacing and pore size distribution.'' They define soil structure as ''that property of the soil which provides for its integrity.'' In addition to the particle spacing and pore size distribution, the structure of soil encompasses the mineralogy and chemistry of the three phases that influence interparticle forces. Although the two terms, fabric and structure, are often used synonymously, the distinction is important in the areas where the physical stability of soils is important. For instance, the sensitive nature of quick soils, collapsing and expanding soils, cannot be explained by mere consideration of fabric alone; the interplay of physicochemical forces between the particles or particle groups must also be considered.

Most of the present conceptualizations of how soil particles group themselves are based on observations using an optical or electron microscope. Figure 2.1 shows, for instance, photomicrographs of six different materials at the same magnification. Because of the subjective judgment involved in observing the patterns, the terminology associated with soil structure is quite extensive. Many of the terms are often used loosely, although their definitions are rigid elsewhere in other disciplines. An excellent example is the usage of *flocculated* and *aggregated*

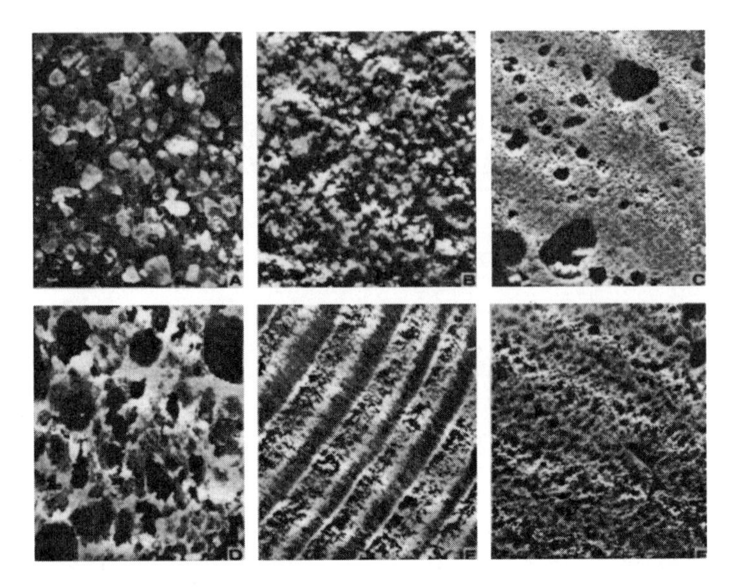

Figure 2.1 Examples of natural porous materials (× 10): (a) beach sand, (b) sandstone, (c) limestone, (d) rye bread, (e) wood, (f) human lung. (*Source*: Collins, 1961)

structures. These terms are used synonymously, although the latter is used to denote a cemented flocculation, and not mere flocculation of particles.

Our interest in the study of soil structure stems from the fact that it plays a significant role in the way contaminants are transported in soils. Porosity and pore-size distribution are directly dependent on soil structure. These parameters govern the mass transfer and transport properties in soils. We will start with a discussion of how the soil structure is conceptualized at different scales of observation and zoom into the elementary particle arrangements which contribute to the unique characteristics of a given soil. In that process, we will find it necessary to study clay mineralogy and the interparticle forces operative at the scale of an individual clay platelet.

2.2 DIFFERENT SCALES OF SOIL STRUCTURE

Soil structure is exhibited in different ways at different scales of observation. Visual observation of the size and shape of the individual clods might characterize a sample from topsoil relatively easily. When the same clods are inspected using a microscope, they present a number of different arrangements of individual particles and their groups, which require additional terminology to characterize.

Soil scientists whose primary focus is the root growth, tillage, and erosion in topsoils term the visible clods of the topsoils as aggregates. These aggregates range in size from millimeters to several centimeters. The U.S. Department of Agriculture (USDA) arrived at a scheme to classify the structure of aggregates based on their size and shape. This classification scheme, shown in Table 2.1, uses primarily three descriptive terms to characterize the size and shape: *platelike, prismlike, and blocklike*. These three basic elements are shown in Figure 2.2. Platelike structure is one where horizontal axes are longer than vertical axes (E in Fig. 2.2), whereas prismlike structure is one where vertical axes dominate. Pillarlike structure without round caps is termed prismatic (A) and with rounded caps is termed columnar (B). Blocky structure is used to denote aggregate cubes up to 10 cm in size, with either angular (C) or subangular (D) planes. It also includes smaller spheroids, which are further subdivided into crumbs, or granules (F) based on whether they appear to be porous or not.

Although the USDA classification of soil structure provides a simple way to characterize the aggregates, it does not permit us to go to scales smaller than the visual size range. It is at small scales that soil structure plays its important role in altering the pore size distribution, and the chemical interactions between pore fluid and solid phase take place. Another difficulty associated with the USDA classification is that it is truly not possible to affix an absolute size to any visible aggregate in the field because the aggregate can always be broken down to smaller groups of particles. The size distribution of the aggregates therefore depends on the mechanical means used to separate them from one another.

This leads us to consider other characterizations of soil structure at smaller scales. Of the several conceptualizations that exist in the literature, the one proposed by Collins and McGown (1974) appears to be extensive. They studied the structures exhibited by normally or lightly overconsolidated clays, silts, and sands from a wide variety of geographic locations. These soils were associated with different environments and depositional processes. Two distinct types of arrangements were apparent from these studies:

- Type 1: Elementary particle arrangements consisting of single particles of clay, silt, and sand
- Type 2: Particle assemblages consisting of one or more forms of elementary particle arrangements or smaller particle assemblages

These two types of arrangements are associated with different scales of pores, discussed later. Figure 2.3 shows a schematic of type 1 arrangements at particle scale. It is very likely that a given soil exhibits more than one of these arrangements side by side. An important observation that strengthens this characterization scheme is that soils of different environments and of different modes of deposition exhibited similar arrangements. Figure 2.4 shows a schematic of type 2 arrangements, which are those of assemblages and are therefore of a

Table 2.1 Types and Classes of Soil Structure

	Type (shape and arrangement of peds)							
				Blocklike; polyhedronlike, or spheroidal, with three dimensions of the same order of magnitude, arranged around a point				
	Platelike with one dimension (the vertical) limited and greatly less than the other two; arranged around a horizontal plane; faces mostly horizontal	Prismlike with two dimensions (the horizontal) limited and considerably less than the vertical; arranged around a vertical line; vertical faces well defined; vertices angular		Blocklike; blocks or polyhedrons having plane or curved surfaces that are casts of the molds formed by the faces of the surrounding peds		Spheroids or polyhedrone having plane or curved surfaces which have slight or no accommodation to the faces of surrounding peds.	
		Without rounded caps	With rounded caps	Faces flattened; most vertices sharply angular	Mixed rounded and flattened faces with many rounded vertices	Relatively nonporous peds	Porous peds
Class	Platy	Prismatic	Columnar	(Anglular) blocky[a]	Subangular blocky[b]	Granular	Crumb
Very fine or very thin	Very thin platy; < 1 mm	Very fine prismatic; <10 mm	Very fine columnar; <10 mm	Very fine angular blocky; <5 mm	Very fine subangular blocky; <5 mm	Very fine granular; <1 mm	Very fine crumb; <1 mm
Fine or thin	Thin platy; 1 to 2 mm	Fine prismatic; 10–20 mm	Fine columnar; 10–20 mm	Fine angular blocky; 5–10 mm	Fine subangular blocky; 5–10 mm	Fine granular; 1–2 mm	Fine crumb; 1–2 mm
Medium	Medium platy; 2–5 mm	Medium prismatic; 20–50 mm	Medium columnar; 20–50 mm	Medium angular blocky; 10–20 mm	Medium subangular blocky; 10–20 mm	Medium granular; 2–5 mm	Medium crumb; 2–5 mm
Coarse or thick	Thick platy; 5–10 mm	Coarse prismatic; 50–100 mm	Coarse columnar; 50–100 mm	Coarse angular blocky; 20–50 mm	Coarse subangular blocky; 20–50 mm	Coarse granular; 5–10 mm	
Very coarse or very thick	Very thick platy; >10 mm	Very coarse prismatic; >100 mm	Very coarse columnar; >100 mm	Very coarse angular blocky; > 50 mm	Very coarse subangular blocky; >50 mm	Very coarse granular; >10 mm	

[a] Sometimes called nut; the word ''angular'' in the name can ordinarily be omitted.
[b] Sometimes called nuciform, nut, or subangular nut. Since the size connotation of these terms is a source of great confusion to many, they are not recommended.
Source: USDA (1951).

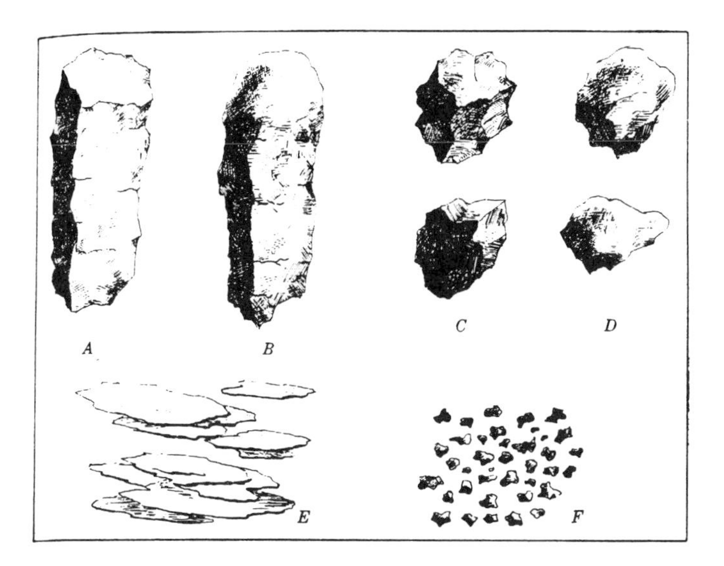

Figure 2.2 Types of soil structure: A, prismatic; B, columnar; C, angular blocky; D, subangular blocky; E, platy; F, granular. (*Source*: USDA, 1951)

slightly higher scale of dimension. Several photomicrographs published by Collins and McGown (1974) seem to suggest that it is possible to delineate boundaries of the assemblages. Again, similar assemblages were found to be present in soils having a wide variety of environment and depositional factors, so one could use this as a general characterization scheme.

2.3 PORE SIZES ASSOCIATED WITH SOIL STRUCTURE

The scale dependence of soil structure reflects in the tremendous variation in the pore sizes of the soil. Characterization of pores associated with a given soil structure is a difficult task. Dullien (1992) notes this difficulty by saying, "While a completely general definition of a pore is probably not possible, a pore is defined as a portion of pore space bounded by solid surfaces and by planes erected where the hydraulic radius of the pore space exhibits minima, analogously as a room is defined by its walls and the doors opening to it." The shape and size of the pores are of utmost importance to us in quantifying the flow and in assessing the entrapment of contaminants. Leaving a detailed discussion of these aspects to the next three chapters, we briefly take up the issue of scale dependence of pore characterization here.

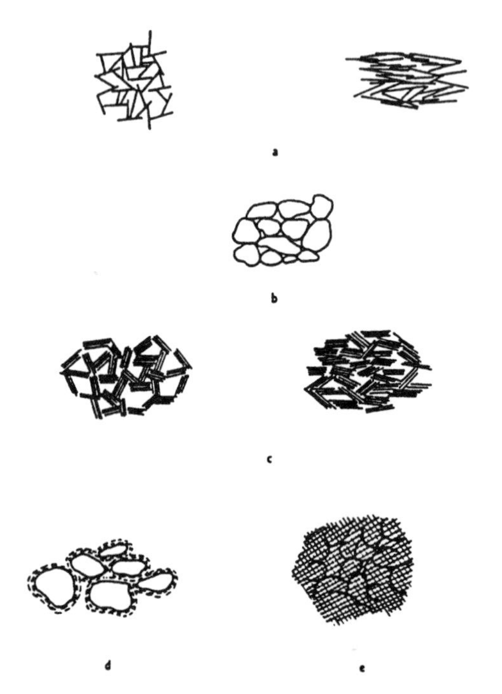

Figure 2.3 Schematic representations of elementary particle arrangements: (a) individual clay platelet interaction; (b) individual silt or sand particle interaction; (c) clay platelet group interaction; (d) clothed silt or sand particle interaction; (e) partly discernible particle interaction. (*Source*: Collins and McGown, 1974)

The existence of soil aggregates with root holes, cracks, and fissures, provides the largest scale at which pores can be characterized. These form an important set of pores, yet are elusive because they cannot be simulated in the laboratory and therefore their effect on contaminant transport in the field cannot be predicted adequately. The next size range of pores is that formed by the arrangement of particle assemblages. Of lesser importance, because of their small sizes, are the pores formed within the elementary particle arrangements. Several classifications exist in the literature for these basic pore arrangements. Olsen (1962), for instance, defined two classes of pores, one intercluster and the other intracluster, where a cluster is defined as a grouping of particles or aggregations into larger units (Fig. 2.5). This classification was used by Olsen to observe the relative contributions of the two sets of pores and to study the effect of particle aggregates on flow rates through soil.

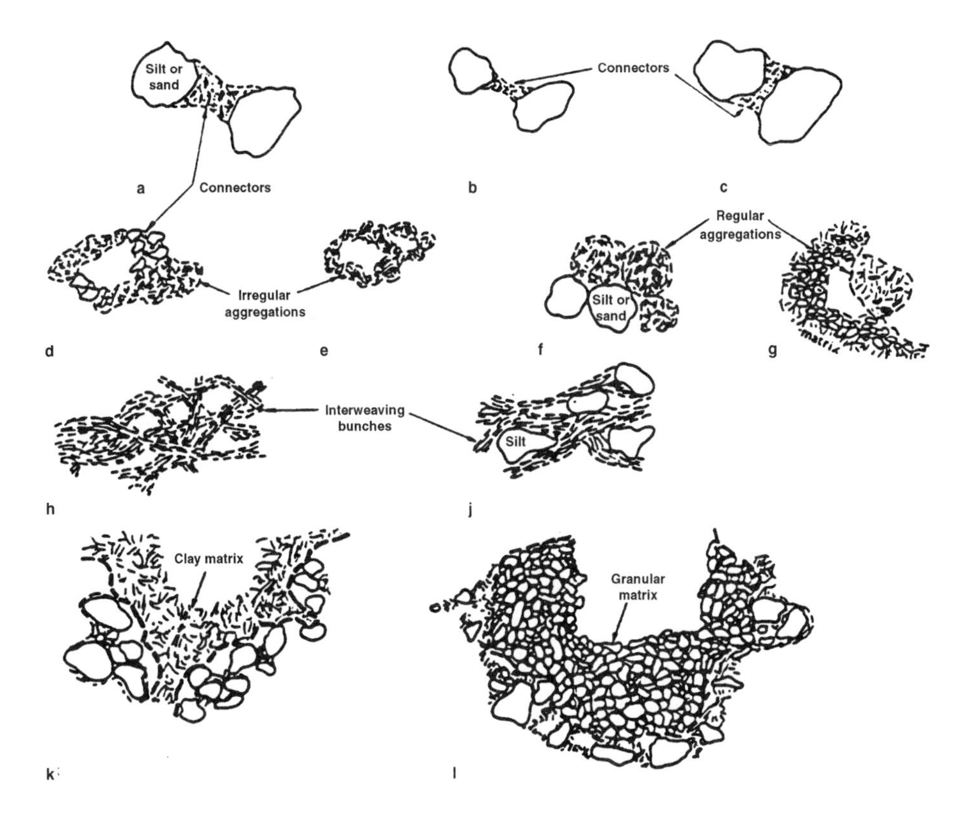

Figure 2.4 Schematic representations of particle assemblages: (a) (b) (c) connectors; (d) irregular aggregations linked by connector assemblages; (e) irregular aggregations forming a honeycomb arrangement; (f) regular aggregations interacting with silt or sand grains; (g) regular aggregation interacting with particle matrix; (h) interweaving bunches of clay; (j) interweaving bunches of clay with silt inclusions; (k) clay particle matrix; (l) granular particle matrix. (*Source*: Collins and McGown, 1974)

Collins and McGown (1974) provided another classification based on observations of soil fabric discussed above. Four classes of pores were suggested in that study:

1. Intraelemental pores occurring within the various elementary particle arrangements, including both interparticle pores and intergroup pores (those occurring between groups of clay plates)
2. Intraassemblage pores existing within particle assemblages or between smaller particle assemblages within a bigger assemblage

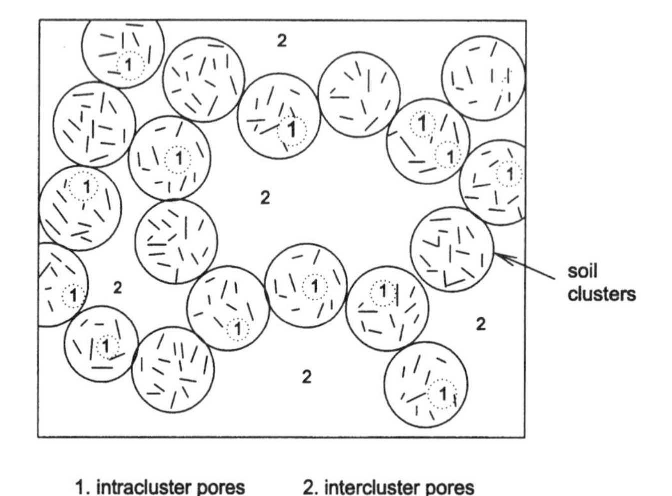

Figure 2.5 Classification of pores according to Olsen (1962).

3. Intraassemblage pores occurring between particle assemblages
4. Transassemblage pores, which are larger in size than the others, traversing several assemblages in the soil fabric

These four types are shown schematically in Figure. 2.6. It is often difficult to distinguish between intraassemblage and interassemblage pores, and similarly between interassemblage and transassemblage pores. For this reason, this grouping may be collapsed further to only three groups for convenience. Mitchell (1993) describes these three in terms of micro-, mini-, and macro-fabric. Microfabric consists of elementary particle arrangements with very small pores between them, minifabric consists of interassemblage pores, and macrofabric refers to transassemblage pores including cracks, fissures, root holes, etc. It is generally acknowledged that fluid flow through a soil is governed primarily by the mini- and macrofabrics of the soils, with little contribution from microfabric.

2.4 SINGLE-PARTICLE ARRANGEMENTS

Having looked at the complexity in characterizing the soil structure and the associated pores, we now proceed to identify the mechanisms involved in particle grouping at the smallest scale. We will see shortly that a host of physicochemical forces operate at the scale of an individual particle to give the visible aggregate of soil its complex structure. It would benefit us to start with sands and silts,

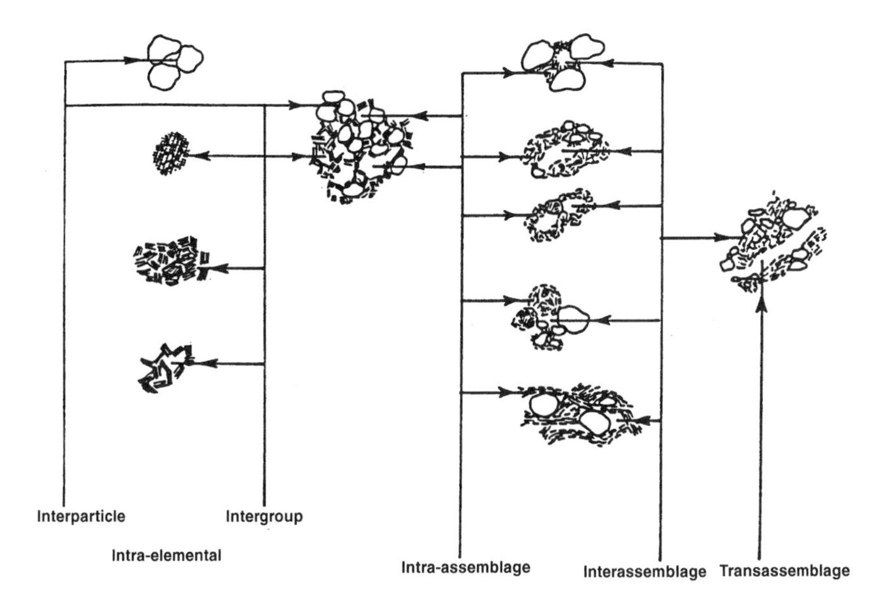

Figure 2.6 Schematic representation of pore space types. (*Source*: Collins and McGown, 1974)

because their arrangements are far simpler than clays. Unlike in clays, there is little chemical interaction in the particle arrangement of sands and silts.

The ways in which spherical grains of the same size can be packed are limited in number. Because of the uniformity with which we can pack these grains, it is possible to predict mathematically the volume occupied by the pore space given the packing arrangement. Figure 2.7 shows five regular packing arrangements of equal spheres. The cubical arrangement shown in case (a) offers the minimal density and maximum porosity, with a coordination number (points of contact per sphere) of 6 and a mean bulk porosity of 47.64%. Pyramidal and tetrahedral arrangements offer the densest states of packing, with a coordination number of 12 and a mean bulk porosity of 25.95%. Table 2.2 shows the properties of intermediate regular packings.

Similar to regular packings, random packings were also found to be associated with a characteristic range of porosities. For instance, Haughey and Beveridge (1969) identified four random packing arrangements with narrow range of porosities exhibited by each of the arrangements: (1) very loose random packing with a porosity of about 0.44, obtained by slowly reducing the velocity in a fluidized bed; (2) loose random packing with a porosity range of 0.40–0.41, obtained by individual random hand packing; (3) poured random packing with a

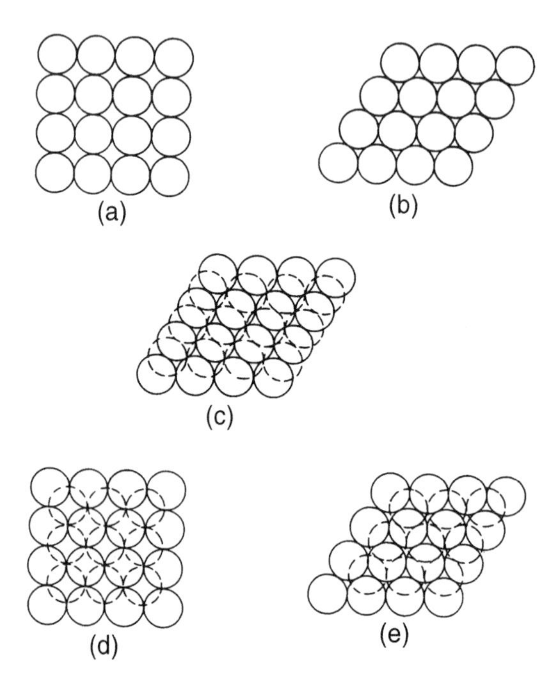

Figure 2.7 Models of regular packing of uniformly sized spheres: (a) simple cubic; (b) cubical tetrahedral; (c) tetragonal sphenoidal; (d) pyramidal; (e) tetrahedral. (*Source*: Deresiewicz, 1958)

Table 2.2 Properties of Regular Packing Arrangements of Uniformly Sized Spheres

Type of packing	Coordination number	Layer spacing (R = radius)	Volume of unit	Porosity (%)	Void ratio
Simple cubic	6	$2R$	$8R^3$	47.64	0.91
Cubical-tetrahedral	8	$2R$	$4\sqrt{3R^3}$	39.54	0.65
Tetragonal-sphenoidal	10	$R\sqrt{3}$	$6R^3$	30.19	0.43
Pyramidal	12	$R\sqrt{2}$	$4\sqrt{2R^3}$	25.95	0.34
Tetrahedral	12	$2R\sqrt{2/3}$	$4\sqrt{2R^3}$	25.95	0.34

Source: Deresiewicz (1958).

porosity range of 0.375 to 0.391, when spheres are poured into a container; and (4) close random packing, with porosities of 0.359 to 0.375, obtained when the packing is vibrated or shaken vigorously. In general, the coordination number is representative of the fabric and may be used to infer the porosity. Ridgway and Tarbuck (1967) give us a regression relationship between coordination number u, and porosity n:

$$n = 1.072 - 0.1193u + 0.004312u^2 \tag{2.1}$$

Similar relationships exist in the literature between porosity and coordination number. One has to be careful in applying these regression equations to polydisperse systems wherein the sphere size varies and therefore the relationships are far more complex. In general, the porosities of natural systems consisting of sands and silts of varying size lie within the theoretical limits of the packing arrangements discussed above.

In the case of clays, the individual particle arrangement varies significantly because of the varied mineralogy of the particles and their rather complex association with the pore fluid. Early studies were focused on the particle arrangements in suspensions only. Although there are additional factors to consider in real soil masses, a knowledge of particle arrangement in clay suspensions is an important first step in understanding the structure of clays. Figure 2.8 shows the possible associations between clay particles in suspensions. The terminology associated with these associations is as follows.

Dispersed: No face-to-face association between particles exists.
Aggregated: Face-to-face association of several particles exists.
Flocculated: Edge-to-face or edge-to-edge association of particles exists.
Deflocculated: No edge-to-face or edge-to-edge association between particles exists.

An understanding of the factors responsible for various associations shown in Fig. 2.8 is essential in order to at least qualitatively predict the structure changes of clays due to environmental factors. This takes us to a study of clay–water interactions because the individual particle arrangement of clay particles occurs through the medium of water films attracted to clay particles. The two primary mechanisms for attraction between clays and water are that (1) the dipolar nature of water molecules brings the molecules close to the negatively charged clay particles, and (2) any cations existing in the vicinity of the clay particle tend to be attracted to the clay particles, bringing with them the water of hydration. Mitchell (1993) subdivides these into four specific mechanisms (Fig. 2.9):

1. Hydrogen bonding between the hydrogen of water molecules and the oxygen or hydroxyl of clay particle surfaces
2. Hydration of cations in the vicinity of the clay particles

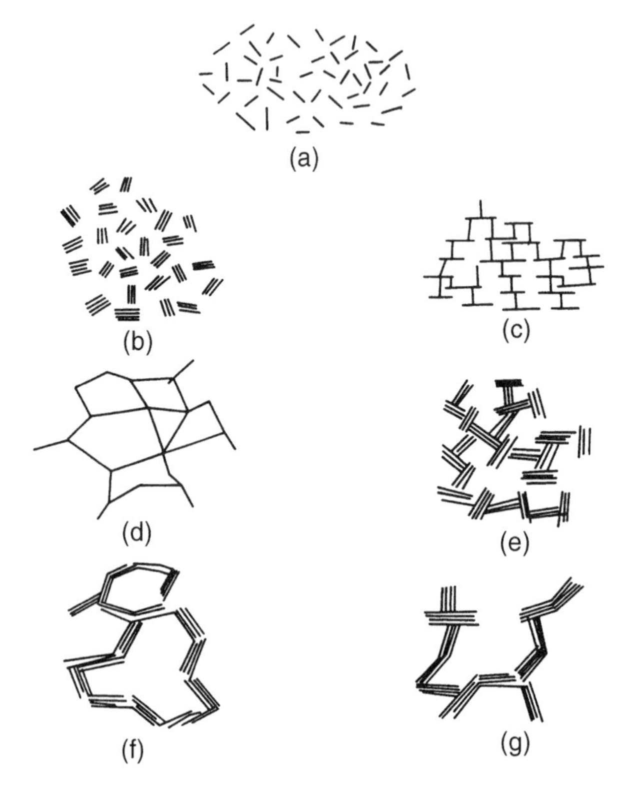

Figure 2.8 Modes of particle association in clay suspensions, and terminology: (a) dispersed and deflocculated; (b) aggregated but deflocculated; (c) edge-to-face flocculated but dispersed; (d) edge-to-edge flocculated but dispersed; (e) edge-to-face flocculated and aggregated; (f) edge-to-edge flocculated and aggregated; (g) edge-to-face and edge-to-edge flocculated and aggregated. (*Source*: van Olphen, 1977)

3. Osmosis attraction, which is the result of increased concentration of cations near the particle surface and the consequent movement of water molecules toward the particle surface to equalize the concentrations
4. Direct attraction between water dipoles and charged particle surfaces because of strong force of hydration of the surface and consequent displacement of cations away from the surface

The force with which water is attracted to the clay particle decreases with distance. Lambe (1958) divides the attracted water into three types on the basis of the relative magnitude of this force: adsorbed water, double-layer water, and free water. The adsorbed water is of the order of only a few molecules thick (of

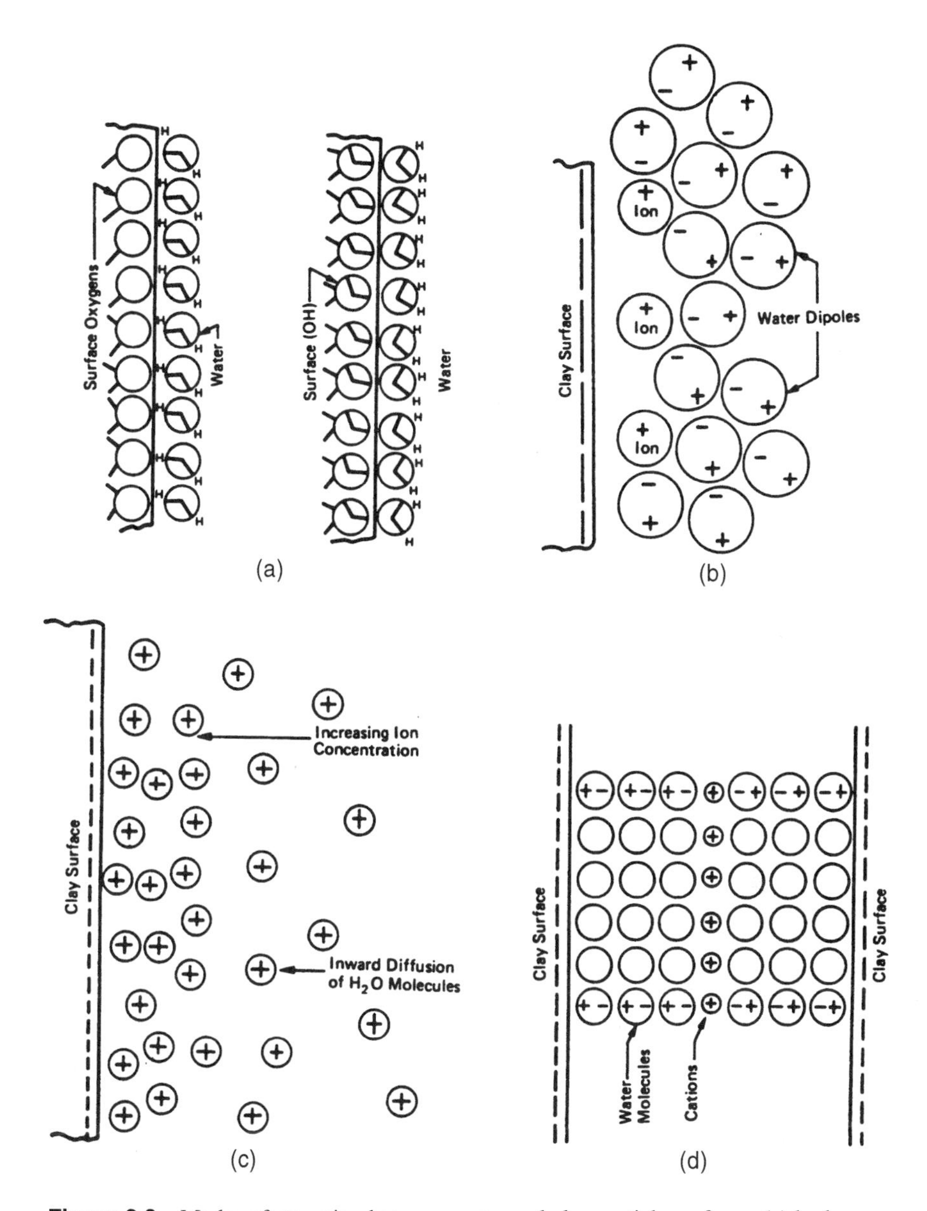

Figure 2.9 Modes of attraction between water and clay particle surfaces: (a) hydrogen bonding; (b) ion hydration; (c) attraction by Osmosis; (d) dipole attraction. (*Source*: Mitchell, 1993)

the order of 10 Å) held by strong forces—so strong that it takes about 10,000 atmospheres to pull the water molecules off the surface. The double-layer water, which is often lumped with adsorbed water in some classification terminologies, is attracted by less strong forces, which make it easy for flow to take place parallel to the particle surface but make it difficult to pull the water without subsequent replacement. Obviously, the thicknesses of these layers depend on the charge density of the particle surfaces, among other parameters. The higher the charge density, the higher is the thickness of these layers, and vice versa. Figures 2.10a and 2.10b demonstrate this dependency of the thickness of water layers on the charge density. It is seen that kaolinite, by virtue of its high charge density, possesses a thicker water layer than montmorillonite. This should not lead us to conclude that the water content of montmorillonite systems is lower than those of kaolinite. Quite the contrary is true, because the specific surface area of montmorillonite is far greater than that of kaolinite. Consideration of the double layer is quite important in soil systems, since most pore water in field soils is contained in that layer.

Contact between individual particles takes place through the water layers. The interaction of forces of the water layers of two approaching particles, coupled with externally applied forces and environmental changes, governs the particle arrangement. Whether two mineral surfaces ever come into physical contact is still an issue of controversy. However, a knowledge of the various factors affecting the forces of attraction and repulsion between two particles is needed to un-

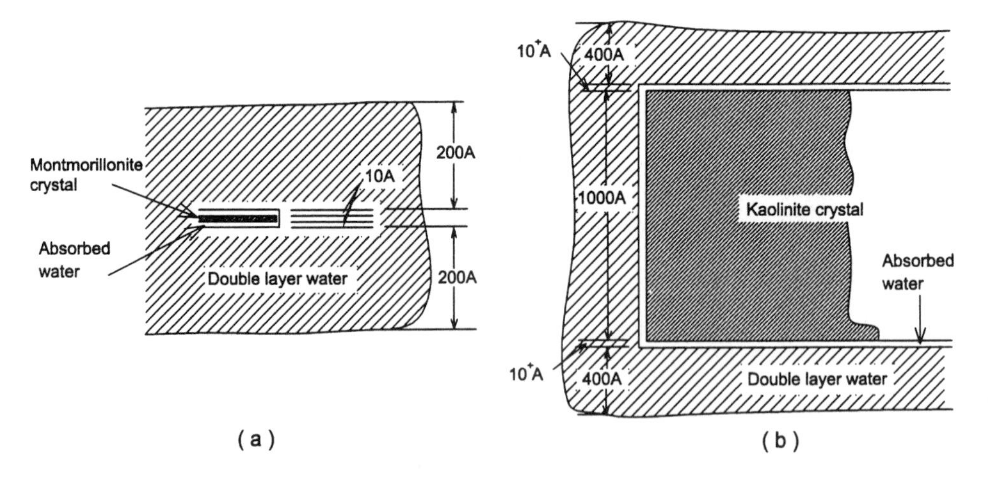

Figure 2.10 Comparison of water layer thicknesses: (a) montmorillonite crystal; (b) kaolinite crystal. (*Source*: Lambe, 1958)

derstand the environmental influences of particle arrangement. Theoretical treatment of the double layer based on electrostatics provides us with this knowledge.

2.5 GOUY-CHAPMAN THEORY OF THE DOUBLE LAYER

Gouy (1910) and Chapman (1913) independently arrived at a theoretical expression for electric potential in the double layer, which enables us to look at the effect of various characteristics of the medium on the thickness of the layer. The theory invokes and combines two well-known equations, one expressing the variation of electric potential with respect to distance known as the Poisson equation, and the other describing the relationship between ion distribution and electric potential known as the Boltzmann equation.

The Poisson equation indicates that the electric potential ψ decreases as the distance from the source, i.e., a clay particle surface with a negative electric potential, increases. The presence of charges or ions, and the dielectric constant of the medium, govern this decrease in potential. The equation is commonly expressed as

$$\frac{d^2\psi}{dx^2} = -\frac{4\pi\rho}{\epsilon} \tag{2.2}$$

where x = distance from the source; ρ = charge density; and ϵ = dielectric constant of the medium. The charge density in the medium is the contribution of the anions and cations in the medium,

$$\rho = c(v_+n_+ - v_-n_-) \tag{2.3}$$

where c = electronic charge; v = valency; n = number of ions per unit volume, and the + and − subscripts indicate cations and anions, respectively. The number of ions in Eq. (2.3) is itself governed by the electric potential ψ. The Boltzmann equation provides us with the relationship between the two for cations and anions:

$$n_+ = n_0 e^{-(vc\psi/kT)} \tag{2.4}$$

$$n_- = n_0 e^{(vc\psi/kT)} \tag{2.5}$$

where n_0 = density of ions far away from the surface, k = Boltzmann constant (1.38×10^{-23} J K^{-1}), and T = temperature in kelvins. For a clay particle, the potential ψ is negative; therefore, the cation population increases and the anion population decreases as the negative potential is increased. Considering that the potential ψ decreases with distance from the charged clay surface, the distribution

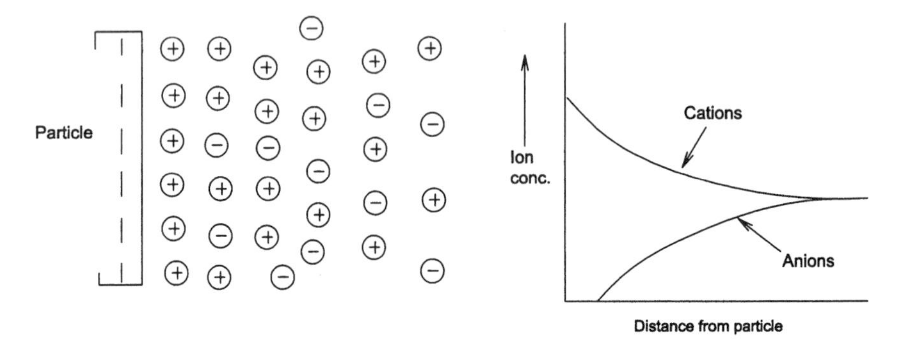

Figure 2.11 Schematic distribution of ions adjacent to a negatively charged clay surface.

of ions may be represented schematically as shown in Figure 2.11. For the simple case of monovalent ions, Eqs. (2.4) and (2.5) may be substituted in Eq. (2.3) to give

$$\rho = -2n_0 c \sinh\left(\frac{c\psi}{kT}\right) \tag{2.6}$$

Equation (2.6), when substituted in (2.2), gives us an expression for the variation of ψ with respect to x:

$$\frac{d^2\psi}{dx^2} = \frac{8\pi n_0 c}{\epsilon} \sinh\left(\frac{c\psi}{kT}\right) \tag{2.7}$$

Equation (2.7) can be simplified as

$$\frac{d^2y}{d\xi^2} = \sinh y \tag{2.8}$$

with the two dimensionless quantities

$$y = \frac{c\psi}{kT} \qquad \text{and} \qquad \xi = Kx \tag{2.9}$$

where

$$K = \sqrt{\frac{8\pi c^2 n_0}{\epsilon kT}} \tag{2.10}$$

We now seek a particular solution for Eq. (2.8) for two specific boundary conditions, one at the surface of the particle and the other at infinite distance from the particle. These conditions can be expressed as

$$y = 0 \qquad \text{and} \qquad \frac{dy}{d\xi} = 0 \qquad \text{at } \xi = \infty \tag{2.11}$$

and

$$y = z = \frac{e\psi_0}{kT} \qquad \text{at } \xi = 0 \tag{2.12}$$

where the potential at the surface of the particle is taken to be ψ_0. For these boundary conditions, the solution of Eq. (2.8) could be written as (Verwey and Overbeek, 1948)

$$e^{y/2} = \frac{e^{z/2} + 1 + (e^{y/2} - 1)e^{-\xi}}{e^{z/2} + 1 - (e^{y/2} - 1)e^{-\xi}} \tag{2.13}$$

Equation (2.13) implies an exponential decay of potential, decreasing from $y = z$ ($\psi = \psi_0$) at the surface of the particle to $y = 0$ ($\psi = 0$) at infinite distance from the particle surface. Several simplifications can be made to this solution for a number of special cases, as discussed by Verwey and Overbeek (1948). Three popular simplifications are for the cases of $z \ll 1$, $z =$ infinity, and $\xi \gg 1$. For these cases, the solutions may be obtained as

$$\psi = \psi_0 e^{-Kx} \qquad \text{for } z \ll 1 \tag{2.14}$$

$$\psi = \frac{2kT}{e} \ln \coth \frac{Kx}{2} \qquad \text{for } z = \infty \tag{2.15}$$

and

$$\psi = \frac{4kT}{e} \left\{ \frac{e^{z/2} - 1}{e^{z/2} + 1} \right\} e^{-\xi} \qquad \text{for arbitrary } z \text{ and } \xi \gg 1 \tag{2.16}$$

Equation (2.14) is often referred as the Debye-Huckel equation, and $1/K$ represents the characteristic length or thickness of the double layer (also known as the Debye length). With the known electric potentials, the ion distributions may now be obtained using Eqs. (2.4) and (2.5).

The Gouy-Chapman theory may be extended for the case of two interacting double layers when two clay surfaces approach each other. This case is shown

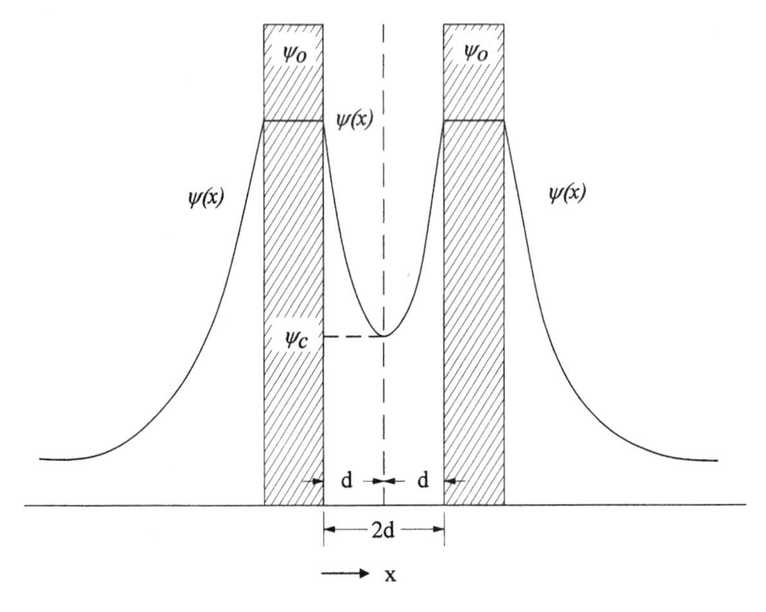

Figure 2.12 Schematic representation of the variation of electric potential between two parallel surfaces in comparison with that of a single surface.

schematically in Figure 2.12. Equation (2.8) again governs the electric potential variation, although the boundary conditions are different. Considering the symmetry of potential variation between the two surfaces, Eq. (2.8) may be solved between $x = 0$ and $x = d$, with the following boundary conditions:

$$y = y_c = \frac{c\psi_c}{kT} \qquad \text{and} \qquad \frac{dy}{d\xi} = 0 \qquad \text{at } x = d \tag{2.17}$$

and

$$y = z = \frac{c\psi_0}{kT} \qquad \text{at } x = 0 \tag{2.18}$$

Solution of Eq. (2.8) for these boundary conditions will lead to the following elliptic integral, which can be solved numerically:

$$\int_z^{y_c} \frac{dy}{\sqrt{2(\text{coshy } y - \cosh y_c)}} = -Kd \tag{2.19}$$

Tables to solve for y_c when z and d are known are given by Verwey and Overbeek (1948). Simplifications again exist for special cases when the interaction is strong (i.e., $y_c \gg 1$), and when the interaction is weak. The latter case in particular lends itself to a very simple solution. The electric potential at the midplane can be approximated in that case as the summation of the electric potentials (y_c') due to the two unperturbed double layers calculated separately. Thus,

$$y_c = 2y_c' \tag{2.20}$$

For weak interactions, ξ is necessarily large. Therefore Eq. (2.16) may be used to calculate y_c' for an individual double layer at a distance d. Substituting (2.16) in (2.20),

$$y_c = 8e^{-Kd}\frac{e^{z/2} - 1}{e^{z/2} + 1} \tag{2.21}$$

Langmuir (1938) provided yet another simplification for obtaining midplane potential assuming that the particle surfaces are so close that no anions are present between them and the potential at the particle surface is infinity. For these conditions, his solution is

$$y_c = 2 \ln \frac{\pi}{Kd} \tag{2.22}$$

This relatively simple solution compares well with the accurate solution in (2.19) beyond $y_c = 1$. For y_c values less than 1, Eq. (2.21) may be used.

2.6 FORCES OF INTERACTION BETWEEN CLAY PARTICLES

The Gouy-Chapman theory discussed in the previous section describes the electric potential variation and the ion distribution in the diffuse double layers, although with ideal configurations. The arrangement of the particles will now depend on how the particles repel or attract each other through these double layers. It is now necessary to relate the results of the Gouy-Chapman theory to forces of interaction between the particles. Because of the like nature of surface charges, the forces due to diffuse double layers are repulsive. The forces of interaction between two charged surfaces in an electrolyte can be understood in terms of the increase in pressure as the two surfaces are brought together from infinity (at which instance the pressure is zero) to a separation distance of D. The increase in pressure depends on the ion density at the midplane and is given by the contact value theorem as (Israelachvili, 1992)

$$P(D) = kT[\rho_c(D) - \rho_c(\infty)] \tag{2.23}$$

where $P(D)$ = pressure between the surfaces when the two surfaces are separated by a distance D, and $\rho_c(D)$ and $\rho_c(\infty)$ = ion densities at the midplane, when the two surfaces are separated by D and infinity, respectively.

To demonstrate the calculation of pressure between charged surfaces, let us take the case of a simple 1:1 electrolyte such as NaCl. Equation (2.6) can be used to obtain charge densities in terms of electric potentials at the midplane, ψ_c. Assuming that the midplane potential is zero when the two surfaces are separated by infinity, and considering both cations and anions, Eq. (2.23) may be expressed by

$$P(D) = kTn_0[(e^{-c\psi_c/kT} - 1) + (e^{c\psi_c/kT} - 1)] \tag{2.24}$$

which can be approximated as

$$P(D) \approx \frac{n_0 c^2 \psi_c^2}{kT} \tag{2.25}$$

Using the result from Eq. (2.21) for ψ_c, which is valid when the interaction between the surfaces is weak, and noting that $D = 2d$, $P(D)$ may be obtained as

$$P(D) = 64kTn_0 e^{-KD} \tan^2 h\left(\frac{c\psi_0}{4kT}\right) \tag{2.26}$$

The interaction between two particles is often expressed in terms of energy per unit area W_R, the numerical value of which is simply the pressure integrated with respect to D. Thus,

$$W_R(D) = 64kTn_0 \frac{e^{-KD}}{D} \tan^2 h\left(\frac{c\psi_0}{4kT}\right) \tag{2.27}$$

The repulsive energy due to double layers is not the only energy acting when two parallel charged surfaces approach each other. There are attractive forces, called van der Waals forces, between any two surfaces due to the mutual influencing of the electronic motion of atoms on the two surfaces. The fluctuating charges due to the orbital movement of electrons causes instantaneous electric moment and the resulting attraction between the two surfaces. For two parallel plates of thickness δ separated by a distance $2d$, the attractive energy due to van der Waals forces, W_A, is expressed as

$$W_A = -\frac{A}{48\pi}\left[\frac{1}{d^2} + \frac{1}{(d+\delta)^2} - \frac{2}{(d+\delta/2)^2}\right] \tag{2.28}$$

where A is known as the Hamaker constant and is of the order of 10^{-20} J. W_A is often approximated for certain limiting dimensions of d and δ. The approximate energies for the limiting cases are

$$W_A = -\frac{\delta^2 A}{32\pi d^4} \quad \text{for } d \gg \delta \tag{2.29}$$

$$W_A = -\frac{A}{48\pi}\left\{\frac{1}{d^2} - \frac{7}{\delta^2}\right\} \quad \text{for } d < \delta \tag{2.30}$$

$$W_A = -\frac{A}{48\pi d^2} \quad \text{for } d \ll \delta \tag{2.31}$$

The resultant of the repulsive (2.27) and attractive (2.28) energies governs the particle arrangement. Because of the inverse relationship to the power of d, van der Waals forces of attraction diminish rapidly as the particle spacing increases. Other kinds of energies, such as those termed Born repulsion, may be operative when the distances d are reduced to the order of an atomic diameter, which prevent interpenetration of the two surfaces. We do not have to be concerned with separations of this order, since for the most part, the particle separations are such that double-layer repulsive and van der Waals attractive energies are dominant.

The net repulsive/attractive force that is operative as the distance between two charged particles is varied is often described quantitatively using DLVO theory, after *D*erjaguin and *L*andau (1941), and *V*erwey and *O*verbeek (1948). This theory outlines the various stages of interaction between two particles, as shown schematically in Figure 2.13. Depending on the concentration of the electrolyte and the electric potential of particle surfaces, the particles may attract each other at certain distances of separation and may repel each other at other distances. Figure 2.13a shows two peaks of the total interaction energy, one on the repulsion side and the other on the attraction side. The peak on the repulsion side at a particle separation distance of d_1 (typically of the order of 1–4 nm in most systems) represents the *energy barrier*, which has to be crossed in order for particle surfaces to be attracted. The magnitude of this peak is significant for systems with thick diffuse layers associated with highly charged surfaces and electrolytes of low concentrations. Beyond the distances representing the energy barrier, weak attractive forces may be present, which are referred as *secondary minima* (to distinguish them from the *primary minima* occurring much closer to the particle surfaces). These are prominent in systems with concentrated electrolyte systems. Many times, the particle separation distances are such that the energy barrier is not crossed, and the particles are attracted to each other by weak forces corresponding to these *secondary minima*. Clay particles often flocculate

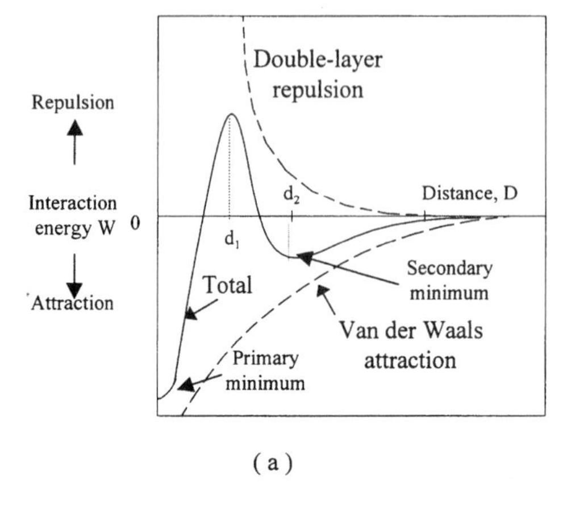

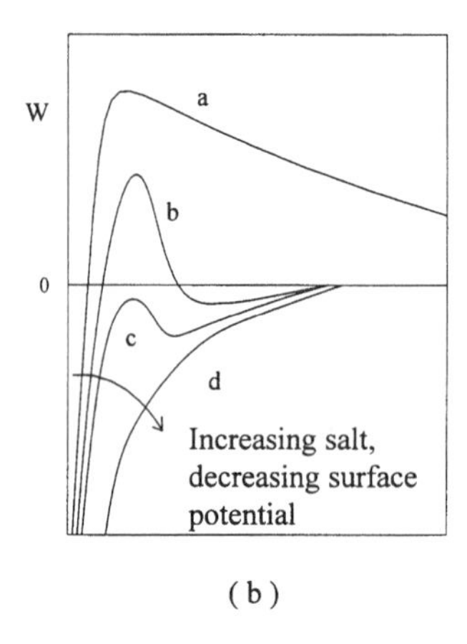

Figure 2.13 Schematic profiles of energy versus distance: (a) DLVO particle interaction energies as the distance between two particles is varied; (b) effects of salt concentration and particle surface potential on interaction energy.

in a slurry because of these weak forces. Upon slight disturbance such as shaking, the attractive forces are easily overcome and the particles are dispersed back.

Figure 2.13b shows the possible interaction curves between the particle surfaces. With particles of weak interaction such as in the cases of low electric potential and high salt concentrations, the energy barrier will become weak too. Thus a "stable" suspension of particles slowly begins to flocculate (from "a" to "b"). A situation may arise when the concentrations of the electrolyte reach "critical coagulation concentrations" beyond which repulsive barrier disappears ("c" and "d") and particles begin to flocculate rapidly.

It is apparent from the above discussion that quantitative analysis of interparticle forces is possible to a certain extent when we idealize the clay particles as parallel plates with their surface charge and the media properties known. In a realistic system, the forces of interaction may be due not only to particle surfaces approaching each other, but also to the particle edges being exposed. For instance, the edges of the particles may expose cations and exhibit double layers containing a swarm of anions. It is therefore possible that double-layer attractive energies prevail in the system between particle edges and surfaces. Quantitative analyses of these forces are also possible using the same general concepts outlined above. However, the uncertainties associated with the magnitudes of charge density on the particle surfaces and edges, and the possible presence of foreign bonding materials between particles, make such analyses less attractive. Organic matter present in the system and certain nonclay materials such as iron oxide, aluminium oxide, and carbonates bond to surfaces of one or more clay particles. These materials participate in bonding via either the stronger chemical bonds or by the van der Waals forces. Forces of bonding due to these cementing agents do not lend themselves to a thorough quantitative analysis at a particle-scale level.

It is also important to know that several differences exist between the clay particle systems in slurry form and the intact soil masses. Soil masses are rarely comprised of a single mineral. The interacting particles may belong to different minerals and as such may not conform to the ideal parallel-plate arrangement. The electric potentials of the particles may also vary significantly. The presence of silt and sand particles embedded within the fine particle matrix poses additional challenges, since the colloidal principles do not apply to those particles. In spite of these limitations, the theories explained above provide us a theoretical basis to judge qualitatively the impact of varying a given system parameter on the particle arrangement and on the soil structure.

2.7 STRUCTURE VARIATIONS DUE TO CONSOLIDATION AND COMPACTION

An external force applied to a fine-grained soil system acts against the repulsive forces between the particles and tends to reduce the average spacing between the particles. Although consolidation and compaction involve application of an external load, an important difference exists between the two processes in terms of the amount of water present in the system. In the case of consolidation, a reduction in void volume takes place under saturated conditions. The particle rearrangement during the pore space reduction process is in general conceptualized as shown in Figure 2.14. To accommodate a reduced void ratio, an increased orientation of clay particles takes place. From an open flocculated structure involving edge-to-face contacts at A, the structure changes to an increased face-to-face arrangement at B. Continued application of load will result in state C, where all clay particles are oriented face to face.

It is questionable whether a load of sufficient magnitude can ever be applied to a soil mass so that the repulsive forces existing between particles can be overcome completely and the particles can be brought to physical contact. The distribution of force from the sample level to the particle level is difficult to assess because of the intricate structure of the soil mass. Indirect attempts were made to study how the pores become compressed as the consolidation stress is increased. Griffiths and Joshi (1989) observed, using mercury porosimetry, the changes in pore size distributions of clays at various stages of consolidation starting from the liquid limit state. The volume intruded by mercury represents the volume of the pore space. Figure 2.15 shows how the volume of the pore space occupied by pores of various sizes changed as the consolidation stress is increased. With reference to the liquid limit state, an increase in consolidation stress led to a

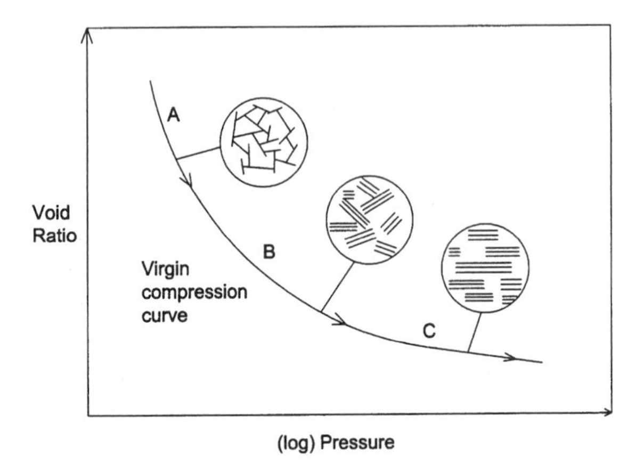

Figure 2.14 Structure changes during consolidation process.

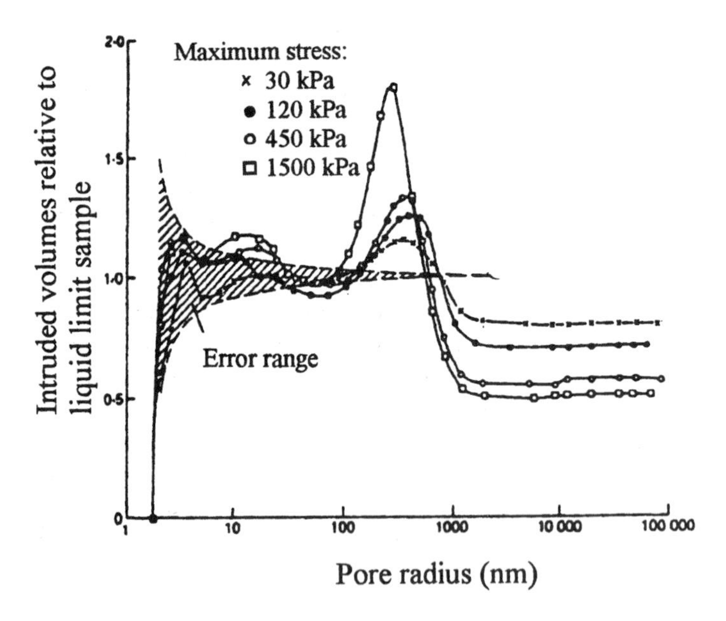

Figure 2.15 Changes in pore volumes corresponding to pores of various sizes as a result of consolidation stress. (*Source*: Griffiths and Joshi, 1989)

decrease in the pore space occupied by pores greater than about 800 nm. Although the consolidation stress results in reduction of the total void space, the volume occupied by pores in the size group of 100 to 800 nm increased as the consolidation stress is increased. This indicates the nonbrittle nature of pores in the sense that larger pores (larger than 800 nm) do not collapse completely but instead are compressed to smaller ones (between 100 and 800 nm). Of significance here is the group of pores smaller than 100 nm. Consolidation stress does not seem to affect this group to a level greater than the error range of the experiments. This indicates that at least up to a consolidation stress of 1500 kPa, the pores below a size of 100 nm could not be noticeably compressed.

In the case of compaction, however, the amount of water present in the system is limited by the molding water content. The available water may not be sufficient for complete development of the double layer. Lambe (1958) provides a conceptualization of the particle arrangement during the compaction process. As shown in Figure 2.16, the low water content at A results in high electrolyte concentration near the particle and in a thin double layer. The interparticle repulsion is reduced and therefore the particles tend to flocculate. An increased molding water content at B causes an expansion in the double layers and therefore a reduced degree of flocculation. This results in a more orderly arrangement and orientation of particles and consequently in a higher compacted density. Further increase in water content results in continued expansion of the double layer, reduction in electrolyte concentrations around the particle surfaces, and a higher degree of particle orientation. However, the density of soil decreases because of

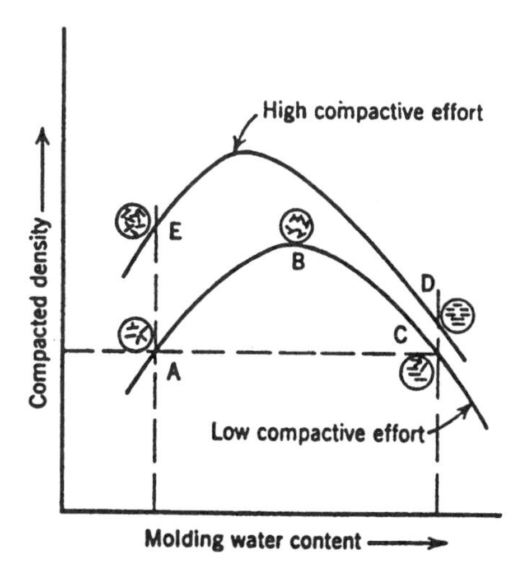

Figure 2.16 Structure changes during compaction process. (*Source*: Lambe, 1958)

dilution of soil particles. This corresponds to the state identified as C in Fig. 2.16. Higher compaction effort results in higher degree of particle orientation and a closer arrangement, because of the additional energy imparted to the system (states D and E in Fig. 2.16).

The above conceptualizations were validated to a limited extent by other investigators who made direct measurements of particle orientations (Bolt, 1955; Quigley and Thompson, 1966; Seed and Chan, 1959). Figure 2.17 shows a summary of the data obtained by Seed and Chan (1959), which concurred with the above conceptualization that the degree of particle orientation increases as the water content is changed from dry-of-optimum to wet-of-optimum.

A change in particle arrangement of soil is in general associated with changes in porosity and pore size distribution. The changes in these parameters directly influence the fluid-conducting properties of soils. Garcia-Bengochea et al. (1979) determined pore size distributions of compacted clays using mercury porosimetry in an attempt to examine how the compaction variables affect the fabric. Pore size distribution curves of a mixture of 90% silt and 10% kaolin compacted at a high effort, obtained from their studies, are shown in Figure 2.18. It is clear from these curves that two different groups of pores exist, one between 10 μm and 1 μm (referred as large pore mode), and the other concentrated around 0.1 μm (referred as small pore mode). This observation validates our earlier grouping of pores between microfabric and minifabric, or intercluster and in-

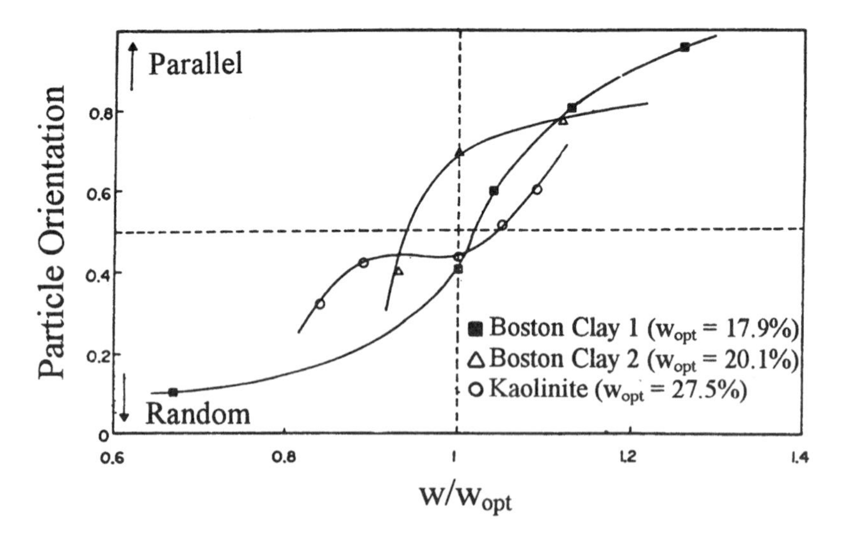

Figure 2.17 Change in particle orientation of compacted clays as the water content is increased from dry-of-optimum to wet-of-optimum. (Data from Seed and Chan, 1959)

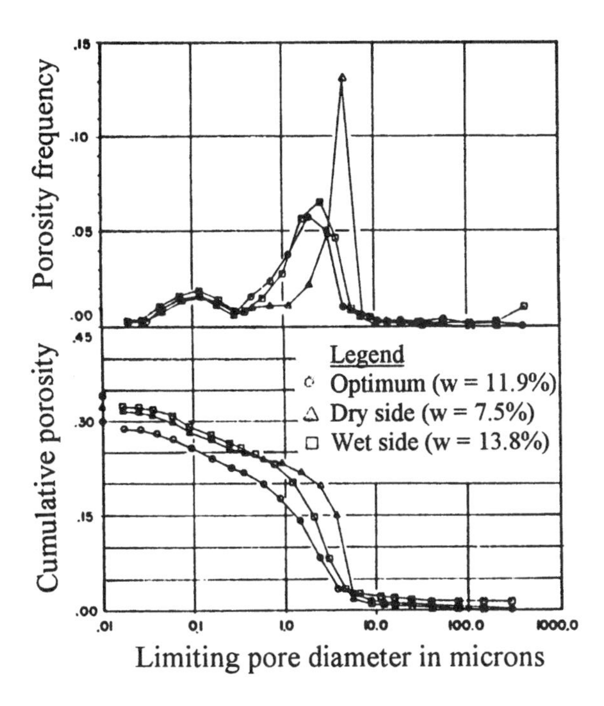

Figure 2.18 Differential pore size distribution curves for 90% silt and 10% kaolin compacted at high effort. (*Source*: Garcia-Bengochea et al., 1979)

tracluster. It may be seen in Figure 2.18 that the small pore mode is relatively the same at all states of compaction, and it is the large pore mode position and frequency that seem to change with compaction. Garcia-Bengochea et al. (1979) indicate that in general, the position and the frequency of the large pore mode are the best indicators of the change in soil fabric caused by varying molding water content and compactive effort for a given soil type. It was also established that the large pore mode controls the fluid conducting nature of the soil. Therefore, the change in soil structure must be an important consideration in any study dealing with fate and transport of contaminants in compacted soils.

2.8 ROLE OF SOIL STRUCTURE IN THE ENGINEERING BEHAVIOR OF SOILS

Soil structure, by virtue of its influence on pore size distribution, plays a key role in the fluid-conducting properties of soils. It will also govern the stability of the porous matrix. The relative ease with which fine colloid-sized particles may be detached from the intact matrix and mobilized in the pore stream has important implications in fate and transport of contaminants. Clay particles possess high surface area and adsorb significant amounts of contaminants. Their mobility implies the mobility of associated contaminants and therefore alters the contaminant transport in the subsurface. This process has been termed *facilitated transport* and has received growing attention in the last decade.

Soil structure also plays a key role in the engineering and mechanical behavior of soils. Geotechnical engineers are concerned primarily with how the soil structure influences the strength of soil. The loss of strength when an undisturbed clay is remolded, often expressed in terms of sensitivity (ratio of undisturbed to remolded strength) of soil, is governed to a large extent by the soil structure. The stress–strain relationship of a sensitive clay, shown schematically in Figure 2.19, illustrates this behavior. The fabric of undisturbed clays typically consists of open flocculated assemblages which when subjected to low strains undergo rupture of some internal bonds. The clay then collapses to a state identical to that produced through remolding of the soil wherein the structural bonds were broken initially.

Another group of clays known as dispersive clays exhibit a highly erodible nature, again because of the stability of soil structure. At high percentages of sodium in the adsorbed layer, soils become susceptible to instantaneous dispersion of particles, even when there is no pore fluid flow. This is fundamentally different from particle erosion that would occur when hydraulic gradients are

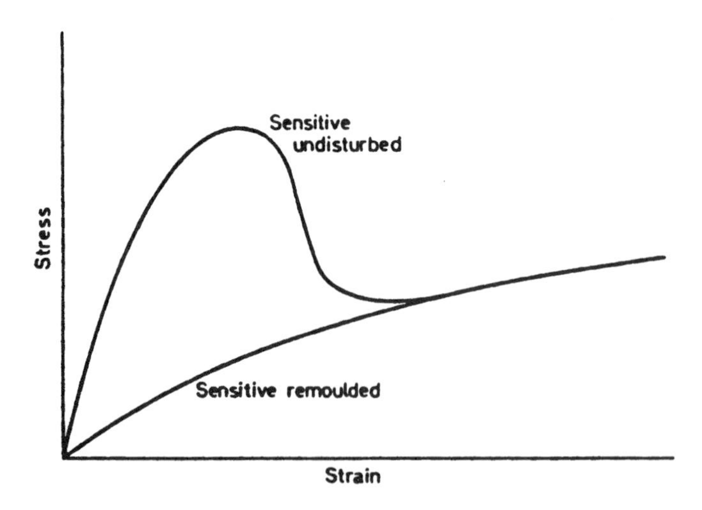

Figure 2.19 Stress–strain relationship for a sensitive clay.

excessive. Two terms, known as *sodium adsorption ratio* (SAR) and *exchangeable sodium percentage* (ESP), are used as indicators of dispersive nature of clays. They are defined as

$$\text{SAR(mEq/liter)}^{1/2} = \left\{ \frac{\text{Na}^+}{[(\text{Ca}^{2+} + \text{Mg}^{2+})/2]^{1/2}} \right\} \tag{2.32}$$

$$\text{ESP} = \frac{(\text{Na}^+)}{\text{total exchange capacity}} \tag{2.33}$$

The concentrations in the SAR refer to the pore solution, whereas the sodium concentration in the ESP refers to the exchange complex of the soil. In general, soils with ESP greater than 2% are considered to be susceptible to dispersion. Sherard et al. (1976) obtained criteria for soil dispersion using a number of experimental observations (Fig. 2.20). Three zones were identified: zone A, where dispersion might be expected; zone B, where dispersion is not likely to occur; and an intermediate zone C, where soils may or may not be dispersive.

A quantitative knowledge of the forces involved in a given structure theoretically allows one to obtain the strength of soils on a given plane. This, however, requires us to postulate the arrangement of interparticle bonds that lie across the plane. Ingles (1962) attempted to estimate the tensile strength of soils directly, in terms of the microscopic forces of interaction. A consideration of the electro-

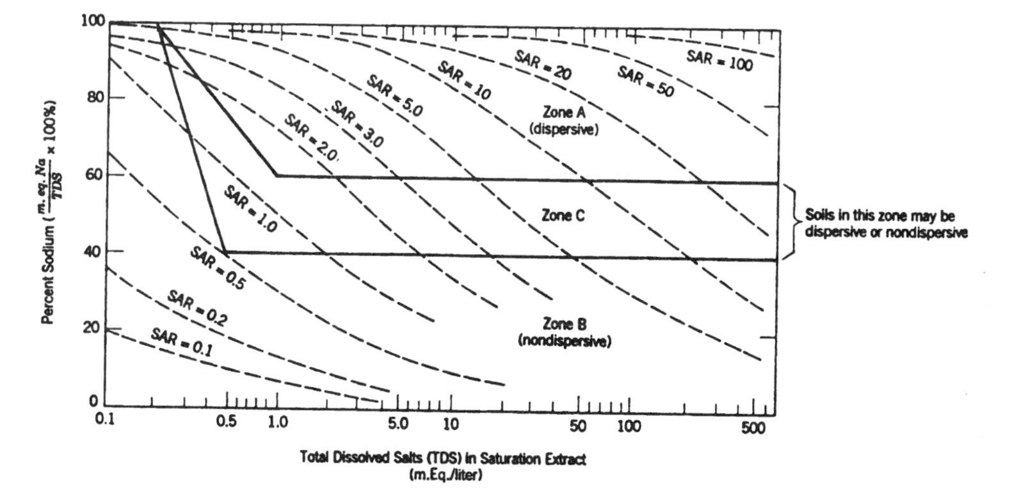

Figure 2.20 Influence of total dissolved salts and percent sodium on dispersibility. (*Source*: Sherard et al., 1976)

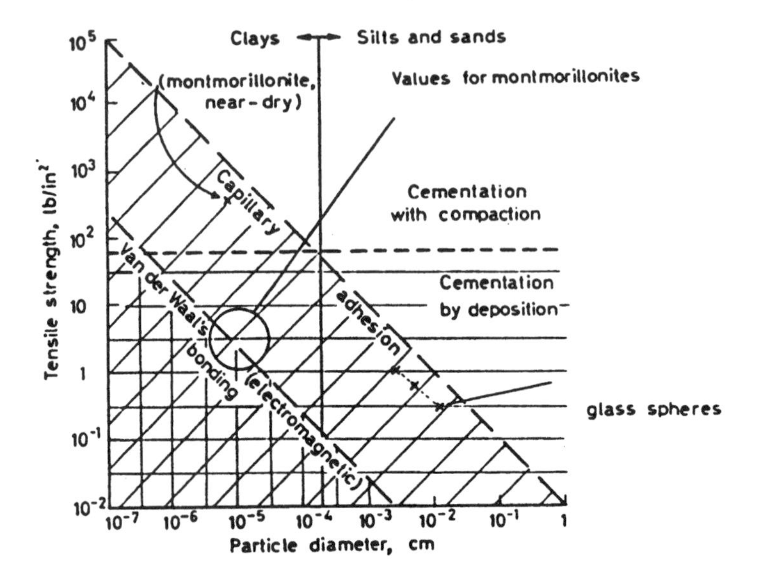

Figure 2.21 Potential contributions of various bonding mechanisms to soil strength. (*Source*: Ingles, 1962)

magnetic, electrostatic, cementation, and capillary adhesion forces permitted him to conclude that the interparticle distance is an important parameter controlling the bonds that contribute to the strength of soils. For soils consisting of fine particles, electrostatic and electromagnetic forces seem to dominate the strength, as shown in Figure 2.21. These forces, however, play a negligible role when compared to cementation forces in the case of soils with coarser particles, such as silts and sands.

REFERENCES

Bolt, G. H. (1955). Analysis of the validity of the Gouy-Chapman theory of the electric double layer. *J. Colloid Sci.*, 10:206.

Chapman, D. L. (1913). A contribution to the theory of electrocapillarity. *Phil. Mag.*, 25(6):475–481.

Collins, R. E. (1961). *Flow of Fluids Through Porous Materials*. Reinhold, New York.

Collins, K., and McGown, A. (1974). The form and function of microfabric features in a variety of natural soils. *Geotechnique*, 24(2):223–254.

Deresiewicz, H. (1958). Mechanics of granular matter. *Adv. Appl. Mech.*, 5:233–306.

Derjaguin, B. V., and Landau, L. (1941). *Acta Physicochim. URSS*, 14:633–662.

Dullien, F.A.L. (1992). Porous Media, Fluid Transport and Pore Structure. Academic Press, London.

Garcia-Bengochea, I., Lovell, C. W., and Altschaeffl, A. G. (1979). Pore distribution and permeability of silty clays. *ASCE J. Geotech. Eng.*, 105GT7:839–856.

Gouy, G. (1910). Sur la constitution de la charge electrique a la surface d'un electrolyte. *Annale Physique (Paris), Ser. 4*, 9:457–468.

Griffiths, F. J., and Joshi, R. C. (1989). Change in pore size distribution due to consolidation of clays. *Geotechnique*, 39(1):159–167.

Haughey, D. P., and Beveridge, G. S. G. (1969). Structural properties of packed beds—A review. *Can. J. Chem. Eng.*, 47:130.

Ingles, O. G. (1962). Bonding forces in soils, part 3: A theory of tensile strength for stabilized and naturally coherent soils. *Proceedings of the First Conference of the Australian Road Research Board*, Vol. 1, pp. 1025–1047.

Israelachvili, J. (1992). *Intermolecular and Surface Forces*, 2nd ed. Academic Press, San Diego.

Lambe, T. W. (1958). Compacted clay: Structure. *ASCE J. Soil Mech. Found. Div.*, Proceedings Paper 1654, pp. 682–706.

Langmuir, I. (1938). Repulsive forces between charged surfaces in water and the cause of the Jones-Ray effect. *Science*, 88:430.

Mitchell, J. K. (1993). *Fundamentals of Soil Behavior*, 2nd ed. Wiley, New York.

Olsen, H. W. (1962). Hydraulic flow through saturated clay. *Proceedings of the Ninth National Conference on Clays and Clay Minerals*, pp. 131–161.

Quigley, R. M., and Thompson, C. D. (1966). The fabric of aniostropically consolidated sensitive marine clay. *Can. Geotech. J.*, 3(2):61–73.

Ridgway, K., and Tarbuck, K. J. (1967). Random packing of spheres. *Bri. Chem. Eng.*, 12(3):384–388.

Seed, H. B., and Chan, C. K. (1959). Structure and strength characteristics of compacted clays. *ASCE J. Soil Mech. Found. Div.*, 85(SM5), Proceedings Paper 2216.

Sherard, J. L., Dunnigan, L. P., Decker, R. S., and Steele, E. F. (1976). Identification and nature of dispersive soils. *ASCE J. Geotech. Eng. Div.*, 102(GT4):187–301.

U.S. Department of Agriculture. (1951). *Soil Survey Manual, Handbook No. 18.*

van Olphen, H. (1977). *An Introduction to Clay Colloid Chemistry*, 2nd ed. Wiley-Interscience, New York.

Verwey, E. J. W., and Overbeek, J. Th. G. (1948). *Theory of the Stability of Lyophobic Colloids*. Elsevier, New York.

Yong, R. N., and Sheeran, D. E. (1973). Fabric unit interaction and soil behavior *Proccedings of the International Symposium on Soil Structure*, Gothenburg, Sweden, pp. 176–183.

3
Flow of Water in Soils

3.1 INTRODUCTION

Water is an important phase in soil for a variety of reasons. A study of flow of water in soils is essential in diverse disciplines. The energy state of water in soil pores dictates the soil strength, and this is studied in soil mechanics using the effective stress principle. Soil physics treats the fundamental aspects of these interactions as they pertain to agricultural needs, generally dealing with topsoils. Hydrology deals with water quantity and its management, particularly at regional and global scales. Groundwater flow at aquifer scale is of primary importance in this discipline. In the context of geoenvironmental engineering, we are interested in how the pore water in soils facilitates transport as well as transformation of contaminants. In this chapter, we will study the general principles of pore water flow necessary to understand contaminant transport in soils. While only water flow is considered in this chapter, the transport of contaminants in the form of dissolved and immiscible phases associated with this flow will be the central theme of Chapters 4 and 5.

The subject of groundwater flow is so broad that it is necessary to divide it into two categories, saturated and unsaturated soils, for the purpose of easier understanding. The flow of water is governed by the same driving mechanisms in both cases; however, the presence of air as the third phase requires additional considerations in the case of unsaturated soils. We will first discuss energy states of water in soils, which provide driving force for flow, before narrowing down to saturated or unsaturated soils. Our objective in this chapter is to lay down the necessary foundations to address the contaminant transport aspects in Chapters 4 and 5.

3.2 ENERGY STATES OF WATER IN SOIL

Water, like any other substance in nature, possesses two types of energy, potential and kinetic. The potential energy is associated with its position and state with reference to some datum conditions, whereas the kinetic energy is associated with its motion. Fortunately, in most cases, the velocities of water flow in soils are not significant enough to warrant kinetic energy considerations. To compensate for this simplicity however, nature has created a variety of sources for soil water to derive its potential energy from. It is the variation in the total potential energy from one location to another that is responsible for water flow in soils. Therefore, it is important for us to understand each of these potential energy sources.

One of the formal definitions for the total potential of soil water, encompassing all the possible sources, was provided by Aslyng (1963):

> [The] amount of work that must be done per unit quantity of pure water in order to transport reversibly and isothermally an infinitesimal quantity of water from a pool of pure water at a specified elevation at atmospheric pressure to the soil water (at the point under consideration).

This definition was merely an attempt to view the soil– water potential in a comprehensive manner. It is, of course, practically impossible to evaluate the potential by transporting water from one point to the other the way it is stated in the definition. Nevertheless, the definition indicates the important factors that cause variations in potential energy, viz., elevation, pressure, and chemical composition (indicated by the word, ''pure'') of water. The definition also uses a reference state (''from a pool of pure water at a specified elevation at atmospheric pressure'') and relates the water potential to that state. By separating the individual contributions, the total potential of soil water, ψ, can be expressed as

$$\psi = \psi_g + \psi_p + \psi_o \tag{3.1}$$

where ψ_g = gravitational potential due to difference in elevation alone, ψ_p = pressure potential due to difference in pressure alone, and ψ_o = osmotic potential due to differences in chemical composition of water, perhaps due to presence of solutes.

The simplest of the three sources is the gravitational potential, ψ_g, which is expressed as the product of unit weight of water, γ_w, and the elevation of the water body above a specified datum. Thus, the potential energy of a unit volume of water at a height z_1 above a reference elevation is simply

$$\psi_g = \gamma_w z_1 \tag{3.2}$$

As shown in Figure 3.1, ψ_g is independent of pressure conditions of soil water and of whether the soil is saturated or unsaturated, and is dependent only on the elevation of water body relative to the datum.

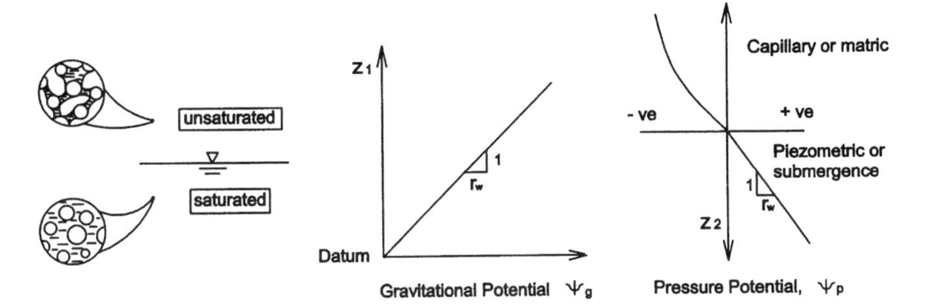

Figure 3.1 Gravitational and pressure potentials.

The pressure potential, ψ_p, can be positive or negative, depending on whether the point in soil under consideration is at hydrostatic pressure greater than atmospheric pressure, or is at a pressure lower than atmospheric pressure (Fig. 3.1). When it is positive, it implies that the point is below a free water surface, and is often referred as piezometric or submergence potential. This positive potential can be expressed as

$$\psi_p = \gamma_w z_2 \tag{3.3}$$

where z_2 = difference in elevation of the point under consideration and water table, where atmospheric conditions prevail.

When the pressure at the point under consideration is subatmospheric, the energy required to transport a unit volume of water from atmospheric conditions is governed by the capillarity principle. The potential in this case is often referred as capillary or matric potential. Water is trapped in the pore space under subatmospheric conditions because of the adsorptive forces of the soil matrix. These forces are better visualized in terms of the menisci formed when narrow capillary tubes are dipped in a body of free water (Fig. 3.2). The occurrence of meniscus in the capillary tube causes a pressure difference to develop across the interface. If the medium above the water is lighter, such as air, the meniscus is concave as shown in Figure 3.2; therefore, the pressure under the meniscus is smaller than the pressure above the meniscus (which is atmospheric in this case). This pressure difference causes the transport (or rise) of water in the capillary tube from the water surface under atmospheric pressure. From the definition of surface tension, the pressure potential can be expressed as

$$\psi_p = \frac{2\sigma}{R_m} \tag{3.4}$$

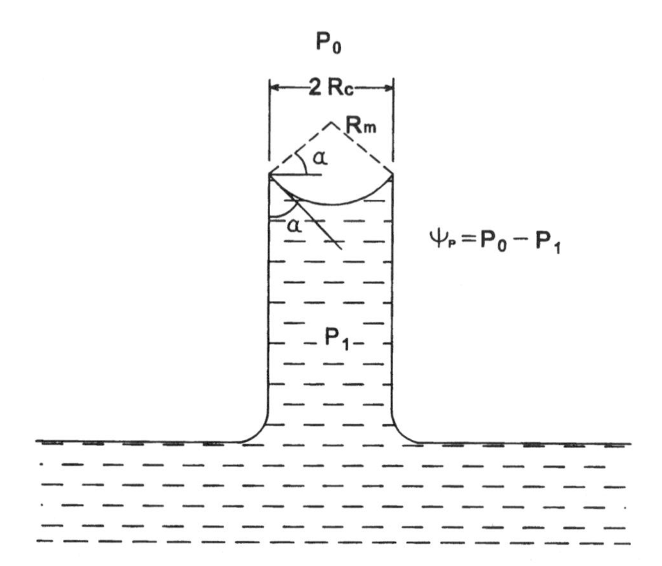

Figure 3.2 Pressure balance in a capillary tube.

where σ = surface tension of water and R_m = radius of curvature of the meniscus. From geometric considerations (see Fig. 3.2),

$$R_m = \frac{R_c}{\cos \alpha} \tag{3.5}$$

which when substituted in Eq. (3.4) yields

$$\psi_p = \frac{2\sigma \cos \alpha}{R_c} \tag{3.6}$$

The potential expressed in Eq. (3.6) is valid only when the geometric considerations assumed in the derivation, viz., cylindrical capillary tubes, prevail in soils. This is rarely the case. Even if capillary tubes can be discerned in certain granular soils, these tubes are not necessarily of constant diameter, and the adsorption of water films on to soil grains may not result in spherical menisci. For this reason, estimation of matric potential in the unsaturated zone is more complicated, as discussed later in Section 3.6. In the case of clays, the adsorptive forces responsible for subatmospheric pressures of water cannot be assessed by the capillarity theory, and considerations of electrochemistry of clay plate surfaces (which were discussed in Chapter 2) are essential. Also, in the case of clays exhibiting shrinkage and swelling behavior, the change in structure of the soil, coupled with the mechanical forces exerted by the matrix on the pore water, make it difficult to assess the matric potential.

In spite of the limitations, the capillarity principle allows us to understand the factors that cause negative potential, and its usefulness goes beyond the mere expression of matric potential. In cases where soil pores are occupied by immiscible fluids other than air, such as gasoline (popularly termed NAPLs for *N*on-*A*queous-*P*hase *L*iquids), the principle allows us to characterize the entrapment of the fluid. These issues will be dealt with in Chapter 5.

In dealing with the pressure potentials, it is important to account for the ambient pressure changes in the air. When the air pressure at the reference (atmospheric) state is different from the air pressure in soil at the point under consideration, the difference in the pressure (termed pneumatic potential) must be added to either the positive or the negative pressure potential.

The differences in osmotic potential of soil water, ψ_o, are due to the changes in chemical composition. The presence of solutes in the soil water lowers the vapor pressure and reduces the potential energy of water to a small extent. This is traditionally explained using the semipermeable membrane concept. When a membrane permeable only to water and not to the solute separates the soil solution and pure water, a process called osmosis causes the water to diffuse into the solution side in an attempt to equilibrate the solute concentrations. The hydrostatic pressure that exists across the barrier as a result of this diffusion process causes osmotic potential. Although this potential is relatively small when compared to the others described above, it is of importance when we consider the chemical transport across barriers in vapor phase. The magnitude of the osmotic potential is dependent on the concentration of the solute and its temperature. For dilute solutions, it can be approximated by (Slatyer, 1967)

$$\psi_o = MCT \tag{3.7}$$

where M = molar concentration of the solute particles, C = universal gas constant (8.32×10^7 erg/mol deg), and T = absolute temperature.

Let us carry the membrane concept further to understand the combined effect of matric potential and osmotic potential. This is best illustrated in Figure 3.3. Notice that the gravitational potential does not exist because of the fact that the elevations of both reference solution (pure water) and the soil solution are the same. The membrane to the immediate left of the soil sample separates the free soil solution with soil sample and therefore is subjected to a pressure difference due to capillary potential alone. The second membrane to the far left of the sample separates pure water and soil solution and is therefore subjected to osmotic potential alone. The combined effect of the two potentials is shown by the single membrane on the right of the sample, which separates the subatmospheric soil solution in the soil and pure water.

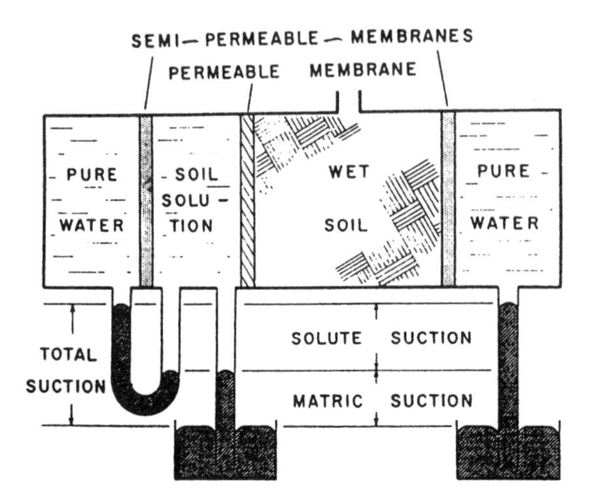

Figure 3.3 Combined and individual matric and osmotic potentials. (*Source*: Richards, 1965)

3.3 PRINCIPLES OF FLOW IN SATURATED SOILS

Flow of water in soils takes place because of the spatial differences in the energy states described in Section 3.2. In some cases, the energy states of water may be artificially varied by inducing electrical or thermal potential differences in water. The processes involved in flow of water due to thermal and electrical potential differences are called thermoosmosis and electroosmosis, respectively. These processes are being used increasingly to manipulate flow in several practical problems dealing with waste containment and site remediation. In most cases dealing with flow assessment in soils, however, the flow is dominated by the differences in only gravitational potential, ψ_g, and pressure potential, ψ_p; hence, we will restrict our subsequent discussion to flow due to these potential differences.

An important difference arises between saturated and unsaturated soils when we attempt to characterize the flow. In the case of saturated soils, it can be generally assumed that the entire pore space is participating in flow (some exceptions arise in compacted clays and other fine-grained soils, where only a portion of pore space is believed to conduct flow). This assumption makes estimation of flow quantities easier. In the case of unsaturated soils, however, only a limited pore space, which is saturated with water, will participate in flow of water. As we will discuss in Section 3.6, there is a strong interdependence between the energy states of water and the volume of conducting pore space, in the case of unsaturated soils. This complication makes it necessary and convenient to treat the saturated and unsaturated flow separately.

Earlier attempts to conceptualize flow in soils assumed that the soil mass is analogous to a bundle of parallel capillary tubes, which are not connected to each other. This assumption enabled application of our knowledge of hydraulics in pipes to characterize flow. We will start with the principles of flow in a single capillary tube (using Poiseuille's law) and work our way up to extrapolating these principles to soils (using Darcy's law). This treatment will give us valuable insight into one of the most important properties of soils—the coefficient of permeability—and some methods to evaluate this property.

3.3.1 Poiseuille's and Darcy's Laws

We will first consider flow through a single capillary tube (Fig. 3.4) due to difference in pressure potential, ΔP. This potential difference between the two ends of the tube is the driving force for flow, which takes place against the frictional resistance offered by the viscosity of water. The force F acting on an annular cylinder of water of radius r due to pressure difference ΔP is

$$F = \Delta P(\pi r^2) \tag{3.8}$$

This force is resisted by the shear stress τ at the interface between the annular ring and the rest of the water body. Equating the two forces for equilibrium,

$$F = \Delta P(\pi r^2) = \tau(2\pi rL) \tag{3.9}$$

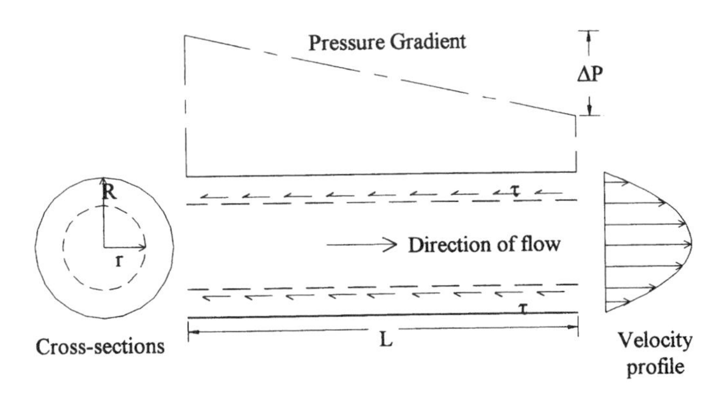

Figure 3.4 Flow through a single capillary tube.

which yields an expression for the shear stress in terms of the potential difference:

$$\tau = \frac{(\Delta P)r}{2L} \tag{3.10}$$

The fundamental law of viscosity given by Newton gives us another expression for the shear stress τ in terms of the radius-dependent velocity of flow $V_t(r)$ in the tube,

$$\tau = -\eta \frac{dV_t(r)}{dr} \tag{3.11}$$

where η = viscosity of water. Combining Eqs. (3.10) and (3.11) and integrating within the limits of r and R will yield

$$V_t(r) = \frac{\Delta P}{4L\eta}(R^2 - r^2) \tag{3.12}$$

In obtaining Eq. (3.12), we used a no-slip boundary condition, which means that at the walls of the tube ($r = R$), $V_t = 0$. Equation (3.12) indicates that the change in velocity with respect to the radius is parabolic in nature. To calculate the water discharge, i.e., the amount of flow per unit time (q_t) Eq. (3.12) must be integrated across the cross-sectional area of the tube. This amounts to the volume of a paraboloid of revolution. Leaving the integration exercise to the reader, we express the end result as

$$q_t = \left(\frac{\pi R^2}{2}\right)\left(\frac{\Delta P R^2}{4L\eta}\right) = \left(\frac{\pi R^4}{8\eta}\right)\left(\frac{\Delta P}{L}\right) \tag{3.13}$$

This is an important equation known as Poiseuille's law. It shows that the discharge through a capillary tube is directly proportional to the difference in pressure potential ΔP across the length of the tube, and to the fourth power of the radius of the tube, R. Equation (3.13) can be rewritten in terms of the area of cross-section of the tube A_t as

$$q_t = \left(\frac{R^2}{8\eta}\right)\left(\frac{\Delta P}{L}\right)(A_t) \qquad \text{where } A_t = \pi R^2 \tag{3.14}$$

which in turn enables us to express the "average" velocity of flow $\overline{V}_t$ in the capillary tube as

$$\overline{V}_t = \frac{q_t}{A_t} = \left(\frac{R^2}{8\eta}\right)\left(\frac{\Delta P}{L}\right) \tag{3.15}$$

The first term on the right-hand side of Eq. (3.15) can now be treated as a constant, expressing the proportionality between flow velocity and the pressure gradient in the capillary tube, $\Delta P/L$.

The flow considerations in a single capillary tube described above make us speculate that when we idealize a soil system as a bundle of parallel capillary tubes, the flow velocities in soils may also exhibit proportionality with respect to the difference in pressure gradient. This was indeed found to be true experimentally by Henry Darcy in 1856. A direct proportionality between the flow rate and the pressure potential difference was observed by him. Expressed mathematically,

$$q = k\left(\frac{\Delta P}{L}\right)A \tag{3.16}$$

where q is the discharge through the soil sample, A is the area of cross section of the sample, and k is a constant of proportionality. Equation (3.16) is one of the most commonly used expressions of Darcy's law for characterization of flow in soils.

The validity of Darcy's law for soil systems was studied in a number of investigations subsequent to that of Darcy. There were some investigations suggesting that the linearity between flow rate and the hydraulic gradient may not be obeyed at very low or at very high gradients. Equation (3.16) suggests that the relationship between q and gradient ($\Delta P/L$) is linear. At low hydraulic gradients, a threshold gradient was noticed in some studies below which flow did not take place in fine-grained soils (Swartzendruber, 1962). Similarly, at high hydraulic gradients corresponding to Reynolds number greater than 10, flow in soils may not obey Darcy's law (Bear, 1972). Fortunately, in a majority of cases of importance in groundwater flow, these conditions do not occur. When the conditions are such that the fluid can be categorized as Newtonian (obeying Newton's law of friction) and the flow conditions are in the laminar flow range, Darcy's law is in general believed to be applicable to soil systems.

3.3.2 Coefficient of Permeability

A comparison of Eqs. (3.14) and (3.16) indicates that the proportionality constant k encompasses the effects of both soil structure (represented by the pore radii) and the viscous property of the liquid on flow rates. The constant of proportionality is known as the coefficient of permeability, and is one of the most researched properties of soils. It exhibits the widest range of variability among all engineering properties of soils, with its value ranging from 10^2 cm/sec in gravels to 10^{-10}

cm/sec in clays. Considering that the discharge q is a product of flow velocity and area of cross section,

$$q = VA \tag{3.17}$$

the velocity of flow in soils, V, can be expressed as

$$V = k\left(\frac{\Delta P}{L}\right) \tag{3.18}$$

We should note that the velocity given by Eq. (3.18) is only superficial, since the gross area A was used to express the discharge q in Eq. (3.17) although only the porous area of the cross section contributes to flow. Recognizing that porosity n could be used to relate total and pore areas in a given cross section, the actual pore velocity, V_a may be expressed as

$$V_a = \frac{V}{n} \tag{3.19}$$

This distinction between superficial and actual velocities is important in our discussion of contaminant transport later on. Common ways of obtaining permeability coefficient typically involve observing the discharge through the soil system under an applied pressure gradient $\Delta P/L$, and using Eq. (3.16). As such, these methods are only indirect. A direct approach to determining permeability coefficient requires a thorough characterization of the pore geometry. Kozeny (1927) and Carman (1956) provided one of the earliest approaches to link permeability of granular soils with their pore geometry. This approach is based on a more rational extrapolation of Poiseuille's law of flow through capillary tubes to soil systems with irregular pore space. It provides a useful link between the coefficient of permeability and the void ratio, which is a macroscopic indicator of the pore space. For this reason, we will briefly discuss below the Kozeny-Carman equation.

We start with flow in a single capillary tube [Eq. (3.14)] and, in anticipation of having to deal with irregular cross sections, we express the size of the pore tube in terms of the hydraulic radius, R_H. Using R_H allows us to generalize the expression for any shape of the tube, and it will eventually yield an expression in terms of void ratio, e. Therefore, for a capillary tube of any geometric shape, flow can be expressed as

$$q_t = C_s \frac{R_H^2}{\eta} ia \tag{3.20}$$

where i = pressure gradient $\Delta P/L$, a = area of cross section of the irregular pore tube, and C_s = shape constant. We now apply Eq. (3.20) to a soil sample and attempt to express a and R_H in terms of measurable parameters. The area of an irregular cross section, a, can be written in terms of the total cross-sectional area, A, as

$$a = nA \tag{3.21}$$

where n = porosity of the soil sample. R_H, on the other hand, may be expressed as

$$R_H = \frac{\text{flow area}}{\text{wetted perimeter}} = \frac{\text{flow volume}}{\text{wetted area}} = \frac{eV_s}{V_s S_o} = \frac{e}{S_o} \tag{3.22}$$

where V_s = volume of solid particles in the sample, and S_o = wetted surface area per unit volume of solid particles. Note that the entire pore volume of soil is assumed to participate in flow in Eq. (3.22). Substitution of Eqs. (3.21) and (3.22) in Eq. (3.20) with the equality $n = e/(1 + e)$ yields

$$q = C_s \left(\frac{1}{\eta S_o^2}\right)\left(\frac{e^3}{1 + e}\right) iA \tag{3.23}$$

By analogy of Eq. (3.23) with Darcy's law [Eq. (3.16)], permeability coefficient k may be expressed as

$$k = C_s \left(\frac{1}{\eta S_o^2}\right)\left(\frac{e^3}{1 + e}\right) \tag{3.24}$$

Equation (3.24) gives us a valuable relationship between k and e; however, its use is possible only when we can quantify the two constants, C_s and S_o, which are difficult to estimate. This is primarily the reason why this equation has remained academic, with little potential for practical application. The proportionality between k and e is, however, a useful result, which led Casagrande to propose a simple and practical expression for k of sands as a function of e:

$$k(e) = 1.4 k_{0.85} e^2 \tag{3.25}$$

where $k_{0.85}$ = permeability at a void ratio of 0.85. In the case of coarse-grained soils such as sands, void ratio is dependent on the representative size of soil grains. This led to another useful expression between permeability and the effective size of soils (Hazen, 1911):

$$k_{(\text{cm/s})} = C D_{10}^2 \tag{3.26}$$

where D_{10} = effective size in centimeter and C = a proportionality constant varying from 90 to 120. Although the proportionality between k and void ratio or effective size of grains is usually obeyed in the case of coarse-grained soils such as sands and silts, serious discrepancies are observed when such equations are applied to clays. Because of the additional and more dominant factors influencing the soil structure, such equations are of limited use in compacted clays, as seen in the next section.

Permeability coefficient k is a lumped parameter encompassing the fluid properties (primarily the viscosity) and the pore space geometry of the media. It is convenient to separate the effects of the fluid and the media parameters. For instance, one may be interested in the effects of pore space geometry alone on permeability of a given fluid. A property defined as ''intrinsic'' or ''absolute'' permeability is often used to address permeability in terms of only media properties. Excluding the viscosity of pore fluid from Eq. (3.24), the intrinsic permeability may be expressed as

$$k_a = \left(\frac{C_s}{S_o^2}\right)\left(\frac{e^3}{1 + e}\right) \tag{3.27}$$

Equation (3.27) contains only soil parameters, and a comparison with Eq. (3.24) indicates that

$$k = \frac{k_a}{\eta} \tag{3.28}$$

While the term ''coefficient of permeability'' refers to the permeation of any fluid in general, the term ''hydraulic conductivity'' is used specifically to denote the permeability of water. This distinction, however, is not followed strictly in the literature. Of the two sets of properties—those of the fluid and the media—the effect of the former on k is easier to evaluate. k is inversely proportional to the viscosity, η [Eq. (3.28)]. For a given fluid, the variation of viscosity with respect to temperature is well known, and therefore permeability values obtained at a given temperature may be translated to those at any other temperature using this variation. When reporting the permeability, it is customary to standardize the values at 20°C (70°F), and have the user transform the standardized permeability to a specific temperature using the relation

$$\frac{k_{20}}{k_t} = \frac{\eta_t}{\eta_{20}} \tag{3.29}$$

where the subscript denotes the temperature in degrees celsius.

The effect of pore space geometry on k is not easy to evaluate, since it is a consequence of a number of variables, viz., grain size distribution, density, and structure of the soil. Soil systems with smaller particle sizes are naturally expected to be associated with smaller pore openings and are likely to exhibit lower permeability. Similarly, an increase in density of the soil, either by compaction or consolidation, will reduce the size of the pore openings and therefore will result in a smaller k. Figure 3.5 shows the effects of particle size and soil density on coefficient of permeability. It is important to notice the wide range of variability in k, 10^{-1} to 10^{-10} cm/s, which is perhaps greater than variability in any other engineering parameter of soils.

Prediction of the influence of soil structure on permeability is perhaps one of the greatest challenges that has not yet been met. This is especially true in the case of compacted clays. The soil structure, discussed in Chapter 2, is governed not only by the physicochemical properties of soils, but also by the mode of preparation in the case of compacted clays. It is possible to compact two soils such that they both have the same void ratio but differ greatly in their fabric and structure. Considering the importance of compacted clays in waste-containment structures dealt with later in this book, it is appropriate to take up this topic at some length.

3.3.3 Permeability of Compacted Clays

The preparation of compacted clays involves choice of molding water content, energy of compaction, and type of compaction—all of which together govern their dry density. A number of studies conducted over the past few decades indicate that the void ratio of compacted clays [which is uniquely related to dry density, see Eq. (1.14)] is not the only factor governing the permeability. This is attributed mainly to the microstructure, which is not unique for a given void ratio. It is difficult to characterize quantitatively the particle orientation and the resulting pore structure in compacted clays from qualitative information such as that shown in Figure 2.17. This makes it difficult to develop a theoretical basis for prediction or estimation of permeability of compacted clays in terms of the changes in soil structure.

The complexity of permeability behavior of compacted clays is best illustrated in Figure 3.6. We see that permeability varies by several orders magnitude although the four clays were prepared at nearly the same conditions of molding water content and maximum dry density. This is typical of compacted clays, and attempts to develop mathematical expressions for permeability have not been completely successful. Research during the last two decades, done in the context of waste containment, however, has resulted in useful information on the important variables affecting permeability of compacted clays. In general, the parameters responsible for such wide variation of permeability in the case of compacted clays can be grouped into three categories:

1. Variables involved in the preparation of compacted clays
2. Chemistry of the permeant
3. Testing methods

The effect of compaction variables on the microstructure and k of compacted clays was first studied by Mitchell et al. (1965) and is shown in Figure 3.7. This figure shows that, in general, reduction of two to three orders of magnitude in

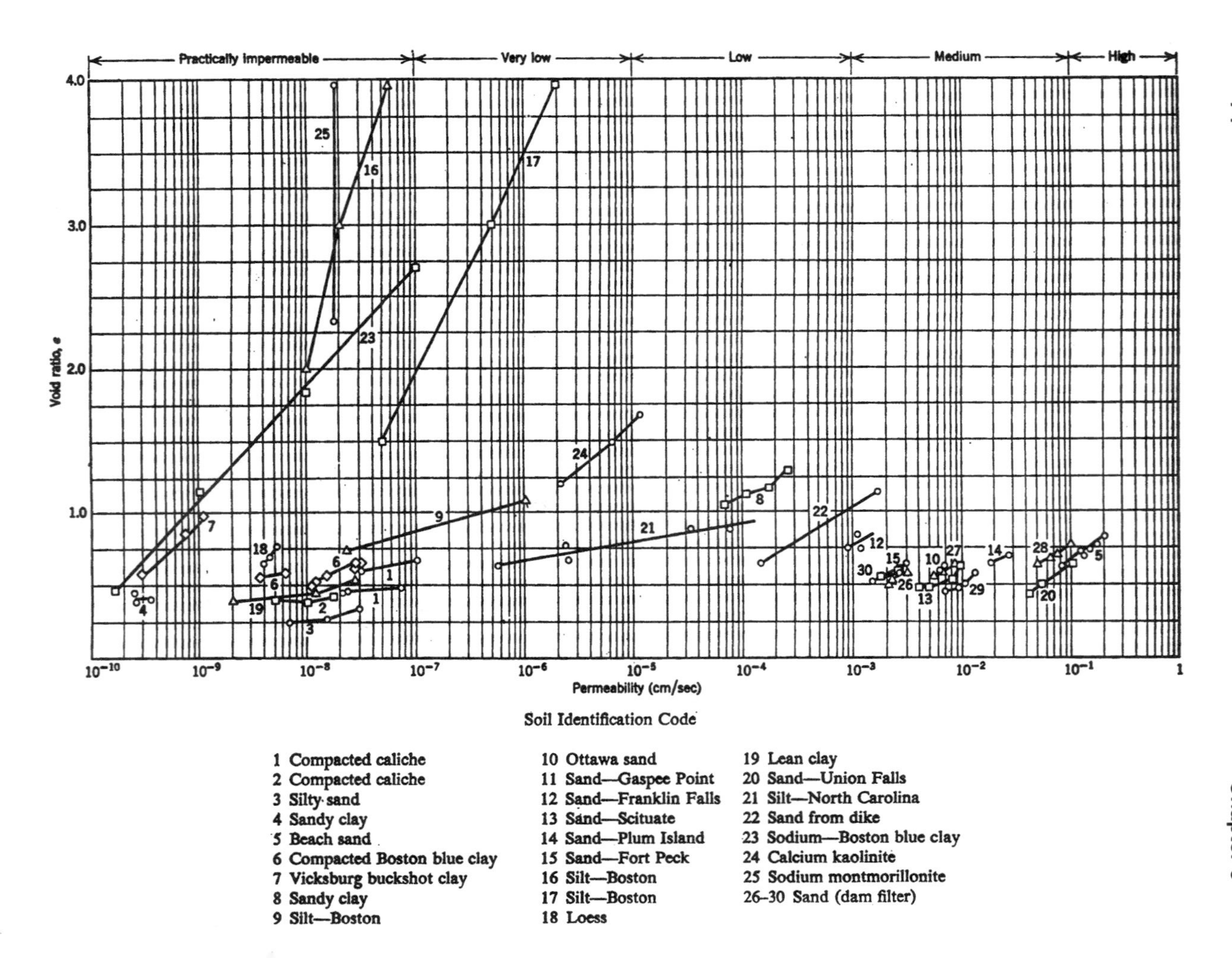
Practically impermeable
Very low
Low
Medium
High
Void ratio, e
4.0
3.0
2.0
1.0
Permeability (cm/sec)
10^{-10}
10^{-9}
10^{-8}
10^{-7}
10^{-6}
10^{-5}
10^{-4}
10^{-3}
10^{-2}
10^{-1}
1
Soil Identification Code
1 Compacted caliche
2 Compacted caliche
3 Silty sand
4 Sandy clay
5 Beach sand
6 Compacted Boston blue clay
7 Vicksburg buckshot clay
8 Sandy clay
9 Silt—Boston
10 Ottawa sand
11 Sand—Gaspee Point
12 Sand—Franklin Falls
13 Sand—Scituate
14 Sand—Plum Island
15 Sand—Fort Peck
16 Silt—Boston
17 Silt—Boston
18 Loess
19 Lean clay
20 Sand—Union Falls
21 Silt—North Carolina
22 Sand from dike
23 Sodium—Boston blue clay
24 Calcium kaolinite
25 Sodium montmorillonite
26–30 Sand (dam filter)

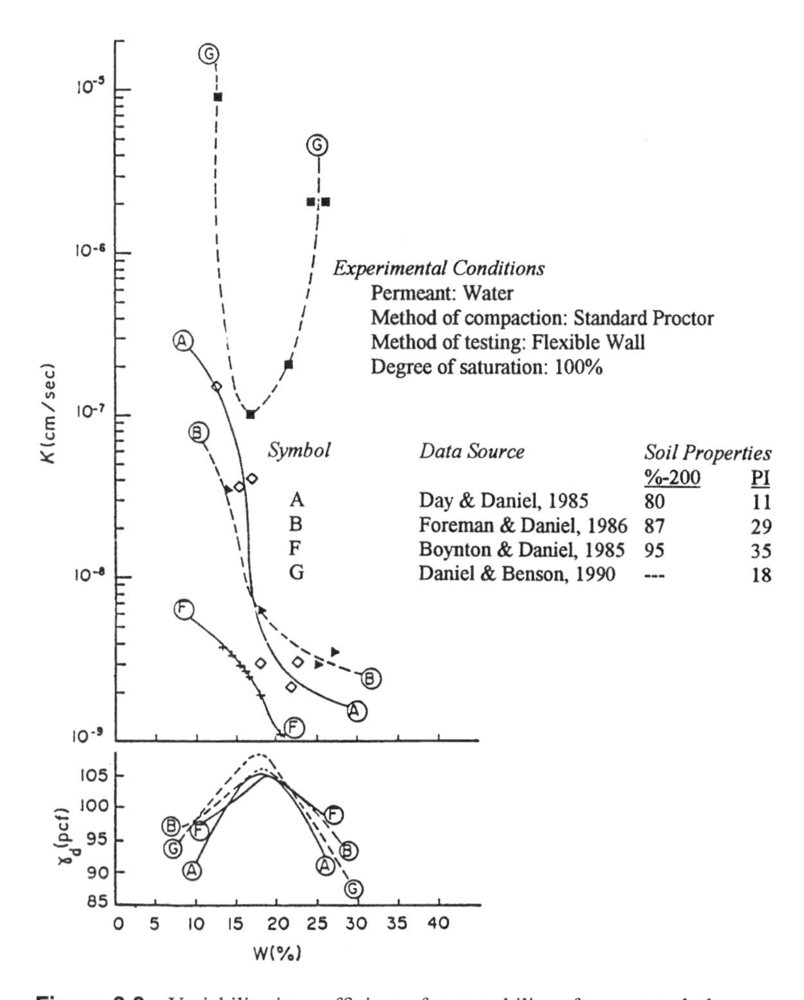

Figure 3.6 Variability in coefficient of permeability of compacted clays.

Figure 3.5 Laboratory results showing the effect of soil type (indicative of particle size) and void ratio (indicative of soil density) on coefficient of permeability. (*Source*: Lambe and Whitman, 1969)

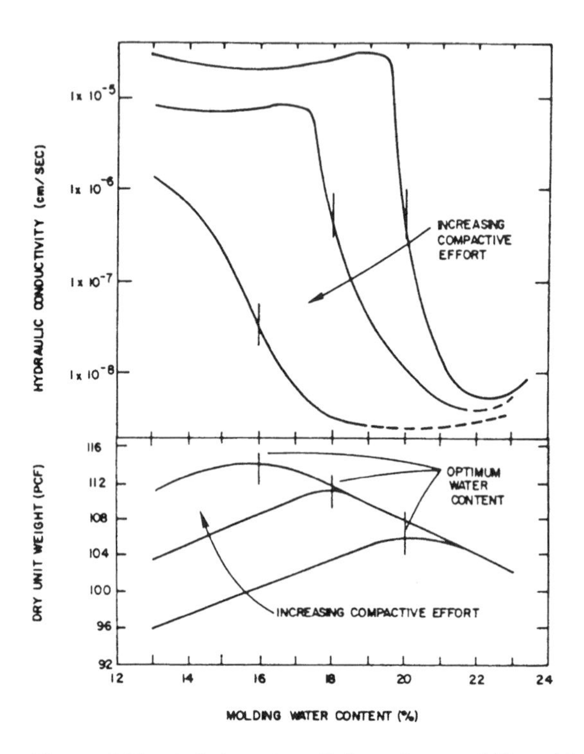

Figure 3.7 Variation of coefficient of permeability with respect to molding water content and compaction effort. (*Source*: Mitchell et al. 1965)

permeability may result as the structure of clay changes from dry-of-optimum to wet-of-optimum. The lowest permeability occurs wet-of-optimum, beyond which a slight rebound is possible. This is the basis for the current practice of preparing compacted clays wet-of-optimum in the construction of clay liners for waste containment. The permeabilities in Figure 3.7 correspond to saturated state with saturation achieved from the initial molding state. A replot of the same data in Figure 3.8 brings out the sharp contrast between sands and compacted clays in that the permeability of compacted clays does not depend uniquely on the dry density (or void ratio). For the molding water content of 19%, the permeability changed almost three orders of magnitude, although the dry density is constant. More recent studies established the effects of such variables as type and energy of compaction and size of clods used in the compaction process. Field compaction of soils can be done using different methods, including impact, static, kneading, and vibratory methods. The compaction method adopted greatly influences the

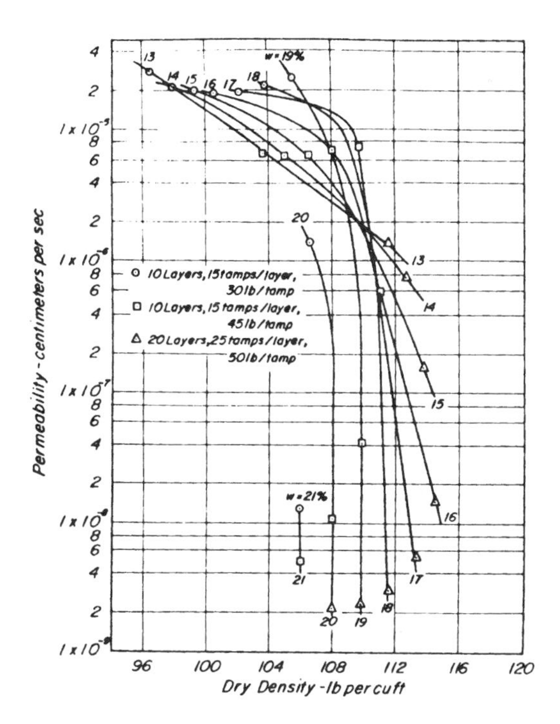

Figure 3.8 Inadequacy of dry density (which is indicative of void ratio) as a macroscopic indicator of permeability in the case of compacted clays. (*Source*: Mitchell et al. 1965)

fabric and hence the permeability of compacted clay. For a detailed description of these issues, the reader is referred to Day and Daniel (1985) and Daniel (1987). The effect of clod sizes on permeability could be understood in terms of Olsen's cluster model (Fig. 2.5). Reduction of clod sizes and elimination of interclod pores would result in smaller minifabric pores, which would in turn result in lower permeability. This has been established experimentally by Benson and Daniel (1990).

Unlike the case of sands, the chemistry of the permeant (in addition to the physical property of viscosity) influences the permeability of compacted clays. This is because of the surface chemistry which clay particles exhibit when interacting with the pore fluid during the molding stages as well as during the permeation stages. The diffuse double-layer theory discussed in Chapter 2 provides a basis on which to draw general conclusions on how the permeant chemistry alters the microstructure of compacted clays. Based on whether the permeant is promoting a flocculated or a dispersed structure, it is possible to infer at least qualitatively the effect on permeability. Factors which cause a decrease in the double-layer thickness, or Debye length [$1/K$, where K is expressed by Eq. (2.10)], promote flocculated fabric, whereas those which cause an increase promote dispersed fabric. In general, a flocculated fabric would result in an increased permeability and a dispersed fabric would cause a decrease in permeability. Based on the effects of pore fluid composition on diffuse double-layer thickness, it is therefore possible to infer the likely changes in permeability. The trends in variation of permeability as a result of changes in common pore fluid parameters are shown in Table 3.1. Experimental research summarized by Mitchell and Madsen (1987) confirm the qualitative trends shown in Table 3.1. However, many of the investigations were limited to pure chemicals and to short periods of permeability testing. The effects of several multispecies and real-life contaminants on soil structure and k have not been established yet, and these issues will continue to be an active area of research in geoenvironmental engineering.

A number of laboratory methods are currently used to measure the permeability of compacted clays. The need for accurate determination of permeability has resulted in tremendous growth of testing methods. The reader is referred to Daniel (1994) for an in-depth treatment of the various laboratory testing methods available for determination of k of compacted clays. In general, the experimental methods are grouped under two broad categories based on the equipment used: rigid wall and flexible wall permeaters. Although good agreement is found between these two methods in general (Boutwell and Rauser, 1990; Daniel 1994), the number of parameters involved in sample preparation and the differences in laboratory and field preparation create tremendous uncertainty in the determination of k. Daniel (1984) showed that field-scale permeabilities were often 10 to 1000 times greater than those estimated using laboratory methods. Accurate determination of k of compacted clays thus remains a significant challenge to geoenvironmental engineers.

Table 3.1 Trends in Permeability Variation as a Result of *Increase* in Common Pore Fluid Parameters

Parameter increased	Trend in diffuse double-layer thickness (1/K)	Resulting soil structure	Trend in permeability
pH	Increase	Dispersion	Decrease
Electrolyte concentration	Decrease	Flocculation	Increase
Cation valence	Decrease	Flocculation	Increase
Cation size	Increase	Dispersion	Decrease
Dielectric Constant	Increase	Dispersion	Decrease

3.4 GOVERNING EQUATION FOR SATURATED FLOW

We now proceed to use Darcy's law and the permeability concept to develop a mathematical equation governing the saturated flow process. We seek a general expression that controls the process in a flow domain of any given shape and under any given set of initial and boundary conditions. As we do with similar processes, we invoke the universal law of mass conservation and couple it with a cause-and-effect relationship, which in this case is Darcy's law. The reader may observe parallelism between this and other processes in solid mechanics (where force equilibrium is coupled with Hooke's law), in heat conduction (where energy conservation is coupled with the law of thermal conductivity), etc.

We apply mass conservation principle to an elemental control volume shown in Figure 3.9. With reference to this volume, the principle dictates:

$$\text{(Rate of mass output)} - \text{(rate of mass input)} = \text{(rate of change in storage in the elemental control volume)} \tag{3.30}$$

When expressed mathematically in all three dimensions, the left-hand side of Eq. (3.30) becomes

$$\left[\rho V_x - \frac{\partial}{\partial x}(\rho V_x)\right] - \rho V_x + \left[\rho V_y \frac{\partial}{\partial y}(\rho V_y)\right] - \rho V_y + \left[\rho V_z - \frac{\partial}{\partial z}(\rho V_z)\right] - \rho V_z$$

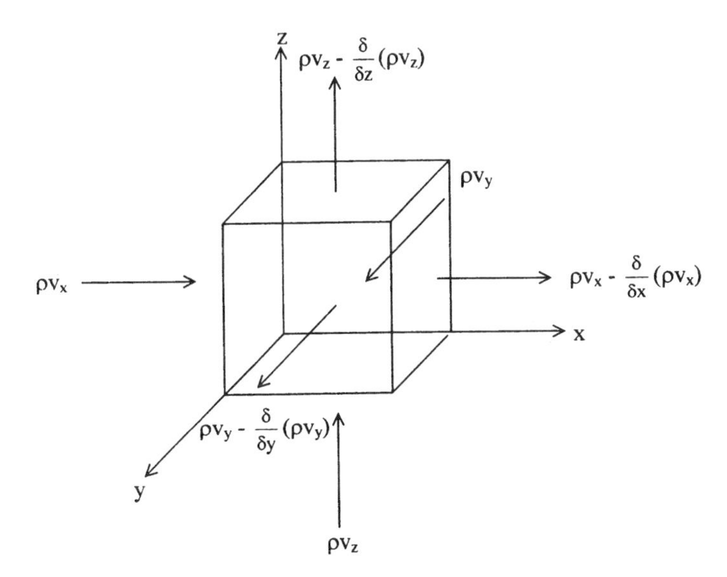

Figure 3.9 Mass conservation in an elemental control volume.

or

$$-\frac{\partial}{\partial x}(\rho V_x) - \frac{\partial}{\partial y}(\rho V_y) - \frac{\partial}{\partial z}(\rho V_z)$$

where ρ = density of the liquid, and V_x, V_y, and V_z = velocities of flow in the x, y, and z dimensions.

Storage in the elemental control volume changes due to either a change in density of the fluid, which causes expansion or contraction of the fluid, or a change in pore volume of the element. The latter is in turn due to time-dependent compressibility of the soil. For fine-grained soils, this is well known to geotechnical engineers as consolidation process. Accounting for these two factors on the right-hand-side of Eq. (3.30), the mass conservation principle may be expressed as

$$\frac{\partial}{\partial x}(\rho V_x) - \frac{\partial}{\partial y}(\rho V_y) - \frac{\partial}{\partial z}(\rho V_z) = \frac{\partial}{\partial t}(\rho n) \tag{3.31}$$

Rate of change in storage is usually expressed in terms of a parameter known as *specific storage*, S_s, which is defined as the volume of water that a unit volume of aquifer releases from storage under a unit decline in hydraulic head. This parameter can be expressed to include the effects of compressibility of the fluid and the medium as (see Freeze and Cherry, 1979)

$$S_s = \rho g(\alpha + n\beta) \tag{3.32}$$

where α = compressibility of the pore medium defined as the change in the bulk soil volume per unit change in effective stress, and β = compressibility of the fluid defined as the change in its volume per unit change in pore water pressure, u. Thus,

$$\alpha = \frac{-\,dV_T/V_T}{d\bar{\sigma}} \tag{3.33}$$

and

$$\beta = \frac{-\,dV_w/V_w}{du} \tag{3.34}$$

As will be seen later, Eq. (3.33) provides a link between saturated flow equation and Terzaghi's consolidation equation.

Expressed in terms of specific storage, the mass conservation principle [Eq. (3.31)] can be rewritten as

$$-\frac{\partial}{\partial x}(\rho V_x) - \frac{\partial}{\partial y}(\rho V_y - \frac{\partial}{\partial z}(\rho V_z) = \rho S_s\frac{\partial h}{\partial t} \tag{3.35}$$

where h= hydraulic head. We now incorporate Darcy's law to write velocity components with pressure potentials expressed in terms of hydraulic head, h:

$$\frac{\partial}{\partial x}\left(k_x \frac{\partial h}{\partial x}\right) - \frac{\partial}{\partial y}\left(k_y \frac{\partial h}{\partial y}\right) - \frac{\partial}{\partial z}\left(k_z \frac{\partial h}{\partial z}\right) = S_s \frac{\partial h}{\partial t} \tag{3.36}$$

where k_x, k_y, and k_z are permeabilities in the x, y, and z dimensions, respectively, ρ is eliminated in Eq. (3.36) because the variation of velocity components is much greater than that of ρ with respect to space coordinates. This would eliminate the $V_x(\partial\rho/\partial x)$, $V_y(\partial\rho/\partial y)$, and $V_z(\partial\rho/\partial z)$ components when the left-hand side is expanded using the chain rule. Assuming that the saturated medium is homogeneous and isotropic ($k_x = k_y = k_z = k$), Eq. (3.36) can be simplified as

$$\frac{\partial^2 h}{\partial x^2} + \frac{\partial^2 h}{\partial y^2} + \frac{\partial^2 h}{\partial z^2} = \frac{S_s}{k}\frac{\partial h}{\partial t} \tag{3.37}$$

Most groundwater-flow modeling activities involve solving Eq. (3.37) in one form or other for a specific domain of interest. While closed-form solutions exist for certain special cases, numerical methods are in general necessary. We will discuss below a few special cases to gain insight into saturated flow problems.

3.5 SPECIAL CASES OF SATURATED FLOW

3.5.1 Steady-State Saturated Flow: Flow Nets

Under steady-state conditions, when there is no change in storage of the porous media, Eq. (3.37) reduces to the well-known Laplace equation:

$$\frac{\partial^2 h}{\partial x^2} + \frac{\partial^2 h}{\partial y^2} + \frac{\partial^2 h}{\partial z^2} = \nabla^2 h = 0 \tag{3.38}$$

Starting with the simplest situation of one-dimensional flow, such as the case in a thin laboratory soil column (Fig. 3.10),

$$\frac{\partial^2 h}{\partial x^2} = 0 \tag{3.39}$$

Integration of Eq. (3.39) twice gives us the general solution,

$$h = C_1 x + C_2 \tag{3.40}$$

where C_1 and C_2 are integration constants. To evaluate these constants, we use the following simple boundary conditions:

$$h = h_1 \quad \text{at} \quad x = 0 \tag{3.41}$$

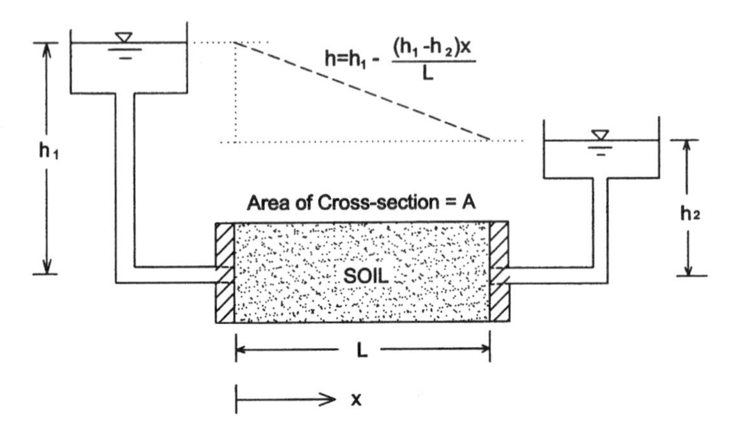

Figure 3.10 One-dimensional steady-state saturated flow.

and

$$h = h_2 \qquad \text{at} \qquad x = L \tag{3.42}$$

where $h_1 > h_2$. For these conditions, the solution becomes

$$h = h_1 - \frac{(h_1 - h_2)x}{L} \tag{3.43}$$

which simply states an intuitive fact—that the pressure potential is dissipated linearly with respect to the length of travel (see Fig. 3.10).

For two-dimensional flow, Eq. (3.38) becomes

$$\frac{\partial^2 h}{\partial x^2} + \frac{\partial^2 h}{\partial y^2} = 0 \tag{3.44}$$

For this situation, one can show that there exist two mathematical functions, $\phi(x, y)$ and $\psi(x, y)$, such that

$$\frac{\partial \phi}{\partial x} = -k\frac{\partial h}{\partial x} \tag{3.45}$$

$$\frac{\partial \psi}{\partial x} = k\frac{\partial h}{\partial y} \tag{3.46}$$

$$\frac{\partial \phi}{\partial y} = -k\frac{\partial h}{\partial y} \tag{3.47}$$

and

$$\frac{\partial \psi}{\partial y} = -k\frac{\partial h}{\partial x} \tag{3.48}$$

which satisfy the governing Eq. (3.44). To understand the usefulness of these arbitrary functions, $\phi(x, y)$ and $\psi(x, y)$, let us consider Eq. (3.45). Integration of both sides of Eq. (3.45) with respect to x yields

$$\phi(x, y) = -kh(x, y) + C_3 \tag{3.49}$$

or

$$h(x, y) = \frac{1}{k}[C_3 - \phi(x, y)] \tag{3.50}$$

where C_3 is a constant of integration. A constant pressure potential $h(x, y)$ indicates that $\phi(x, y) =$ constant. Thus, the curve represented by $\phi(x, y) =$ constant would correspond to a constant pressure potential. Such a curve is called an *equipotential line*. A number of equipotential lines can be drawn in a flow field by using different constants for $h(x, y)$. Similarly, $\psi(x, y) =$ constant represents another set of curves on the (x, y) plane. These are called *flow lines*, because the slope of these curves in the (x, y) plane is in the same direction as the resultant velocity. This follows from a simple chain-rule differentiation of $\psi(x, y) =$ constant:

$$\left(\frac{dy}{dx}\right)_{\psi} = \frac{V_y}{V_x} \tag{3.51}$$

Similarly, we can show, by differentiating $\phi(x, y) =$ constant, that

$$\left(\frac{dy}{dx}\right)_{\phi} = -\frac{V_x}{V_y} \tag{3.52}$$

From Eqs. (3.51) and (3.52), we note that $\phi(x, y)$ and $\psi(x, y)$ yield mutually perpendicular sets of curves. This mathematical outcome gave us the powerful technique to solve saturated flow problems using the graphical method of drawing flow nets. It is a simple technique, which involves drawing two sets of lines perpendicular to each other in the flow domain consistent with boundary conditions. The flow net offers a useful method to estimate pressure potentials in a flow domain and also to estimate the seepage quantities. We demonstrate this

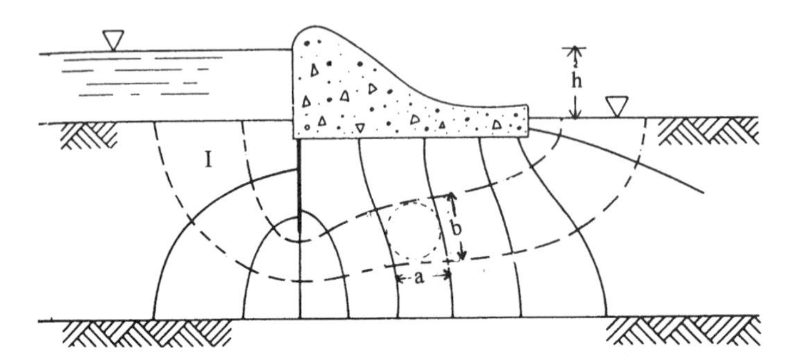

Figure 3.11 Flow net for a dam (solid and dashed lines correspond to equipotential lines and flow lines, respectively).

using the simple flow net, shown in Figure 3.11, for the case of a concrete dam with a sheet pile at the upstream end. The flow net is drawn such that the flow elements (bounded by the equipotential and flow lines) are curvilinear squares (i.e., $a = b$), as is commonly done. The drop in pressure potential between two consecutive equipotential lines, Δh, is equal to h/N_d, where h is the total difference in pressure potential and N_d = number of potential drops. For a unit length of the dam, the flow rate Δq through the flow channel I may be expressed using Darcy's law as

$$\Delta q = k\frac{\Delta h}{a}(b) = k\,\Delta h = k\frac{h}{N_d} \tag{3.53}$$

If the total number of channels in the net is N_f, the total seepage rate through the domain is

$$q = kh\frac{N_f}{N_d} \tag{3.54}$$

The construction of a flow net is independent of the permeability in an homogeneous, isotropic domain. In an anisotropic domain, however, transformation of axes is necessary in accordance with the directional differences in permeability. For a detailed explanation of flow nets in these and other, more involved cases, the reader is referred to specialized sources (Bear, 1979; Cedergren, 1989). A few examples of flow nets which are of relevance in geoenvironmental engineering are shown in Figure 3.12.

3.5.2 Transient Saturated Flow: Terzaghi's Consolidation Equation

Consolidation of porous media is the time-dependent change in void ratio due to flow of water. The gradients driving the flow are created by externally imposed

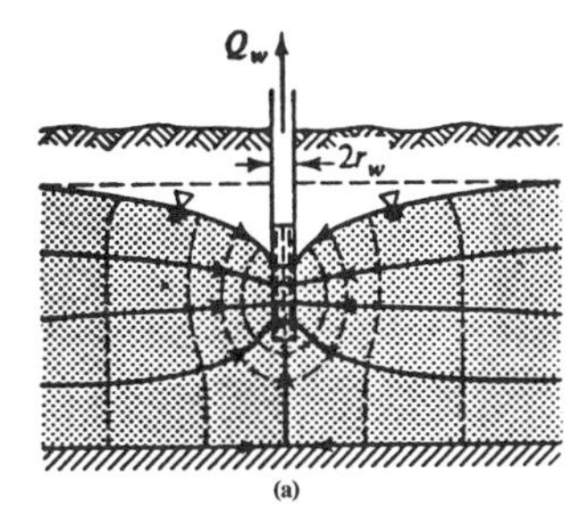

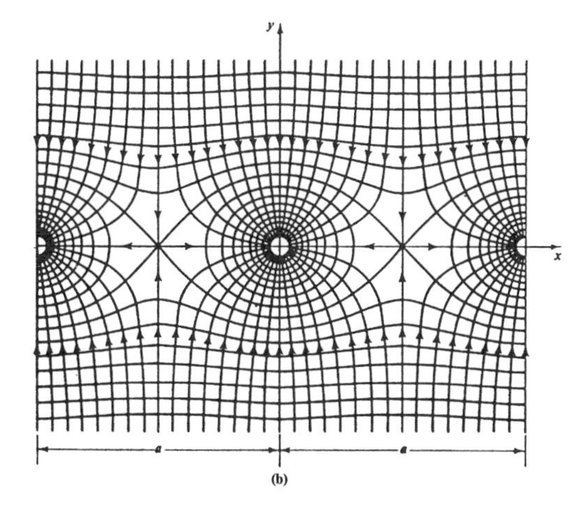

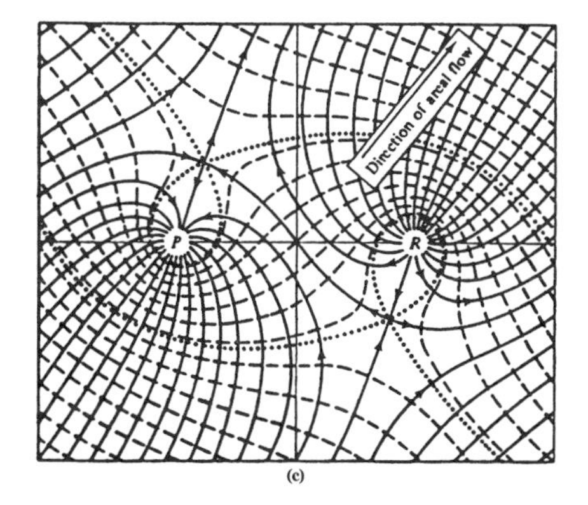

Figure 3.12 Examples of flow nets in groundwater flow problems: (a) flow toward a well in a phreatic aquifer; (b) flow in an aquifer with an infinite array of wells; (c) a region of uniform flow with a pair of pumping and recharge wells. (*Source*: Bear, 1979)

pressures in excess of existing hydrostatic pressures. The time dependence is due to the gradual transfer of excess pressure from pore water to solid grains. Since it involves flow with simultaneous change in storage (void ratio), the consolidation process can be treated as a special case of saturated flow. Using Eq. (3.32) for S_s, in Eq. (3.37), the saturated flow for a compressible soil structure can be expressed as

$$\frac{\partial^2 h}{\partial x^2} + \frac{\partial^2 h}{\partial y^2} + \frac{\partial^2 h}{\partial z^2} = \frac{\rho g \alpha}{k} \frac{\partial h}{\partial t} \tag{3.55}$$

Note that we assumed the fluid to be incompressible (i.e., $\beta = 0$). The compressibility of the medium α was defined earlier [see Eq. (3.33)]. The change in bulk volume V_T can be taken to be equal to the change in pore volume, since it is reasonable to assume that the solid particles are incompressible. α may therefore be expressed as

$$\alpha = \frac{- dV_v/V_T}{d\bar{\sigma}} = -\frac{de}{(1 + e_0)\, d\bar{\sigma}} \tag{3.56}$$

where e_0 = initial void ratio of the medium. The change in void ratio per unit change in effective stress is typically obtained in the laboratory by conducting consolidation experiments on soil samples. This is the slope of the e versus $\bar{\sigma}$ curve, a_v, shown in Figure 3.13.

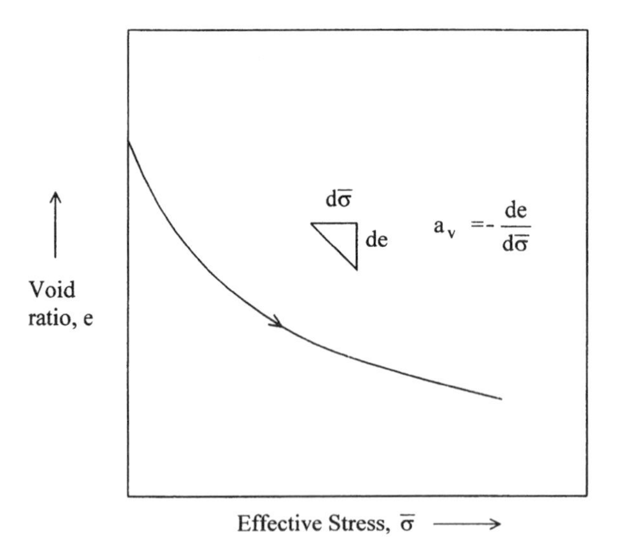

Figure 3.13 Schematic variation of void ratio with respect to effective stress.

Considering a one-dimensional consolidation process in the vertical dimension z, Eq. (3.55) may now be expressed as

$$\frac{\partial^2 h}{\partial z^2} = \frac{a_v}{1 + e_0}\left(\frac{\rho g}{k}\right)\frac{\partial h}{\partial t} \tag{3.57}$$

We recognize here that the total hydraulic head can be divided into two components, hydrostatic head and head due to excess pore pressures. Therefore,

$$h = h_0 + \frac{u}{\gamma_w} \tag{3.58}$$

where h_0 = hydrostatic head and u = excess pore pressure. Limiting ourselves to problems where h_0 is constant,

$$\frac{\partial^2 h}{\partial z^2} = \frac{1}{\gamma_w}\frac{\partial^2 u}{\partial z^2} \tag{3.59}$$

and

$$\frac{\partial h}{\partial t} = \frac{1}{\gamma_w}\frac{\partial u}{\partial t} \tag{3.60}$$

Substituting Eqs. (3.59) and (3.60) in Eq. (3.57),

$$\frac{1}{\gamma_w}\frac{\partial^2 u}{\partial z^2} = \frac{a_v}{(1 + e_0)k}\frac{\partial u}{\partial t} \tag{3.61}$$

or, in its most familiar form,

$$\frac{\partial u}{\partial t} = C_v\frac{\partial^2 u}{\partial z^2} \tag{3.62}$$

where C_v is known as the coefficient of consolidation and is given by

$$C_v = \frac{k(1 + e_0)}{a_v \gamma_w} \tag{3.63}$$

Equation (3.62) is the consolidation equation originally studied by Karl Terzaghi (1925). It is similar in form to the heat conduction equation, and a closed-form solution is possible using Laplace transforms for simple initial and boundary

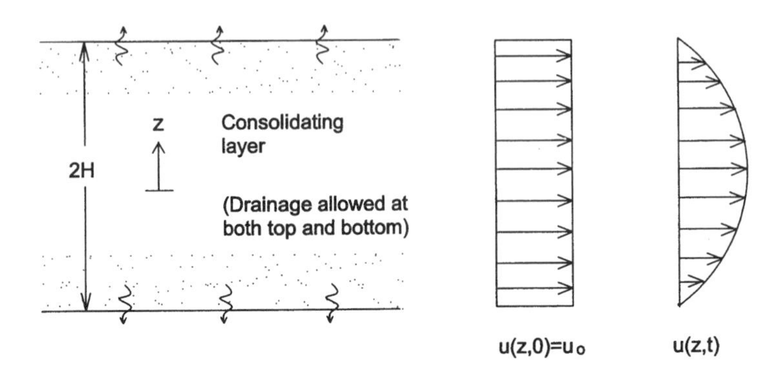

Figure 3.14 Consolidation of a soil layer drained from both top and bottom.

conditions. In a soil layer of height $2H$ (Fig. 3.14), drained at both top and bottom, the solution is symmetric about the midplane. For initial and boundary conditions given by

$$u(z, 0) = u_0(0 < z < H) \tag{3.64}$$

$$u(H, t) = 0(t > 0) \tag{3.65}$$

$$\frac{\partial u}{\partial z}(0, t) = 0(t \geq 0) \tag{3.66}$$

this becomes a problem commonly solved in geotechnical engineering. Its solution is as follows:

$$u(z, t) = \sum_{m=0}^{m=\infty} \left(\frac{2u_0}{M} \sin \frac{Mz}{H} \right) e^{-M^2 T_v} \tag{3.67}$$

in which $M = \pi(2m + 1)/2$, m is an integer, and $T_v = C_v t/H^2$, which is a dimensionless number called the time factor.

3.6 PRINCIPLES OF FLOW IN UNSATURATED SOILS

Consistent with our treatment of saturated soils, we adopt the capillary tube concept to characterize unsaturated flow. We again restrict our attention to only one driving mechanism, i.e., pressure potential, which is called matric potential when referring to the unsaturated zone. We noted that for a single capillary tube, the matric potential is inversely proportional to the radius of the tube [Eq. (3.6)].

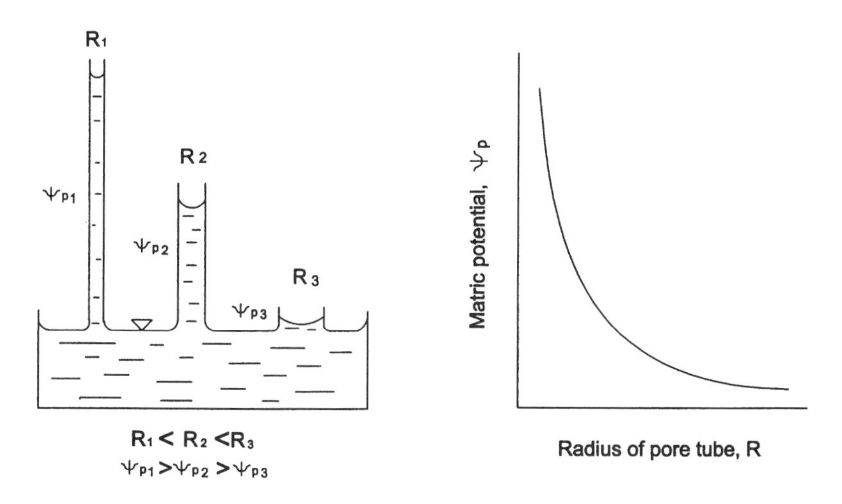

Figure 3.15 Inverse relation between matric potential and radius of capillary tube.

The inverse relation, shown in Figure 3.15, indicates that when pores in soil mass are idealized as capillary tubes, water in smaller pores possesses much higher matric potential than that in larger pores. This implies that when a saturated soil begins to become unsaturated, the larger pores are drained first, followed by smaller pores. Therefore, the volumetric water content θ, at a given unsaturated state of soil, is a reflection of the energy state, or matric potential, of water. This interdependence between water content and matric potential is an important one, and it makes the unsaturated flow more difficult to characterize than saturated flow. While a constant volume of water equal to porosity is mobilized as a result of applied difference in pressure potential in the saturated zone, the water content changes continuously with pressure potential in the case of unsaturated zone.

3.6.1 Moisture Characteristic Curves

When the information in Figure 3.15 is coupled with a knowledge of the relationship between sizes and the corresponding water-holding capacities of soil pores, we get what are known as the moisture (or soil–water) characteristic curves. These curves indicate the relationship between water content and matric potential, and are well researched in soil science. A schematic of these curves is shown in Figure 3.16. In general, coarse-grained soils contain much of their water in large pores, which can be drained at relatively modest suctions. Fine-grained soils on the other hand, have their water distributed in a range of relatively smaller pores, which require high suctions to be drained. In either case, there is a water content

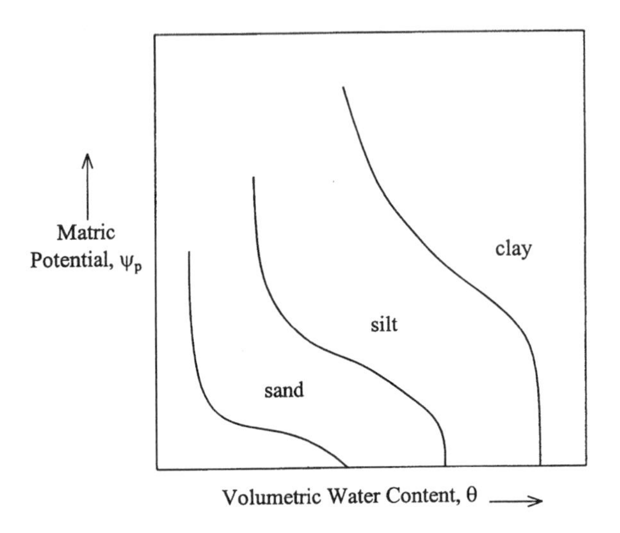

Figure 3.16 Schematic of moisture characteristic curves.

(known as irreducible or residual water content) below which water cannot be practically drained, and the moisture characteristic curve goes asymptotic to the matric potential axis. A number of relationships have been suggested in the soil science literature for moisture characteristic curves. The expressions commonly used are listed in Table 3.2. The important parameters of these expressions are the water content at complete saturation, the irreducible water content (that water which remains immobilized in the form of thin rings and films even at high suctions), and air-entry suction, which is the minimum suction necessary for air to penetrate the soil system. The empirical constants in these expressions obviously depend on the distribution of the pore sizes, since it is the pore size distribution that indicates the water-holding capacity of different sizes of pores. This is where the pore size distributions of soils discussed in Chapter 2 play an important role in governing the moisture characteristic curves. The x axis (pore sizes) of a pore-size distribution curve corresponds to the matric potential via the capillarity principle, and the y axis (proportion of a given size of pores in the soil) corresponds to the volumetric water content.

One of the complications associated with tracing the moisture characteristic curve is that it exhibits hysteresis; i.e., the shape of the curve depends on whether the soil is wetting or drying, as shown schematically in Figure 3.17. This is primarily because the relationship between water content and matric potential depends strongly on the properties of the air–water interface as discussed in Section 3.2. The contact angle of the interface with the grains tends to be different in

Table 3.2 Common Empirical Expressions for Moisture Characteristic Curves

Equation	Description of parameters	Reference
$\psi = a(n - \theta)^b/\theta^c$	n = max water content (porosity); a, b, c = empirical constants	Visser (1966)
$\left(\frac{\theta - \theta_r}{n - \theta_r}\right) = \left(\frac{\psi_a}{\psi}\right)^\lambda$	θ_r = residual water content; ψ_a = air-entry suction; λ = empirical parameter	Brooks and Corey (1966)
$\psi = f\theta^{-g}$	f, g = empirical constants	Gardner et al. (1970)
$\frac{\theta}{n} = \left(\frac{\psi_a}{\psi}\right)^{1/l}$	l = empirical constant	Campbell (1974)
$\left(\frac{\theta - \theta_r}{n - \theta_r}\right) = \frac{1}{[1 + (\alpha\lvert\psi\rvert)^p]^m}$	α = inverse of air-entry pressure; m, n = empirical constants ($m = 1 - 1/p$)	Van Genuchten (1978)

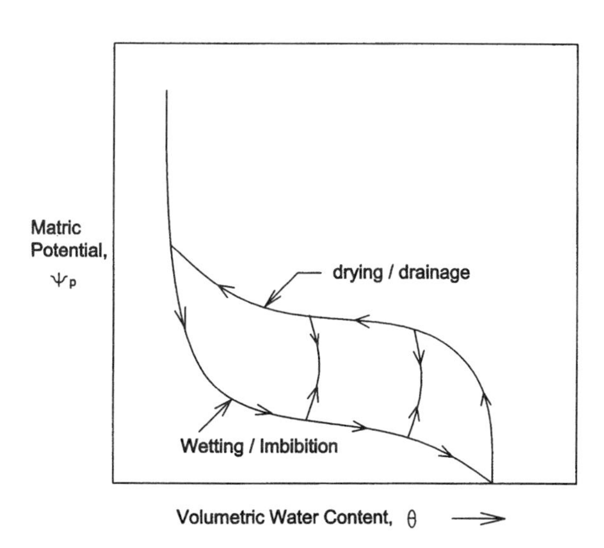

Figure 3.17 Nature of hysteresis in moisture characteristic curves.

the drying and the wetting phases, and the pores tend to be interconnected in different ways in the two phases. The characteristic curve is called a *drying* or *drainage curve* when the soil is incrementally dried from a saturated state, and a *wetting* or *imbibition curve* when the soil is wetted from an initially dry state.

3.6.2 Variation of Coefficient of Permeability with Water Content

At a given state of unsaturated flow, not all capillary tubes participate in flow. The permeability of the unsaturated zone depends on the number and the sizes of the pore tubes participating in flow. Since water content θ is a direct reflection of these parameters, it is obvious that the permeability of unsaturated soils depends on θ. From Poiseulle's law discussed in Section 3.3.1, the flow rate through a capillary tube is directly proportional to the fourth power of radius of the tube. This means that flow rates are larger when large pores are conducting flow than when small pores are conducting. Since large pores are associated with higher water contents, it is clear that permeability increases with an increase in water content. A schematic variation of unsaturated permeability with respect to water content, for various soil types, is shown in Figure 3.18. At complete saturation (corresponding to $\theta = n$), coarser soils such as sands have higher permeability because of larger pore tubes. However, after these large pores are drained, perme-

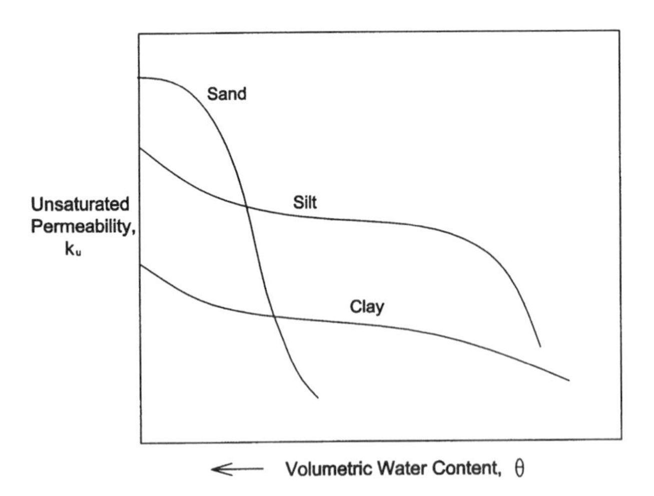

Figure 3.18 Schematic variation of unsaturated permeability with respect to volumetric water content.

ability decreases rapidly. The decrease in permeability is more gradual in the case of fine-grained soils, such as clays.

We may represent the flow in a single capillary tube as we did in Section 3.3.1, and sum up the contributions of all conducting tubes to obtain a relation for unsaturated permeability. Assuming that there are K different radii of capillary tubes of length L, all of which are conducting flow, the total flow rate may be expressed as

$$q = \frac{\pi}{8\eta}\left(\frac{\Delta P}{L}\right)\sum_{i=1}^{K} N_i R_i^4 \tag{3.68}$$

where N_i = the number of tubes of radius R_i per unit area of cross section. To know the number and size of tubes contributing to flow at a given unsaturated state, we use the moisture characteristic curve. The matric potential of an unsaturated soil can be transformed to pore tube radius using the capillarity equation (3.6). Consider the state of soil at a moisture content of θ_1, which is less than the saturated water content (n) by a small increment, $\Delta\theta$. The volumetric water content θ_1 corresponds to the state when $\Delta\theta$ has been drained through tubes of radius R_1, which is given by $2\sigma\cos\alpha/\psi_p$ [Eq. (3.6)]. For a unit length of pore tubes, $\Delta\theta$ may be expressed as

$$\Delta\theta = N_1 \pi R_1^2 \tag{3.69}$$

which gives us an expression for the number of pores of radius R_1 as

$$N_1 = \frac{\Delta\theta}{\pi R_1^2} \tag{3.70}$$

Using this process, the number of pores contributing to flow can be obtained for an unsaturated soil at any state dryer than θ_1 using small increments of $\Delta\theta$. At water content θ_1, pores of radius R_1 have all drained, and only those smaller than R_1 conduct flow. Expressing the number of pore tubes N_i in terms of $\Delta\theta$ and R_i, the total flow rate corresponding to water content θ_1 may be expressed as follows using Eq. (3.68):

$$q_u(\theta_1) = \frac{1}{8\eta}\left(\frac{\Delta P}{L}\right)\Delta\theta\sum_{i=2}^{K} R_i^2 \tag{3.71}$$

By analogy with Darcy's law, the permeability of the unsaturated soil, k_u, may be expressed as

$$k_u(\theta_1) = \frac{\Delta\theta}{8\eta}\sum_{i=2}^{K} R_i^2 \tag{3.72}$$

The radii can be transformed to the matric potentials [Eq. (3.6)], and the permeability can be expressed in terms of matric potentials as

$$k_u(\psi_{p1}) = \frac{\sigma^2 \Delta\theta}{2\eta} \sum_{i=2}^{K} \frac{1}{\psi_{pi}^2} \tag{3.73}$$

Thus, a knowledge of the moisture characteristic curve can be theoretically used to obtain an expression for the unsaturated permeability. As an alternative, the Kozeny-Carmen equation derived in Section 3.3.2 may be used to obtain an expression for unsaturated permeability in terms of void ratio and degree of saturation. Expressing Eqs. (3.21) and (3.22) in terms of the degree of saturation S, to account for the reduced flow area, the reader can deduce that

$$k_u \propto S^3 \tag{3.74}$$

Again, the above equations are more of theoretical importance because of the capillary tube analogy used in their derivation. Similar to moisture characteristic

Table 3.3 Common Empirical Relationships for Unsaturated Permeability

Function	Source
$k_u(\psi) = a(b + \psi^n)^{-1}$	Childs and Collis-George (1950)
$k_u(\psi) = \dfrac{k_s}{\left[1 + \left(\dfrac{\psi}{\psi_c}\right)^n\right]}$	Gardner (1958)
$k_u(\theta) = k_s\left(\dfrac{\theta - \theta_r}{\theta_s - \theta_r}\right)^n$	Brooks and Corey (1966)
$k_u(\psi) = \dfrac{a}{\psi}$	Baver et al. (1972)
$k_u(\psi) = k_s \exp(a\psi)$	Mualem (1976)
$k_u(\theta) = a(\theta)^n$	Marshall and Holmes (1979)
$k_u(\theta) = k_s\sqrt{\dfrac{\theta - \theta_r}{\theta_s - \theta_r}}\left\{1 - \left[1 - \left(\dfrac{\theta - \theta_r}{\theta_s - \theta_r}\right)^{1/m}\right]^m\right\}^2$	Van Genuchten (1980)

Note: k_s = saturated permeability; θ_s = volumetric water content at saturation (equal to porosity); ψ_c = matric potential for which $k_u = k_s/2$; θ_r = residual water content; and a, b, m, and n are empirical constants.

relations, several empirical relationships exist to describe the variation of k_u with either θ or ψ. Common examples of these are listed in Table 3.3. The power relation between the unsaturated permeability and the degree of saturation, predicted by the Kozeny-Carman equation, is consistent with these empirical relationships.

3.7 GOVERNING EQUATION FOR UNSATURATED FLOW

Similar to our treatment of saturated flow, we will express unsaturated flow mathematically by coupling mass conservation and Darcy's law. Because hydraulic conductivity is not constant and is highly dependent on water content, certain simplifications that were possible earlier cannot be made in the case of unsaturated flow. Referring back to the elemental control volume of Figure 3.9, its storage constantly changes as a result of flow, although the fluid and the medium may be incompressible. Therefore, unsaturated flow is always transient. Since the water content is always less than the total porosity, and can be expressed as a product of degree of saturation S and porosity n, the right-hand side of Eq. (3.31) may be modified to reflect unsaturated flow as

$$\begin{aligned} -\frac{\partial}{\partial x}(\rho V_x) - \frac{\partial}{\partial y}(\rho V_y) - \frac{\partial}{\partial z}(\rho V_z) &= \frac{\partial}{\partial t}(\rho S n) \\ &= \rho S \frac{\partial n}{\partial t} + S n \frac{\partial \rho}{\partial t} + \rho n \frac{\partial S}{\partial t} \end{aligned} \tag{3.75}$$

The effects due to compressibility of the pore fluid and the medium are in general considered to be negligible compared with the change in degree of saturation. Equation (3.75) can then be simplified as

$$\frac{\partial}{\partial_x}(V_x) + \frac{\partial}{\partial_y}(V_y) + \frac{\partial}{\partial_z}(V_z) = \frac{\partial}{\partial_t}(Sn) = \frac{\partial \theta}{\partial t} \tag{3.76}$$

Incorporating Darcy's law,

$$\frac{\partial}{\partial x}\left[k_u(\psi)\frac{\partial \psi}{\partial x}\right] + \frac{\partial}{\partial y}\left[k_u(\psi)\frac{\partial \psi}{\partial y}\right] + \frac{\partial}{\partial z}\left[k_u(\psi)\frac{\partial \psi}{\partial z}\right] = \frac{\partial \theta}{\partial t} \tag{3.77}$$

Unlike in the case of saturated flow, further simplification of Eq. (3.77) is not possible because of the strong dependence of k_u on ψ. Analogous to the specific storage term in saturated flow problem, a term called specific moisture capacity $C(\psi)$ is used to express the right-hand-side of Eq. (3.77). It is defined as the change in moisture content per unit change in pressure head, i.e.,

$$C(\psi) = \frac{d\theta}{d\psi} \tag{3.78}$$

$C(\psi)$ is equivalent to the slope of moisture characteristic curve. Choosing to write the right-hand-side of Eq. (3.77) in terms of matric potential ψ and substituting Eq. (3.78) yields:

$$\frac{\partial}{\partial x}\left[k_u(\psi)\frac{\partial \psi}{\partial x}\right] + \frac{\partial}{\partial y}\left[k_u(\psi)\frac{\partial \psi}{\partial y}\right] + \frac{\partial}{\partial z}\left[k_u(\psi)\frac{\partial \psi}{\partial z}\right] = C(\psi)\frac{\partial \psi}{\partial t} \tag{3.79}$$

Equation (3.79) is the well-known Richard's equation. Because of its high nonlinearity, numerical methods are commonly used for its solution. Analytical solutions pioneered by Philip (1957) do exist, however, for some simplified initial and boundary conditions. In certain problems where infiltration of water into the subsurface is of importance rather than prediction of matric potentials throughout the unsaturated zone, simpler conceptualizations exist. Solution of Eq. (3.79) is not necessary in such cases. The water intake at the ground surface is controlled primarily by the method of introduction (ponding, sprinkling, etc.) and the available input, and antecedent soil moisture conditions. Infiltration rates into a dry soil, for instance, are initially large; however, as top soil becomes saturated, the gradients driving the flow decrease and the rates approach saturated hydraulic conductivity with time. A schematic of infiltration rate variation with respect to time is shown in Figure 3.19. Some common empirical expressions for estimation of infiltration rate are listed in Table 3.4.

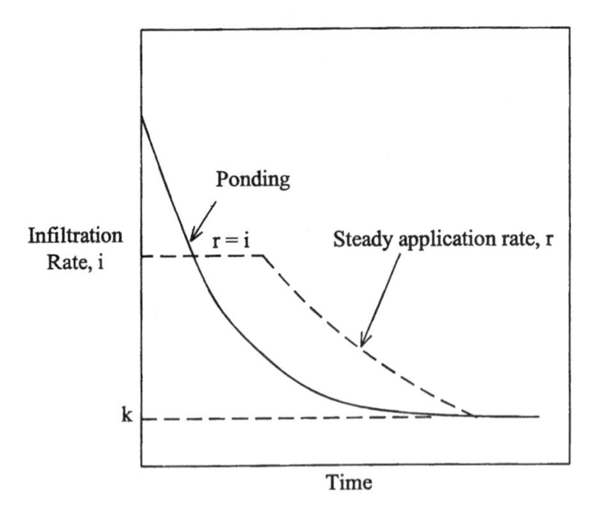

Figure 3.19 Schematic of infiltration rate variation with respect to time.

Table 3.4 Empirical Expressions for Estimating Infiltration Rates into Unsaturated Soil

Equation	Source
$i = i_c + (b/I)$	Green and Ampt (1911)
$i = Bt^{-n}$	Kostiakov (1932)
$i = i_c + (i_o - i_e)e^{-kt}$	Horton (1940)
$i = i_e + (3/2)t^{-1/2}$	Philip (1957)
$i = i_c + a(M - I)^n;\ I < M$	Holton (1961)
$i = i_c;\ I > M$	

Note: i is the infiltration rate (cm^3 of water per cm^2 area per hour); I is the cumulative volume (cm^3) of water infiltrated in time, t; a, B, M, S, n, and k are constants; i_c is the steady-state infiltration rate; i_e is the initial infiltration rate; and i_o is the final infiltration rate.
Source: National Research Council (1990)

3.8 ANALYTICAL SOLUTIONS OF STEADY AND TRANSIENT FLOW IN SOILS

In some cases, simple analytical solutions for saturated flow can be obtained by integrating the mass conservation equation directly. A good number of problems dealing with discharge and recharge wells can be solved by simplifying the problem in a single Cartesian dimension, or in radial dimensions. The Dupuit assumption is especially useful in such cases. This assumption states that when the change in slope of phreatic surface (that of the groundwater table at atmospheric pressure) is small, the equipotential and flow surfaces can be treated as vertical

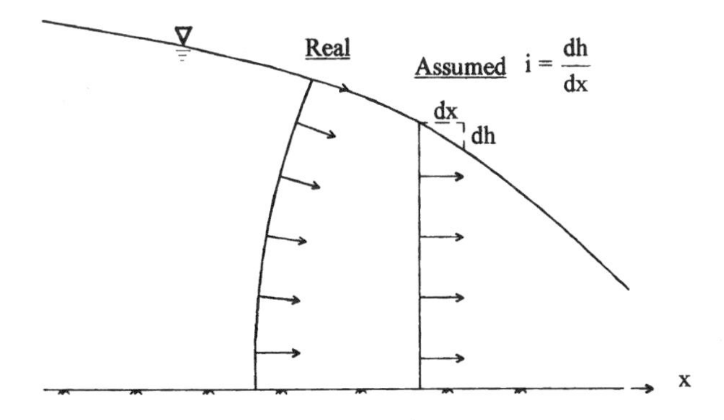

Figure 3.20 The Dupuit assumption.

Table 3.5 Analytical Solutions for Steady and Transient Flow in Soils

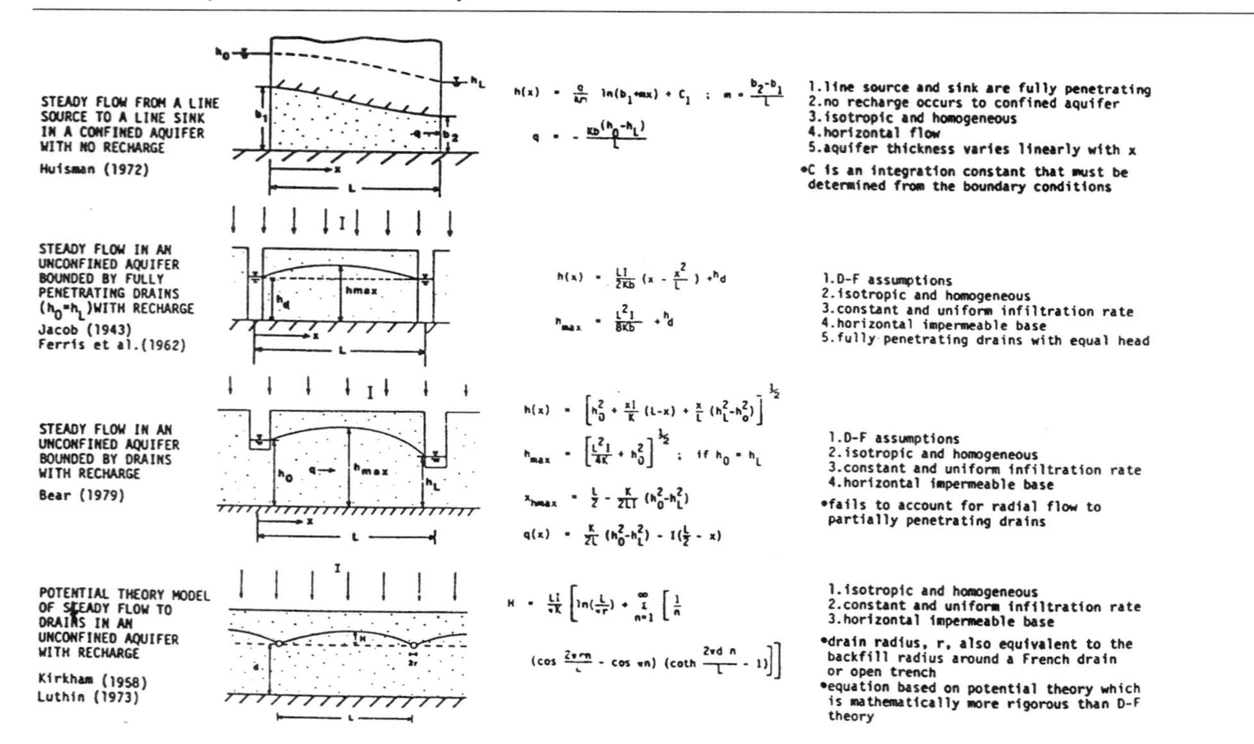

Description	Equations	Assumptions
STEADY FLOW FROM A LINE SOURCE TO A LINE SINK IN A CONFINED AQUIFER WITH NO RECHARGE Huisman (1972)	$h(x) = -\frac{q}{Km}\ln(b_1+mx) + C_1 \;;\; m = \frac{b_2-b_1}{L}$ $q = -\frac{Kb(h_0-h_L)}{L}$	1. line source and sink are fully penetrating 2. no recharge occurs to confined aquifer 3. isotropic and homogeneous 4. horizontal flow 5. aquifer thickness varies linearly with x •C is an integration constant that must be determined from the boundary conditions
STEADY FLOW IN AN UNCONFINED AQUIFER BOUNDED BY FULLY PENETRATING DRAINS ($h_0=h_L$) WITH RECHARGE Jacob (1943) Ferris et al. (1962)	$h(x) = \frac{LI}{2Kb}\left(x - \frac{x^2}{L}\right) + h_d$ $h_{max} = \frac{L^2 I}{8Kb} + h_d$	1. D-F assumptions 2. isotropic and homogeneous 3. constant and uniform infiltration rate 4. horizontal impermeable base 5. fully penetrating drains with equal head
STEADY FLOW IN AN UNCONFINED AQUIFER BOUNDED BY DRAINS WITH RECHARGE Bear (1979)	$h(x) = \left[h_0^2 + \frac{xI}{K}(L-x) + \frac{x}{L}(h_L^2-h_0^2)\right]^{1/2}$ $h_{max} = \left[\frac{L^2 I}{4K} + h_0^2\right]^{1/2} \;;\; \text{if } h_0 = h_L$ $x_{hmax} = \frac{L}{2} - \frac{K}{2LI}(h_0^2-h_L^2)$ $q(x) = \frac{K}{2L}(h_0^2-h_L^2) - I\left(\frac{L}{2} - x\right)$	1. D-F assumptions 2. isotropic and homogeneous 3. constant and uniform infiltration rate 4. horizontal impermeable base •fails to account for radial flow to partially penetrating drains
POTENTIAL THEORY MODEL OF STEADY FLOW TO DRAINS IN AN UNCONFINED AQUIFER WITH RECHARGE Kirkham (1958) Luthin (1973)	$H = \frac{LI}{\pi K}\left[\ln\left(\frac{L}{\pi r}\right) + \sum_{n=1}^{\infty}\left[\frac{1}{n}\left(\cos\frac{2\pi rn}{L} - \cos \pi n\right)\left(\coth\frac{2\pi d\, n}{L} - 1\right)\right]\right]$	1. isotropic and homogeneous 2. constant and uniform infiltration rate 3. horizontal impermeable base •drain radius, r, also equivalent to the backfill radius around a French drain or open trench •equation based on potential theory which is mathematically more rigorous than D-F theory

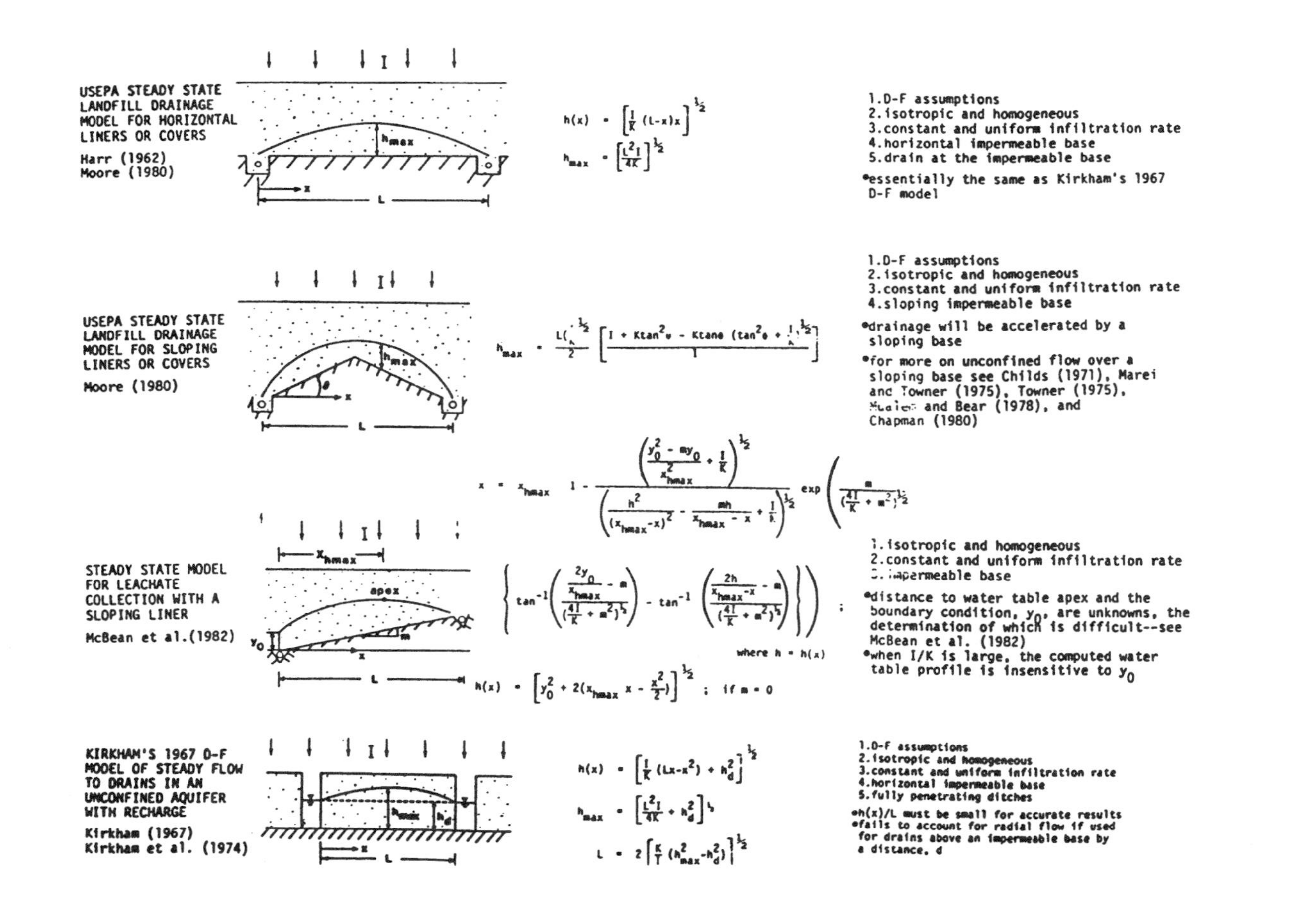

Model	Equations	Assumptions / Comments
USEPA STEADY STATE LANDFILL DRAINAGE MODEL FOR HORIZONTAL LINERS OR COVERS Harr (1962) Moore (1980)	$h(x) = \left[\frac{I}{K}(L-x)x\right]^{1/2}$ $h_{max} = \left[\frac{L^2 I}{4K}\right]^{1/2}$	1. D-F assumptions 2. isotropic and homogeneous 3. constant and uniform infiltration rate 4. horizontal impermeable base 5. drain at the impermeable base •essentially the same as Kirkham's 1967 D-F model
USEPA STEADY STATE LANDFILL DRAINAGE MODEL FOR SLOPING LINERS OR COVERS Moore (1980)	$h_{max} = \frac{L(\frac{I}{K})^{1/2}}{2}\left[\frac{I + K\tan^2\phi - K\tan\phi\,(\tan^2\phi + \frac{I}{K})^{1/2}}{I}\right]$	1. D-F assumptions 2. isotropic and homogeneous 3. constant and uniform infiltration rate 4. sloping impermeable base •drainage will be accelerated by a sloping base •for more on unconfined flow over a sloping base see Childs (1971), Marei and Towner (1975), Towner (1975), Mualem and Bear (1978), and Chapman (1980)
STEADY STATE MODEL FOR LEACHATE COLLECTION WITH A SLOPING LINER McBean et al. (1982)	$x = x_{hmax}\left[1 - \frac{\left(\frac{y_0^2 - my_0}{x_{hmax}^2} + \frac{I}{K}\right)^{1/2}}{\left(\frac{h^2}{(x_{hmax}-x)^2} - \frac{mh}{x_{hmax}-x} + \frac{I}{K}\right)^{1/2}}\exp\left(\frac{m}{(\frac{4I}{K}+m^2)^{1/2}}\left\{\tan^{-1}\left(\frac{\frac{2y_0}{x_{hmax}} - m}{(\frac{4I}{K}+m^2)^{1/2}}\right) - \tan^{-1}\left(\frac{\frac{2h}{x_{hmax}-x} - m}{(\frac{4I}{K}+m^2)^{1/2}}\right)\right\}\right)\right]$; where $h = h(x)$ $h(x) = \left[y_0^2 + 2(x_{hmax}x - \frac{x^2}{2})\right]^{1/2}$; if $m = 0$	1. isotropic and homogeneous 2. constant and uniform infiltration rate 3. impermeable base •distance to water table apex and the boundary condition, y_0, are unknowns, the determination of which is difficult--see McBean et al. (1982) •when I/K is large, the computed water table profile is insensitive to y_0
KIRKHAM'S 1967 D-F MODEL OF STEADY FLOW TO DRAINS IN AN UNCONFINED AQUIFER WITH RECHARGE Kirkham (1967) Kirkham et al. (1974)	$h(x) = \left[\frac{I}{K}(Lx - x^2) + h_d^2\right]^{1/2}$ $h_{max} = \left[\frac{L^2 I}{4K} + h_d^2\right]^{1/2}$ $L = 2\left[\frac{K}{I}(h_{max}^2 - h_d^2)\right]^{1/2}$	1. D-F assumptions 2. isotropic and homogeneous 3. constant and uniform infiltration rate 4. horizontal impermeable base 5. fully penetrating ditches •h(x)/L must be small for accurate results •fails to account for radial flow if used for drains above an impermeable base by a distance, d

Table 3.5 Continued

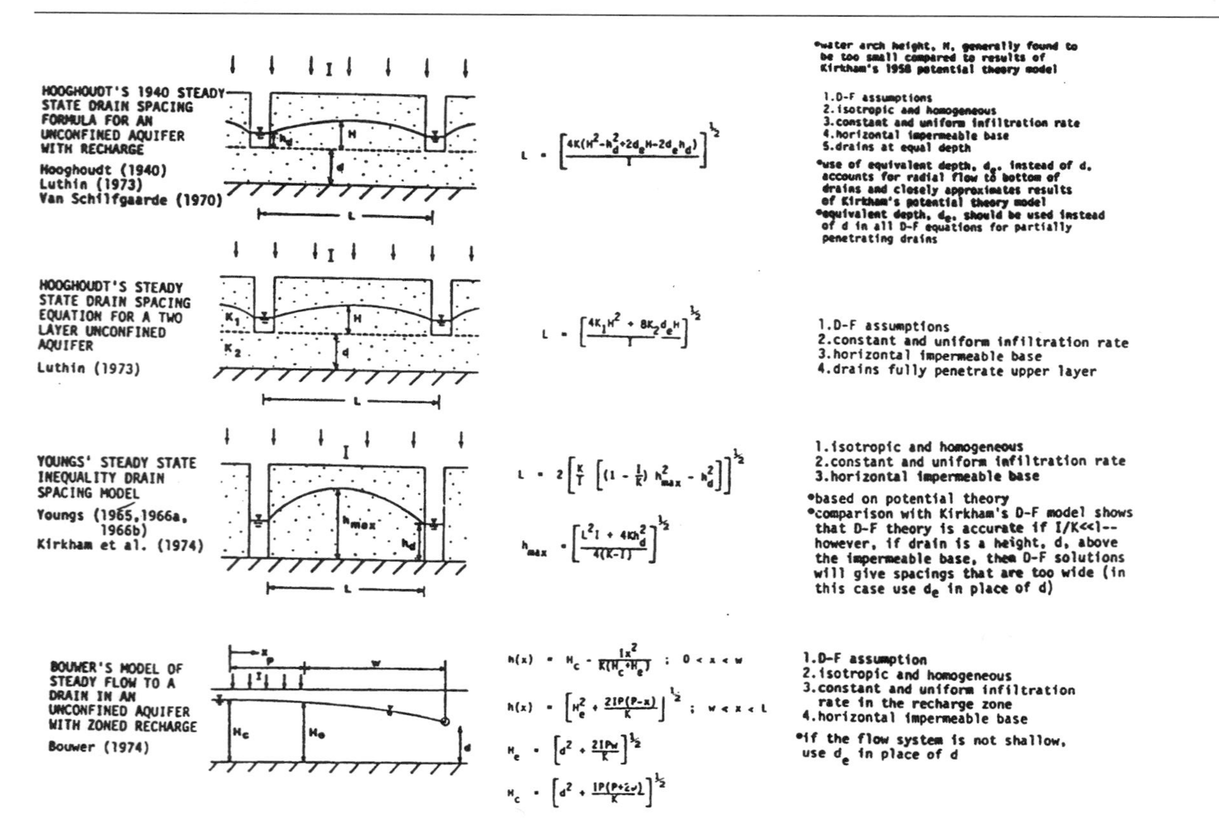

Model	Equation	Assumptions / Comments
HOOGHOUDT'S 1940 STEADY STATE DRAIN SPACING FORMULA FOR AN UNCONFINED AQUIFER WITH RECHARGE Hooghoudt (1940) Luthin (1973) Van Schilfgaarde (1970)	$L = \left[\frac{4K(H^2 - h_d^2 + 2d_eH - 2d_eh_d)}{I}\right]^{1/2}$	•water arch height, H, generally found to be too small compared to results of Kirkham's 1958 potential theory model 1. D-F assumptions 2. isotropic and homogeneous 3. constant and uniform infiltration rate 4. horizontal impermeable base 5. drains at equal depth •use of equivalent depth, d_e, instead of d, accounts for radial flow to bottom of drains and closely approximates results of Kirkham's potential theory model •equivalent depth, d_e, should be used instead of d in all D-F equations for partially penetrating drains
HOOGHOUDT'S STEADY STATE DRAIN SPACING EQUATION FOR A TWO LAYER UNCONFINED AQUIFER Luthin (1973)	$L = \left[\frac{4K_1H^2 + 8K_2d_eH}{I}\right]^{1/2}$	1. D-F assumptions 2. constant and uniform infiltration rate 3. horizontal impermeable base 4. drains fully penetrate upper layer
YOUNGS' STEADY STATE INEQUALITY DRAIN SPACING MODEL Youngs (1965, 1966a, 1966b) Kirkham et al. (1974)	$L = 2\left[\frac{K}{I}\left[\left(1 - \frac{I}{K}\right)h_{max}^2 - h_d^2\right]\right]^{1/2}$ $h_{max} = \left[\frac{L^2I + 4Kh_d^2}{4(K-I)}\right]^{1/2}$	1. isotropic and homogeneous 2. constant and uniform infiltration rate 3. horizontal impermeable base •based on potential theory •comparison with Kirkham's D-F model shows that D-F theory is accurate if I/K<<1--however, if drain is a height, d, above the impermeable base, then D-F solutions will give spacings that are too wide (in this case use d_e in place of d)
BOUWER'S MODEL OF STEADY FLOW TO A DRAIN IN AN UNCONFINED AQUIFER WITH ZONED RECHARGE Bouwer (1974)	$h(x) = H_c - \frac{Ix^2}{K(H_c + H_e)}$; $0 < x < w$ $h(x) = \left[H_e^2 + \frac{2IP(P-x)}{K}\right]^{1/2}$; $w < x < L$ $H_e = \left[d^2 + \frac{2IPw}{K}\right]^{1/2}$ $H_c = \left[d^2 + \frac{IP(P+2w)}{K}\right]^{1/2}$	1. D-F assumption 2. isotropic and homogeneous 3. constant and uniform infiltration rate in the recharge zone 4. horizontal impermeable base •if the flow system is not shallow, use d_e in place of d

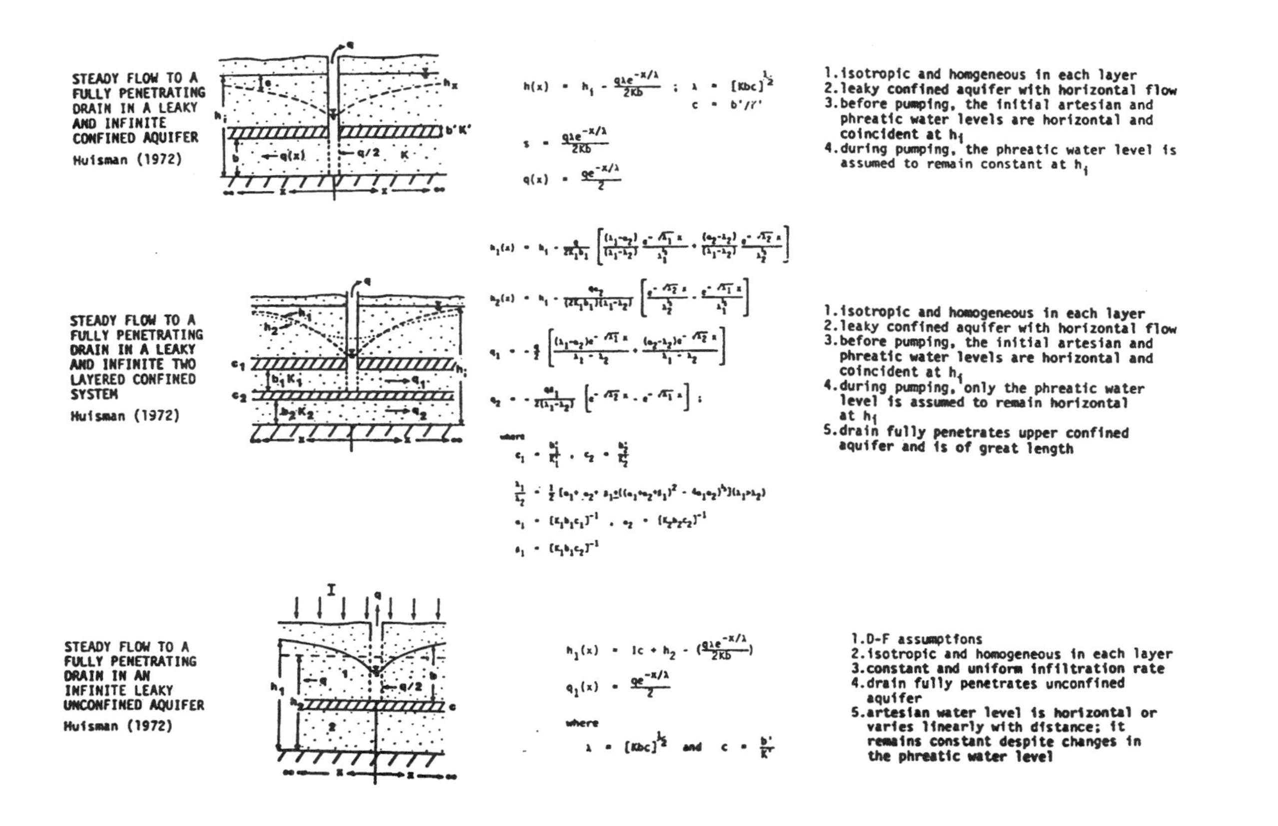

Case	Equations	Assumptions
STEADY FLOW TO A FULLY PENETRATING DRAIN IN A LEAKY AND INFINITE CONFINED AQUIFER Huisman (1972)	$h(x) = h_i - \frac{q\lambda e^{-x/\lambda}}{2Kb}$; $\lambda = [Kbc]^{\frac{1}{2}}$ $c = b'/K'$ $s = \frac{q\lambda e^{-x/\lambda}}{2Kb}$ $q(x) = \frac{qe^{-x/\lambda}}{2}$	1. isotropic and homgeneous in each layer 2. leaky confined aquifer with horizontal flow 3. before pumping, the initial artesian and phreatic water levels are horizontal and coincident at h_i 4. during pumping, the phreatic water level is assumed to remain constant at h_i
STEADY FLOW TO A FULLY PENETRATING DRAIN IN A LEAKY AND INFINITE TWO LAYERED CONFINED SYSTEM Huisman (1972)	$h_1(x) = h_i - \frac{q}{2K_1b_1}\left[\frac{(\lambda_1-a_2)}{(\lambda_1-\lambda_2)}\frac{e^{-\sqrt{\lambda_1}x}}{\lambda_1^{\frac{1}{2}}} + \frac{(a_2-\lambda_2)}{(\lambda_1-\lambda_2)}\frac{e^{-\sqrt{\lambda_2}x}}{\lambda_2^{\frac{1}{2}}}\right]$ $h_2(x) = h_i - \frac{qa_2}{(2K_1b_1)(\lambda_1-\lambda_2)}\left[\frac{e^{-\sqrt{\lambda_2}x}}{\lambda_2^{\frac{1}{2}}} - \frac{e^{-\sqrt{\lambda_1}x}}{\lambda_1^{\frac{1}{2}}}\right]$ $q_1 = -\frac{q}{2}\left[\frac{(\lambda_1-a_2)e^{-\sqrt{\lambda_1}x}}{\lambda_1-\lambda_2} + \frac{(a_2-\lambda_2)e^{-\sqrt{\lambda_2}x}}{\lambda_1-\lambda_2}\right]$ $q_2 = -\frac{qa_1}{2(\lambda_1-\lambda_2)}\left[e^{-\sqrt{\lambda_2}x} - e^{-\sqrt{\lambda_1}x}\right]$; where $c_1 = \frac{b_1'}{K_1'}$, $c_2 = \frac{b_2'}{K_2'}$ $\frac{\lambda_1}{\lambda_2} = \frac{1}{2}[a_1 + a_2 + \beta_1 \pm ((a_1+a_2+\beta_1)^2 - 4a_1a_2)^{\frac{1}{2}}](\lambda_1 > \lambda_2)$ $a_1 = [K_1b_1c_1]^{-1}$, $a_2 = [K_2b_2c_2]^{-1}$ $\beta_1 = [K_1b_1c_2]^{-1}$	1. isotropic and homogeneous in each layer 2. leaky confined aquifer with horizontal flow 3. before pumping, the initial artesian and phreatic water levels are horizontal and coincident at h_i 4. during pumping, only the phreatic water level is assumed to remain horizontal at h_i 5. drain fully penetrates upper confined aquifer and is of great length
STEADY FLOW TO A FULLY PENETRATING DRAIN IN AN INFINITE LEAKY UNCONFINED AQUIFER Huisman (1972)	$h_1(x) = Ic + h_2 - (\frac{q\lambda e^{-x/\lambda}}{2Kb})$ $q_1(x) = \frac{qe^{-x/\lambda}}{2}$ where $\lambda = [Kbc]^{\frac{1}{2}}$ and $c = \frac{b'}{K'}$	1. D-F assumptions 2. isotropic and homogeneous in each layer 3. constant and uniform infiltration rate 4. drain fully penetrates unconfined aquifer 5. artesian water level is horizontal or varies linearly with distance; it remains constant despite changes in the phreatic water level

Table 3.5 Continued

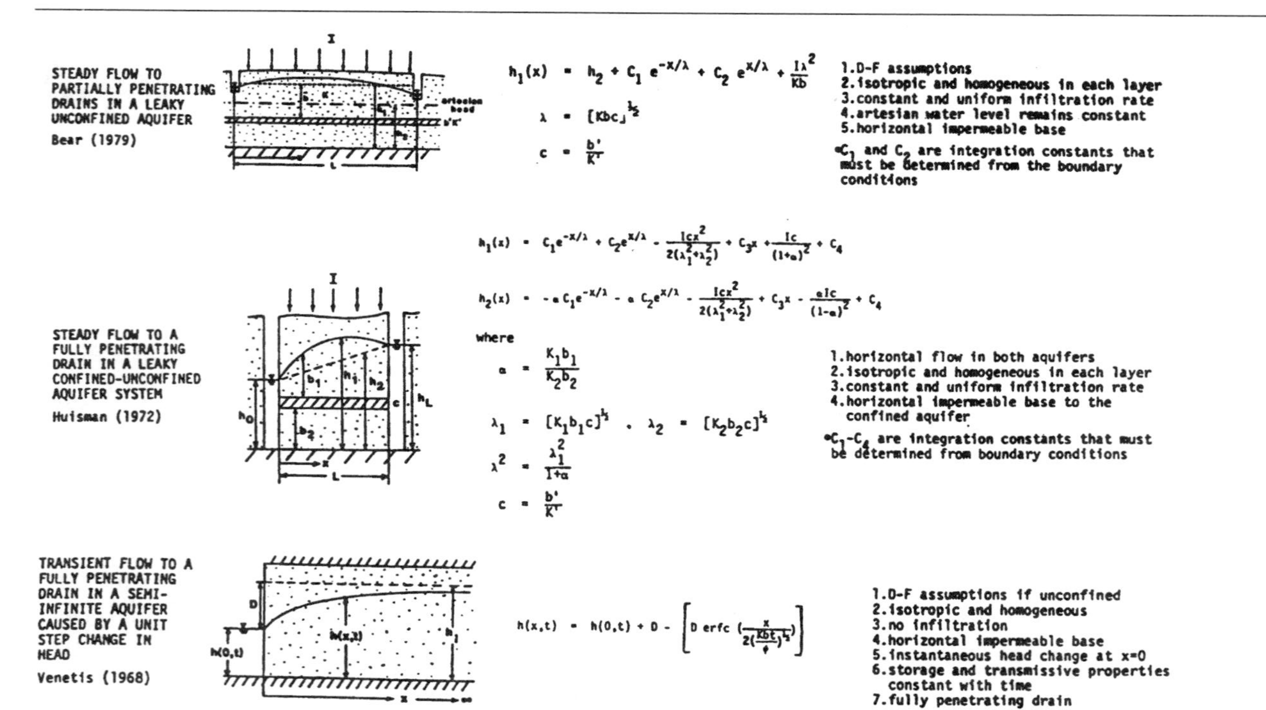

Case	Equation	Assumptions
STEADY FLOW TO PARTIALLY PENETRATING DRAINS IN A LEAKY UNCONFINED AQUIFER Bear (1979)	$h_1(x) = h_2 + C_1 e^{-x/\lambda} + C_2 e^{x/\lambda} + \frac{I\lambda^2}{Kb}$ $\lambda = [Kbc_1]^{1/2}$ $c = \frac{b'}{K'}$	1. D-F assumptions 2. isotropic and homogeneous in each layer 3. constant and uniform infiltration rate 4. artesian water level remains constant 5. horizontal impermeable base * C_1 and C_2 are integration constants that must be determined from the boundary conditions
STEADY FLOW TO A FULLY PENETRATING DRAIN IN A LEAKY CONFINED-UNCONFINED AQUIFER SYSTEM Huisman (1972)	$h_1(x) = C_1 e^{-x/\lambda} + C_2 e^{x/\lambda} - \frac{Icx^2}{2(\lambda_1^2+\lambda_2^2)} + C_3 x + \frac{Ic}{(1+\alpha)^2} + C_4$ $h_2(x) = -\alpha C_1 e^{-x/\lambda} - \alpha C_2 e^{x/\lambda} - \frac{Icx^2}{2(\lambda_1^2+\lambda_2^2)} + C_3 x - \frac{\alpha Ic}{(1-\alpha)^2} + C_4$ where $\alpha = \frac{K_1 b_1}{K_2 b_2}$ $\lambda_1 = [K_1 b_1 c]^{1/2}$, $\lambda_2 = [K_2 b_2 c]^{1/2}$ $\lambda^2 = \frac{\lambda_1^2}{1+\alpha}$ $c = \frac{b'}{K'}$	1. horizontal flow in both aquifers 2. isotropic and homogeneous in each layer 3. constant and uniform infiltration rate 4. horizontal impermeable base to the confined aquifer * C_1-C_4 are integration constants that must be determined from boundary conditions
TRANSIENT FLOW TO A FULLY PENETRATING DRAIN IN A SEMI-INFINITE AQUIFER CAUSED BY A UNIT STEP CHANGE IN HEAD Venetis (1968)	$h(x,t) = h(0,t) + D - \left[D \operatorname{erfc}\left(\frac{x}{2(\frac{Kbt}{\phi})^{1/2}}\right) \right]$	1. D-F assumptions if unconfined 2. isotropic and homogeneous 3. no infiltration 4. horizontal impermeable base 5. instantaneous head change at x=0 6. storage and transmissive properties constant with time 7. fully penetrating drain

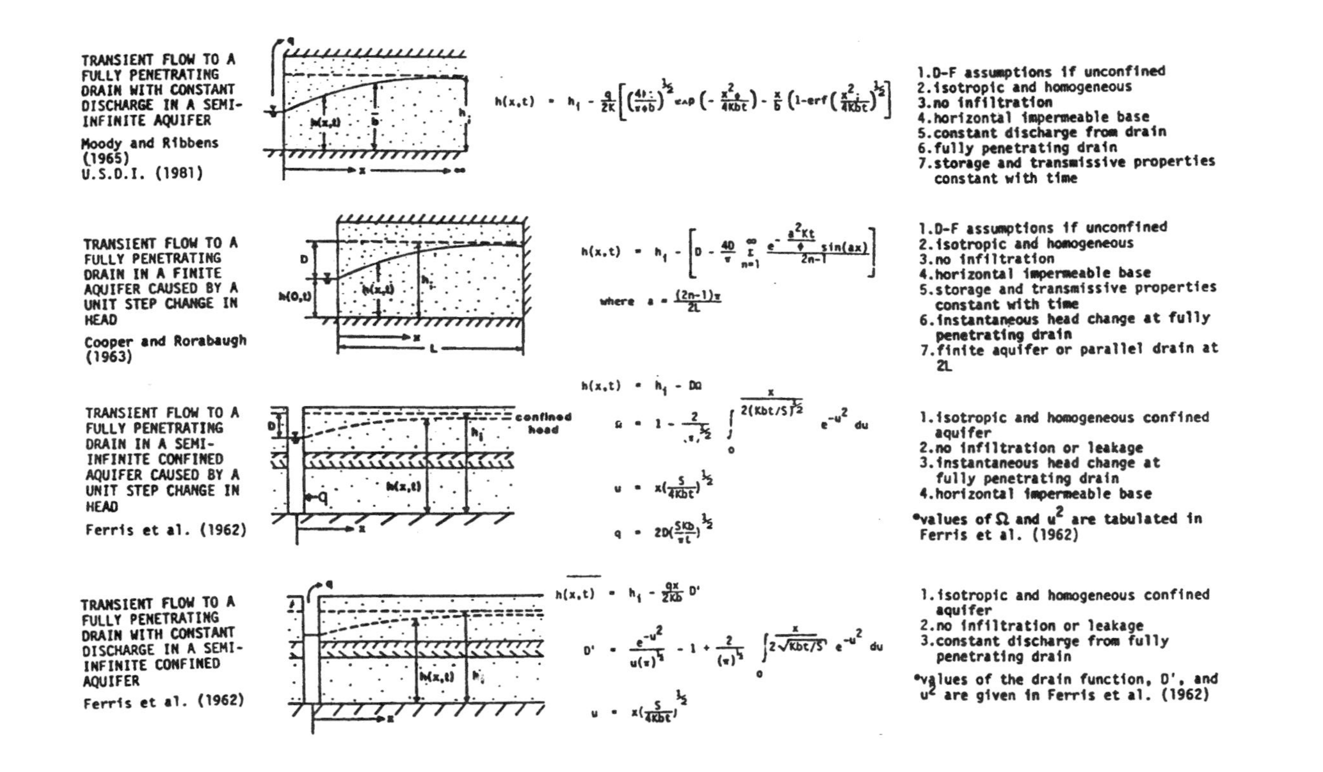

Flow condition / Reference	Equation	Assumptions
TRANSIENT FLOW TO A FULLY PENETRATING DRAIN WITH CONSTANT DISCHARGE IN A SEMI-INFINITE AQUIFER Moody and Ribbens (1965) U.S.D.I. (1981)	$h(x,t) = h_i - \frac{q}{2K}\left[\left(\frac{4t}{\pi\phi b}\right)^{1/2}\exp\left(-\frac{x^2\phi}{4Kbt}\right) - \frac{x}{b}\left(1-\mathrm{erf}\left(\frac{x^2\phi}{4Kbt}\right)^{1/2}\right)\right]$	1.D-F assumptions if unconfined 2.isotropic and homogeneous 3.no infiltration 4.horizontal impermeable base 5.constant discharge from drain 6.fully penetrating drain 7.storage and transmissive properties constant with time
TRANSIENT FLOW TO A FULLY PENETRATING DRAIN IN A FINITE AQUIFER CAUSED BY A UNIT STEP CHANGE IN HEAD Cooper and Rorabaugh (1963)	$h(x,t) = h_i - \left[D - \frac{4D}{\pi}\sum_{n=1}^{\infty} e^{-\frac{a^2Kt}{\phi}}\frac{\sin(ax)}{2n-1}\right]$ where $a = \frac{(2n-1)\pi}{2L}$	1.D-F assumptions if unconfined 2.isotropic and homogeneous 3.no infiltration 4.horizontal impermeable base 5.storage and transmissive properties constant with time 6.instantaneous head change at fully penetrating drain 7.finite aquifer or parallel drain at 2L
TRANSIENT FLOW TO A FULLY PENETRATING DRAIN IN A SEMI-INFINITE CONFINED AQUIFER CAUSED BY A UNIT STEP CHANGE IN HEAD Ferris et al. (1962)	$h(x,t) = h_i - D\Omega$ $\Omega = 1 - \frac{2}{\pi^{1/2}}\int_0^{\frac{x}{2(Kbt/S)^{1/2}}} e^{-u^2}\,du$ $u = x\left(\frac{S}{4Kbt}\right)^{1/2}$ $q = 2D\left(\frac{SKb}{\pi t}\right)^{1/2}$	1.isotropic and homogeneous confined aquifer 2.no infiltration or leakage 3.instantaneous head change at fully penetrating drain 4.horizontal impermeable base *values of Ω and u^2 are tabulated in Ferris et al. (1962)
TRANSIENT FLOW TO A FULLY PENETRATING DRAIN WITH CONSTANT DISCHARGE IN A SEMI-INFINITE CONFINED AQUIFER Ferris et al. (1962)	$h(x,t) = h_i - \frac{qx}{2Kb}D'$ $D' = \frac{e^{-u^2}}{u(\pi)^{1/2}} - 1 + \frac{2}{(\pi)^{1/2}}\int_0^{\frac{x}{2\sqrt{Kbt/S}}} e^{-u^2}\,du$ $u = x\left(\frac{S}{4Kbt}\right)^{1/2}$	1.isotropic and homogeneous confined aquifer 2.no infiltration or leakage 3.constant discharge from fully penetrating drain *values of the drain function, D', and u^2 are given in Ferris et al. (1962)

Table 3.5 Continued

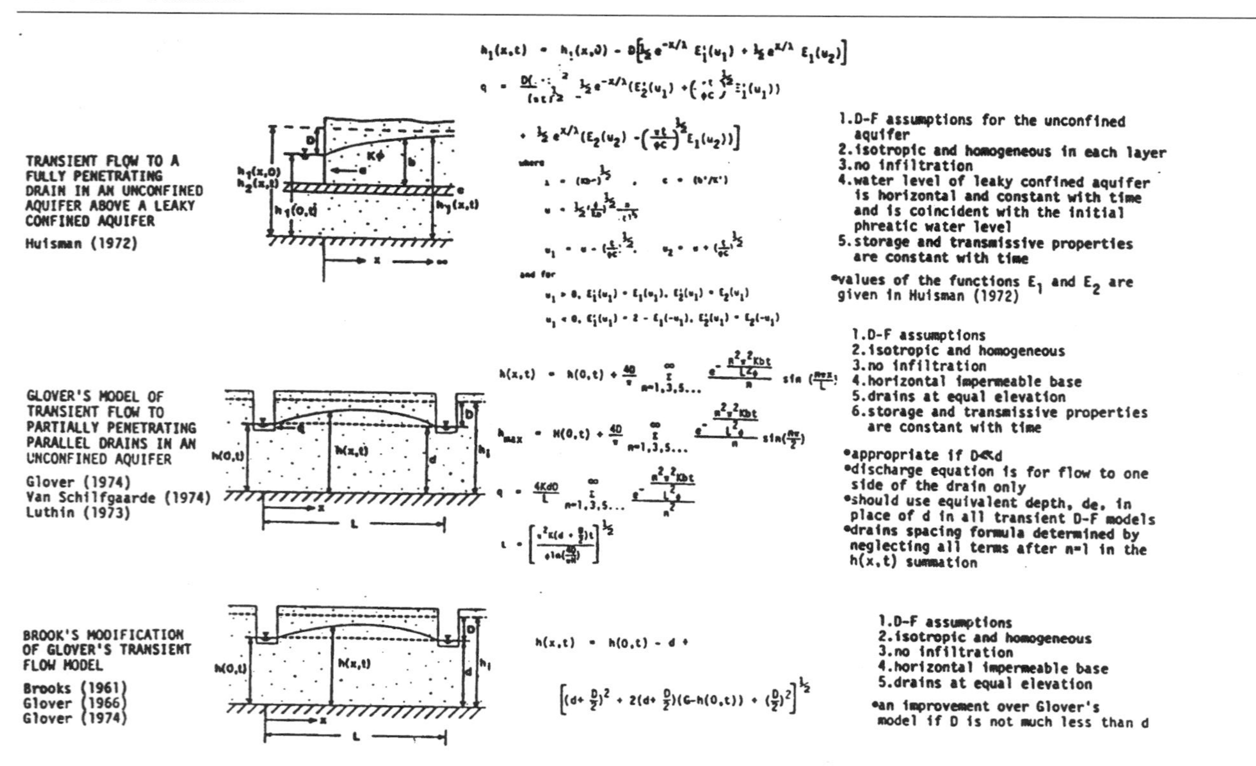

Model	Equations	Assumptions
TRANSIENT FLOW TO A FULLY PENETRATING DRAIN IN AN UNCONFINED AQUIFER ABOVE A LEAKY CONFINED AQUIFER Huisman (1972)	$h_1(x,t) = h_1(x,0) - D\left[\tfrac{1}{2}e^{-x/\lambda}E_1^*(u_1) + \tfrac{1}{2}e^{x/\lambda}E_1(u_2)\right]$ $q = \frac{D(\cdot)}{(\pi t)^{1/2}}\left[\tfrac{1}{2}e^{-x/\lambda}\left(E_2^*(u_1) + \left(\frac{\pi t}{\phi c}\right)^{1/2}E_1^*(u_1)\right) + \tfrac{1}{2}e^{x/\lambda}\left(E_2(u_2) - \left(\frac{\pi t}{\phi c}\right)^{1/2}E_1(u_2)\right)\right]$ where $\lambda = (KDc)^{1/2}$, $c = (b'/K')$ $u = \tfrac{1}{2}\left(\frac{\phi}{KD}\right)^{1/2}\frac{x}{t^{1/2}}$ $u_1 = u - \left(\frac{t}{\phi c}\right)^{1/2}$, $u_2 = u + \left(\frac{t}{\phi c}\right)^{1/2}$ and for $u_1 > 0$, $E_1^*(u_1) = E_1(u_1)$, $E_2^*(u_1) = E_2(u_1)$ $u_1 < 0$, $E_1^*(u_1) = 2 - E_1(-u_1)$, $E_2^*(u_1) = E_2(-u_1)$	1. D-F assumptions for the unconfined aquifer 2. isotropic and homogeneous in each layer 3. no infiltration 4. water level of leaky confined aquifer is horizontal and constant with time and is coincident with the initial phreatic water level 5. storage and transmissive properties are constant with time • values of the functions E_1 and E_2 are given in Huisman (1972)
GLOVER'S MODEL OF TRANSIENT FLOW TO PARTIALLY PENETRATING PARALLEL DRAINS IN AN UNCONFINED AQUIFER Glover (1974) Van Schilfgaarde (1974) Luthin (1973)	$h(x,t) = h(0,t) + \frac{4D}{\pi}\sum_{n=1,3,5\ldots}^{\infty}\frac{e^{-\frac{n^2\pi^2Kdt}{L^2\phi}}}{n}\sin\left(\frac{n\pi x}{L}\right)$ $h_{max} = h(0,t) + \frac{4D}{\pi}\sum_{n=1,3,5\ldots}^{\infty}\frac{e^{-\frac{n^2\pi^2Kdt}{L^2\phi}}}{n}\sin\left(\frac{n\pi}{2}\right)$ $q = \frac{4KdD}{L}\sum_{n=1,3,5\ldots}^{\infty}\frac{e^{-\frac{n^2\pi^2Kdt}{L^2\phi}}}{n^2}$ $L = \left[\frac{\pi^2K(d + \frac{D}{2})t}{\phi\ln(\frac{4D}{\pi H})}\right]^{1/2}$	1. D-F assumptions 2. isotropic and homogeneous 3. no infiltration 4. horizontal impermeable base 5. drains at equal elevation 6. storage and transmissive properties are constant with time • appropriate if D≪d • discharge equation is for flow to one side of the drain only • should use equivalent depth, de, in place of d in all transient D-F models • drains spacing formula determined by neglecting all terms after n=1 in the h(x,t) summation
BROOK'S MODIFICATION OF GLOVER'S TRANSIENT FLOW MODEL Brooks (1961) Glover (1966) Glover (1974)	$h(x,t) = h(0,t) - d + \left[(d+\frac{D}{2})^2 + 2(d+\frac{D}{2})(G-h(0,t)) + (\frac{D}{2})^2\right]^{1/2}$	1. D-F assumptions 2. isotropic and homogeneous 3. no infiltration 4. horizontal impermeable base 5. drains at equal elevation • an improvement over Glover's model if D is not much less than d

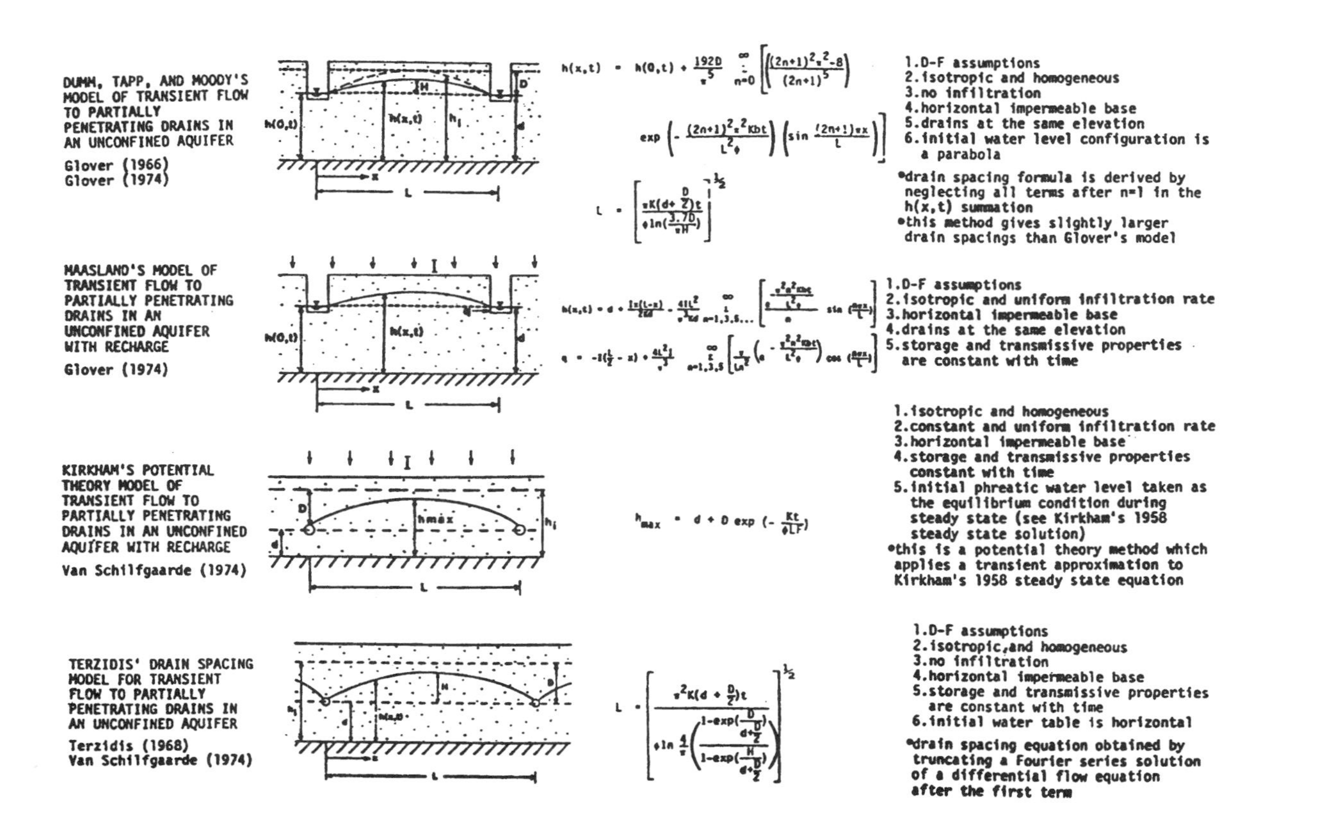

Model	Equation	Assumptions and remarks
DUMM, TAPP, AND MOODY'S MODEL OF TRANSIENT FLOW TO PARTIALLY PENETRATING DRAINS IN AN UNCONFINED AQUIFER Glover (1966) Glover (1974)	$h(x,t) = h(0,t) + \frac{192D}{\pi^5} \sum_{n=0}^{\infty} \left[\left(\frac{(2n+1)^2\pi^2 - 8}{(2n+1)^5} \right) \exp\left(-\frac{(2n+1)^2\pi^2 Kbt}{L^2\phi} \right) \left(\sin \frac{(2n+1)\pi x}{L} \right) \right]$ $L = \left[\frac{\pi K(d+\frac{D}{2})t}{\phi \ln(\frac{3.7D}{\pi H})} \right]^{\frac{1}{2}}$	1. D-F assumptions 2. isotropic and homogeneous 3. no infiltration 4. horizontal impermeable base 5. drains at the same elevation 6. initial water level configuration is a parabola •drain spacing formula is derived by neglecting all terms after n=1 in the h(x,t) summation •this method gives slightly larger drain spacings than Glover's model
MAASLAND'S MODEL OF TRANSIENT FLOW TO PARTIALLY PENETRATING DRAINS IN AN UNCONFINED AQUIFER WITH RECHARGE Glover (1974)	$h(x,t) = d + \frac{Ix(L-x)}{2Kd} - \frac{4IL^2}{\pi^3 Kd} \sum_{n=1,3,5...}^{\infty} \left[\frac{e^{-\frac{\pi^2 n^2 Kdt}{L^2\phi}}}{n} \sin\left(\frac{n\pi x}{L}\right) \right]$ $q = -I(\frac{L}{2} - x) + \frac{4L^2 I}{\pi^3} \sum_{n=1,3,5}^{\infty} \left[\frac{\pi}{Ln^2} \left(e^{-\frac{\pi^2 n^2 Kdt}{L^2\phi}} \right) \cos\left(\frac{n\pi x}{L}\right) \right]$	1. D-F assumptions 2. isotropic and uniform infiltration rate 3. horizontal impermeable base 4. drains at the same elevation 5. storage and transmissive properties are constant with time
KIRKHAM'S POTENTIAL THEORY MODEL OF TRANSIENT FLOW TO PARTIALLY PENETRATING DRAINS IN AN UNCONFINED AQUIFER WITH RECHARGE Van Schilfgaarde (1974)	$h_{max} = d + D \exp\left(-\frac{Kt}{\phi LF}\right)$	1. isotropic and homogeneous 2. constant and uniform infiltration rate 3. horizontal impermeable base 4. storage and transmissive properties constant with time 5. initial phreatic water level taken as the equilibrium condition during steady state (see Kirkham's 1958 steady state solution) •this is a potential theory method which applies a transient approximation to Kirkham's 1958 steady state equation
TERZIDIS' DRAIN SPACING MODEL FOR TRANSIENT FLOW TO PARTIALLY PENETRATING DRAINS IN AN UNCONFINED AQUIFER Terzidis (1968) Van Schilfgaarde (1974)	$L = \left[\frac{\pi^2 K(d+\frac{D}{2})t}{\phi \ln \frac{4}{\pi} \left(\frac{1-\exp(-\frac{D}{d+\frac{D}{2}})}{1-\exp(-\frac{H}{d+\frac{D}{2}})} \right)} \right]^{\frac{1}{2}}$	1. D-F assumptions 2. isotropic, and homogeneous 3. no infiltration 4. horizontal impermeable base 5. storage and transmissive properties are constant with time 6. initial water table is horizontal •drain spacing equation obtained by truncating a Fourier series solution of a differential flow equation after the first term

Table 3.5 Continued

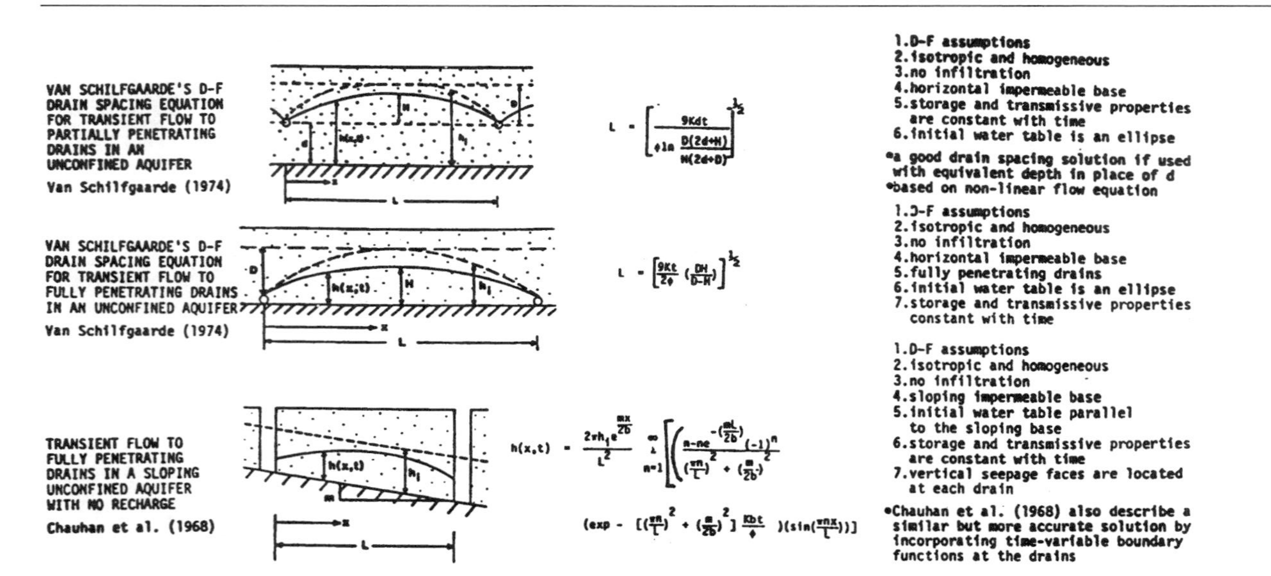

VAN SCHILFGAARDE'S D-F DRAIN SPACING EQUATION FOR TRANSIENT FLOW TO PARTIALLY PENETRATING DRAINS IN AN UNCONFINED AQUIFER Van Schilfgaarde (1974)		$L = \left[\dfrac{9Kdt}{\phi \ln \dfrac{D(2d+H)}{H(2d+D)}}\right]^{\frac{1}{2}}$	1.D-F assumptions 2.isotropic and homogeneous 3.no infiltration 4.horizontal impermeable base 5.storage and transmissive properties are constant with time 6.initial water table is an ellipse •a good drain spacing solution if used with equivalent depth in place of d •based on non-linear flow equation
VAN SCHILFGAARDE'S D-F DRAIN SPACING EQUATION FOR TRANSIENT FLOW TO FULLY PENETRATING DRAINS IN AN UNCONFINED AQUIFER Van Schilfgaarde (1974)		$L = \left[\dfrac{9Kt}{2\phi}\left(\dfrac{DH}{D-H}\right)\right]^{\frac{1}{2}}$	1.D-F assumptions 2.isotropic and homogeneous 3.no infiltration 4.horizontal impermeable base 5.fully penetrating drains 6.initial water table is an ellipse 7.storage and transmissive properties constant with time
TRANSIENT FLOW TO FULLY PENETRATING DRAINS IN A SLOPING UNCONFINED AQUIFER WITH NO RECHARGE Chauhan et al. (1968)		$h(x,t) = \dfrac{2\pi h_1 e^{\frac{mx}{2b}}}{L^2} \sum_{n=1}^{\infty} \left[\left(\dfrac{n - ne^{-(\frac{mL}{2b})}(-1)^n}{(\frac{\pi n}{L})^2 + (\frac{m}{2b})^2}\right)\left(\exp - \left[(\frac{\pi n}{L})^2 + (\frac{m}{2b})^2\right]\frac{Kbt}{\phi}\right)\left(\sin(\frac{\pi nx}{L})\right)\right]$	1.D-F assumptions 2.isotropic and homogeneous 3.no infiltration 4.sloping impermeable base 5.initial water table parallel to the sloping base 6.storage and transmissive properties are constant with time 7.vertical seepage faces are located at each drain •Chauhan et al. (1968) also describe a similar but more accurate solution by incorporating time-variable boundary functions at the drains

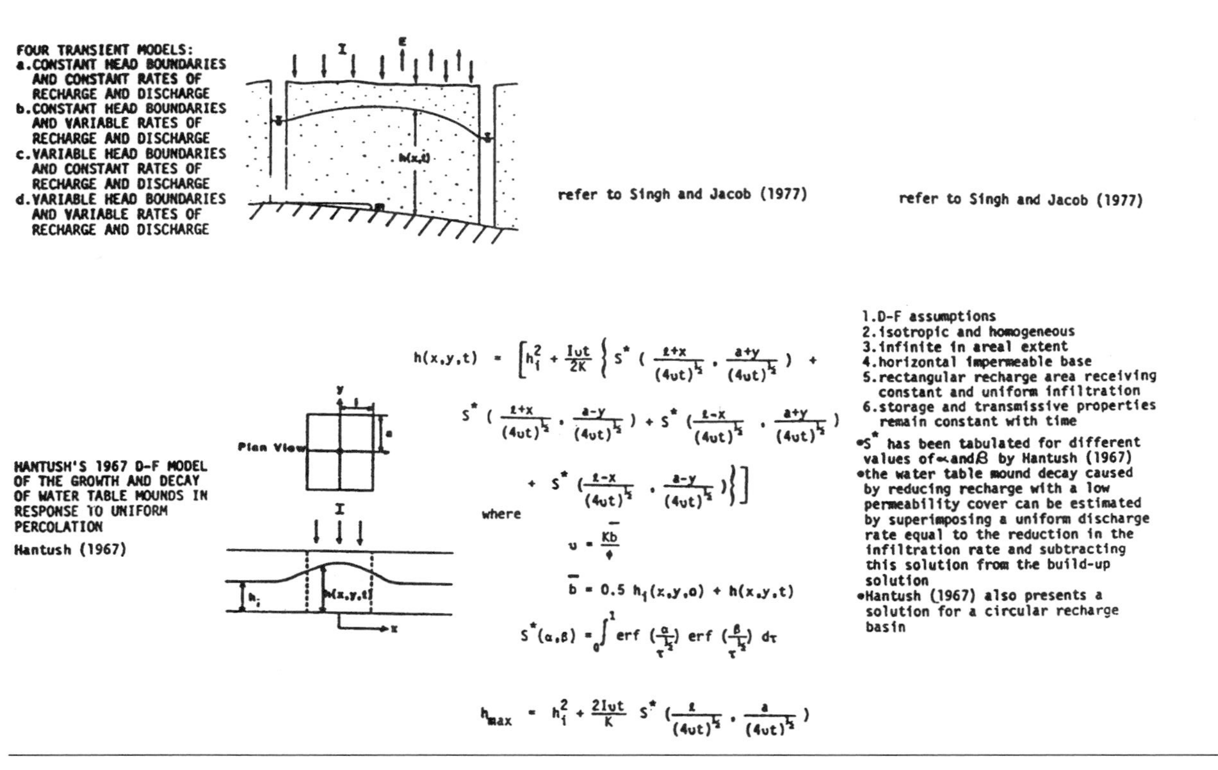

Model	Equation	Remarks
FOUR TRANSIENT MODELS: a. CONSTANT HEAD BOUNDARIES AND CONSTANT RATES OF RECHARGE AND DISCHARGE b. CONSTANT HEAD BOUNDARIES AND VARIABLE RATES OF RECHARGE AND DISCHARGE c. VARIABLE HEAD BOUNDARIES AND CONSTANT RATES OF RECHARGE AND DISCHARGE d. VARIABLE HEAD BOUNDARIES AND VARIABLE RATES OF RECHARGE AND DISCHARGE	refer to Singh and Jacob (1977)	refer to Singh and Jacob (1977)
HANTUSH'S 1967 D-F MODEL OF THE GROWTH AND DECAY OF WATER TABLE MOUNDS IN RESPONSE TO UNIFORM PERCOLATION Hantush (1967)	$h(x,y,t) = \left[h_i^2 + \frac{Ivt}{2K}\left\{S^*\left(\frac{\ell+x}{(4vt)^{1/2}}, \frac{a+y}{(4vt)^{1/2}}\right) + S^*\left(\frac{\ell+x}{(4vt)^{1/2}}, \frac{a-y}{(4vt)^{1/2}}\right) + S^*\left(\frac{\ell-x}{(4vt)^{1/2}}, \frac{a+y}{(4vt)^{1/2}}\right) + S^*\left(\frac{\ell-x}{(4vt)^{1/2}}, \frac{a-y}{(4vt)^{1/2}}\right)\right\}\right]$ where $v = \frac{K\bar{b}}{\phi}$ $\bar{b} = 0.5\, h_i(x,y,0) + h(x,y,t)$ $S^*(\alpha,\beta) = \int_0^1 \mathrm{erf}\left(\frac{\alpha}{\tau^{1/2}}\right) \mathrm{erf}\left(\frac{\beta}{\tau^{1/2}}\right) d\tau$ $h_{max} = h_i^2 + \frac{2Ivt}{K} S^*\left(\frac{\ell}{(4vt)^{1/2}}, \frac{a}{(4vt)^{1/2}}\right)$	1. D-F assumptions 2. isotropic and homogeneous 3. infinite in areal extent 4. horizontal impermeable base 5. rectangular recharge area receiving constant and uniform infiltration 6. storage and transmissive properties remain constant with time • S^* has been tabulated for different values of α and β by Hantush (1967) • the water table mound decay caused by reducing recharge with a low permeability cover can be estimated by superimposing a uniform discharge rate equal to the reduction in the infiltration rate and subtracting this solution from the build-up solution • Hantush (1967) also presents a solution for a circular recharge basin

Source: Cohen and Miller (1983).

and horizontal, respectively (Fig. 3.20). Bear (1972) has shown that the error due to this assumption is small when $i^2 \ll 1$, where i is the slope of the phreatic surface. The scenarios where this assumption would lead to simple and useful solutions are shown in Table 3.5. Many of the solutions listed in Table 3.5 were obtained assuming that the velocity of flow in the aquifer is proportional to the slope of the water table. Flow in the unsaturated zone is not considered in these solutions, and the infiltration to the groundwater table is generally considered as equal to the rate of application at the ground surface. We will find many of these solutions useful in site remediation activities and in designing drainage facilities for waste containment systems, as discussed later in this book.

REFERENCES

Aslyng, H. C. (1963). Soil physics terminology. *In. Soc. Soil Sci. Bull.*, 23:7.

Baver, L. D., Gardner, W. H., and Gardner, W. R. (1972). *Soil Physics*. Wiley, New York.

Bear, J. (1972). *Dynamics of Fluids in Porous Media*. American Elsevier, New York.

Bear, J. (1979). *Hydraulics of Ground Water*. McGraw-Hill, New York.

Benson, C. H., and Daniel, D. E. (1990). Influence of clods on hydraulic conductivity of compacted clay. ASCE J. Geotech. Eng., 116(8):1231–1248.

Boutwell, G. P., and Rauser, C. L. (1990). Clay liner construction. Geotechnical Engineering in Today's Environment, conference sponsored by the Central Pennsylvania Section American Society of Civil Engineers, Hershey, PA, April 10–11, 1990.

Bouwer, H. (1974). Design and operation of land treatment systems for minimum contamination of ground water. *Ground Water*, 12(3):140–147.

Boynton, S. S., and Daniel, D. E. (1985). Hydraulic conductivity tests on compacted clay. ASCE J. Geotech. Eng. 111(4):465–477.

Brooks, R. H. (1961). Unsteady flow of ground water into drain tile. *ASCE Proc.*, 87:27–37.

Brooks, R. H., and Corey, A. T. (1966). Properties of porous media affecting fluid flow. *Proc. Am. Soc. Civ. Eng., J. Irrigation Drainage Div.*, IR2:61–88.

Campbell, G. S. (1974). A simple method for determining unsaturated conductivity from moisture retention data. *Soil Sci.* 117(6):311–314.

Carman, P. C. (1956). *Flow of Gases Through Porous Media*. Academic Press, New York.

Cedergren, H. R. (1989). *Seepage, Drainage, and Flow Nets*, 3rd ed. J Wiley, New York.

Chauhan, H. S., Schwab, G. O., and Hamdy, M. Y. (1968). Analytical and computer solutions of transient water tables for drainage of sloping land. *Water Resources Res.* 4(3):573–579.

Childs, E. C., and Collis-George, N. (1950). The permeability of porous materials. *Proc. R. Soc. London*, A201:392–405.

Cohen, R. M., and Miller, W. J. (1983). Use of analytical models for evaluating corrective actions at hazardous waste disposal facilities. *Proceedings of the Third National Symposium on Aquifer Restoration and Ground Water Monitoring*. National Water Well Association, Worthington, OH, pp. 85–97.

Cooper, H. H., Jr., and Rorabaugh, M. I. (1963). Groundwater movements and bank storage due to flood stages in surface streams. *USGS Water Supply Paper 1536-J*, pp. 352–357.

Daniel, D. E. (1984). Predicting hydraulic conductivity of clay liners. *ASCE J.* Geotech. Eng., 110:285–300.

Daniel, D. E. (1987). Earthen liners for land disposal facilities. In: R. D. Woods (ed.), *Geotechnical Practice for Waste Disposal*, Geotechnical Special Publication No. 13, ASCE, pp. 21–39.

Daniel, D. E. (1994). State-of-the-Art: Laboratory Hydraulic Conductivity tests for saturated soils. In: D. E. Daniel and S. J. Trautwein (eds.), *Hydraulic Conductivity and Waste Contaminant Transport in Soil,* STP1142, ASTM, pp. 30–78.

Daniel, D. E., and Benson, C. H. (1990). Water content-density criteria for compacted soil liners. *ASCE J. Geotech. Eng*., 116(12):1811–1830.

Day, S. R., and Daniel, D. E. (1985). Hydraulic conductivity of two prototype clay liners. *ASCE J. Geotech. Eng*., 111(8):957–970.

Ferris, J. G., Knowles, D. B., Brown, R. H., and Stallman, R. W. (1962). Theory of aquifer tests. *USGS Water Supply Paper 1536E.*

Foreman, D. E., and Daniel, D. E. (1986). Permeation of compacted clay with organic chemicals. ASCE J. Geotech. Eng., 112(7).

Freeze, R. A., and Cherry, J. A. (1979). *Groundwater*. Prentice Hall, Englewood Cliffs, NJ.

Gardner, W. R. (1958). Some steady-state solutions of the unsaturated moisture flow equation with application to evaporation from a water table. *Soil Sci*. 85(4):228–232.

Gardner, W. R., Hillel, D., and Benyamini, Y. (1970). Post irrigation movement of soil water. I. Redistribution. *Water Resources Res*., 6(3):851–861; II. Simultaneous redistribution and evaporation. *Water Resources Res*., 6(4):1148–1153.

Glover, R. E. (1966). *Theory of Ground-Water Movement*. U. S. Bureau of Reclamation Engineering Monograph No. 31.

Glover, R. E. (1974). *Transient Ground-Water Hydraulics*. Water Resources Publications, Fort Collins, Co.

Green, W. H., and Ampt, G. A. (1911). Studies on soil physics, I. Flow of air and water through soils. *J. Agr. Sci*., 4:1–24.

Hantush, M. S. (1967). Growth and decay of ground-water mounds in response to uniform percolation. *Water Resources Res*. 3(1):227–234.

Harr, M. E. (1962). *Ground-Water and Seepage*. McGraw-Hill, New York.

Hazen, A. (1911). Discussion of "Dams on Sand Foundation" by A. C. Koenig, *Trans. ASCE*, 73:199.

Holton, H. N. (1961). *A Concept of Infiltration Estimates in Watershed Engineering*, Agricultural Research Service Publication 41–51. U. S. Department of Agriculture, Washington, DC.

Hooghoudt, S. B. (1940). Bijdragen tot de kennis van eenige natuurkundige grootheden van den grond, 7. Algemeene beschouwing van het probleem van de detail ontwatering en de infiltratie door middel van parallel loopende drains, greppels, slooten, en kanalen. *Versl. Landbouwk. Ond*., 46:515–707.

Horton. (1940). An approach toward a physical interpretation of infiltration capacity. *Soil Sci. Soc. Am. Proc*., 5:399–417.

Huisman, L. (1972). *Ground-Water Recovery*. Winchester Press, New York.
Jacob, C. E. (1943). Correlation of ground-water levels with precipitation on Long Island, New York. *Ann. Geophys. Union Trans.*, pt. 2:564–573.
Kirkham, D. (1958). Seepage of steady rainfall through soil into drains. *Am. Geophys. Union Trans.*, 39:892–908.
Kirkham, D. (1967). Explanation of paradoxes in the Dupuit-Forchheimer seepage theory. *Water Resources Res.*, 3:609–622.
Kirkham, D., Toksoz, S., and van der Ploeg, R. R. (1974). Steady flow to drains and wells. In J. van Schilfgaarde (ed.) *Drainage for Agriculture*, pp. 203–244. American Society of Agronomy, Madison, WI.
Kostiakov, A. N. (1932). On the dynamics of the coefficient of water percolation in soils and on the necessity of studying it from a dynamic point of view for purposes of amelioration. *Transactions Com. International Soil Science*, 6th, Moscow, Russia, Part A, pp. 17–21.
Kozeny, J. (1927). Ueber kapillare Leitung des Wassers im Boden, Wien. *Akad. Wiss.*, 136, pt 2a:271.
Lambe, T. W., and Whitman, R. V. (1969). *Soil Mechanics*. Wiley, New York.
Luthin, J. N. (1973). *Drainage Enginering*. R. E. Krieger, Huntington, NY.
Marshall, T. J., and Holmes, J. W. (1979). *Soil Physics*. Cambridge University Press, New York.
McBean, E. A., Poland, R., Rovers, F. A., and Crutcher, A. J. (1982). Leachate collection design for containment landfills. *ASCE J. Env. Eng. Div.*, 108(1):204–209.
Mitchell, J. K., Hooper, D. R., and Campanella, R. G. (1965). Permeability of compacted clay. ASCE J. Soil Mech. Found. Div., 91(SM4), Proceedings Paper 4392.
Mitchell, J. K., and Madsen, F. T. (1987). Chemical effects on clay hydraulic conductivity. In Woods R. D. (ed.), *Geotechnical Practice for Waste Disposal*, Geotechnical Special Publication No. 13, ASCE, pp. 87–116.
Moody, W. T., and Ribbens, R. W. (1965). Ground water—Tetrama-Colusa canal reach no. 3, Sacramento Canal Units, Central Valley Project. Memorandum to Chief, Canals Branch. Bureau of Reclamation, Office of Chief Engineer, Denver, Co.
Moore, C. A. (1980). *Landfill and Surface Impoundment Performance Evaluation*. U.S. EPA SW-869.
Mualem, Y. (1976). Hysteretical models for prediction of the hydraulic conductivity of unsaturated porous media. *Water Resources Res.*, 12:1248–1254.
National Research Council (1990). Groundwater models: Scientific and regulatory applications. National Academy Press, Washington, DC.
Philip, J. R. (1957). The theory of infiltration, 4. Sorptivity and algebraic infiltration equations. *Soil Sci.*, 84:257–264.
Richards, S. J. (1965). Soil suction measurements with tensiometers. In Methods of Soil Analysis, Monograph 9, pp. 153–163. American Society of Agronomy, Madison, WI.
Singh, S. R., and Jacob, C. M. (1977). Transient analysis of phreatic aquifers lying between two open channels. *Water Resources Res.*, 13(2) 411–419.
Slatyer, R. O. (1967). *Plant–Water Relationships*. Academic Press, New York.
Swartzendruber, D. (1962). Non-Darcy flow behavior in liquid-saturated porous media. *J. Geophys. Res.*, 67:5205–5213.

Terzaghi, K. (1925). *Erdbaumechanik auf Boden-physicalischen Grundlagen*. Deuticke, Vienna.

Terzidis, G. (1968). Discussion, falling water table between tile drains. *ASCE Proc*. 94: 159.

U. S. Department of the Interior. (1981). *Ground-Water Manual*. U. S. Department of the Interior.

Van Genuchten, R. (1978). Calculating the unsaturated hydraulic conductivity with a new closed-form analytical model. Publication of the Water Resourses Program, Department of Civil Engineering, Princeton University Princeton, NJ.

Van Genuchten, M. Th. (1980). A closed-form equation for predicting the hydraulic conductivity of unsaturated soils. *Soil Sci. Soc. Am. J.* 44:892–898.

Van Schilfgaarde, J. (1970). Theory of flow to drains. In *Advances in Hydroscience*, vol. 6, pp. 43–106. Academic Press, New York.

Van Schilfgaarde, J. (1974). Nonsteady flow to drains. In J. van Schilfgaarde ed., Drainage for Agriculture, pp. 245–270. American Society of Agronomy, Madison, WI.

Venetis, C. (1968). On the impulse response of an aquifer. Bull. Int. Assoc. Sci. Hydrol. 13:136–139.

Visser, W.C. (1996). Progress in the knowledge about the effect of soil moisture content on plant production. Inst. Land Water Management, Wageningen, Netherlands, Tech. Bull. 45.

Youngs, E. G. (1965). Horizontal seepage through unconfined aquifers with hydraulic conductivity varying with depth. *J. Hydrol.*, 3:448–451.

Youngs, E. G. (1966a). Horizontal seepage through unconfined aquifers with nonuniform hydraulic conductivity. *J. Hydrol.*, 4:91–97.

Youngs, E. G. (1966b). Exact analysis of certain problems of ground-water flow with free surface conditions. *J. Hydrol.*, 4:227–281.

4

Mass Transport and Transfer in Soils

4.1 INTRODUCTION

In this chapter, we will address the issue of how contaminants migrate in the subsurface relative to pore water while simultaneously interacting with each other. We note that contaminants participate in several physical, chemical, and biological transformation processes during the course of their travel. These transformation processes should be taken into account in the assessment of soil and groundwater quality. As we will see later in this book, knowledge of contaminant transport and transfer in soils is crucial in site remediation and waste containment efforts.

As noted in Chapter 3, any formulation for transport in porous media essentially involves application of the mass conservation principle and cause-and-effect relationships in the form of constitutive equations. The mass conservation principle applied to contaminants in an elemental volume of porous media dictates that

$$\begin{aligned}\text{(Rate of mass input)} &- \text{(rate of mass output)}\\ &\pm \text{(rate of mass production or consumption)}\\ &= \text{rate of mass accumulation}\end{aligned} \tag{4.1}$$

where the addition and subtraction on the left-hand side (l.h.s.) apply to mass production and consumption, respectively. Mass transfer processes contribute to mass consumption or production in porous media and thus directly relate to the sink/source terms in Eq. (4.1). Mass transport processes, on the other hand, enter as constitutive relations in our formulation. We will see that the transport processes involve not only the application of Darcy's law (elaborated in Chapter 3), but also Fick's law discussed subsequently.

We need to make an important distinction between contaminants that are soluble in water and those that are insoluble or immiscible in water. The two types of contaminants differ in the way their transport takes place relative to pore water. The transport of soluble contaminants is more closely linked to the flow of pore water than that of immiscible contaminants, which is governed by a host of pore-scale and field-scale mechanisms not related to flow of water. We will defer discussion of immiscible contaminants to the next chapter and take up the dissolved contaminants first.

4.2 MASS TRANSPORT MECHANISMS

The transport of dissolved contaminants follows that of water via advection and is therefore related to the velocity of water flow. The direction of hydraulic gradients dictates to a large extent the direction of dissolved contaminant transport. If advection is the only mechanism of transport, the pore velocity (Darcy velocity divided by porosity) is an indicator of the transport of dissolved contaminants. Thus, when a chemical with a concentration of C_0 is introduced into a saturated soil system as shown in Figure 4.1a, it migrates as a sharp front at a velocity equal to the pore velocity, V_a. In reality, however, there are other mechanisms augmenting advection. The saturated soil system possesses concentration gradients in addition to hydraulic gradients because of the localized presence of the

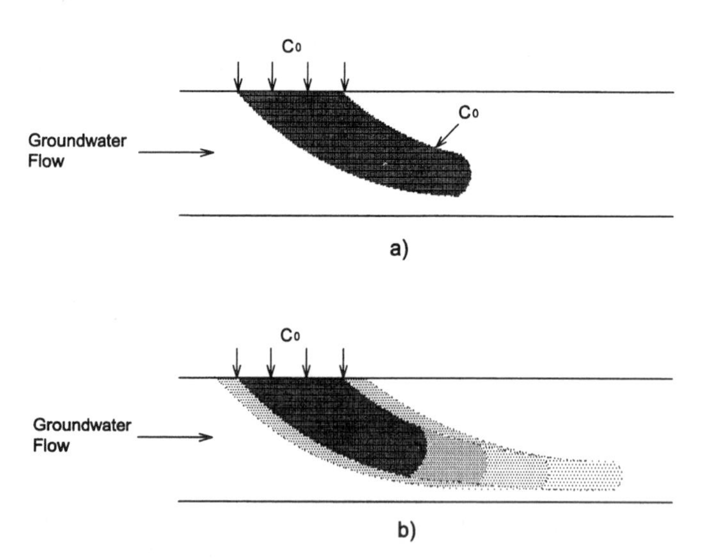

Figure 4.1 Chemical transported in groundwater: (a) via advection; (b) via advection and diffusion.

dissolved chemical. These concentration gradients exercise kinetic activity and provide for an additional mechanism of transport namely, diffusion. As shown schematically in Figure 4.1b, the effect of diffusion is to cause the contaminant to be spread out in all directions in response to concentration gradients. The relative contributions of advection and diffusion are therefore dependent on the magnitudes of velocity and the concentration gradients. In systems where velocities are low (by virtue of low permeability, such as in the case of waste-containment barriers), one can expect diffusion process to dominate. On the other hand, in systems where flow velocities are high, such as in coarse-grained soils, advection dominates. Leaving the quantitative analysis of these relative contributions to a later stage, we start with a discussion of the diffusion process.

Fick's first law provides us with a formal representation of the diffusion process. Analogous to Darcy's law, it states that the flux of diffusing chemical, J, (mass of solute per unit area per unit time, M/L^2T) is directly proportional to the concentration gradient:

$$J = -D^* \frac{dC}{dx} \tag{4.2}$$

where C is the solute concentration (M/L^3), D^* is the diffusion coefficient for the soil medium (L^2/T), and dC/dx is the concentration gradient, which is negative in the direction of diffusion. D^* is thus a coefficient of proportionality analogous to the coefficient of permeability k in Eq. (3.16). Carrying the analogy between the coefficients of permeability and diffusion further, the measurement of diffusion coefficients is also often indirect. Concentration gradients are applied over a soil sample and the observed chemical flux is used in Fick's first law to determine the diffusion coefficient D^*. Efforts to obtain a direct expression for diffusion coefficient in terms of the soil and pore fluid parameters were based on extrapolations of free-solution diffusion, as discussed below.

The fundamental basis for deriving an expression for the free-solution diffusion coefficient is the mobility of ions as determined by their mass, radius, and valence, and by the properties of the liquid, such as viscosity, dielectric constant, and temperature. Three expressions derived from fundamental principles are given below.

$$D_o = \frac{uRT}{N} \quad \text{Nernst-Einstein equation} \tag{4.3}$$

$$D_o = \frac{RT\lambda_o}{F^2\,|z|} \quad \text{Nernst equation} \tag{4.4}$$

$$D_o = \frac{RT}{6\pi N\eta r} \quad \text{Einstein-Stokes equation} \tag{4.5}$$

where D_o is the free-solution diffusion coefficient; u = absolute mobility of the solute; R = universal gas constant; T = absolute temperature; N = Avogadro's number (6.022×10^{23} mol^{-1}); λ_o = limiting ionic conductivity; F = Faraday constant (96,500 C/Eq); $|z|$ = absolute value of the ionic valence; η = viscosity of the fluid; and r = radius of the molecular or hydrated ion. These equations were all based on the same principles of ionic mobility. The absolute mobility of the solute, u, is expressed in terms of ionic conductivity λ_o in Eq. (4.4) and in terms of the size of the ion and the viscosity of the fluid in Eq. (4.5). The diffusion of species in a free solution is thus a complex function of a number of variables, viz, ionic radius, valence, temperature, viscosity of the diffusing medium, etc. Typical values of D_o for selected anions and cations are shown in Table 4.1. Values for other species may be found in Lerman (1979), and in Li and Gregory (1974).

The diffusion coefficient for a species in porous medium will be smaller than that in free solution, primarily because of the limited domain available for ionic flow. An effective diffusion coefficient, D^*, in soils may therefore be expressed as

$$D^* = \omega D_o \tag{4.6}$$

where ω is an empirical constant (typically ranging from 0.5 to 0.01) accounting for the presence of soil solids in the medium. More rigorous ways of accounting for the various factors involved in the diffusion process in soils are reviewed by

Table 4.1 Free-Solution Diffusion Coefficients for Selected Species[a]

Cation	$D_0 \times 10^{-10}$ (m^2/s)	Anion	$D_0 \times 10^{-10}$ (m^2/s)
H^+	93.1	OH^-	52.8
Na^+	13.3	Cl^-	20.3
K^+	19.6	F^-	14.7
NH_4^+	19.8	Br^-	20.8
Mg^{2+}	7.1	I^-	20.4
Ca^{2+}	7.9	NO_3^-	19.0
Fe^{2+}	7.2	HCO_3^-	11.8
Cu^{2+}	7.3	SO_4^{2-}	10.7
Zn^{2+}	7.2	CO_3^{2-}	9.6

[a] Values correspond to infinite dilution in water at 25°C.
Sources: Li and Gregory (1974); Lerman (1979).

Shackelford and Daniel (1991a). In general, one can group all the factors responsible for differences in diffusion between free solution and soil medium and express D^* as

$$D^* = D_o \tau \alpha \gamma \theta \tag{4.7}$$

where τ = tortuosity factor accounting for the tortuous pathway of ions; α = fluidity or mobility factor to account for the increased viscosity of water adjacent to the surface of solid particles relative to that of bulk pore fluid; γ = anion exclusion factor to account for the possible exclusion of anions from smaller pores in case of low-porosity soils; and θ = volumetric water content accounting for the reduced flow domain for ionic travel. These factors are often lumped together for simplification purposes. Two such simplifications are

$$D^* = D_o f_i \theta \qquad \text{Nye (1979)} \tag{4.8}$$
$$D^* = D_o t_r \qquad \text{Olsen et al. (1965)} \tag{4.9}$$

where f_i is an ''impedance factor'' and t_r is a ''transmission factor.'' Shackelford and Daniel (1991a) warn us about the usage in literature of ''effective diffusion coefficient'' and caution us to interpret carefully the factors that were taken into account in the specific context where the term was used.

The diffusion of chemicals in soils described above is typically grouped with another important transport mechanism known as mechanical dispersion. The mechanical dispersion is the effect of advective velocities which, when sufficiently high, cause a mixing of the chemical in the porous medium. The extent of mechanical mixing of the substance is a direct function of the advective velocity. To account for the combined effect of diffusion and dispersion, a dispersion coefficient, D, is used:

$$D = D^* + D_m \tag{4.10}$$

where D_m = coefficient of mechanical dispersion and is commonly expressed as a linear function of the pore velocity, V_a:

$$D_m = \alpha V_a \tag{4.11}$$

where α is the dispersivity [L]. Through laboratory experiments on tracer migration in saturated granular media, Perkins and Johnston (1963) identified the relative importance of diffusion and mechanical dispersion in soils. Based on the experimental results, the α (in meters) in Eq. (4.11) was expressed as

$$\alpha = 1.75d \tag{4.12}$$

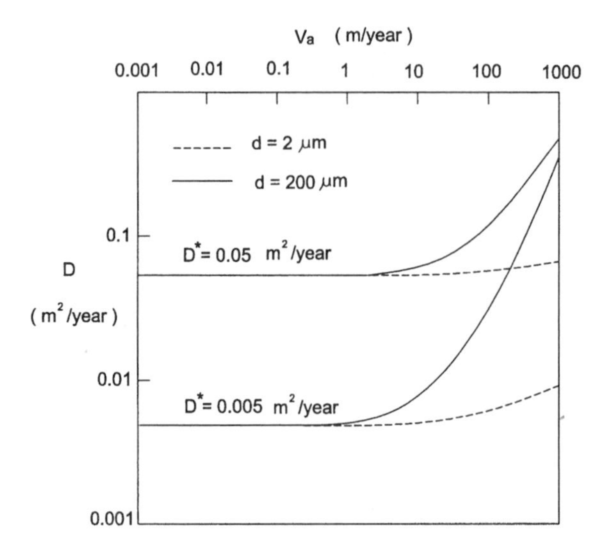

Figure 4.2 Relative effects of molecular diffusion and mechanical dispersion on the dispersion coefficient, D. (*Source*: Perkins and Johnston, 1963)

where d is the mean grain diameter of the soil in meters. Figure 4.2 shows the variation of D with respect to the actual velocity, V_a, for the two extreme conditions of D^* and d. In fine soils with high diffusion, D^* alone controls the dispersion coefficient; on the other hand, in coarse soils with a low diffusion, D is controlled by mechanical dispersion for velocities greater than 1 m per year. It is customary in practice to substitute D for D* in Eq. (4.2); however, the reader may question the validity of lumping mechanical dispersion with a constant of proportionality between mass flux and concentration gradient.

In the case of engineered systems such as compacted clays, the above principles are generally taken to be valid. Unlike the permeability aspect, the diffusion behavior of compacted clays has been addressed in only a few studies (Quigley et al., 1987; Shackelford and Daniel, 1991b). These studies indicate that diffusion plays an important role in the breakthrough of chemicals through compacted clays. They suggest that the experimental errors far outweigh any variability of effective diffusion coefficient with respect to molding water content and type of compaction. Diffusion in compacted clays is still an area of active research, and future studies may lead to further understanding of the process.

4.3 MASS TRANSFER MECHANISMS

A specific chemical element or compound may exist in groundwater in any of the following forms (Johnson et al., 1989):

1. As "free" ions surrounded by water molecules
2. As insoluble species, e.g., Ag_2S, $BaSO_4$
3. As metal–ligand complexes, e.g., $Al(OH)^{2+}$
4. As adsorbed species
5. As species held on a surface by ion exchange
6. As species that differ by oxidation state, e.g., Fe^{2+}, Fe^{3+}; Cr^{3+}, Cr^{4+}

The various chemicals existing in one or more of these forms, while participating in reactions with each other to achieve equilibrium, may transfer from one phase to the other among the three phases—solids, liquids, and gases. In other words, the reactions can be homogeneous (occurring in only one phase) or heterogeneous (involving phase transfer). The transformation of a chemical from its initial state is an intricate process, which constantly influences and is influenced by the transport mechanisms. Furthermore, change in state of one species may alter the mass transfer mechanisms of other species. It is here that one would find it necessary to augment a mass transport model with a geochemical model, which considers the myriad changes that a chemical in the subsurface undergoes during the course of its migration.

While some transfer mechanisms are better understood than others, many of the mechanisms are still actively being researched in the field of geochemistry. Our purpose in the following subsections is to obtain knowledge of the individual mass transfer mechanisms to the extent needed to understand the transport phenomena in soils. An important consideration in transport modeling is the rate at which the various transfer processes occur in soils. The time period required for a specific transfer process relative to that of advection and diffusion processes will govern the significance of that process in transport modeling. Thus, the transfer processes, which take place at time scales far greater than those required by advection and diffusion, can safely be ignored in transport modeling. Unfortunately, this is not true for a majority of the processes, which have reaction rates much greater than the typical rates of groundwater flow—hence the need for a detailed discussion of these processes. We will take up the issue of reaction times after discussing briefly the individual mechanisms.

It is convenient to treat the mass transfer mechanisms under two groups: (1) abiotic processes; and (2) biotic processes. The abiotic processes refer to those that are nonbiological in nature. The biotic processes involve mass consumption of the chemicals by microorganisms, often referred as biodegradation. The population growth/decline of microorganisms in the presence of various chemicals is of primary importance in these processes. In reality, both abiotic and biotic processes occur concurrently. The microorganisms involved in the biotic processes act as catalysts for some of the reactions in abiotic processes.

4.3.1 Abiotic Processes

The major mechanisms under this classification are: (1) acid–base reactions, (2) hydrolysis, (3) oxidation–reduction (redox) reactions, (4) complexation, (5) precipitation and dissolution, (6) exsolution and volatilization, (7) radioactive decay, and (8) sorption.

Acid–Base Reactions

Acid–base reactions are the most fundamental processes involving the exchange of hydrogen ions (H^+). These reactions take place almost instantaneously between substances tending to lose H^+ ions (acids) and those tending to gain the same (bases). They alter the pH of the system, which in turn affects other mechanisms such as sorption. Very often, it is this indirect effect of acid–base reactions (through changes in pH) that is of importance in transport modeling. Because of the pH changes, the chemicals participating in these reactions may influence transport of contaminants, although the chemicals by themselves may not be the contaminants of concern. The following are two examples of these reactions:

$$HCO_3^- + H_2O = H_3O^+ + CO_3^{2-} \tag{4.13}$$

$$NH_3 + H_2O = NH_4^+ + OH^- \tag{4.14}$$

In the first reaction, water serves as the base accepting a proton, and in the second reaction, it serves as the acid donating a proton. The acid–base reactions are often referred as weak or strong depending on the extent of proton transfer. Several examples of acid–base reactions common to groundwater systems may be found in Morel (1983) and in Domenico and Schwartz (1990).

Hydrolysis

Hydrolysis is a mass transfer process that is a result of substitution between an organic compound and water. It may be represented as

$$(R - X) + H_2O \rightarrow (R - OH) + X^- + H^+ \tag{4.15}$$

where R refers to the main part of organic molecule and X represents the attached halogen, carbon, phosphorous, or nitrogen. The introduction of hydroxyl into the organic molecule may make the reaction product more soluble and susceptible to biodegradation. Thus, hydrolysis may be an important transfer mechanism in remediation efforts, since it may transform an organic compound, originally resistant to biodegradation, to a degradable compound. However, not all organic compounds participate in hydrolysis reactions. Those which are susceptible to hydrolysis include alkyl halides, amides, carbamates, carboxylic acid esters, epoxides and lactones, phosphoric acid esters, and sulfonic acid esters (Neely, 1985).

Oxidation–Reduction (Redox) Reactions

Oxidation–reduction reactions involve exchange of electrons, similar to exchange of protons in the case of acid–base reactions. Oxidation and reduction refer to the removal and acceptance of electrons, respectively. Examples of these reactions are

$$Fe^{2+} = Fe^{3+} + e^- \tag{4.16}$$

$$Fe^{3+} + e^- = Fe^{2+} \tag{4.17}$$

Equation (4.16) involves oxidation of Fe^{2+}, which donates an electron, and Eq. (4.17) involves reduction of Fe^{3+}, which accepts an electron. These reactions essentially alter the oxidation number of elements. The oxidation number of iron, for instance, changes from 2 to 3 in the reaction shown in (4.16). It is important to note, however, that no free electrons result from these reactions, since electrons given up by one compound are gained by another compound. The following describes a complete redox reaction:

$$O_2 + 4Fe^{2+} + 4H^+ = 2H_2O + 4Fe^{3+} \tag{4.18}$$

wherein iron and oxygen participate as the electron donor (reductant) and electron acceptor (oxidant), respectively. The oxidation number of oxygen changes from 0 to -2 and that of iron changes from 2 to 3. In general, any element that can have different valencies may participate in redox reactions. Several trace elements have variable oxidation numbers. Common elements found in variable oxidation states, and examples of ions formed from these elements, are shown in Table 4.2.

An important feature that distinguishes redox reactions from others is that they can be mediated by microorganisms, which act as catalysts to these reactions. The energy that is released when the electrons are donated in the redox reactions becomes a source for the microbial cells to sustain and grow. The following is an example of a biologically mediated redox reaction, which involves oxidation of an organic compound, CH_2O, and reduction of oxygen:

$$\frac{1}{4}CH_2O + \frac{1}{4}O_2 = \frac{1}{4}CO_2 + \frac{1}{4}H_2O \tag{4.19}$$

Table 4.2 Examples of Elements Found in More Than One Oxidation State and Ions Formed from Those Elements

Element and oxidation state in parentheses	Ion or solid formed from the element
Fe(+III)	Fe^{3+}, $Fe(OH)_3$
Fe(+II)	Fe^{2+}
Cr(+IV)	CrO_4^{2-}, $Cr_2O_7^{2-}$
Cr(+III)	Cr^{3+}, $Cr(OH)_3$
C(+IV)	HCO_3^-, CO_3^{2-}
C(0)	CH_2O, C
C(−IV)	CH_4
N(+V)	NO_3^-
N(+III)	NO_2^-
N(0)	N
N(−III)	NH_4^+, NH_3
S(+VI)	SO_4^{2-}
S(+V)	$S_2O_6^{2-}$
S(+II)	$S_2O_3^{2-}$
S(−II)	H_2S, HS^-

Source: Morel (1983).

The transfer of electrons from CH_2O to O_2 provides the necessary energy for cell growth. Reactions similar to (4.19) occur with oxygen substituted by NO_3^-, SO_4^{2-}, CO_2, and others. When these substitutions take place, the biotransformation reactions result in denitrification (involving NO_3^-), sulfate reduction (involving SO_4^{2-}), and methane formation (involving CO_2).

Complexation

Complexation involves reaction between simple cations (usually metallic), and anions called ligands. The ligands might be inorganic, such as Cl^-, F^-, Br^-, SO_4^{2-}, PO_4^{3-}, and CO_3^{2-}, or organic, such as amino acid. Examples of complexation reactions are

$$Mn^{2+} + Cl^- = MnCl^+ \tag{4.20}$$

$$Cu^{2+} + H_2O = CuOH^+ + H^+ \tag{4.21}$$

Complexation reactions may also occur in series, with complex of one reaction participating with ligand in another reaction. An example is the reaction involving Cr^{3+} (Domenico and Schwartz, 1990):

$$Cr^{3+} + OH^- = Cr(OH)^{2+} \tag{4.22}$$

$$Cr(OH)^{2+} + OH^- = Cr(OH)_2^{\ +} \tag{4.23}$$

$$Cr(OH)_2^{\ +} + OH^- = Cr(OH)_3^{\ 0} \tag{4.24}$$

The speciation of Cr might be controlled by pH of the pore fluid and thus by the acid–base reactions. Figure 4.3 shows the dominance of various chromium hydroxide complexes as a function of pH. Complexation reactions are important in studying the transport of metals in the subsurface. In assessing the total metal concentration in the pore fluid, the complexes must be considered in addition to the free metal ion concentration. Because of the complexation reactions, the actual quantities of metal transported downstream in the pore fluid might be more than what the concentrations of metal ions alone might indicate.

Precipitation and Dissolution

These processes are fundamental in nature and are responsible for the vast amounts of mass transfer occurring in the subsurface environment. Water is an excellent solvent for several chemicals in gas, solid, as well as liquid phases. Dissolution refers to complete solubilizing of a given element in groundwater. This mechanism alone might be responsible for bringing contaminants into the pore fluid at the source. Natural minerals also dissolve in pore fluid in a process similar to weathering discussed in Chapter 1. Many times, dissolution involves only partial solubilizing of certain elements in the mineral, leaving behind certain secondary minerals. Examples of dissolution occurring in quartz and kaolinite are as follows:

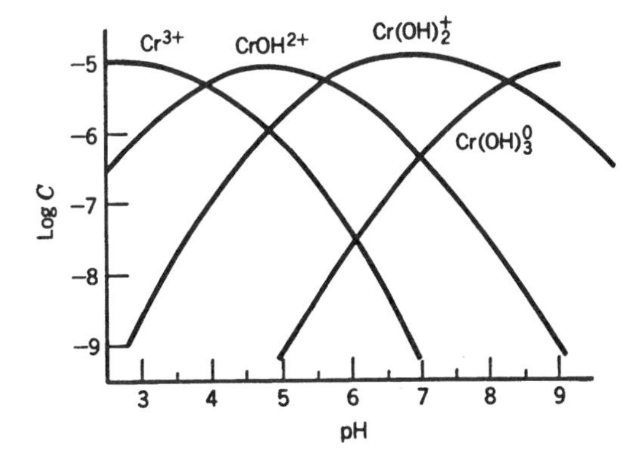

Figure 4.3 Effect of pH on complexation of chromium hydroxide. (*Source*: Domenico and Schwartz, 1990)

$$SiO_2 + H_2O = H_2SiO_3 \tag{4.25}$$

$$\text{Kaolinite} + 5H_2O = 2Al(OH)_3 + 2H_4SiO_4 \tag{4.26}$$

The dissolved components might be precipitated again due to reactions with other dissolved species. Some examples of chemicals that could be reduced to lower concentrations in the pore fluid through the formation of precipitates include lead and silver, which enter into reactions with sulfides, carbonates, or chlorides. Precipitates may also be formed as the products of hydrolysis reactions (precipitates of iron, copper, chromium, and zinc) or redox reactions (precipitates of chromium and arsenic). It is possible in certain cases that both precipitation and dissolution continue one after another as the plume moves downgradient.

Because of the pH dependence of speciation of metals (see Fig. 4.3), the solubility of the metal oxides and hydroxides is also dependent on pH. Figure 4.4 shows as examples the solubilities of zinc and copper as a function of pH. The lines on the graphs show the equilibrium conditions between solid and soluble forms of the metals. Thus the shaded portions of the plots show the cumulative solubility for all the species of a given metal. The figure demonstrates that there exists a pH at which the solubilities are minimum and beyond which the solubility increases again in the anionic form. The pH dependence of solubility is also related to the specific type of the metal and the form of the precipitate. Figure 4.5 shows the solubilities of certain metal oxides and sulfides, which shows the difference in pH dependence between the solubilities of hydroxides and sulfides.

The solubility of organic solutes is highly variable. In general, charged species of organic compounds or those containing oxygen or nitrogen are most soluble. An example is alcohols, which participate in hydrogen bonding with

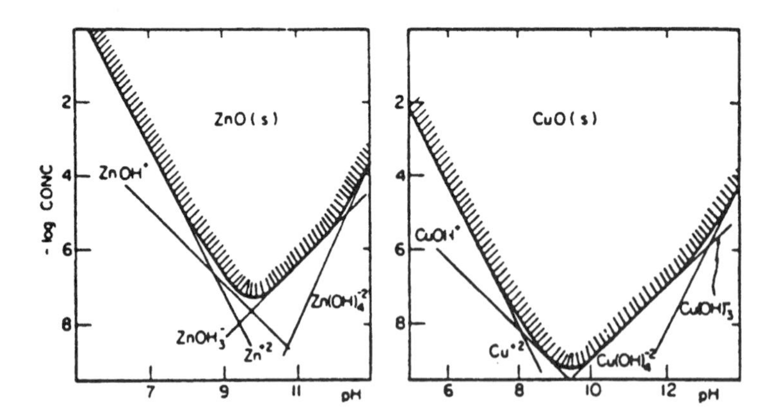

Figure 4.4 Solubilities of zinc and copper as a function of pH. (*Source*: Stumm and Morgan, 1981)

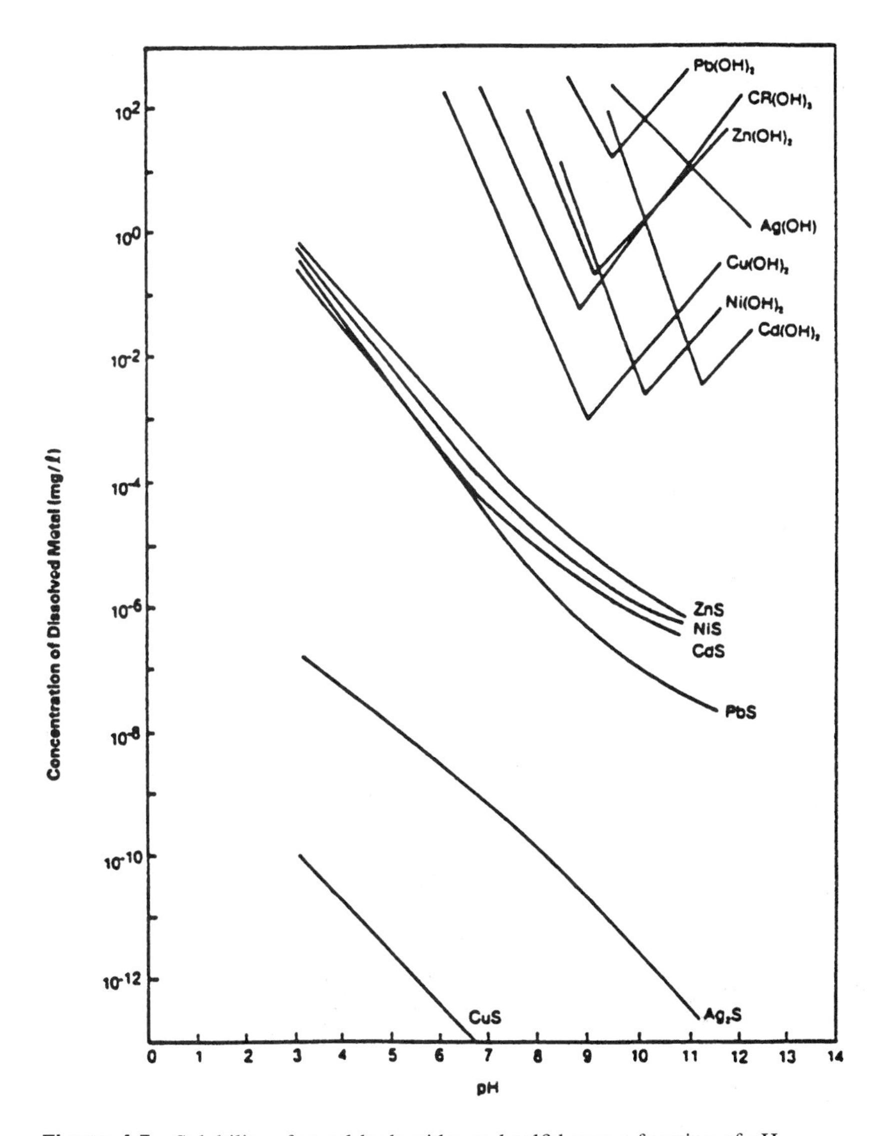

Figure 4.5 Solubility of metal hydroxides and sulfides as a function of pH.

water and fit into its structure. Nonpolar molecules require high energies to enter into the structure of water and therefore are sparingly soluble or completely hydrophobic. In general, larger organic molecules are less soluble because they cannot be accommodated in the structure of water. The solubility is in general inversely proportional to the mass of the organic molecules (Mackay and Leinonen, 1975). The octanol/water partition coefficient (K_{ow}) is commonly used to indicate the solubility of an organic solute in water. K_{ow} is a dimensionless constant, which expresses partitioning of the organic solute between octanol and water. Therefore, the higher the K_{ow}, the lower is the water solubility of the organic solute. Figure 4.6 shows the relationship between octanol–water partition coefficients and the water solubilities for several organic compounds.

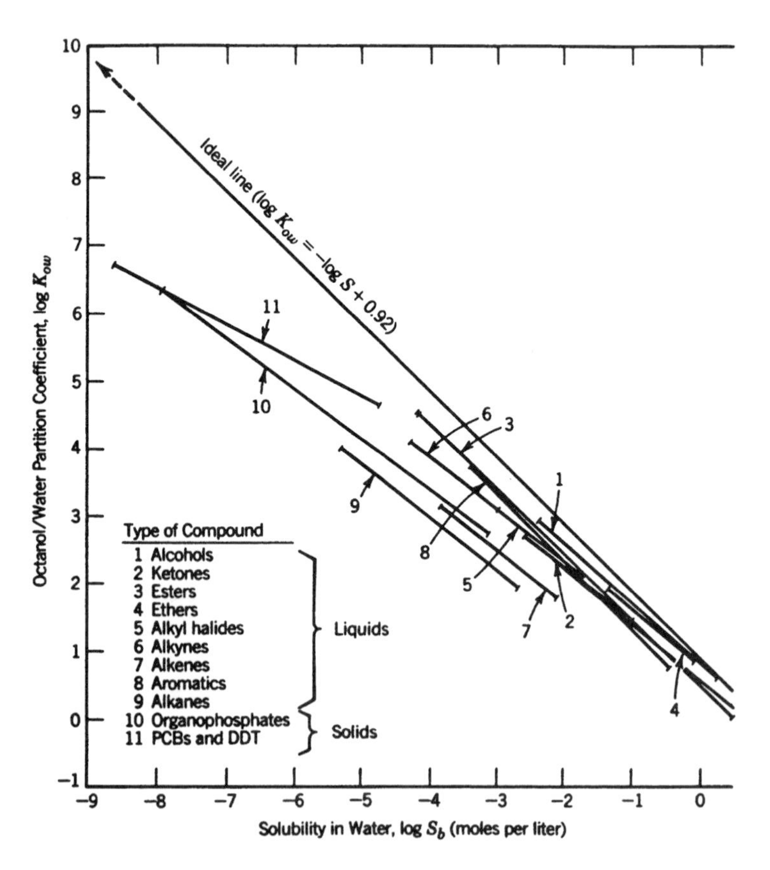

Figure 4.6 Relationship between solubility in water and octanol-water partition coefficients. (*Source*: Chiou et al., 1982).

Exsolution and Volatilization

These processes involve mass transfer between gaseous and either liquid or solid phases. Similar to precipitation, these processes remove mass from the pore water into the gaseous phase. They are controlled by the vapor pressure, which is the pressure of the gas in equilibrium with respect to the liquid or solid at a given temperature. Vapor pressure reflects the solubility of a compound in gas and is therefore an indicator of the compound's tendency to evaporate. The process is governed by two well-known laws, Raoults' law and Henry's law. The former is used to relate the equilibrium partial pressure of a volatile organic in the atmosphere to the pure constituent vapor pressure. In other words,

$$P_i = x_i P_i^0 \tag{4.27}$$

where P_i = partial pressure of the vapor in the gas phase; x_i = mole fraction of the organic solvent; and P_i^0 = vapor pressure of the pure organic solvent. Equation (4.27) is applicable only when the mole fraction of the solvent is greater than 0.9, and is therefore valid to represent volatilization of pure solvents. To describe volatilization of many solutes dissolved in water, Henry's law is used instead. It relates the equilibrium partial pressure of a constituent directly to the mole fraction of the constituent in the aqueous phase. It is expressed as

$$P_i = H_i X_i \tag{4.28}$$

where P_i = partial pressure of the vapor in the gas phase; X_i = mole fraction of the constituent in the aqueous phase, and H_i = Henry's constant specific for the constituent. As seen in Eq. (4.28), Henry's constant is an indicator of the ratio of constituent concentration in the vapor phase to that in the aqueous phase. Table 4.3 lists the vapor pressures and Henry's constants for selected group of compounds.

Radioactive Decay

Radioactive decay is the process whereby unstable isotopes (atoms of the same element that differ in their mass) decay to form new ones. It involves emission of particles from the element's nucleus. The process is termed α decay or β decay depending on whether the element loses an α particle (helium) or a β particle (electron). The decay is an irreversible process, with the parent element continuously decaying over time while the daughter isotope increases in quantity. Of importance in geoenvironmental engineering are the radioactive species released into groundwater from such activities as mining, milling, and storage of wastes. The movement of radioactive isotopes such as uranium, plutonium, cesium, and selenium, away from high-level radioactive waste repositories and defense installations, is of utmost importance.

Table 4.3 Henry's Law Constants for Selected Chemicals

Compound	Henry's law constant (atm m^3/mol)
Acetone	3.97×10^{-5} (25°C)
Benzene	5.48×10^{-3} (25°C)
Carbon tetrachloride	2.4×10^{-2} (20°C)
Ethylbenzene	8.68×10^{-3} (25°C)
Fluorene	2.1×10^{-4}
n-Hexane	1.184 (25°C)
Isopropylbenzene	1.47×10^{-2} (25°C)
Methyl alcohol	4.66×10^{-6} (25°C)
Naphthalene	4.6×10^{-4}
n-Octane	3.225 (25°C)
PCB-1232	8.64×10^{-4}
Phenol	3.97×10^{-7} (25°C)
Propane	7.06×10^{-1} (25°C)
Styrene	2.61×10^{-3}
Tetrachloroethylene	0.0153
Toluene	0.00674 (25°C)
Trichloroethylene	0.0091
Trifluralin	4.84×10^{-5} (23°C)
Vinyl chloride	2.78
o-Xylene	5.35×10^{-3} (25°C)

Sources: Montgomery and Welkom (1990); Montgomery (1991).

Sorption

Sorption is the phase-transfer process describing the transfer of contaminants from liquid to solid state and is a dominant process affecting almost all dissolved species in groundwater. *Sorption* is a term used to include all the processes responsible for the mass transfer—*absorption, adsorption*, and *ion exchange*. Absorption is the process whereby the chemical is incorporated into the interior of the solid, whereas adsorption is the process of attraction of chemicals to the surface. Ion exchange occurs because of the substitution of cations in solids as discussed in Chapter 1, and may therefore be considered a special case of absorption. Many sorption reactions are completely or partially reversible, and *desorption* is used to indicate this reverse process.

Sorption reactions are in general classified as either *sorbent driven* or *solvent driven* (Weber, 1972). When they are sorbent driven, the solid phase is the active one attracting the chemical, perhaps because of the presence of cation-exchange sites on the solids. When they are solvent driven, the chemicals (usually nonpolar and organic) are hydrophobic in nature and therefore tend to associate with the nonpolar phase such as the organic content of the porous matrix. Thus, in the former case, opposites attract, and in the latter, likes interact with likes. These two types encompass all the possible sorption reactions and allow us to identify the following parameters governing sorption:

- Contaminant and pore fluid characteristics
  - pH of the pore fluid
  - Water solubility of the contaminant
  - Polar nature of the contaminant
- Solids and porous media characteristics
  - Mineralogy: surface charge and surface area
  - Organic carbon content
  - Homogeneity
  - Texture
  - Porosity and permeability of the porous medium

The extent of sorption of a given constituent is often estimated using sorption isotherms. These isotherms allow partitioning of the constituent between the solid and liquid phases. They are obtained in the laboratory using batch tests, which involve mixing water containing the constituent at a specific concentration with the solid medium and allowing the mixture to equilibrate. By repeating the tests at other concentrations of the constituent at the same temperature (hence the name, isotherm), a relation between sorbed mass of the constituent and the equilibrium concentration is obtained. Such isotherms may assume several shapes—linear, concave, or convex, or any combination of these. Theoretical equations are generally used to fit these experimental isotherms. Three of the common equations are

Linear isotherm: $$S = K_d C \tag{4.29}$$

Freundlich isotherm: $$S = K\,C^n \tag{4.30}$$

Langmuir isotherm: $$S = \frac{\alpha\beta C}{1 + \alpha C} \tag{4.31}$$

where S = quantity of mass sorbed on the surface (mg/g or μg/g), C = equilibrium concentration of the solution, K_d = slope of the linear sorption isotherm, also known as the distribution coefficient (ml/g), K = partition coefficient indicating the extent of sorption, n is the Freundlich exponent, usually ranging between 0.7 and 1.2, α = coefficient known as Langmuir constant, and β = maximum sorptive capacity of the solid surface (mg/kg). The linear isotherm is thus

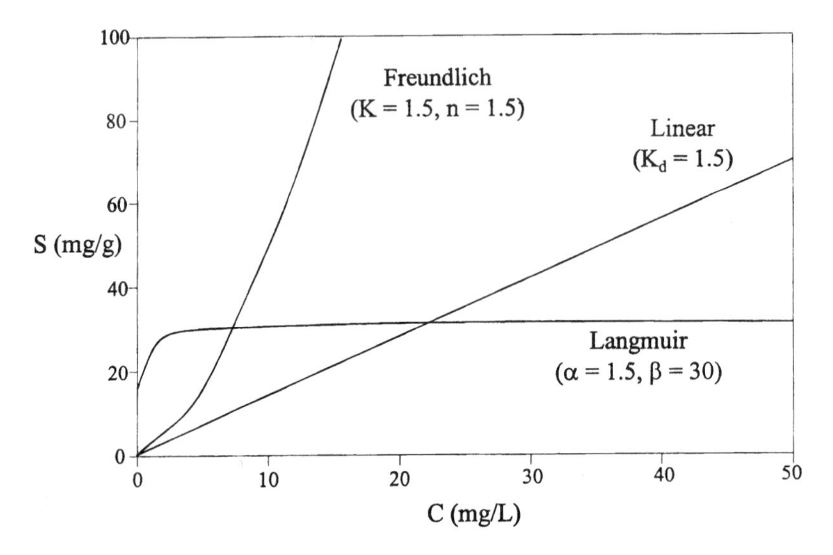

Figure 4.7 Examples of linear, Freundlich, and Langmuir isotherms.

a special case of the Freundlich isotherm with the exponent $n = 1$. Figure 4.7 shows the possible sorption reactions as represented by these three equations using representative parameters. A wide variation in sorption behavior is apparent from these relationships. Of the three, the linear isotherm offers the simplest relationship to account for sorption reactions, and is therefore commonly used in simulating mass transport.

When more than one constituent is present in pore fluid, a selectivity sequence is used to identify the contaminant preferred over other species by the solid surface. The preference of one constituent over the other by the solid surface, referred as affinity or selectivity, is a consequence of the differences in the characteristics (positive charges, size of the ions, etc.) of the cations, which compete for the exchange sites on the solid surface. Table 4.4 presents a summary of the adsorption selectivity of some common heavy metals in different types of soils under different pH conditions.

In the case of organic compounds, the sorption is a result of their hydrophobic nature. The sorption of these compounds is considered to take place exclusively onto the organic carbon fraction, f_{oc} if the soil contains at least 1% of the organic carbon on a weight basis (Karickhoff et al., 1979). The sorption of the hydrophobic compounds is also modeled using a linear isotherm assuming K_d to be directly proportional to f_{oc},

$$K_d = K_{oc} f_c \tag{4.32}$$

Table 4.4 Adsorption Selectivity of Heavy Metals in Different Soils

Material	Selectivity order	References
Kaolinite clay (pH 3.5–6)	Pb > Ca > Cu > Mg > Zn > Cd	Farrah and Pickering (1977)
Kaolinite clay (pH 5.5–7.5)	Cd > Zn > Ni	Puls and Bohn (1988)
Illite clay (pH 3.5–6)	Pb > Cu > Zn > Ca > Cd > Mg	Farrah and Pickering (1977)
Montmorillonite clay (pH3.5–6)	Ca > Pb > Cu > Mg > Cd > Zn	Farrah and Pickering (1977)
Montmorillonite clay (pH5.5–7.5)	Cd = Zn > Ni	Puls and Bohn (1988)
Al oxides (amorphous)	Cu > Pb > Zn > Cd	Kinniburgh et al. (1976)
Mn oxides	Cu > Zn	Murray (1975)
Fe oxides (amorphous)	Pb > Cu > Zn > Cd	Benjamin and Leckie (1981)
Goethite	Cu > Pb > Zn > Cd	Forbes et al. (1974)
Fulvic acid (pH 5.0)	Cu > Pb > Zn	Schnitzer and Skinner (1967)
Humic acid (pH 4–6)	Cu > Pb > Cd > Zn	Stevenson (1977)
Japanese dominated by volcanic parent material	Pb > Cu > Zn > Cd > Ni	Biddappa et al. (1981)
Mineral soils (pH 5.0), (with no organics)	Pb > Cu > Zn > Cd	Elliot et al. (1986)
Mineral soils (containing 20–40 g/kg organics)	Pb > Cu > Cd > Zn	Elliot et al. (1986)

Source: Yong et al. (1992).

where K_{oc} = partition coefficient of a compound between organic carbon and water (Karickhoff et al., 1979). In several studies, K_{oc} was found to be related to K_{ow}, the octanol/water partition coefficient, because of the similarity in the adsorption behavior of organic carbon and octanol. Some regression equations relating the two parameters are

Karickhoff et al., 1979 $\log K_{oc} = -0.21 + \log K_{ow}$ (4.33)

Kenaga and Goring, 1980 $\log K_{oc} = 1.377 + 0.544 \log K_{ow}$ (4.34)

Schwarzenbach and Westall, 1981 $\log K_{oc} = 0.49 + 0.72 \log K_{ow}$ (4.35)

Hassett et al., 1983 $\log K_{oc} = 0.088 + 0.909 \log K_{ow}$ (4.36)

4.3.2 Equilibrium and Kinetic Models of Reactions

The processes discussed in the preceding section occur at different rates to achieve chemical equilibrium. A simplistic approach would be to assume that the reactions would come to equilibrium instantaneously. This is justified in the case of acid–base reactions and a few others, and is termed the equilibrium model of reaction. However, in majority of the reactions, the reaction process is transient similar to the transport processes of advection and dispersion. Some of the reactions, such as radioactive decay and biologically mediated redox reactions, may take several months or years. It is essential in these cases to adopt a kinetic model and treat the reaction as time dependent. Figure 4.8 shows the ranges of half-times (time periods for half-life of the reaction processes) for some common reactions occurring in the subsurface. Radioactive decay, which is not shown in Figure 4.8, has a wide range of half-times.

The equilibrium model of reaction is based on the law of mass action, which allows one to describe the equilibrium distributions of mass between reactants and products. Consider a reversible reaction in which reactants A and B participate to produce C and D.

$$aA + bB = cC + dD \tag{4.37}$$

where a, b, c, and d represent the number of moles of the individual constituents. At equilibrium, the mass law dictates that

$$K = \frac{[C]^c[D]^d}{[A]^a[B]^b} \tag{4.38}$$

where K is known as the equilibrium constant, and [ ] indicates a thermodynamic property equivalent to aqueous-phase concentration for dissolved species, or to the partial pressure for a gas. The equilibrium constant K is termed the dissociation constant in the case of acid–base reactions, the solubility constant in the case of dissolution reactions, the stability or complexation constant in the case

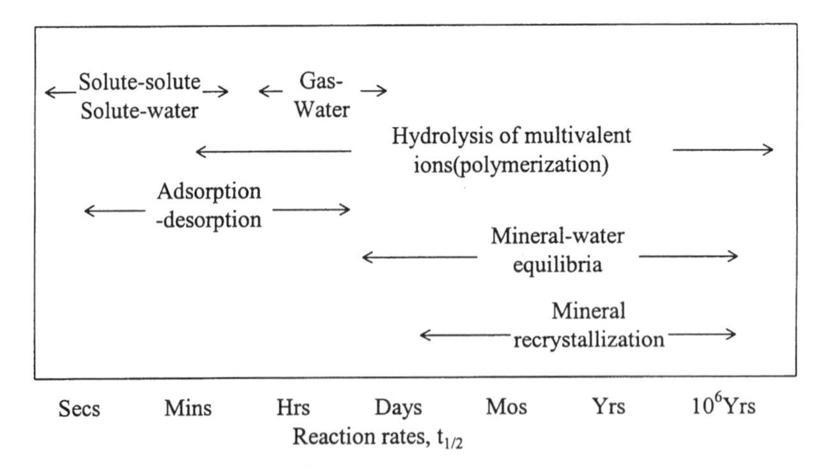

Figure 4.8 Half-time ranges for common reactions occurring in the subsurface. (*Source*: Langmuir and Mahoney, 1984)

of complexation reactions, the substitution constant in the case of substitution/hydrolysis, and the adsorption constant for surface reactions. Equation (4.38) is also used to identify the departure from equilibrium of a given reaction. For instance, if the right-hand side (r.h.s.) of Eq. (4.38), also known as the ion activity product (IAP), is less than K, the reaction is proceeding forward; and if it is greater than K, the reaction is proceeding backward, reducing the products and increasing the reactants.

A kinetic model for the reaction in Eq. (4.37) involves a rate law of the form

$$r_A = -k_1[A]^{x_1}[B]^{x_2} + k_2[C]^{y_1}[D]^{y_2} \tag{4.39}$$

where r_A is the rate of mass accumulation of the constituent A, k_1 and k_2 are the rate constants for the forward and reverse reactions, and x_1, x_2, y_1, and y_2 are empirical or stoichiometric coefficients and represent the order of the reaction. The first term of the r.h.s. in Eq. (4.39) indicates the rate at which the species A is consumed in the forward direction and the second term indicates the rate at which it is produced in the reverse direction. Equation (4.39) yields a sink-source term for a given constituent and can be incorporated directly in the mass transport equation as discussed in the next section. In a number of cases, the reactions can be represented by simplified forms of rate laws such as those given below (Langmuir and Mahoney, 1984),

$$r_C = -k_1[^{14}C] \tag{4.40}$$

$$r_{Fe} = -k_1[Fe^{2+}][O_2] \tag{4.41}$$

where r_C and r_{Fe} are the rates of mass changes of the irreversible decay of ^{14}C and that of Fe^{2+} in the following radioactive decay and oxidation reactions:

$$^{14}C \rightarrow {}^{14}N + e \tag{4.42}$$

$$Fe^{2+} + \frac{1}{4}O_2 + \frac{1}{2}H_2O \rightarrow FeOH^{2+} \tag{4.43}$$

Often, a number of reactions may be proceeding in parallel, making it unwieldy to express the kinetics of a given constituent in a single equation. Also, when the reactions are of higher order, the solution of mass transport equations tends to be complicated. For these reasons, simplifications are often sought when describing geochemical reactions in mass transport phenomena.

The National Research Council (1990) notes that, in general, the equilibrium models are well established only for oxidation/reduction and acid/base reactions. The kinetic models are well established for radioactive reactions, dissolution, complexation, and substitution/hydrolysis; however, the rate constants needed for these models are poorly defined for several relevant species in the subsurface. Thus, one may expect increased research efforts in future on several of these mass transfer mechanisms.

4.3.3 Biotic Processes

Soils provide habitat for a vast range of microorganisms. Typically, the microbial population is of the order of 1×10^6 cells per gram dry weight in natural systems. These microorganisms act as catalysts for the abiotic transformations described above, in particular for the redox reactions. The redox reactions involve release of energy when electrons are donated in the oxidation process of a chemical compound. Much like human respiration, the majority of organisms utilize oxygen during respiration while oxidizing organic matter. The oxidation of organic matter to form carbon dioxide is an example of aerobic respiration. However, oxygen may be toxic for some organisms, in which case growth of organisms takes place under anaerobic conditions. For these organisms, oxygen is substituted by another organic compound. An excellent example of a biological process involving these organisms is the process known as denitrification, in which nitrate is reduced to nitrogen gas.

In our treatment of mass transfer of contaminants in the subsurface, our interest lies in the consumption of a certain chemical by microorganisms. In dealing with biotic reactions, we are thus faced with a need to know the existence and population growth/decline of the microbial population. How the population could be metabolically stimulated will depend on the presence of energy sources, nutrients (primarily nitrogen and phosphorous), and suitable environmental conditions (temperature, pH, etc). Stimulated growth of the organisms and increased uptake of the contaminant of concern by these organisms are desirable outcomes in site remediation. A remediation technology based on this concept is bioremediation.

The various phases of microbial growth typically occurring in an environment limited by supply of nutrients and electron acceptors are shown in Figure 4.9. In general, there are six phases. In the first phase, known as the lag phase, the microbes adopt to the new environmental conditions and the available substrate for energy and carbon sources. Following this phase is an acceleration phase when the microbial population grows at an increasing rate. The maximum rate of growth is sustained in the exponential growth phase, when the substrate is consumed significantly. This rate of growth may be followed by declining and stationary phases, perhaps due to the limited availability of the substrate, insufficient nutrients, or depleting electron acceptor. The cells die in these phases, leading ultimately to the endogenous phase, where the rate of cell death exceeds the rate of cell growth.

The various phases of microbial growth, shown in Figure 4.9, make it a complicated problem to assess the rate at which a chemical in the pore stream

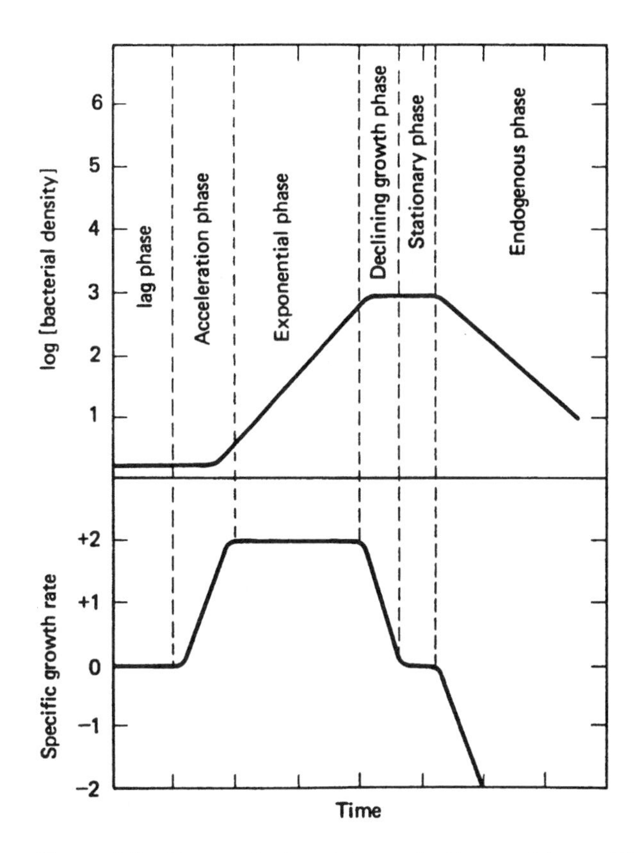

Figure 4.9 Phases of microbial growth. (*Source*: Monod, 1949)

is consumed by microorganisms. A simplistic view is to simulate the degradation of the chemical using an exponential relationship,

$$C(t) = C_0 e^{-kt} \tag{4.44}$$

where $C(t)$ = concentration at time t, C_0 = initial concentration, k = rate of consumption. Such approaches may perhaps adequately simulate the chemical consumption during the exponential phase of microbial growth, but they do not consider the limiting aspects of growth. The most popular expression for microbial utilization of substrates is based on a hyperbolic saturation function proposed by Monod (1942):

$$J = \frac{k_{\max} M C}{Y(C_k + C)} \tag{4.45}$$

where J = rate of change in substrate concentration, k_{max} = maximum growth rate of microbes, M = microbial mass, C = concentration of the growth-limiting substrate, Y = yield coefficient indicating the organisms formed per substrate utilized, and C_k = substrate concentration at which growth rate is $k_{max}/2$. As seen in Eq. (4.45), smaller values of C_k and larger values of k_{max} are associated with maximum rate of substrate consumption. It is important to note that microbial mass M, is itself not constant, so Eq. (4.45) must be coupled with another equation describing the rate of change in M. This is usually accomplished using Monod kinetics with a decay term that accounts for cell death. Thus,

$$\frac{dM}{dt} = k_{max}\, MY \frac{C}{C_k + C} - bM \tag{4.46}$$

where b = coefficient representing the rate at which microorganisms decay. The interdependency between the microbial mass and substrate concentrations has led to several biodegradation models, most of which use the above principles based on Monod kinetics. The biofilm model (McCarty et al., 1984) is one such model which accounts for the processes involved in the transport of substrate chemical to the microbes. It conceptualizes that the microorganisms bind to the surface of the solid matrix in the form of a biofilm, deriving the needed substrate from the pore liquid (Fig. 4.10). A diffusion layer is identified between the bulk

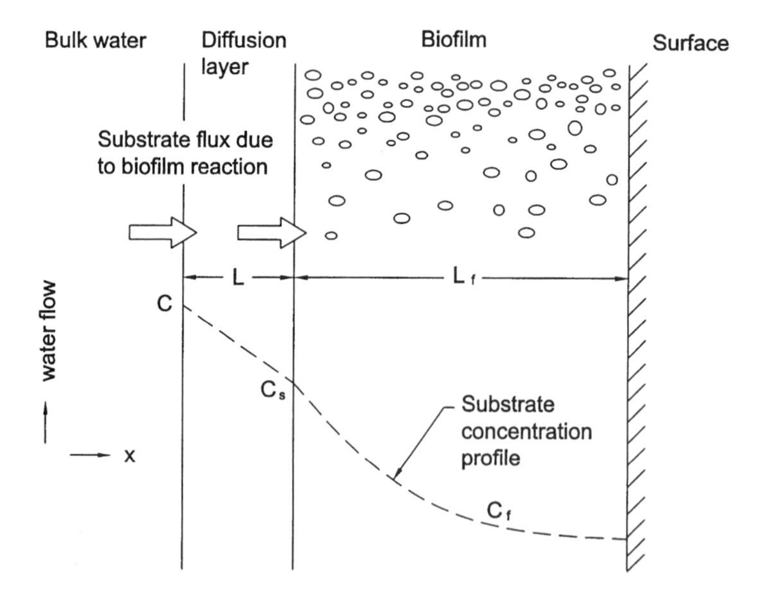

Figure 4.10 Conceptualization of substrate flux into the biofilm. (*Source*: McCarty et al., 1984)

pore liquid and the biofilm. Within the biofilm itself, the substrate is transported by molecular diffusion while being utilized by the microorganisms. The rate at which this diffusion occurs is given by Fick's second law,

$$J_f = D_f \frac{d^2C_f}{dx^2} \tag{4.47}$$

where D_f = molecular diffusion coefficient of the substrate in the biofilm, x = dimension normal to the surface of the film. Note the difference between Fick's first and second laws. The second law involves a combination of the equation of continuity and the first law [Eq. (4.2)]. Under steady state conditions of substrate utilization, the rates given by Eqs. (4.45) and (4.47) are equal. Therefore,

$$\frac{d^2C_f}{dx^2} = \frac{k_{\max}M}{D_f} \frac{C_f}{Y(C_k + C_f)} \tag{4.48}$$

For the substrate to reach the biofilm, it must be transported through water in the diffusion layer, and this occurs according to Fick's first law,

$$r = D\frac{dC}{dx} = D\frac{C - C_s}{L} \tag{4.49}$$

Simultaneous solution of Eqs. (4.48) and (4.49) will therefore yield the steady-state concentrations of the substrate across the diffusion and biofilm layers. r in Eq. (4.49) can then be used to represent substrate being consumed by the microorganisms. The problem of biomass estimation still remains, and a coupled equation for microbial growth needs to be solved in order to estimate the extent of biofilm over which r in Eq. (4.49) is applicable.

4.4 GOVERNING EQUATION FOR MASS TRANSPORT

The mass transport and transfer processes discussed in the preceding sections may now be combined to obtain a general equation describing the transport of contaminants. To make the problem simple, we will assume that the porous medium is homogeneous, isotropic, and saturated. Similar to our treatment of flow process (Section 3.4), we will invoke the universal law of mass conservation and couple it with a cause-and-effect relationship, which in the present problem is supplied by Fick's first law.

Consider the elemental control volume shown in Figure 4.11. The mass conservation equation for this volume, stated in Eq. (4.1), may be expressed mathematically as

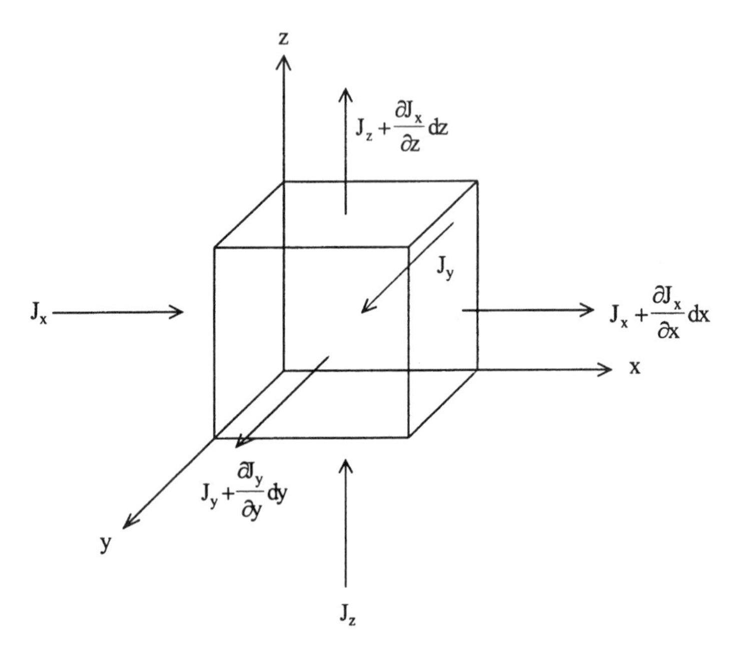

Figure 4.11 Elemental control volume for mass flux.

$$-\left[\frac{\partial J_x}{\partial x} + \frac{\partial J_y}{\partial y} + \frac{\partial J_z}{\partial z}\right] \pm r = \frac{\partial (nC)}{\partial t} \tag{4.50}$$

where J = mass flux of solute per unit cross-sectional area transported in the direction indicated by the subscript x, y, or z; r = rate of mass production/consumption given by the kinetic model of reaction, n = porosity, and C = solute concentration expressed as mass of solute per unit volume of solution.

The two mass transport processes of advection and dispersion govern J in Eq. (4.50). In the x direction, J due to advection and dispersion may be expressed as

$$J_{\text{due to advection}} = V_x nC \tag{4.51}$$

and

$$J_{\text{due to dispersion}} = -nD_x \frac{\partial C}{\partial x} \tag{4.52}$$

where V_x = pore velocity in the x direction. Note that D_x in Eq. (4.52) includes the two components of molecular diffusion and mechanical dispersion. Summing up the contributions from advection and dispersion, the mass fluxes in the three dimensions become

$$J_x = V_x nC - nD_x \frac{\partial C}{\partial x} \tag{4.53}$$

$$J_y = V_y nC - nD_y \frac{\partial C}{\partial y} \tag{4.54}$$

$$J_z = V_z nC - nD_z \frac{\partial C}{\partial z} \tag{4.55}$$

Substitution of Eqs. (4.53), (4.54), and (4.55) in (4.50) yields

$$-\left[\frac{\partial}{\partial x}\left(V_x nC - nD_x \frac{\partial C}{\partial x}\right)\right] - \left[\frac{\partial}{\partial y}\left(V_y nC - nD_y \frac{\partial C}{\partial y}\right)\right] - \left[\frac{\partial}{\partial z}\left(V_z nC - nD_z \frac{\partial C}{\partial z}\right)\right] \pm r = \frac{\partial (nC)}{\partial t} \tag{4.56}$$

Assuming that the velocities are steady and uniform and that the dispersion coefficients do not vary in space, Eq. (4.56) may be simplified as

$$\left[D_x \frac{\partial^2 C}{\partial x^2} - V_x \frac{\partial C}{\partial x}\right] + \left[D_y \frac{\partial^2 C}{\partial y^2} - V_y \frac{\partial C}{\partial y}\right] + \left[D_z \frac{\partial^2 C}{\partial z^2} - V_z \frac{\partial C}{\partial z}\right] \pm \frac{r}{n} = \frac{\partial C}{\partial t} \tag{4.57}$$

It is also assumed in Eq. (4.57) that the porosity of the medium is constant in time and space. In one dimension, Eq. (4.57) reduces to the well-known advection-dispersion equation (ADE):

$$D_x \frac{\partial^2 C}{\partial x^2} - V_x \frac{\partial C}{\partial x} \pm \frac{r}{n} = \frac{\partial C}{\partial t} \tag{4.58}$$

The exact form of the ADE depends on the mass transfer processes accounted for in the term r. Depending on the nature of r, analytical solutions exist for the ADE for simple initial and boundary conditions. The following section presents some common analytical solutions for the ADE.

4.5 SOLUTIONS FOR SPECIAL CASES OF MASS TRANSPORT

We classify the existing analytical solutions for mass transport under three categories:

1. One-dimensional mass transport involving nonreactive constituents, therefore $r = 0$
2. One-dimensional mass transport involving reactive constituents, therefore one or more of the mass transfer processes should be accounted for through r
3. Two- and three-dimensional mass transport involving nonreactive and reactive constituents for simple flow conditions

4.5.1 Nonreactive Constituents ($r = 0$)

Pure Diffusion Conditions

Pure diffusion conditions prevail when groundwater velocity, V_x, is so low that advective transport process can be ignored. Waste containment barriers such as clay or synthetic liners fall into this category, since they possess low permeabilities and diffusion tends to dominate advection. The ADE [Eq. (4.58)] then reduces to Fick's second law:

$$\frac{\partial C}{\partial t} = D_x \frac{\partial^2 C}{\partial x^2} \tag{4.59}$$

Given the following initial and boundary conditions,

$$C(x, 0) = C_i \qquad x \geq 0$$
$$C(0, t) = C_0 \qquad t > 0$$
$$C(\infty, t) = C_i \qquad t \geq 0$$

the solution for Eq. (4.59) may be expressed as

$$C(x, t) = C_i + (C_0 - C_i)\,\mathrm{erfc}\left(\frac{x}{\sqrt{4D_x t}}\right) \tag{4.60}$$

where "erfc" refers to the complement of the error function, "erf," which is expressed in terms of its argument β as

$$\mathrm{erfc}(\beta) = 1 - \mathrm{erf}(\beta) = 1 - \left(\frac{2}{\sqrt{\pi}}\right)\int_0^{\beta} e^{-u^2}\,du \tag{4.61}$$

The error function and its complement are listed in Table 4.5 for positive values of β. For the special condition when $C_i = 0$, i.e., the medium is initially free of the solute, Eq. (4.60) reduces to

$$C(x, t) = C_0\,\mathrm{erfc}\left(\frac{x}{\sqrt{4D_x t}}\right) \tag{4.62}$$

Table 4.5 Error Function (erf) and Complementary Error Function (erfc)

β	erf(β)	erfc(β)
0	0	1.0
0.05	0.056372	0.943628
0.1	0.112463	0.887537
0.15	0.167996	0.832004
0.2	0.222703	0.777297
0.25	0.276326	0.723674
0.3	0.328627	0.671373
0.35	0.379382	0.620618
0.4	0.428392	0.571608
0.45	0.475482	0.524518
0.5	0.520500	0.479500
0.55	0.563323	0.436677
0.6	0.603856	0.396144
0.65	0.642029	0.357971
0.7	0.677801	0.322199
0.75	0.711156	0.288844
0.8	0.742101	0.257899
0.85	0.770668	0.229332
0.9	0.796908	0.203092
0.95	0.820891	0.179109
1.0	0.842701	0.157299
1.1	0.880205	0.119795
1.2	0.910314	0.089686
1.3	0.934008	0.065992
1.4	0.952285	0.047715
1.5	0.966105	0.033895
1.6	0.976348	0.023652
1.7	0.983790	0.016210
1.8	0.989091	0.010909
1.9	0.992790	0.007210
2.0	0.995322	0.004678
2.1	0.997021	0.002979
2.2	0.998137	0.001863
2.3	0.998857	0.001143
2.4	0.999311	0.000689
2.5	0.999593	0.000407
2.6	0.999764	0.000236
2.7	0.999866	0.000134
2.8	0.999925	0.000075
2.9	0.999959	0.000041
3.0	0.999978	0.000022

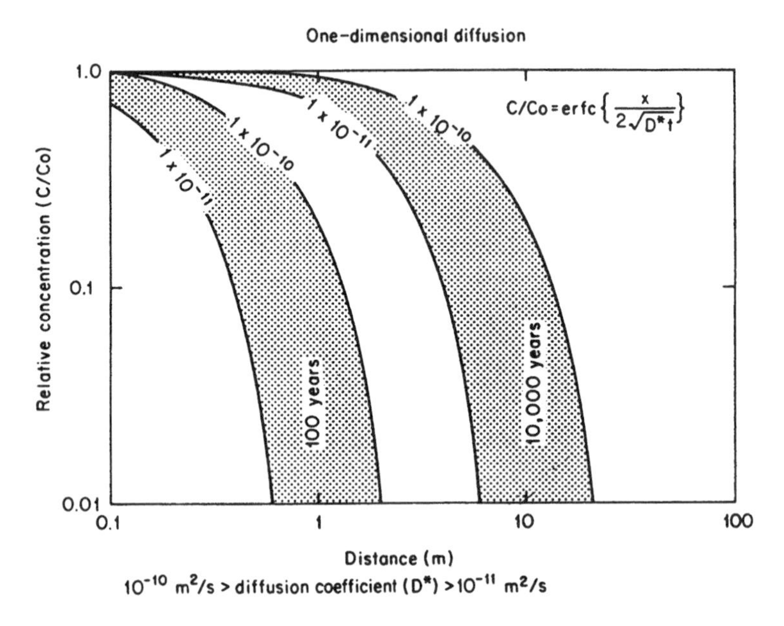

Figure 4.12 Mass transport by molecular diffusion. (*Source*: Freeze and Cherry, 1979)

Figure 4.12 shows this solution for a range of D_x, typical of nonreactive chemical species in clayey geologic deposits. It is seen that diffusion alone can cause contaminants to travel over distances typical of the thickness of a waste-containment barrier (1–2 m) during typical lifespans of waste-containment structures.

Advection and Dispersion for Continuous Source

For a column initially free of solute and subjected to a continuous source C_0 at the inlet, the initial and boundary conditions may be expressed as

$$\begin{aligned} C(x, 0) &= 0 \qquad x \geq 0 \\ C(0, t) &= C_0 \qquad t > 0 \\ C(\infty, t) &= 0 \qquad t \geq 0 \end{aligned}$$

The solution of the ADE under these conditions is given by (Ogata, 1961, 1970)

$$\frac{C}{C_0} = \frac{1}{2}\left[\operatorname{erfc}\left(\frac{x - V_x t}{2\sqrt{D_x t}}\right) + \exp\left(\frac{V_x x}{D_x}\right)\operatorname{erfc}\left(\frac{x + V_x t}{2\sqrt{D_x t}}\right)\right] \tag{4.63}$$

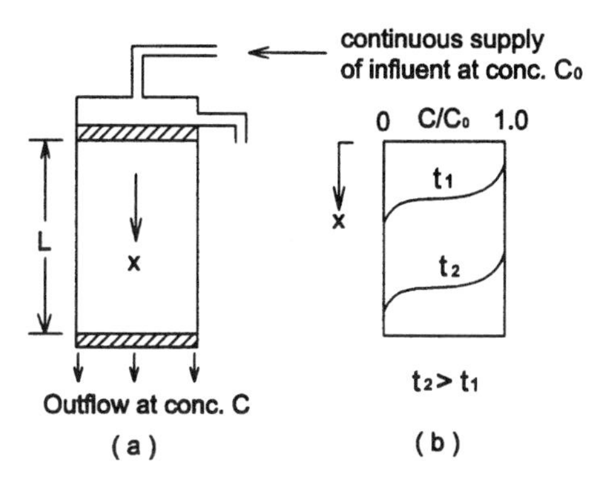

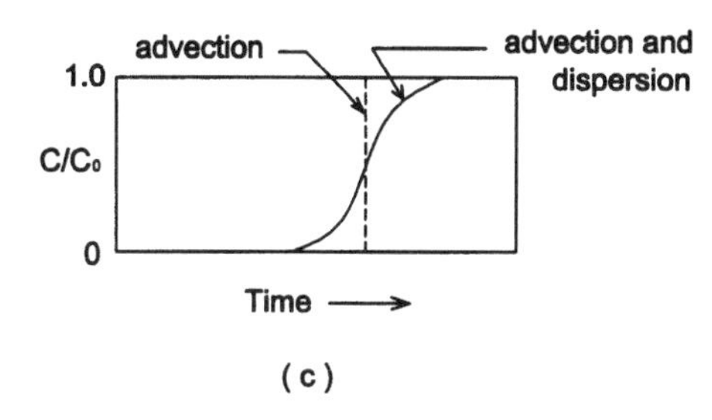

Figure 4.13 One-dimensional transport under advection and dispersion for continuous source: (a) column permeated continuously at concentration C_0; (b) relative concentrations as a function of x; (c) breakthrough curves under advection alone, and under advection and dispersion.

The solution given in Eq. (4.63) is shown schematically in Figure 4.13. The plot of C/C_0 with respect to time is known as the breakthrough curve. A comparison with the solution under plug-flow conditions ($D_x = 0$) illustrates the significance of dispersion. The impact of dispersion on mass transport is to make the solute appear before the arrival of advection front at a given cross section and to delay the instant when full source concentrations are reached at that cross section.

Advection and Dispersion for Instantaneous Source

This case represents the injection of a pulse of tracer at $x = 0$ with the background concentration equal to zero in the rest of the domain. If M is the injected mass per unit cross-sectional area, the solution may be expressed as

$$C(x, t) = \frac{M}{\sqrt{4\pi D_x t}} \exp\left\{-\frac{(x - V_x t)^2}{4 D_x t}\right\} \tag{4.64}$$

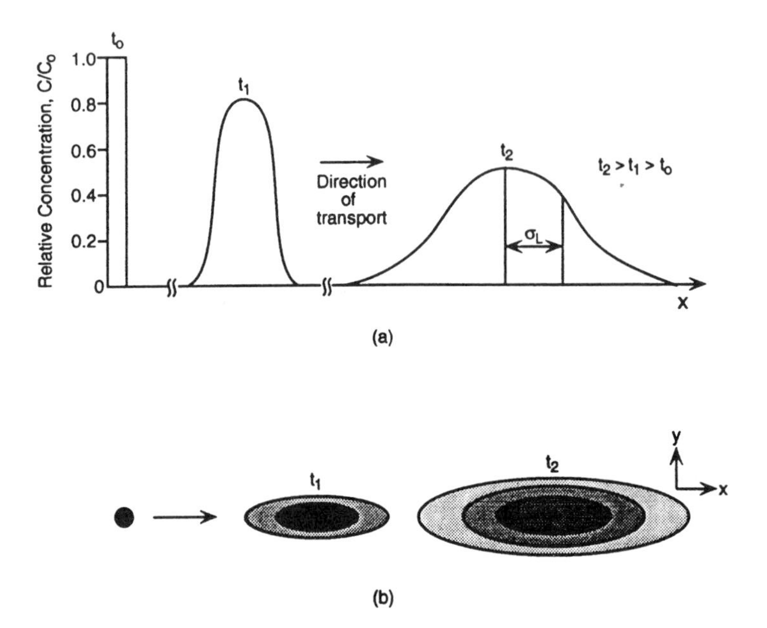

Figure 4.14 Mass transport of an instantaneous source under advection and dispersion. (*Source*: Bedient et al. 1994)

This solution is shown schematically in Fig. 4.14. Unlike the previous case, the concentrations at a given location x will gradually increase and then decrease because of the limited source. Since the tracer is nonreactive, the area encompassed by the two curves in Figure 4.14a must be equal. We should also note that Eq. (4.58) represents normal distribution with mean equal to $V_x t$ and standard deviation equal to $\sqrt{2D_x t}$. This indicates that the diffusion is essentially a Gaussian process.

4.5.2 Reactive Constituents

Advection and Dispersion with Sorption

One of the dominant mass transfer mechanisms occurring during mass transport is sorption. The simplest way of incorporating sorption is to use the linear isotherm discussed in Section 4.3.1:

$$S = K_d C \tag{4.29}$$

where S = quantity of mass sorbed on the surface of solids and K_d = distribution coefficient. The rate expression r is equal to the product of time derivative of S and dry mass density, ρ_b. Thus,

$$r = \rho_b \frac{\partial S}{\partial t} = K_d \rho_b \frac{\partial C}{\partial t} \tag{4.65}$$

Substituting Eq. (4.65) in the ADE [Eq. 4.58)],

$$D_x \frac{\partial^2 C}{\partial x^2} - V_x \frac{\partial C}{\partial x} - \frac{\rho_b}{n} K_d \frac{\partial C}{\partial t} = \frac{\partial C}{\partial t} \tag{4.66}$$

Rearrangement of terms in Eq. (4.66) yields

$$D_x \frac{\partial^2 C}{\partial x^2} - V_x \frac{\partial C}{\partial x} = \frac{\partial C}{\partial t}\left(1 + \frac{\rho_b K_d}{n}\right) \tag{4.67}$$

or, in its familiar form,

$$\frac{D_x}{R}\frac{\partial^2 C}{\partial x^2} - \frac{V_x}{R}\frac{\partial C}{\partial x} = \frac{\partial C}{\partial t} \tag{4.68}$$

where $R = (1 + \rho_b K_d/n)$ is known as the retardation factor since it has the effect of retarding the transport of adsorbed species relative to the advection front. Given the initial and boundary conditions,

$$\begin{aligned} C(x, t) &= 0 && \text{for } t = 0 \text{ and } x > 0 \\ C(x, t) &= C_0 && \text{for } x = 0 \\ C(x, t) &= 0 && \text{for } x = \infty \end{aligned}$$

the solution of Eq. (4.68) is (Ogata and Banks, 1961, 1970)

$$C(x, t) = \frac{C_0}{2}\left[\operatorname{erfc}\left(\frac{Rx - V_x t}{2\sqrt{RD_x t}}\right) + \exp\left(\frac{V_x x}{D_x}\right)\operatorname{erfc}\left(\frac{Rx + V_x t}{2\sqrt{RD_x t}}\right)\right] \tag{4.69}$$

Note that Eq. (4.69) reduces to Eq. (4.63) when $R = 1$, i.e., the species is nonreactive and $K_d = 0$.

Advection and Dispersion with Sorption and Zero-Order Production or Decay

When the reaction process is independent of species concentration, a constant sink/source term γ, can be used for r. Although this is a rare occurrence, it provides simple approximates in a number of mass transport problems. With a zero-order production term, the ADE becomes

$$\frac{D_x}{R}\frac{\partial^2 C}{\partial x^2} - \frac{V_x}{R}\frac{\partial C}{\partial x} + \frac{\gamma}{R} = \frac{\partial C}{\partial t} \tag{4.70}$$

Analytical solutions for Eq. (4.70) are available for simple initial and boundary conditions (van Genuchten and Alves, 1982). One solution for a specific set of initial and boundary conditions is given here as an example.

For

$$C(x, 0) = C_i,$$

$$C(0, t) = \begin{cases} C_0 & 0 < t \le t_0 \\ 0 & t > t_0 \end{cases}$$

and

$$\frac{\partial C}{\partial x}(\infty, t) = \text{finite},$$

$$C(x,t) = \begin{cases} C_i + (C_0 - C_i)A(x,t) + B(x,t) & 0 < t \le t_0 \\ C_i + (C_0 - C_i)A(x,t) + B(x,t) - C_0 A(x, t - t_0) & t > t_0 \end{cases} \tag{4.71}$$

where

$$A(x, t) = \frac{1}{2}\operatorname{erfc}\left(\frac{Rx - V_x t}{2\sqrt{D_x R t}}\right) + \frac{1}{2}\exp\left(\frac{V_x x}{D_x}\right)\operatorname{erfc}\left(\frac{Rx + V_x t}{2\sqrt{D_x R t}}\right)$$

and

$$B(x, t) = \frac{\gamma}{R}\left[t + \frac{(Rx - V_x t)}{2V_x}\operatorname{erfc}\left(\frac{Rx - V_x t}{2\sqrt{D_x R t}}\right) - \frac{(Rx - V_x t)}{2V_x}\exp\left(\frac{V_x x}{D_x}\right)\operatorname{erfc}\left(\frac{Rx + V_x t}{2\sqrt{D_x R t}}\right)\right]$$

Advection and Dispersion with Sorption, Zero-Order Production, and First-Order Decay

Many of the mass transfer processes including radioactive decay, biodegradation, and hydrolysis can be adequately described using first-order kinetics. In these cases,

$$r = \frac{d(nC)}{dt} = -\lambda n C \tag{4.72}$$

where λ is the first-order decay rate [T^1]. The ADE of Eq. (4.70) will now become

$$\frac{D_x}{R}\frac{\partial^2 C}{\partial x^2} - \frac{V_x}{R}\frac{\partial C}{\partial x} - \frac{\lambda C}{R} + \frac{\gamma}{R} = \frac{\partial C}{\partial t} \tag{4.73}$$

Analytical solutions for Eq. (4.73) are listed in van Genuchten and Alves (1982) for several sets of initial and boundary conditions. For the following set of initial and boundary conditions,

$$C(x, 0) = C_i$$

$$C(0, t) = \begin{cases} C_0 & 0 < t \leq t_0 \\ 0 & t > t_0 \end{cases}$$

$$\frac{\partial C}{\partial x}(\infty, t) = 0$$

the solution for Eq. (4.73) is

$$C(x,t) = \begin{cases} \frac{\gamma}{\lambda} + \left(C_i - \frac{\gamma}{\lambda}\right)A(x,t) + \left(C_0 - \frac{\gamma}{\lambda}\right)B(x,t) & 0<t\leq t_0 \\ \frac{\gamma}{\lambda} + \left(C_i - \frac{\gamma}{\lambda}\right)A(x,t) + \left(C_0 - \frac{\gamma}{\lambda}\right)B(x,t) - C_0 B(x, t-t_0) & t>t_0 \end{cases}$$

where

$$A(x, t) = \exp\left(-\frac{\lambda t}{R}\right)\left\{1 - \frac{1}{2}\,\mathrm{erfc}\left(\frac{Rx - V_x t}{2\sqrt{D_x R t}}\right) - \frac{1}{2}\exp\left(\frac{V_x x}{D_x}\right)\mathrm{erfc}\left[\frac{Rx + V_x t}{2\sqrt{D_x R t}}\right]\right\}$$

$$B(x, t) = \frac{1}{2}\exp\left[\frac{(V_x - U)x}{2D_x}\right]\mathrm{erfc}\left(\frac{Rx - U_x t}{2\sqrt{D_x R t}}\right) + \frac{1}{2}\exp\left[\frac{(V_x + U)x}{2D_x}\right]\mathrm{erfc}\left(\frac{Rx + Ut}{2\sqrt{D_x R t}}\right)$$

and

$$U = V_x\left[1 + \frac{4\lambda R D}{V_x^2}\right]^{1/2} \tag{4.74}$$

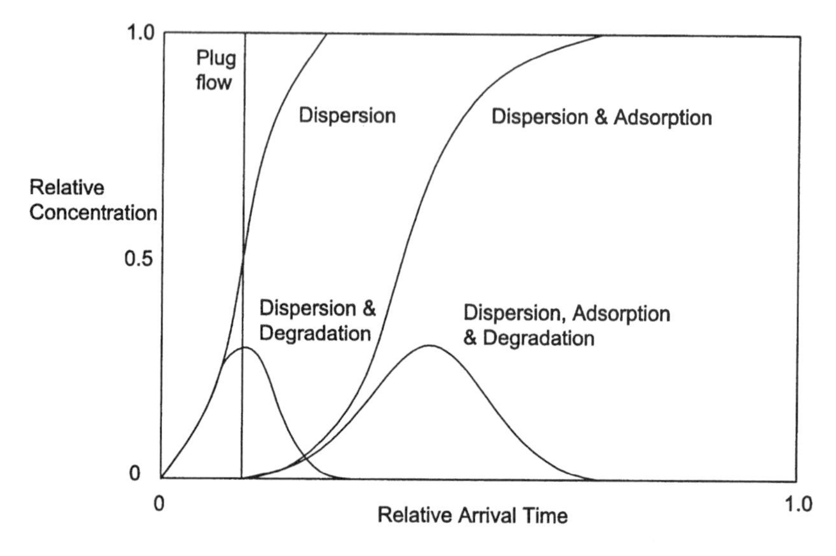

Figure 4.15 Effects of various processes on the breakthrough curve.

It is necessary to have a qualitative understanding of the effects of the various mass transfer processes in dealing with mass transport phenomena. Often, the nature of the breakthrough curve from a laboratory experiment indicates the predominance of one or more of the transport and transfer processes. Figure 4.15 illustrates the individual effects of the processes on the breakthrough curve of a chemical in a column experiment. In summary, the dispersion process causes the chemical to be transported with a front characterized by relative concentration ranging gradually from 0 to 1, as opposed to the advection process which exhibits $C/C_0 = 1$ at the wetting front. The effect of sorption is to retard the overall transport process. The effect of degradation and decay reactions is to consume the chemical species, which may also be a retarded occurrence if sorption is present.

4.5.3 Two- and Three-Dimensional Transport

For a more realistic representation of mass transport phenomena, consideration of transport in y and z dimensions often becomes necessary. Fortunately, in a majority of the cases, groundwater flow is one-dimensional, and it would suffice to include higher dimensionality only in the case of dispersion component. For instance, for an aquifer of uniform thickness, mass transport can be represented on a horizontal plane neglecting velocity components in the vertical dimension. The ADE under such circumstances may be written as

$$\left(D_x \frac{\partial^2 C}{\partial x^2} + D_y \frac{\partial^2 C}{\partial y^2}\right) - V_x \frac{\partial C}{\partial x} \pm \frac{r}{n} = \frac{\partial C}{\partial t} \tag{4.75}$$

De Josselin De Jong (1958) provided a solution for the mass transport in a two-dimensional field for the specific case of pulse injection over a finite line source, such as a well point, into a horizontal flow field. For a pulse tracer with initial concentration C_0 injected at a point (x_0, y_0) in the horizontal field, the solution was given as

$$C(x, y, t) = \frac{C_0 A}{4\pi t(D_x D_y)^{1/2}} \exp\left\{-\frac{[(x - x_0) - V_x t]^2}{4D_x t} - \frac{(y - y_0)^2}{4D_y t}\right\} \tag{4.76}$$

where A = area over which the pulse is injected. This solution can be extended to include dispersion component in the z dimension (Baetsle, 1969). For a source with an initial concentration of C_0 and volume V_0, its spread in space and time coordinates was given as

$$C(x, y, z, t) = \frac{C_0 V_0}{8(\pi t)^{3/2}(D_x D_y D_z)^{1/2}} \exp\left[-\frac{(x - V_x t)^2}{4D_x t} - \frac{y^2}{4D_y t} - \frac{z^2}{4D_z t} - \lambda t\right] \tag{4.77}$$

As seen in Eq. (4.77), first-order decay of the species (represented by the decay constant λ) was accounted for in this solution. A close examination of Eq. (4.77) indicates that the source progresses as a Gaussian plume with its center at $x = V_x t$, $y = 0$, $z = 0$, and the standard deviations in the three dimensions equal to

$$\sigma_x = \sqrt{2D_x t} \tag{4.78a}$$

$$\sigma_y = \sqrt{2D_y t} \tag{4.78b}$$

$$\sigma_z = \sqrt{2D_z t} \tag{4.78c}$$

The concentration of the species is always maximum at the center of the plume, and is given by

$$C_{\max} = C(V_x t, 0, 0) = \frac{C_0 V_0 e^{-\lambda t}}{8(\pi t)^{3/2}(D_x D_y D_z)^{1/2}} \tag{4.79}$$

Wilson and Miller (1978) provided another set of analytical solutions for mass transport in two dimensions, accounting for the processes of advection in the x dimension (the direction of flow), dispersion in the x and y dimensions, and including sorption and first-order decay. Their solutions for continuous and pulse sources, respectively, are

$$C(x, y, t) = \frac{f'_m \exp(x/B)}{4\pi n\sqrt{D'_x D'_y}} W\left(u, \frac{\alpha}{B}\right) \tag{4.80}$$

$$C(x, y, t) = \frac{m'}{4\pi nt\sqrt{D'_x D'_y}} \exp\left[-\frac{(x - V'_x t)^2}{4D'_x t} - \frac{y^2}{4D'_y t} - \lambda t\right] \tag{4.81}$$

where

f'_m = continuous rate of chemical injection per vertical unit aquifer
m' = injected contaminant mass per vertical unit aquifer

$$B = \frac{2D'_x}{V'_x}$$

$$u = \frac{\alpha^2}{4\beta D'_x t} \qquad \beta = 1 + \frac{2B\lambda}{V'_x} \qquad \alpha = \left[\left(x^2 + \frac{D'_x y^2}{D'_y}\right)\beta\right]^{1/2}$$

$$V'_x = \frac{V_x}{R} \qquad D'_x = \frac{D_x}{R} \qquad D'_y = \frac{D_y}{R}$$

$W\left(u, \frac{\alpha}{B}\right)$ = leaky well function, which is plotted in Fig. 4.16 against $1/u$

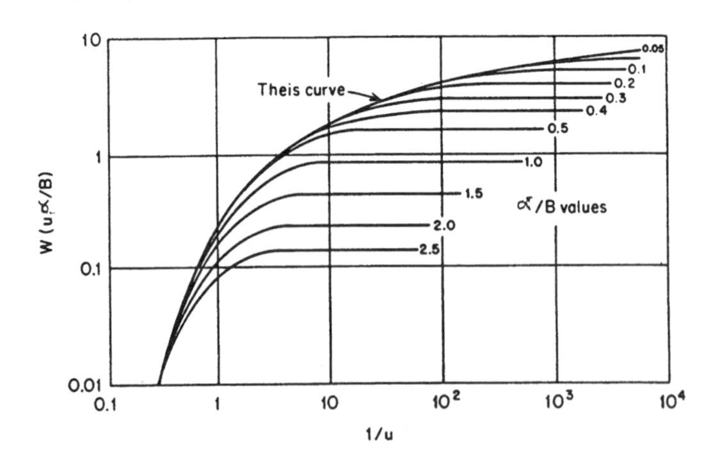

Figure 4.16 Curves of $W(u, \alpha/B)$ versus 1/u. (*Source*: Walton, 1960)

We note that analytical solutions for mass transport equation are possible only for simple initial and boundary conditions. They are suitable for specialized cases where the velocity is steady and the porous medium is homogeneous and isotropic in its hydrogeological parameters. These limitations often restrict the use of the simple solutions presented above. One would then resort to numerical solutions, which implement finite-difference or finite-element methods. A number of software packages are available for the purpose of studying mass transport in porous media, and the next section provides a brief summary of the capabilities of these packages.

4.6 SURVEY OF COMPUTER SOFTWARE FOR MASS TRANSPORT AND TRANSFER MODELING

The Groundwater Quality Technical Committee of the Environmental Engineering Division of the ASCE (American Society of Civil Engineers) launched a group of activities in 1991 in an attempt to summarize the state of the art of contaminated-groundwater modeling practice and to provide guidelines for model users. Based on a survey conducted by the committee as well as by other agencies including the U.S. Environmental Protection Agency (EPA), Geraghty and Miller, and Brookhaven National Laboratory, it was reported that a total of about 54 different models were used by state agencies and consultants. The most frequently used models simulating mass transport phenomena included (American Society of Civil Engineers, 1996), in alphabetical order, are:

1. CFEST
2. DYNTRACK
3. MIGRATE
4. MOC
5. MT3D
6. POLLUTE
7. RANDOMWALK
8. RESSQ
9. RITZ
10. SOLUTE
11. SUTRA

Table 4.6 lists the common practical applications of these models. In general, the popular applications are (1) to estimate contaminant plume migration, and (2) to determine sensitivity of contaminant migration to various hydrogeological and source parameters. The various transport processes simulated by these models are listed in Table 4.7. The models essentially solve the advection-dispersion equation in one form or another, using either analytical, semianalytical (potential theory, finite-layer technique), or numerical (finite-difference, finite-element) methods. All of the models are capable of simulating mass transport due to advection and molecular diffusion, and only RITZ does not simulate transport due to hydrodynamic dispersion. RITZ was a model developed to investigate land treatment alternatives in the disposal of oil sludges, primarily originating

Table 4.6 Practical Applications of Frequently Used Transport Models

Applications	Models
Estimate contaminant plume migration plume migration.	All
Determine sensitivity of migration to various parameters.	All
Assess leachate collection systems below waste sites.	MIGRATE, POLLUTE
Design barriers in a landfill.	MIGRATE, POLLUTE
Simulate pollutant transport through clayey soils.	MIGRATE, POLLUTE
Locate landfill, hazardous waste, and nuclear waste sites.	CFEST, MOC, MT3D, RNDWALK, SUTRA DYNTRACK, RESSQ
Help calibrate flow model.	CFEST, MOC, MT3D, RNDWALK, SUTRA DYNTRACK, RESSQ
Investigate contaminant source distribution and loading history:	CFEST, MOC, MT3D, RNDWALK, SUTRA DYNTRACK, RESSQ
Investigate remedial action alternatives (interceptor drains and withdrawal/injection wells) for cleanup or containment of contaminated water	CFEST, MOC, MT3D RNDWALK DYNTRACK, RESSQ
Simulate saltwater intrusion.	CFEST, SUTRA
Simulate energy transport.	CFEST, SUTRA
Land treatment and cleanup criteria of petroleum refinery wastes.	RITZ
Analytical solutions to several practical problems.	POLLUTE
Stimulate multiphase flow.	RITZ

Source: American Society of Civil Engineers (1996).

from petroleum refinery wastes, and is the only model that can simulate multiphase flow (the next chapter is devoted to issues of multiphase flow). All of the models except RITZ can simulate the effects of radioactive decay, and only CFEST, MOC, and SOLUTE do not simulate the effects of chemical biodegradability. Sorption process is simulated in all the models through a user-defined retardation factor; however, most of the models can also simulate sorption and/or other chemical reactions indirectly, using parameters such as bulk density, organic carbon content, solubility, sorption constant, cation-exchange capacity, reaction rate constants, and partition coefficient.

The various categories of input parameters for these models were identified as follows:

Table 4.7 Transport Processes Simulated in the Frequently Used Models

Processes	Models									
Advection	CF	DY	MI	MO	MT	PO	RI	RW	SO	SU
Molecular diffusion	CF	DY	MI	MO	MT	PO	RI	RW	SO	SU
Hydrodynamic dispersion	CF	DY	MI	MO	MT	PO	—	RW	SO	SU
Density-driven	CF	—	—	—	—	—	—	—	—	SU
Adsorption and/or Chemical Reactions										
Retardation factor	CF	DY	MI	MO	MT	PO	RI	RW	SO	SU
Other	CF[1]	—	MI[2]	MO[3]	MT[4]	PO[2]	RI[5]	RW[6]	SO[7]	SU[8]
Radioactive decay	CF	DY	MI	MO	MT	PO	—	RW	SO	SU
Biodegradation	—	DY	MI	—	MT	PO	RI	RW	—	SU
Volatilization	—	—	—	—	—	—	RI	—	—	—

CF = CFEST DY = DYNTRACK MI = MIGRAT MO = MOC
MT = MT3D PO = POLLUTE RI = RITZ RW = RNDWALK
SO = SOLUTE SU = SUTRA

Adsorption and/or chemical reactions can be determined using user-specified values of:
1. Bulk density, organic carbon content, solubility and sorption constant
2. Bulk density, organic carbon content and cation exchange capacity
3. Bulk density, cation exchange capacity, and sorption constant
4. Bulk density, sorption constant, reaction rate constant, and partition coefficient
5. Bulk density, organic carbon content, and partition coefficient
6. Bulk density, organic carbon content, cation exchange capacity, solubility, sorption Constant, reaction rate constant, and partition coefficient
7. Bulk density, organic carbon content, and partition coefficient
8. Bulk density, reaction rate constant, and sorption constant

Source: American Society of Civil Engineers (1996).

Geological parameters, such as aquifer properties
Mass transport and transfer process-related parameters
Initial and boundary conditions
Dimensionality issues such as spatial and temporal discretization
Solution technique parameters such as convergence criteria and acceleration parameters

Several geochemical models also simulate the mass transformations in porous media. Most of these models, including MINTEQ (Felmy et al., 1984), GEOCHEM (Sposito and Mattigod, 1980), WATEQF (Plummer et al., 1976), and PHREEQE (Parkhurst et al., 1980), are based on solving a set of equations representing simultaneous transformations occurring among the chemical species. The computer codes keep track of the mass balance of each of the chemical species, and the reaction processes are simulated until a thermodynamic equilibrium is achieved. Most codes at present, however, consider only equilibrium models of reactions, which precludes their applicability to field scenarios where a kinetic reaction model might be more appropriate.

REFERENCES

American Society of Civil Engineers. (1996). *Quality of Ground Water, Guidelines for Selection and Application of Frequently Used Models*, ASCE Manuals and Reports on Engineering Practice No. 85. ASCE, New York.

Baetsle, L. H. (1969). Migration of radionuclides in porous media. In A. M. F. Duhamel (ed.), *Progress in Nuclear Energy Series XII, Health Physics*, pp. 707–730. Pergamon Press, Elmsford, NY.

Bedient, P. B., Rifai, H. S., and Newell, C. J. (1994). *Ground Water Contamination, Transport and Remediation*. Prentice-Hall, Englewood Cliffs, NJ.

Benjamin, M. M., and Leckie, J. O. (1981). Multiple-site adsorption of Cd, Zn, and Pb on amorphous iron oxyhydroxide. *J. Colloid Interface Sci.*, 79:209–221.

Biddappa, C. C., Chino, M., and Kumazawa, K. (1981). Adsorption, desorption, potential and selective distribution of heavy metals in selected soils of Japan. *J. Environ. Sci. Health B*, 156:511–528.

Chiou, C. T., Schmedding, D. W., and Manes, M. (1982). Partitioning of organic compounds in octanol-water systems. *Environ. Sci. Technol.*, 16:4–10.

De Josselin De Jong, G. (1958). Longitudinal and transverse diffusion in granular deposits. *Trans. Am. Geophys. Union*, 39(1):67.

Domenico, P. A., and Schwartz, F. W. (1990). *Physical and Chemical Hydrogeology*. Wiley, New York.

Elliott, H. A., Liberati, M. R., and Huang, C. P. (1986). Competitive adsorption of heavy metals by soils. *J. Environ. Qual.*, 15:214–219.

Farrah, H., and Pickering, W. F. (1977). The sorption of lead and cadmium species by clay minerals. *Austral. J. Chem.*, 30:1417–1422.

Felmy, A. R., Brown, S. M., Onishi, Y., Yabusaki, S. B., Argo, R. S., Girvin, D. C., and Jenne, E. A. (1984). Modeling the transport, speciation, and fate of heavy metals in aquatic systems, EPA Project Summary, EPA-600/53-84-033. U.S. Environmental Protection Agency, Athens, GA.

Forbes, E. A., Posner, A. M., and Quirk, J. P. (1974). The specific adsorption of inorganic Hg (II) species and Co (II) complex ions on geothite. *J. Colloid Interface Sci.*, 49: 403–409.

Freeze, R. A., and Cherry, J. A. (1979). *Groundwater*. Prentice-Hall, Englewood Cliffs, NJ.

Hassett, J. J., Banwart, W. L., and Griffin, R. A. (1983). Correlation of compound properties with sorption characteristics of nonpolar compounds by soils and sediments: Concepts and limitations. In C. W. Francis an S. I. Auerbach (eds.), *Environment and Solid Wastes: Characterization, Treatment and Disposal*, pp. 161–178. Butterworth, Stoneham, MA.

Johnson, R. L., Palmer, C. D., and Fish, W. (1989). Subsurface chemical processes. In

Transport and Fate of Contaminants in the Subsurface, EPA/625/4-89/019, pp. 41–56. U.S. Environmental Protection Agency, Cincinnati, OH, and Ada, OK.

Karickhoff, S. W., Brown, D. S., and Scott, T. A. (1979). Sorption of hydrophobic pollutants on natural sediments. *Water Res.*, 13:241–248.

Kenaga, E. E., and Goring, C. A. I. (1980). *ASTM Special Technical Publication 707*. American Society for Testing Materials, Washington, DC.

Kinniburgh, D. G., Jackson, M. L., and Syers, J. K. (1976). Adsorption of alkaline earth, transition and heavy metal cations by hydrous oxide gels of iron and aluminum. *Soil Sci. Soc. Am. J.*, 40:796–799.

Langmuir, D., and Mahoney, J. (1984). Chemical equilibrium and kinetics of geochemical processes in groundwater studies. In B. Hitchon and E. I. Walliek (eds.) *Proceedings of the First Canadian/American Conference on Hydrogeology*, pp. 69–95. National Water Well Association, Dublin, OH.

Lerman, A. (1979). *Geochemical processes—Water and sediment environments*. Wiley, New York.

Li, Y-H., and Gregory, S. (1974). Diffusion of ions in sea water and in deep-sea sediments. *Geochim. Cosmochim. Acta*, 38(5):703–714.

Mackay, D. M., and Leinonen, P. J. (1975). Rate of evaporation of low-solubility contaminants from water bodies to atmosphere. *Environ. Sci. Technol.*, 9(8):1178–1180.

McCarty, P. L., Rittman, B. E., and Bouwer, E. J. (1984). Microbiological processes affecting chemical transformations in groundwater. In G. Bitton and C. P. Gerba (eds.), *Groundwater Pollution Microbiology*, pp. 89–115. Wiley, New York.

Monod, J. (1942) *Recherches sur la croissance des cultures bacteriennes*. Herman & Cie, Paris.

Monod, J. (1949). The growth of bacterial cultures. *Ann. Rev. Microbiol.* 3:371.

Montgomery, J. H. (1991). *Groundwater Chemicals Desk Reference*, Vol. 2. Lewis Publishers, Chelsea, MI.

Montgomery, J. H. and Welkom, L. M. (1990). *Groundwater Chemicals Desk Reference*. Lewis Publishers, Chelsea, MI.

Morel, F. M. M. (1983). *Principles of Aquatic Chemistry*. J Wiley, New York.

Murray, J. W. (1975). The interaction of metal ions at the manganese dioxide solution interface. *Geochim. Cosmochim. Acta*, 39:505–519.

National Research Council. (1990). *Ground Water Models: Scientific and Regulatory Applications*. National Academy Press, Washington, DC.

Neely, W. B. (1985). Hydrolysis. In W. B. Neely and G. E. Blau (eds.), *Environmental Exposure from Chemicals*, Vol. I, pp. 157–173. CRC Press, Boca Raton, FL.

Nye, P. H. (1979). Diffusion of ions and uncharged solutes in soils and soil clays. *Adv. Agron.* 31:25–272.

Ogata, A. (1970). Theory of dispersion in a granular medium. U.S. Geol. Surv. Prof. Paper 411–I.

Ogata, A., and Banks, R. B. (1961). A solution of the differential equation of longitudinal dispersion in porous media. U.S. Geol. Surv. Prof. Paper 411–A.

Olsen, S. R., Kemper, W. D., and van Schaik, J. C. (1965). Self-diffusion coefficients of phosphorous in soil measured by transient and steady-state methods. *Proc. Soil Sci. Soc. Am.*, 29(2):154–158.

Parkhurst, D. L., Thorstenson, D. C., and Plummer, L. N. (1980). PHREEQE—A computer program for geochemical calculations. U.S. Geol. Surv. Water Resources Investigation 80–96.

Perkins, T. K., and Johnston, O. C. (1963). A review of diffusion and dispersion in porous media. *Soc. Petroleum Eng. J.*, 3:70–84.

Plummer, L. N., Jones, B. F., and Truesdell, A. H. (1976). WATEQF—A Fortran IV version of WATEQ, a computer code for calculating chemical equilibria of natural waters. U.S. Geol. Surv. Water Resources Investigation 76–13.

Puls, R. W., and Bohn, H. L. (1988). Sorption of cadmium, nickel, and zinc by kaolinite and montmorillonite suspensions. *Soil Sci. Soc. Am. J.*, 52:1289–1292.

Quigley, R. M., Yanful, E. K., and Fernandez, F. (1987). Ion transfer by diffusion through clayey barriers. In R. Woods (ed.), *Geotechnical Practice for Waste Disposal*, pp. 137–158. Geotechnical Special Publication No. 13.

Schnitzer, M., and Skinner, S. I. M. (1967). Organio-metallic interaction in soils: 7. Stability constants of Pb^{++}-, Ni^{++}-, Co^{++}- and Mg^{++}-fluvic acid complexes. *Soil Sci.*, 103:247–252.

Schwarzenbach, R. P., and Westall, J. (1981). Transport of nonpolar organic compounds rom surface water to groundwater. Laboratory studies. *Environ. Sci. Technol.*, 15: 1300–1367.

Shackelford, C. D., and Daniel, D. E. (1991a). Diffusion in saturated soil. I: Background. ASCE J. Geotech. Eng., 117(3):467–484.

Shackelford, C. D., and Daniel, D. E. (1991b). Diffusion in saturated soil. II: Results for compacted clay. ASCE J. Geotech. Eng., 117(3):485–506.

Sposito, G., and Mattigod, S. V. (1980). GEOCHEM: A computer program for the calculation of chemical equilibria in soil solutions and other natural water systems. Department of Soils and Environment Report, University of California, Riverside.

Stevenson, F. J. (1975). Stability constants of Cu^{2+}, Pb^{2+}, and Cd^{2+} complexes with humic acids. *Soil Sci. Soc. Am. Proc.*, 40:665–672.

Stumm, W. and Morgan, J. J. (1981). *Aquatic Chemistry*, 2nd ed. Wiley, New York.

van Genuchten, M. T., and Alves, W. J. (1982). Analytical solutions of the one-dimensional convective-dispersive solute transport equation. U.S. Department of Agriculture Technical Bulletin No. 1661. Agricultural Research Service, Washington, DC.

Walton, W. C. (1960). Leaky artesian aquifer conditions in Illinois. Illinois State Water Surv. Rept. Invest. 39.

Weber, W. J., Jr. (1972). *Physicochemical Processes for Water Quality Control*. Wiley, New York.

Wilson, J.L., and Miller P. J. (1978). Two-dimensional plume in uniform groundwater flow. *ASCE J. Hydraulics Div.*, 104:503–514.

Yong, R. N., Mohamed, A. M. O., and Warkentin, B. P. (1992). *Principles of Contaminant Transport in Soils*. Developments in Geotechnical Engineering 73. Elsevier, Amsterdam.

5
Nonaqueous-Phase Liquids in Soils

5.1 INTRODUCTION

Nonaqueous-phase liquid (NAPL) is a term used to denote any liquid which is immiscible with water. In geoenvironmental engineering, this term is usually associated with petroleum hydrocarbons such as those leaking from underground oil storage tanks or accidental spills of organic chemicals. By virtue of the immiscibility, an interface or a physical dividing surface between the two bulk liquids exists in the pore spaces of soils saturated with water and NAPL. In the case of unsaturated soils, the presence of air creates additional interfaces. In spite of their immiscible nature, NAPLs do dissolve in water, and mass transfer occurs across these interfaces. Although this dissolution is rather slow and is not extensive in most cases, even slight amounts of dissolved NAPLs in potable water supplies may result in detrimental health effects. The slow dissolution also causes a long-term source of subsurface contamination. Hence the need to study the fate and transport of NAPLs in the subsurface.

It is customary to divide the vast range of NAPLs commonly encountered in geoenvironmental engineering into two general categories, light nonaqueous-phase liquids (LNAPLs) and dense nonaqueous-phase liquids (DNAPLs). This categorization is based on their specific gravity. LNAPLs have a specific gravity less than water, and DNAPLs have a specific gravity greater than water. Tables 5.1 and 5.2 list common types of LNAPLs and DNAPLs, respectively, along with some of their properties, which will be referred in subsequent sections. As we will see in a later section, the difference in density of a NAPL with respect to water governs some important aspects of its transport in the subsurface. Therefore, it is useful to recognize this categorization at the outset.

Throughout this chapter, it will benefit the reader to keep in mind that many of the principles involved in NAPL transport and entrapment in soils are similar

Table 5.1 Chemicals Typically Associated with LNAPLs

LNAPLs	Sp. gr.	Water solubility (mg/liter)	Interfacial tension (dynes/cm)
Benzene	0.88	1.75×10^3	35.0
Ethylbenzene (phenylethane)	0.87	1.52×10^2	35.5
Styrene (ethenylbenzene)	0.91	3.00×10^2	35.5
Toluene (methylbenzene)	0.86	5.35×10^2	36.1
Methyl ethyl ketone	0.81		
m-Xylene	0.86	1.30×10^2	36.4
o-Xylene	0.88	1.75×10^2	36.06
p-Xylene	0.86	1.98×10^2	37.8
Vinyl chloride	0.91		
Crude oil	0.7–0.98		
Diesel fuels	0.80–0.85		50
Gasoline (automative)	0.73		50
Fuel oils (kerosene, jet fuel)	0.81–0.85		48
Mineral oil	0.82		47
Petroleum distillates	0.71–0.75		50
n-Heptane	0.68		50.2
n-Hexane	0.66		51.0

Table 5.2 Chemicals Typically Associated with DNAPLs

DNAPLs	Sp. gr.	Water solubility (mg/liter)	Interfacial tension (dynes/cm)
Carbon tetrachloride	1.56	7.57×10^2	45.0
Chloroform	1.48	8.20×10^3	32.8
Methylene chloride	1.33	2.00×10^4	28.3
Ethylene chloride	1.24		
Bromobenzene	1.49	4.46×10^2	36.5
Chlorobenzene	1.11	4.66×10^2	37.4
Hexachlorobenzene	1.60		
Chlorotoluene	1.10	3.30×10^3	30
Trichloroethylene (TCE)	1.46	1.10×10^3	34.5
1,1,1-Trichloroethane (TCA)	1.34	1.50×10^3	45
Tetrachloroethylene	1.62	1.50×10^2	44.4
Phenol	1.07		
2-Chlorophenol	1.26	2.90×10^4	
Pentachlorophenol	1.98		
Naphthalene	1.03		
Creosote (coal tar)	1.05–1.10		20
1,2-Dichloroethane	1.24	8.52×10^3	30

to those treated in unsaturated flow (Section 3.6). In an unsaturated soil, the existence and transport of air phase is in competition with that of water. In a water-saturated soil penetrated by NAPLs, the NAPLs behave similar to air in their attempt to compete for pore space in the soil. Extending this further to a water-saturated soil penetrated by air and NAPL, all of the three phases compete for the pore space. The relative quantities of the three phases in soil at equilibrium are governed by the same capillarity principle which controls the behavior of unsaturated soil.

We will look first at the principles of NAPL entrapment in soils at a pore scale and then expand our study to a field scale. These principles will lead us to a conceptual understanding of the mass transport and transfer processes of NAPLs. Quantitative modeling of the fate and transport of NAPLs is still at an early stage of development, and is beyond the scope of this book. Our objective in this chapter is to gain a qualitative understanding of the mass transport and transfer characteristics pertinent to NAPL-penetrated soils.

5.2 PRINCIPLES OF NAPL ENTRAPMENT IN SOILS

Fate and transport of NAPLs in the subsurface depends to a large extent on the quantities of entrapment in the soil pore space. The relative competition among the three phases, i.e., water, air, and NAPLs, to occupy the pore space can be understood in terms of

1. Interfacial tension between these phases
2. Wettability of soils with these phases
3. Capillarity principle

In the following sections, we will treat these individually, for a conceptual understanding of the NAPL entrapment first at a pore scale and then at field scale.

5.2.1 Interfacial Tension

Homogeneous fluids contain molecules bound by van der Waals attractive forces. The magnitudes of these forces differ from fluid to fluid. When two immiscible fluids are brought together, there exists an interface which divides the two sets of molecules (Fig. 5.1). To maintain the force equilibrium, the interface is pulled by the fluid whose molecules are bound by greater attractive forces. Thus, either a convex or a concave meniscus is formed. The stretched interface behaves much like an elastic membrane subjected to tension. The tension existing at the interface, known as interfacial tension σ, is a result of the internal pressure difference

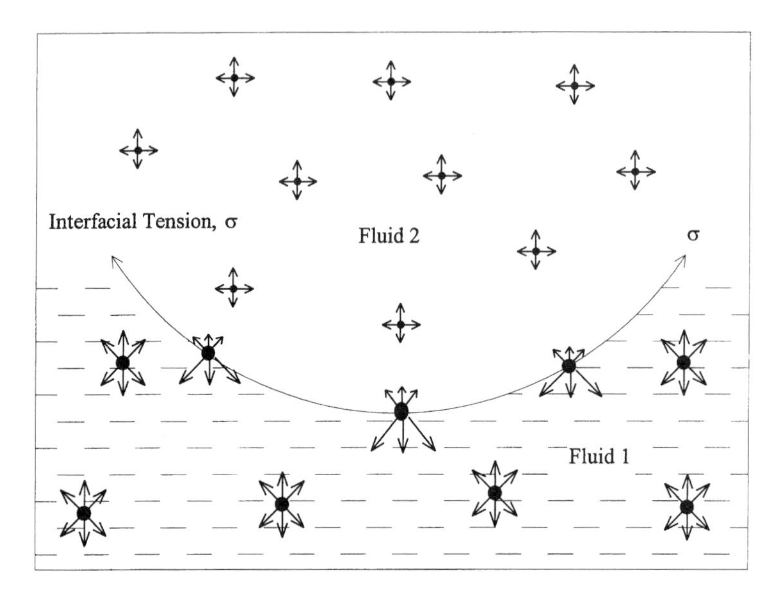

Figure 5.1 Molecular attraction forces in the individual phases and at the interface of two phases.

of the two adjoining phases. One of the two immiscible fluids in Figure 5.1 could be a gas containing the vapor phase of a liquid, in which case the interfacial tension is known as surface tension.

Interfacial tension characterizes the work done by the phases to satisfy force balance. Therefore, it is expressed in terms of energy per unit area, Nm/m^2, which reduces to the units of force per unit length, N/m. Since the internal pressures of fluids vary with temperature, interfacial tensions depend on temperature. The magnitudes of interfacial tension for some common NAPLs are presented in Tables 5.1 and 5.2.

5.2.2 Wettability of Soils

When two immiscible fluids coexist on a solid surface, the interfaces between the solid and the fluid phases control the phase configuration. The configuration of the two phases on the solid surface will be such that the energy is minimized and the force balance is obeyed. Consider two immiscible fluids existing on the

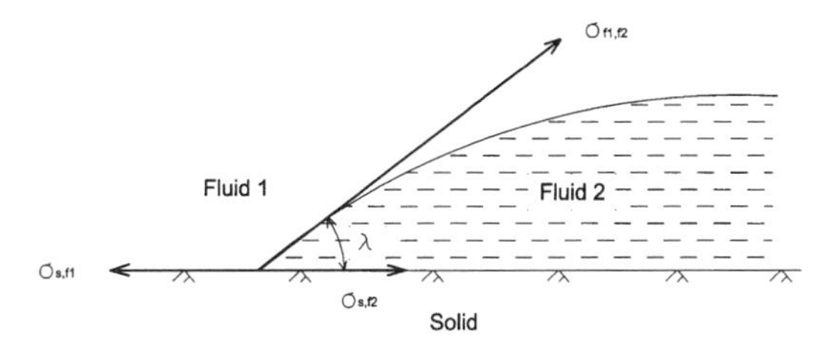

Figure 5.2 Forces acting along interfaces between the three phases at equilibrium.

top of a solid surface as shown in Figure 5.2. Three interface boundaries can be delineated in this configuration: (1) interface between solid and fluid 1, (2) interface between solid and fluid 2, and (3) interface between the two fluids. The stable configuration will be one where the tensions associated with these three interfaces are in equilibrium. At equilibrium,

$$\sigma_{s,f1} = \sigma_{s,f2} + \sigma_{f1,f2} \cos \lambda \tag{5.1}$$

or

$$\lambda = \cos^{-1}\left[\frac{\sigma_{s,f1} - \sigma_{s,f2}}{\sigma_{f1,f2}}\right] \tag{5.2}$$

where the subscripts denote the two phases corresponding to the interface, and s, $f1$, $f2$ denote the solid and the two fluid phases. The angle λ between the surface of the solid phase and the fluid 1–fluid 2 interface is a measure of wettability of the solid surface with respect to fluid 2, and is known as the *contact angle*. Perhaps a common observation of the contact angle is when a drop of water is placed on a glass surface, in which case the two fluids are water and air.

The relative wettability of soils with respect to various fluids is reflected in three possible ranges of contact angle λ. Figure 5.3 shows these three ranges and the associated configurations of the fluids. Figure 5.3a refers to complete wetting of the solid surface by fluid 2. If fluid 2 is water, the solid surface is hydrophilic. An incomplete wetting of the solid surface occurs in the range, $0 < \lambda < 90°$ (Fig. 5.3b). Figure 5.3c indicates that the surface is relatively nonwetting with respect to fluid 2. If fluid 2 is water, the solid surface in this case will be hydrophobic. With regard to a soil mass whose pores are filled with

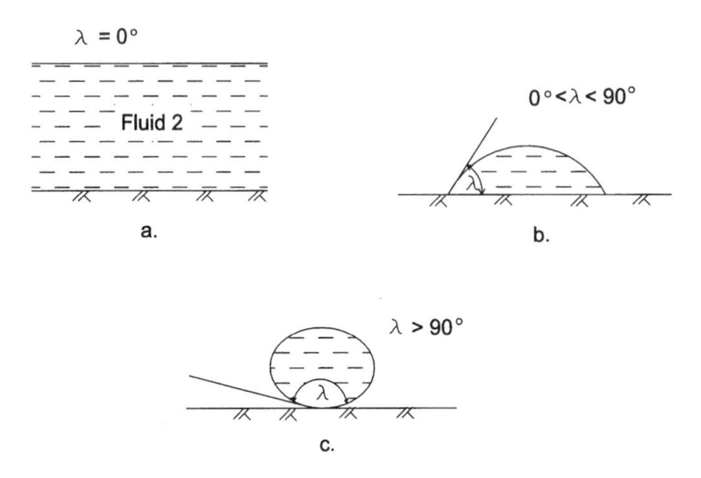

Figure 5.3 Possible contact angles between fluids and solid surfaces.

two fluids 1 and 2, the angle λ therefore dictates which fluid coats the solid particles.

Generally, a porous medium with NAPL and water as fluids 1 and 2 is considered to be water-wet if λ (between water and the solids) $< 70°$, NAPL-wet if $\lambda > 110°$, and neutral if λ is between 70° and 110° (Anderson, 1986). In most cases with NAPL and water, the porous medium is commonly treated as water-wet. In an unsaturated medium where NAPL and air are present, the medium is treated as NAPL-wet.

It is important to note that the contact angle may not be definable microscopically. The solid surfaces in a soil mass may exhibit tremendous heterogeneity in their surface characteristics. The surfaces of solid particles may belong to either silica or clay minerals, which may in turn be coated with iron oxides or macromolecules of soil humus. Considering this heterogeneity, it may be possible to define only a composite value for the contact angle. Also, the contact angle may not be unique for a given surface since it is known to manifest hysteresis, i.e., it depends on the movement of the fluid phase wetting the surface. When a drop of fluid 2 moves along a solid surface (Fig. 5.4), the contact angle λ_1 at the receding end is in general less than the contact angle λ_2 at the proceeding end of the drop. This is the primary reason for hysteresis behavior in soil–water retention discussed in Section 3.6. The hysteretic behavior of contact angle can be observed by pushing and pulling a capillary tube in a liquid. The meniscii observed in the two stages will exhibit different contact angles.

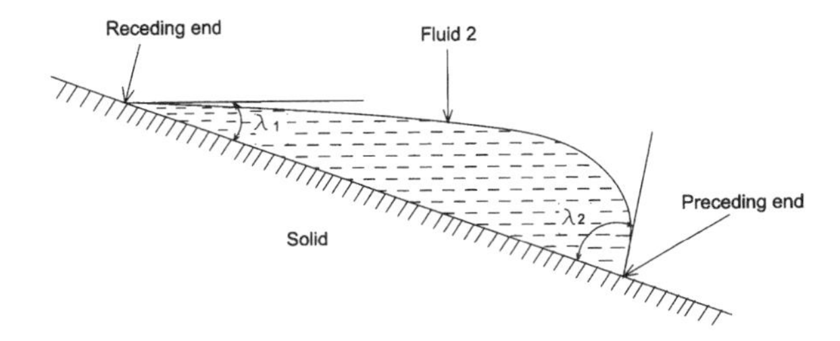

Figure 5.4 Nonuniqueness of contact angle for a moving liquid phase.

5.2.3 Capillarity

Capillarity may be broadly defined as that property of a porous medium which makes it draw the wetting fluid into its pores and repels the nonwetting fluid. It is the difference between the internal pressures of the two immiscible fluids, known as the *capillary pressure*, that causes this preferential filling of soil pores. For a water–NAPL system, the capillary pressure P_c may be defined as

$$P_c = P_n - P_w \tag{5.3}$$

where P_n = NAPL pressure and P_w = water pressure. For a small tube inserted in a water-NAPL system, the capillary pressure causes the wetting fluid to be drawn in (Fig. 5.5a). The height h_c to which the wetting fluid (water) is drawn in is directly related to the capillary pressure. The capillary pressure is also related to the interfacial tension, contact angle, and pore size in accordance with Laplace's law:

$$P_c = \frac{2\sigma_{nw} \cos \lambda}{r} \tag{5.4}$$

where σ_{nw} = interfacial tension between NAPL and water phases, λ = contact angle of the water with the surface of the tube, and r = radius of the tube. Note that Eq. (5.4) can be arrived at by considering force equilibrium on the water column of height h_c (see Fig. 5.5b). Capillary pressures representative of the three soil types, sand, silt, and clay, for three pairs of immiscible fluids are shown in Table 5.3.

Consider now that a soil mass is comprised of cylindrical pore tubes of different radii as shown in Figure 5.6a. It is obvious that the degree to which these pore tubes are filled with wetting or nonwetting phases is dependent on the capillary pressure. To demonstrate how the water and NAPL phases compete for the pore space, consider that all of the pores in Figure 5.6a are initially filled with water, that is, P_c = 0. Let us now introduce the NAPL phase and either

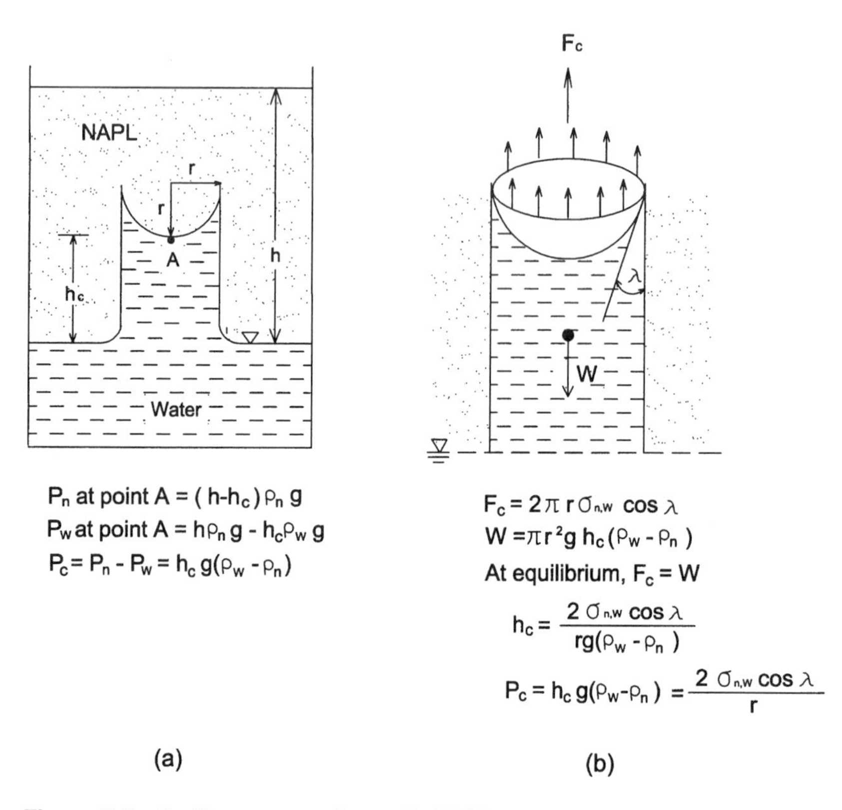

Figure 5.5 Capillary pressure in a soil–NAPL system. (Modified from Luckner and Schestakow, 1991).

Table 5.3 Range of Capillary Pressures for Sands, Silts, and Clays

	Pore radius, r (in m)	Air/water, P_c (in kPa)	Oil/water, P_c (in kPa)	Air/oil, P_c (in kPa)
Sands	10^{-4}–10^{-5}	1.4–14	0.6–6	0.7–7
Silts	10^{-5}–10^{-6}	14–140	6–60	7–70
Clays	10^{-6}–10^{-7}	140–1400	60–600	70–700

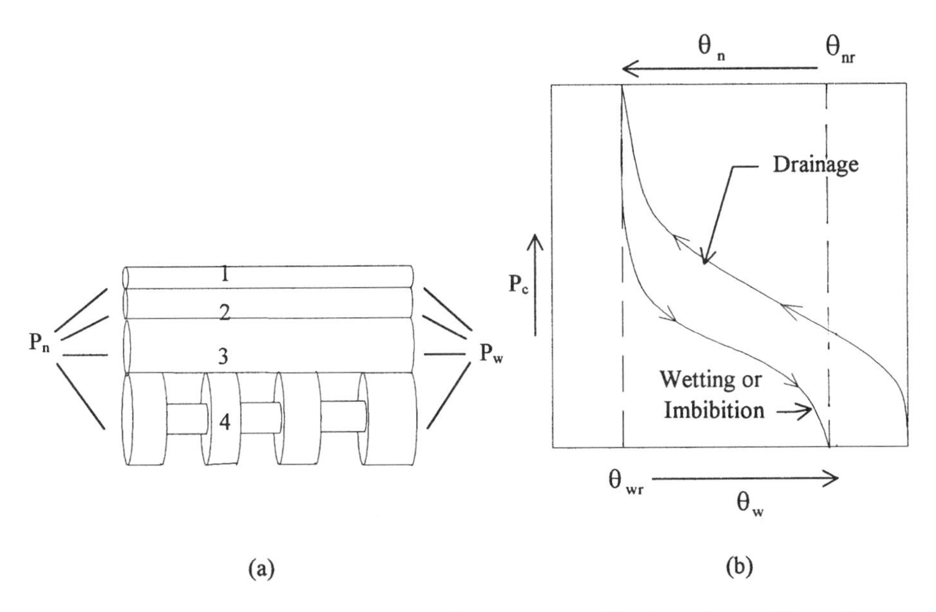

Figure 5.6 Drainage and wetting of soil: (a) conceptualization using capillary tubes; (b) hysteresis behavior of fluid retention. (Modified from Luckner and Schestakow, 1991)

increase P_n or decrease P_w (by providing suction) slowly. P_c therefore increases [Eq. (5.3)]. In accordance with Eq. (5.4), pores of larger radius r are associated with smaller P_c and will thus be first penetrated by the NAPL. The NAPL replaces water in tubes 3 and 4. However, in the case of tube 4, the advancing NAPL stops at the first bottleneck, since to advance further a higher capillary pressure is needed [in accordance with Eq. (5.4)]. The bottlenecks and the pore tubes 2 and 1 will be filled only at higher capillary pressures. This process of NAPL replacing the water phase is often referred as the drainage phase, since the wetting phase is drained. If the volume of pore space occupied by water (θ_w) is tracked during this process, the variation between θ_w and P_c will be as shown in Figure 5.6b.

Let us now reverse the above process in an attempt to wet the NAPL-filled pore system. This is known as the wetting or imbibition phase. It can be accomplished by reducing P_c, that is, either by increasing P_w or by decreasing P_n. It has been established through experimental studies that the resulting path on the θ_w–P_c space does not coincide with the drainage path. Smaller pressures are needed to achieve the same volumetric water contents as in the drainage phase. The cause of this hysteretic retention of fluids is the discontinuous distribution of the wetting fluid at high P_c values, and the hysteretic behavior of the contact angle of the wetting fluid. It is seen from Figure 5.6b that the wetting fluid cannot be drained completely, and a residual water content, θ_{wr}, remains even as P_c approaches infinity. Similarly, during the imbibition phase the nonwetting fluid cannot be displaced completely even when $P_c = 0$. The residual content of the NAPLs that remains during the wetting or imbibition phase is of particular importance in remediation operations. It is usually entrapped in small pores, and the pressures required to overcome the capillary pressures corresponding to these small pores are very high. Therefore, once entrapped, the discontinuous phase of NAPLs is difficult to mobilize.

For a given soil mass, the NAPL pressure P_n that is required to displace the water from the pores is known as the threshold or displacement entry pressure. The variation of capillary pressure represented by Eq. (5.4) is shown in Figure 5.7. Clearly, soils with small pores are associated with higher threshold pressures than those with large pores. This means that it takes much greater thickness of a NAPL pool to penetrate a clayey soil than a sandy soil. Some useful relationships for the pressure required for NAPL penetration into soils, based on the capillarity principle, are given in Table 5.4.

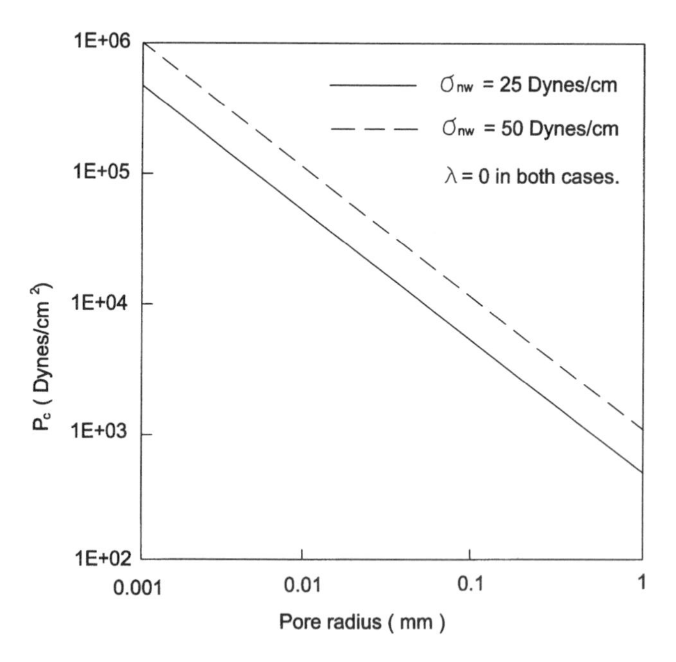

Figure 5.7 Capillary pressure as a function of interfacial tension and pore radius.

Table 5.4 Relationships for Capillary Pressures and Conditions of NAPL Penetration

Capillary pressure exerted on the surface of a nonwetting NAPL sphere	$P_c = P_n - P_w = \frac{2\sigma_{nw}\cos\lambda}{r}$
Capillary pressure exerted on the surface of NAPL in a fracture plane where b is the fracture aperture	$P_c = \frac{2\sigma_{nw}\cos\lambda}{b}$
Critical NAPL height, z_n, for DNAPL penetration into a water saturated geologic medium	$z_n = \frac{2\sigma_{nw}\cos\lambda}{rg(\rho_n - \rho_w)}$
Critical NAPL height for downward DNAPL or upward LNAPL penetration from a coarse-grained medium into a finger-grained, water-saturated medium, where r_t and r_p are the pore throat and pore radii, respectively	$z_n = \frac{2\sigma_{nw}\cos\lambda}{g\lvert\rho_n - \rho_w\rvert}\left(\frac{1}{r_t} - \frac{1}{r_p}\right) = \frac{P_{c(\text{fine})} - P_{c(\text{coarse})}}{g\lvert\rho_n - \rho_w\rvert}$
Estimated critical NAPL height for downward DNAPL or upward LNAPL penetration into a water-saturated medium	$z_n = \frac{2\sigma_{nw}\cos\lambda}{r_t g\lvert\rho_n - \rho_w\rvert}$
Estimated critical NAPL height for downward migration in the vadose zone	$z_n = \frac{2\sigma_{nw}\cos\lambda}{r_t g\rho_n}$

Source: Mercer and Cohen (1990).

5.3 CONCEPTUALIZATION OF FIELD-SCALE TRANSPORT OF NAPLs

The principles discussed in the previous section will enable us to formulate conceptually how NAPLs move into the subsurface. Because of some important effects of density of NAPLs (relative to that of water) on the transport phenomena, it is convenient to treat the LNAPLs and DNAPLs separately.

5.3.1 Light Nonaqueous-Phase Liquids

When a LNAPL is spilled at the ground surface, it enters the unsaturated zone into the soil–water–air system under the influence of gravity. When the soil is completely dry, the LNAPL will be the wetting fluid relative to air; therefore, it will advance into the pores relatively easily. On the other hand, when the soil is relatively wet, the LNAPL will be only partially wetting (wetting with respect to air but nonwetting with respect to water). Therefore, it has to have sufficient pressure to overcome the capillary forces in order to advance into the finer pores filled with water. The volumetric content and distribution of NAPLs in the unsaturated zone are governed by the LNAPL pressure, the initial volumetric contents of air and water phases, and the pore size distribution of the soils. If sufficient LNAPL is available, it advances deep into the unsaturated zone, as shown in Figure 5.8a.

If the groundwater table is shallow or if the LNAPL source is abundant, the product advances to the top of the capillary fringe. Since it is lighter than water, it tends to float on the top of the capillary fringe (Fig. 5.8b). As the pressure of the product released increases, the water table begins to be depressed, with the LNAPL accumulating in the depression. If the LNAPL source at the ground surface is exhausted, the infiltrating LNAPL in the unsaturated zone continues to move by gravity leaving behind thin films and ganglia in the unsaturated zone (Fig. 5.8c). The residual quantities of the LNAPLs thus left behind often amounts to 10–20% of the pore volume. Meanwhile, the infiltrating product reaches the pool at the capillary fringe and begins to spread laterally, forming what is known as a pancake. The lateral spreading of the product relieves the head of the LNAPL pool, causing a slight rebound of the water table. The rebounding water table, however, does not carry with it all the LNAPL, since part of it will be trapped in the pores as residual saturation due to capillarity. This residual LNAPL trapped below the water table will dissolve slowly over several years, often decades, and thus forms a long-term threat to the water quality.

Fluctuations of groundwater table resulting from seasonal variations or recharge operations will cause the pool to move up and down. When the pool drops, the saturated zone is occupied with the free product. Subsequent rising of the pool causes a portion of the LNAPL trapped below the water table at residual saturations. The fluctuations in the water table thus cause a "smear zone" and can spread the LNAPL over a greater thickness of the aquifer in the form of residual saturations. This is of concern in remediation operations because recovery of immobilized and discontinuous blobs of LNAPL is relatively more difficult than recovery of free product at the water table.

The free product thickness, the relative volumetric contents of water, LNAPL, and air in the vicinity of the pool, can be estimated using the capillarity principles discussed in the previous section. We defer details of this estimation to a later section and proceed with a conceptual model for movement of DNAPLs in the subsurface.

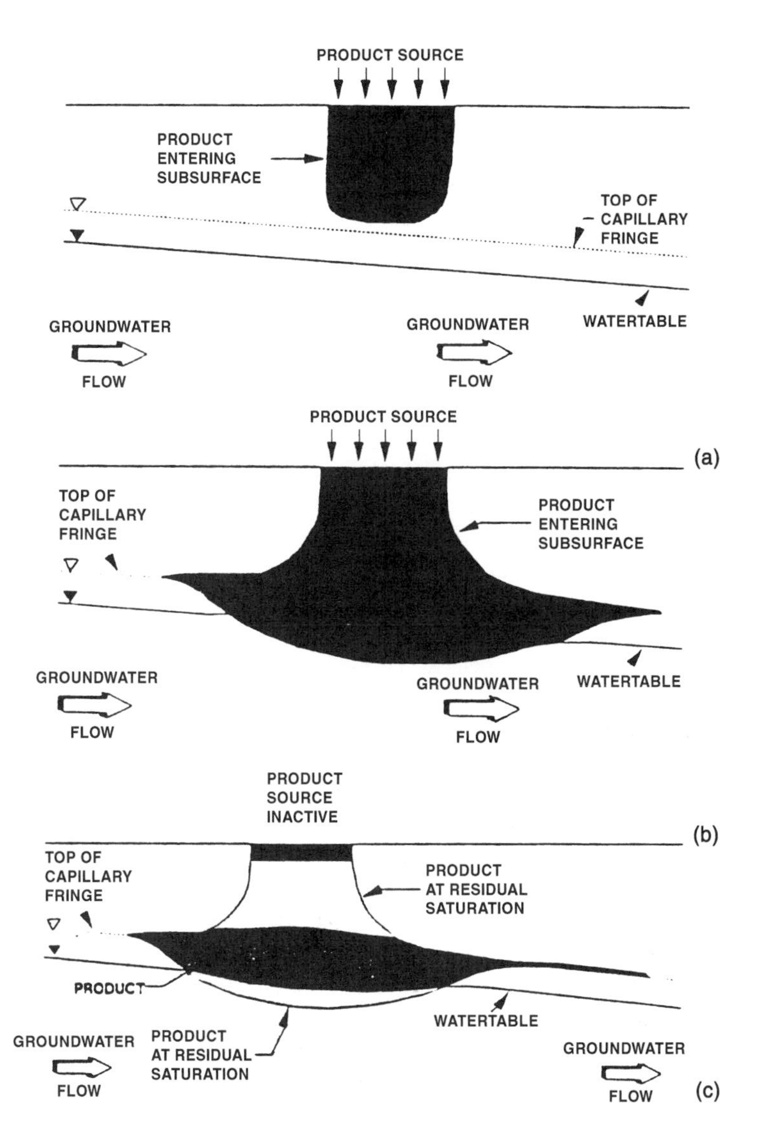

Figure 5.8 Movement of LNAPLs into the subsurface: (a) distribution of LNAPL after small volume has been spilled; (b) depression of the capillary fringe and water table; (c) rebounding of the water table as LNAPL drains from overlying pore space (Palmer and Johnson, 1989).

5.3.2 Dense Nonaqueous-Phase Liquids

Compared to the LNAPLs, the DNAPLs are more complex in the way they migrate in the subsurface. Only during the last decade have they been studied in detail and their movement conceptualized. The mobility of DNAPLs is greater than that of LNAPLs because they are denser than water and most of them have low viscosity. The relatively high density of the DNAPLs carries the product deep into the aquifers, hundreds of feet below the ground surface in some cases.

Similar to LNAPLs, the DNAPLs enter into the soil–water–air system of the unsaturated zone under the influence of gravity. However, unlike LNAPLs, the DNAPLs are known to exhibit fingering behavior (Fig. 5.9a) when the unsaturated soil is relatively wet. This finguring phenomenon has been established to be the result of a high-density, low-viscosity fluid (DNAPL) displacing a lower-density, higher-viscosity fluid (water) (Kueper and Frind, 1988). No finguring occurs, however, if the soil is dry. When a sufficient quantity of DNAPL is released, it advances to the water table, and unlike LNAPLs, it penetrates into the aquifer. The DNAPL is, however, still a nonwetting fluid compared to water, therefore it must overcome capillary forces in order to penetrate into the pores of the soil–water system. As discussed earlier, the DNAPL must possess an entry or threshold pressure sufficient to overcome the capillary forces. The higher density of the DNAPLs contributes to this entry pressure. If the DNAPL pressure is higher than the entry pressure, the DNAPL proceeds into the soil–water system until the source is exhausted. A finer-grained stratum in the unsaturated zone may sometimes stop the DNAPL migration (Fig. 5.9b), because finer-grained materials possess finer pores, which are associated with higher entry pressure. If the DNAPL source is larger, it may penetrate the entire thickness of the aquifer and settle down on the impermeable stratum at the bottom of the aquifer (Fig. 5.9c). In cases where the impermeable stratum is sloping, the DNAPL tends to flow down the dip of the stratum, which may be in a direction opposite to that of the groundwater flow in the aquifer. This causes problems in characterizing the source of the DNAPL, since the dissolved phase may be transported in a direction different from that of the free product.

The transport of DNAPLs is tremendously complicated by field-scale heterogeneities. The examples shown in Figure 5.10 demonstrate the challenges associated with characterizing DNAPL penetration in the subsurface. Figure 5.10a shows a scenario where a rock stratum exists between a surficial sand layer and the water table. The rock stratum may be fractured with cracks of various lengths, widths, and apertures. These cracks convey the DNAPLs either into the dead-end zones or, through interconnectedness, directly to the water table. Characterization of DNAPL penetration into a fractured rock is a challenge that the scientific community has only begun to face. Figure 5.10b shows the DNAPL transport

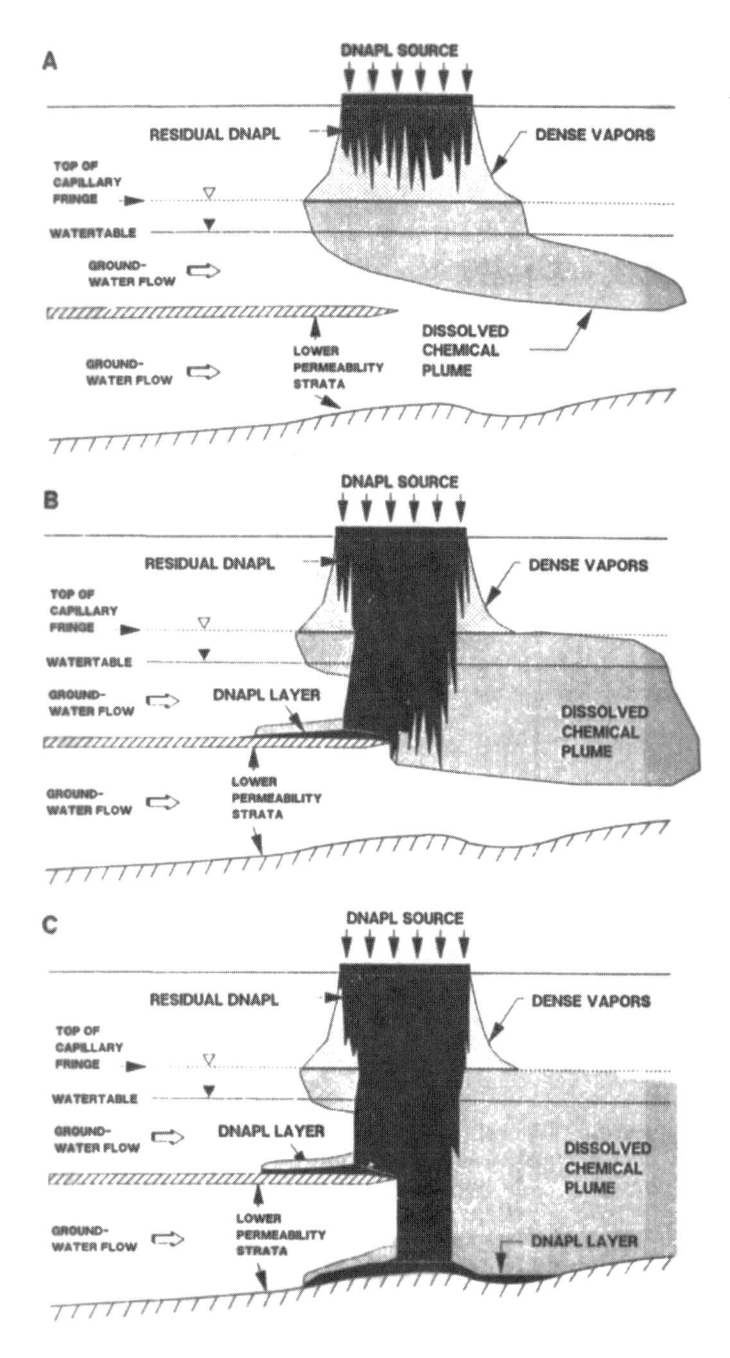

Figure 5.9 Movement of DNAPLs into the subsurface: (a) distribution of DNAPL after small volume has been spilled; (b) distribution of DNAPL after moderate volume has been spilled; (c) distribution of DNAPL after large volume has been spilled (Feenstra and Cherry, 1987).

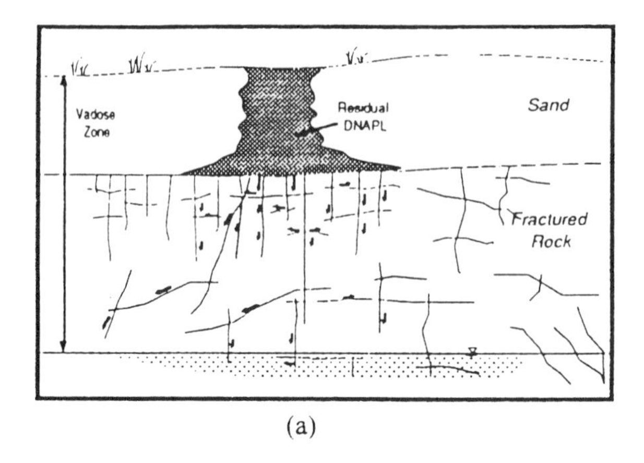

(a)

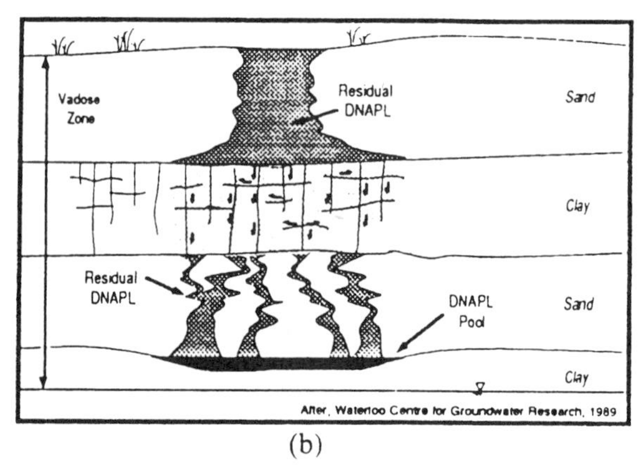

(b)

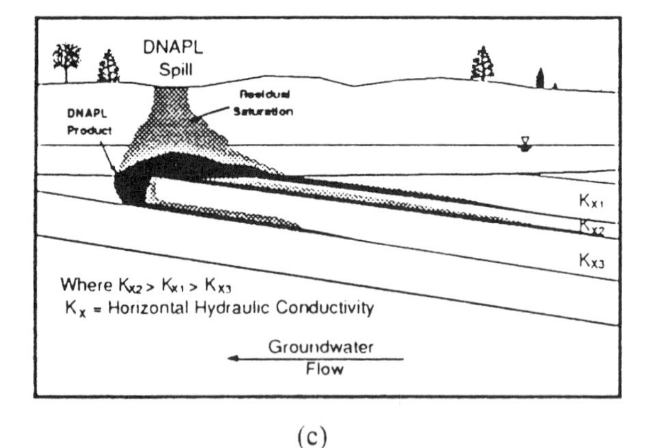

(c)

Figure 5.10 Challenges associated with characterizing DNAPL penetration: (a) spill into fractured rock system; (b) spill into a low permeable formation with preferential pathways; and (c) spill into a system of nonhorizontal stratigraphic units with variable hydraulic conductivity (Huling and Weaver, 1991).

through a layered system where the clay layer may contain a heterogeneous distribution of preferential pathways. Entry pressures based on the homogeneous pore-size distribution of the clay may be high, yet the DNAPL can migrate through the preferential pathways and convey the DNAPL into strata below the clay layer. Figure 5.10c shows the possibility of DNAPLs flowing in directions different from that of the groundwater flow. Inclined strata of different hydraulic conductivities may convey DNAPLs to considerable distances upstream of the location of the spill.

5.4 PHASE DIAGRAM FOR SOIL–WATER–LNAPL–AIR SYSTEMS

Estimation of the volumetric contents of the three phases of air, water, and NAPL is an essential component in modeling the transport of NAPLs in the subsurface. The relationships between capillary pressure and volumetric water contents of NAPL, water, and air will form the constitutive expressions in transport modeling. Substantial advances have been made recently to develop these relationships for LNAPLs. Lenhard and Parker (1987, 1990) and Parker and Lenhard (1987) extended moisture rentention expressions in the unsaturated flow to account for the presence of the NAPL phase. The resulting expressions, which will be described below, were validated in laboratory investigations (Lenhard and Parker, 1988) and were used to calibrate field-scale observations in monitoring wells (Ostendorf et al., 1993).

The phase composition in the vicinity of groundwater table is illustrated in Figure 5.11, where d_w and d_l represent the depths to the water table and to the top of the free product, respectively. The thickness of the LNAPL pool, which accumulates as a pancake at the top of the water table, is therefore equal to $(d_w - d_l)$. If a monitoring well is drilled in a LNAPL-contaminated site, free LNAPL product will be observed up to a height of $(d_w - d_l)$ above the water table. In the range, $d_w > d > d_l$, only LNAPL and water exist in the pore space. Immediately above this range, the free LNAPL exists in combination with water and air. Above a depth $d = d_m$, the free product is not found and only water and air coexist. Apart from air, water, and free LNAPL, a fourth phase is also present in the form of trapped LNAPL ganglia encompassed by free water. As explained earlier, this phase is a result of the fluctuating water table.

At a given depth, the interfacial tensions between air, LNAPL, and water govern the radii of pore throats into which the various phases can penetrate. This follows directly from Eq. (5.4). Because of capillarity, small pores are always filled with the wetting fluid. The interfacial tension between air and LNAPL usually corresponds to a large radius of curvature. If this radius is r_1, all pores greater than size r_1 are filled with air and those smaller than r_1 are filled with

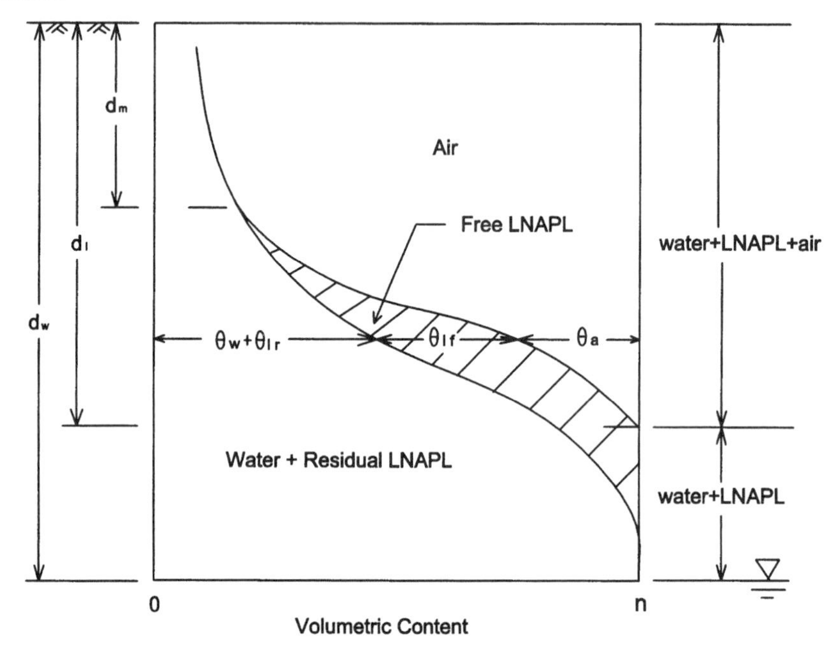

Figure 5.11 Composition of fluid phases near water table (Ostendorf et al., 1993).

either LNAPL or water. For the pores smaller than r_1, the interfacial tension between the LNAPL and water controls the presence of LNAPL and water. If r_2 is the radius of curvature corresponding to the interfacial tension between LNAPL and water, pores smaller than r_2 are filled with water whereas those greater than r_2 (but less than r_1) are filled with the LNAPL.

To present expressions for determining the phase distribution, we define the volumetric fluid contents of the various phases as follows:

$$\theta_w = \frac{\text{total volume of water}}{\text{total pore volume}} \tag{5.5a}$$

$$\theta_l = \frac{\text{total volume of LNAPL}}{\text{total pore volume}} \tag{5.5b}$$

and

$$\theta_a = \frac{\text{total volume of air}}{\text{total pore volume}} = n - \theta_w - \theta_l \tag{5.5c}$$

where n = porosity of the soil. Considering that both LNAPL and water exist in free and residual forms,

$$\theta_w = \theta_{wr} + \theta_{wf} \tag{5.6a}$$

and

$$\theta_l = \theta_{lr} + \theta_{lf} \tag{5.6b}$$

Effective saturation of water (S_w) and that of the total liquids (water and LNAPL), S, may be defined in terms of the above volumetric contents as

$$S_w = \frac{(\theta_w + \theta_{lr}) - \theta_{wr}}{n - \theta_{wr}} \tag{5.7a}$$

$$S = \frac{(\theta_w + \theta_{lf} + \theta_{lr}) - \theta_{wr}}{n - \theta_{wr}} \tag{5.7b}$$

Note that the effective saturation of water includes the residual content of LNAPL. The effective saturations of LNAPL in its free and residual form may be defined similarly as

$$S_{lf} = \frac{\theta_{lf}}{n - \theta_{wr}} = S - S_w \tag{5.8a}$$

and

$$S_{lr} = \frac{\theta_{lr}}{n - \theta_{wr}} \tag{5.8b}$$

The effective saturations S_w and S, defined in Eq. (5.7), are governed by the interfacial tensions between LNAPL and air, and between LNAPL and water. These are analogous to the effective saturation of water in soil–water–air systems, where the moisture content above the water table is governed by the surface tension of water. Therefore, the moisture retention functions, dealt with in Section 3.6, can be adopted here for the purpose of relating S and S_w to the capillary pressures. Using van Genuchten's (1980) function for water retention, Lenhard and Parker (1990) expressed S as

$$S = \{1 + [\beta_l(d_l - d)]^{\alpha}\}^{(1/\alpha)-1} \qquad (d_l > d > d_m) \tag{5.9a}$$

and

$$S = 1 \qquad (d > d_l) \tag{5.9b}$$

where α = empirical coefficient dependent on the pore size distribution of soils, and β_l is a scaling factor, which is related to the interfacial tension between LNAPL and water and the mean pore radius, by

$$\beta_l = \frac{\rho_l g r_m}{2\sigma_{la}} \tag{5.10}$$

where ρ_l = density of the LNAPL, g = acceleration due to gravity, r_m = mean pore radius, and σ_{la} = interfacial tension between LNAPL and air. Note that Eq. (5.10) is the capillary equation with contact angle assumed to be zero and the pores assumed to be cylindrical. Similarly, the effective saturation of water, S_w, is expressed as

$$S_w = \{1 + [\beta_w(d_w - d)]^\alpha\}^{(1/\alpha)-1} \qquad (d_w > d > d_m) \tag{5.11a}$$

$$S_w = 1 \qquad (d > d_w) \tag{5.11b}$$

and

$$S_w = \left\{1 + \left[\frac{\beta_l \sigma_{la}}{\sigma_{wa}}\left(d_l - d_w + \frac{\rho_w}{\rho_l}\left\langle d_w - d\right\rangle\right)\right]^\alpha\right\}^{(1/\alpha)-1} \qquad (d_m > d) \tag{5.11c}$$

where σ_{wa} = interfacial tension between water and air, i.e., the surface tension, ρ_w = density of water, and β_w is a scaling factor given by

$$\beta_w = \beta_l \frac{[(\rho_w/\rho_l) - 1]\sigma_{la}}{\sigma_{wl}} \tag{5.12}$$

where σ_{wl} = interfacial tension between the water and the LNAPL phases. The depth d_m to which the LNAPL free product rises above the water table can be obtained by equating S_w[Eq. (5.11a)] and S [Eq. (5.9a)], which will result in

$$d_m = \frac{\beta_l d_l - \beta_w d_w}{\beta_l - \beta_w} \tag{5.13}$$

Once S and S_w are determined through Eqs. (5.9) and (5.11), S_{lf} can be determined using Eq. (5.8a). The only other quantity that needs to be determined is S_{lr}, in order to estimate the total volume of LNAPL in the soil with respect to depth.

As discussed earlier, the reason for residual entrapment of LNAPLs is the fluctuating water table, which causes changes in capillary pressures of the fluids at a given point. Following Ostendorf et al. (1993) and Parker and Lenhard (1987), the hysteretic entrapment of LNAPLs in water is shown conceptually in Figure 5.12. Consider the changes in capillary pressure at point A, located at a depth d below the ground surface, as the water table fluctuates between the depths of $d_{w,\min}$ and $d_{w,\max}$ (Fig. 5.12a). We choose the depth d such that $d > d_m$, since residual entrapment of LNAPLs is possible only when free product is available. The corresponding retention curves are shown in Figure 5.12b. As the water table is lowered from B to C, the water content at point A reduces along the path BC

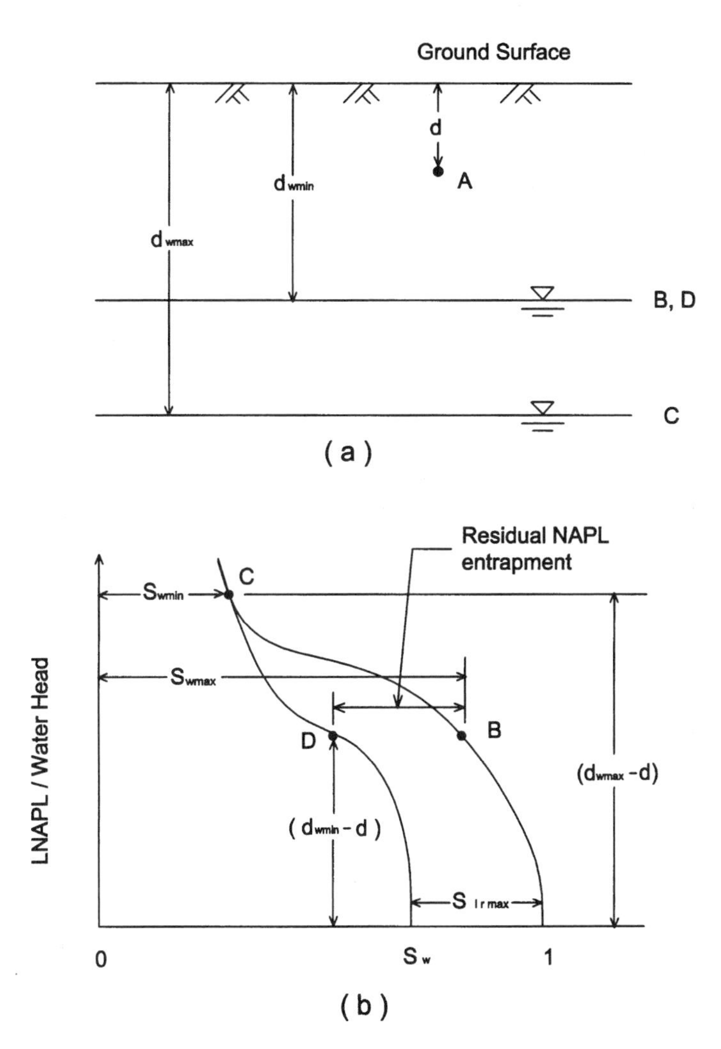

Figure 5.12 Hysteretical trapping of LNAPL: (a) range of water table fluctuations; (b) changes in S_w as a result of water table movement (Ostendorf et al., 1993).

in Figure 5.12b. This is because water is drained from point A as a result of increase in suction, equal in magnitude to $(d_{w,\max} - d_{w,\min})$. The minimum and maximum effective saturations of water, $S_{w,\min}$ and $S_{w,\max}$ respectively, at point A are given in Eq. (5.11a),

$$S_{w,\min} = \{1 + [\beta_w(d_{w,\max} - d)]^{\alpha}\}^{(1/\alpha)-1} \tag{5.14a}$$

and

$$S_{w,\max} = \{1 + [\beta_w(d_{w,\min} - d)]^{\alpha}\}^{(1/\alpha)-1} \tag{5.14b}$$

When the water table moves upward from C to D, the moisture characteristic path does not coincide with CB (Fig. 5.12b) because of hysteresis; instead, it follows the path CD. Some of the free LNAPL is trapped during this stage, the magnitude of which is equal to BD in Figure 5.12b. If in fact the water table rises past the point A under consideration, the amount of residual LNAPL entrapment at the point is equal to $S_{lr,\max}$. This is the maximum entrapment that can occur at point A and is given by Parker and Lenhard (1987)

$$S_{lr,\max} = \frac{1 - S_{w,\min}}{1 + \gamma(1 - S_{w,\min})} \tag{5.15}$$

where γ is a hysteretical trapping factor defined by the hysteresis in the contact angle and the radii of curvature at the interfaces. It is not possible to obtain values of γ directly. Published estimates of its magnitude were obtained by calibration procedures. Using residual LNAPL saturation data available at an aviation gasoline spill site in Traverse City, Michigan, Ostendorf et al. (1993) obtained an estimate of $\gamma = 40$. The soils at the site were characterized as medium-grained uniform sands with a mean particle diameter of 3.8×10^{-4} m.

In order to estimate residual saturations of LNAPLs at elevations higher than the water table, such as BD in Figure 5.12b, Parker and Lenhard (1987) proposed the following empirical relationships:

$$S_{lr} = S_{lr,\max}\left(1 - \frac{1 - S_{w,\max}}{1 - S_{w,\min}}\right) \qquad d_{w,\min} > d \tag{5.16a}$$

$$S_{lr} = S_{lr,\max} \qquad d_{w,\max} > d > d_{w,\min} \tag{5.16b}$$

$$S_{lr} = 0 \qquad 0 < d < d_m \tag{5.16c}$$

and

$$S_{lr} = 0 \qquad d > d_{\max} \tag{5.16d}$$

Equations (5.8a), (5.9), (5.11), and (5.16) completely describe the phase diagram and enable estimation of the volumetric contents of each of the phases in the vicinity of the water table.

5.5 MODELING TRANSPORT OF NAPLs IN SOILS

Modeling the transport of NAPLs in soils is complicated because of the multiple phases involved and the vast number of parameters associated with these phases.

As a result, prediction of NAPL migration in soils has remained as an academic and research endeavor only. We will present below essential principles involved in NAPL transport, restricting the scope to a qualitative treatment of the problem.

5.5.1 Relative Permeability Concept

The transport of NAPLs is analogous to that of flow of water in unsaturated soils in two respects: (1) the effective saturations of the phases keep on changing, and (2) the changes in the effective saturations in turn cause changes in the driving force for fluid flow. Darcy's law is again invoked as the constitutive relationship governing the flow of each of the three fluid phases, i.e., NAPL, water, and air. In order to use the Darcy's law to express flow of air and NAPL, the hydraulic conductivity term is replaced by the intrinsic permeability (with the units of area), or

$$V_\alpha = -\frac{k\rho_\alpha g}{\mu_\alpha}\nabla\phi \tag{5.17}$$

where V_α = velocity of the phase α, k = intrinsic permeability of the medium, g = acceleration due to gravity, ρ_α, μ_α = density and dynamic viscosity of phase α, respectively, and $\nabla\phi$ = potential gradient of the phase α.

As discussed in Chapter 3, the permeability of a fluid varies with the volumetric content of that fluid. When NAPL, water, and air coexist in the pores, the permeability of the medium with respect to any individual fluid is less than when the pore space is entirely occupied by that fluid. This reduction in permeability is dependent on the effective saturation of that fluid, and is described in terms of *relative permeability*, $k_{r\alpha}$, which is defined as

$$k_{r\alpha} = \frac{k_\alpha(S_\alpha)}{k_{s\alpha}} \tag{5.18}$$

where $k_\alpha(S_\alpha)$ = permeability of the soil with respect to phase α at effective saturation S_α, and $k_{s\alpha}$ = permeability of the soil at complete saturation with phase α. Thus, the relative permeability varies from 1.0 at complete saturation of the phase to 0 near 0% saturation. In between 0 and 1.0, the fluids hinder the flow of each other in the pore space.

Considering a two-phase system with only water and NAPL, the variation of the relative permeability with effective saturation is shown in Figure 5.13. As the effective saturation of water decreases from 100%, its relative permeability also decreases. The relative permeability of the soil with respect to water becomes zero at a point where the water phase is no longer continuous and is at irreducible saturation. Similarly, as the effective saturation of the NAPL decreases from 100%, its relative permeability also decreases, ultimately becoming zero at its

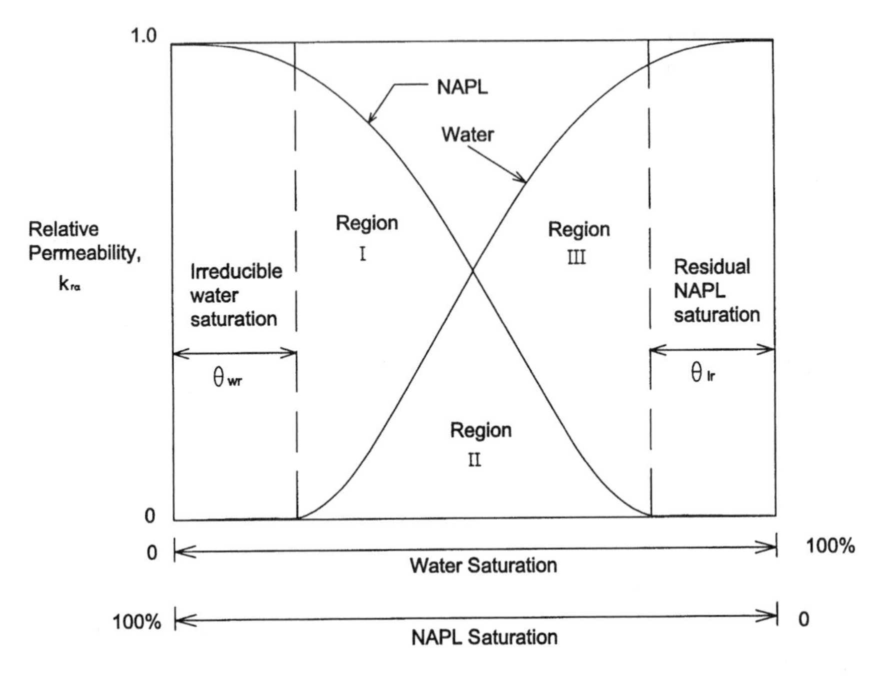

Figure 5.13 Relative permeabilities of water and NAPL as a function of the degree of saturation.

residual saturation. Because of the interference between the phases, the sum of the relative permeabilities (with respect to water and NAPL) do not equal 1. Thus, the overall permeability of the medium is reduced because of this interference. Three regions of flow can be delineated in Figure 5.13. In region I, NAPL dominates the pore space, and the relative permeability with respect to water is nearly zero. In region II, both NAPL and water exist in a continuous form and flow together. In this region, the interferences from each other are maximum, and the permeability of the medium as a whole is greatly reduced. In region III, water dominates the pore space, and the relative permeability with respect to NAPL is nearly zero because the NAPL is present in discontinuous form.

The above concept can be extended to a three-phase system when air is present in addition to NAPL and water. This is usually presented in a ternary diagram as shown in Figure 5.14. The process is considerably more complicated than a two-phase system, although the same principles govern. It can be noted in Figure 5.14 that there are large areas where at least one of the fluids becomes immobile. Also, all three fluids can flow simultaneously only in an extremely

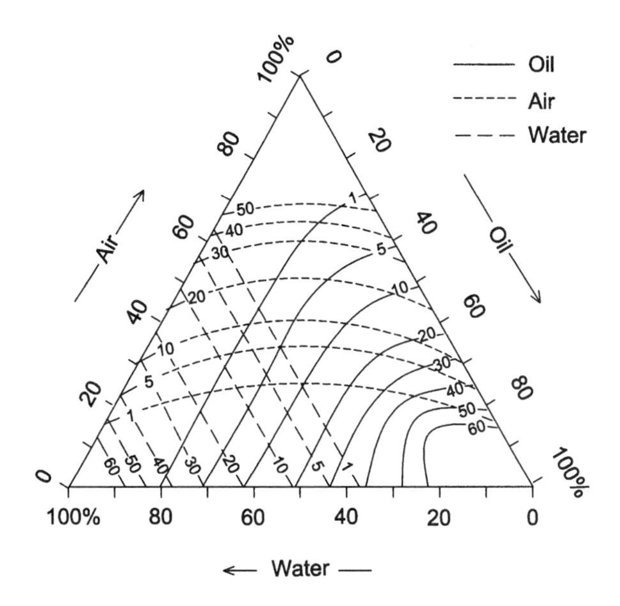

Figure 5.14 Relative permeabilities for three-phase flow (Van Dam, 1967).

limited saturation region. Note that air also has a residual saturation, which hinders the permeability of the other two phases.

In most cases of NAPLs, data describing the relative permeability variation with respect to effective saturations are not available. Both theoretical and empirical equations were developed in the literature describing the variations (Stone, 1973; Lenhard and Parker, 1987). Mathematical details of these equations are beyond the scope of this fundamental treatment.

5.5.2 Equations Governing Multiphase Transport

As in the transport of water (Chapter 3) and dissolved solutes (Chapter 4), the transport of NAPLs involves combination of the mass conservation principle and constitutive (or cause-and-effect) relationships. However, the NAPL transport modeling is far more complicated because of the multiple number of phases involved and possible interphase transfer of fluids. The transport of each of the phases is coupled with capillary pressures and effective saturations of the other phases. In general, governing equations for each of the three phases are developed individually and then they are coupled together using capillary pressure and effective saturations. Although simple in principle, NAPL transport modeling requires a vast number of parameters, not all of which can be obtained through routine experiments.

Abriola and Pinder (1985a, 1985b) and Abriola (1988) presented a comprehensive approach to modeling multiphase transport. Following this approach, we begin with a comprehensive statement of the transport problem by expressing the mass balance of a chemical species i in the α phase as

$$\frac{\partial}{\partial t}(\rho_\alpha \varepsilon_\alpha \omega_\alpha^i) + \nabla \cdot (\rho_\alpha \varepsilon_\alpha V_\alpha \omega_\alpha^i) - \nabla \cdot J_\alpha^i = A_\alpha^i + B_\alpha^i \tag{5.19}$$

where

ρ_α = mass density of the α phase
ε_α = fraction of pore volume occupied by the α phase
ω_α^i = mass fraction of species i in the α phase
V_α = velocity of the α phase
J_α^i = nonadvective flux of the species i in the α phase
A_α^i = exchange of mass of species i due to interphase diffusion and/or phase change
B_α^i = external supply of species i to the α phase
∇ = differential operator

The first term represents the time rate of change in mass of species i in phase α. The second and third terms reflect mass transport of species i due to advection and nonadvective effects (diffusion and dispersion), respectively. The terms on the right-hand side of the equation represent sink/source terms due to phase changes, and chemical and biological reactions leading to destruction or creation of the species.

In order to express total mass balance of the species i in the system, a summation of mass balances in the individual phases should be carried out, or

$$\sum_\alpha \left[\frac{\partial}{\partial t}(\rho_\alpha \varepsilon_\alpha \omega_\alpha^i) + \nabla \cdot (\rho_\alpha \varepsilon_\alpha V_\alpha \omega_\alpha^i) - \nabla \cdot J_\alpha^i\right] = \sum_\alpha B_\alpha^i \tag{5.20}$$

Note that the above equation does not contain the term A_α^i representing exchange of species mass among the phases, since mass lost by one phase is gained by the other phases, i.e.,

$$\sum_\alpha A_\alpha^i = 0 \tag{5.21}$$

Also, the summation should be carried out including soil solids as a separate phase to account for the compressibility of the porous medium and species mass present in the adsorbed form. Equations such as (5.20) are known as species-balance or compositional equations. One such equation has to be solved for each of the species present in the system.

Equations (5.19) and (5.20) may be simplified by assuming $\nabla \cdot J_{\alpha}^{i} = 0$, in the case of those species which maintain a sharp wetting front and which may not be transported by nonadvective means such as diffusion and dispersion. Assuming that Darcy's law is a valid constitutive relation, V_{α} may be expanded in Eq. (5.19) and the resulting set of equations for the water, NAPL, and air phases may be expressed for a single-species chemical as

$$\frac{\partial}{\partial t}(\rho_w n S_w \omega_w) = \nabla \cdot \frac{k \rho_w k_{rw}}{\mu_w}(\nabla \cdot P_w - \rho_w g \nabla \cdot D_w) + A_w + B_w \quad (5.22a)$$

$$\frac{\partial}{\partial t}(\rho_n n S_n \omega_n) = \nabla \cdot \frac{k \rho_n k_{rn}}{\mu_n}(\nabla \cdot P_n - \rho_n g \nabla \cdot D_n) + A_n + B_n \quad (5.22b)$$

and

$$\frac{\partial}{\partial t}(\rho_a n S_a \omega_a) = \nabla \cdot \frac{k \rho_a k_{ra}}{\mu_a}(\nabla \cdot P_a - \rho_a g \nabla \cdot D_a) + A_a + B_a \quad (5.22c)$$

where subscripts w, n, and a, stand for water, NAPL, and air phases, respectively, D_a = elevation head of the phase α, P_{α} = pressure head of the phase α, and all other terms are as defined earlier. Solution of Eqs. (5.22) involves 18 variables: ρ_w, ρ_n, ρ_a, S_w, S_n, S_a, k_{rw}, k_{rn}, k_{ra}, μ_w, μ_n, μ_a, ω_w, ω_n, ω_a, P_w, P_n, and P_a. Therefore, in addition to the three equations above, 15 more equations are needed for complete solution of the problem. These may be obtained as follows.

1. Densities (ρ_w, ρ_n, ρ_a) and viscosities (μ_w, μ_n, μ_a) may be expressed as known functions of pressures (P_w, P_n, P_a), yielding six equations.
2. Effective saturations (S_w, S_n, S_a) may be expressed as functions of pressures (P_w, P_n, P_a), yielding three equations.
3. Relative permeabilities (k_{rw}, k_{rn}, k_{ra}) may be expressed as functions of effective saturations, yielding three equations.
4. Mass fractions of species in all the phases may be assumed to be at local equilibrium, which implies that the time scales are such that adjoining phases reach a thermodynamic equilibrium. This will yield two expressions of the form

$$\omega_{\alpha} = \kappa^{\alpha\beta} \omega_{\beta} \quad (5.23)$$

where $\kappa^{\alpha\beta}$ is known as the partition coefficient of the species between α and β phases. These coefficients are functions of phases compositions and pressures and may be determined using Henry's law and Raoult's law constants. Therefore, two equations of the type (5.23) together with $\Sigma_{\alpha}\omega_{\alpha} = 1$ will give three equations in total.

Note that the soil phase is ignored in Eq. (5.22). The soil is assumed to be incompressible and porosity is assumed to be constant. Further simplification of the above system of equations is possible when the gas phase remains at atmospheric pressure, in which case Eq. (5.22c) can be reduced to a simpler form.

5.6 MOBILIZATION OF RESIDUAL NAPLs

In modeling the transport of NAPLs in Section 5.5, we ignored the discontinuous/residual phase of NAPLs. This is justifiable because the residual content of NAPLs is relatively immobile. In remediation operations, however, the residual content of NAPLs often amounts to significant quantities, and its mobilization and recovery are important. Unlike free product removal, residually trapped ganglia are difficult to mobilize. Research on mobilization of NAPL ganglia was first initiated in the petroleum engineering discipline in the context of enhancement of recoveries from subsurface oil reservoirs. The principles developed in that discipline were later adopted in environmental engineering in the context of subsurface remediation. Due to recent emphasis on in-situ remediation, several research efforts have focused lately on the mechanics of entrapment and the mobilization of ganglia.

Entrapment of ganglia is a pore-scale phenomenon, therefore an understanding of their transport must begin at pore scale. Although several modes of ganglion entrapment have been conceptualized in the literature, two modes are commonly accepted for a soil–water–NAPL system: snap-off and bypassing (Chatzis et al., 1983). Snap-off occurs when the NAPL is displaced from a pore body into a pore throat by water (Fig. 5.15). Water moves along the sides of the pores, where it displaces the NAPL, and reaches the end of the pore body before NAPL can exit. As water begins to enter the pore throat, NAPL "snaps-off" and remains in the pore as a trapped ganglion. As shown in Figure 5.15a, this mode of entrapment is likely in cases where the aspect ratio of the pores (the ratio of the size of the pore body to that of the pore throat) is high. When the aspect ratio is small, complete displacement of the NAPL is possible, as shown in Figure 5.15b. The likelihood of snap-off also depends on the wettability of the two liquids. Snap-off occurs in systems where the media are water-wet. If they are intermediate-wet ($\theta = 90°$), snap-off may not occur.

Bypassing, on the other hand, occurs when the NAPL is displaced by water in a complex pore structure and is described using a pore doublet model (Chatzis et al., 1983). As shown in Figure 5.16a, the pore doublet model is visualized using two arms of a pore. Bypassing does not occur when the two arms have similar geometry. However, as shown in Figure 5.16b, when the downstream node of the upper arm is smaller than that of the lower arm, the advancing water will bypass the NAPL in the lower arm. The greater capillary force associated

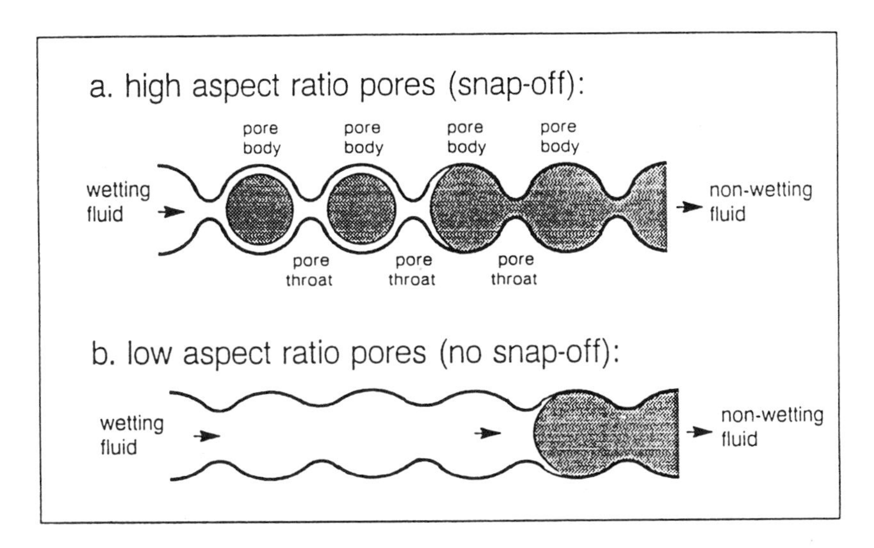

Figure 5.15 Effect of pore aspect ratio on the NAPL trapping in a tube of nonuniform diameter (Chatzis et al., 1983).

with that node will not allow a stable interface of the two liquids to form there. Such bypassing may occur at larger scales extending over many pore bodies, and ganglia of complex shapes may form as a result. Finally, snap-off and bypassing modes can occur together, as shown in Figure 5.16c, yielding ganglia of various sizes and configurations. The sizes of the isolated ganglia can range from one or two pore diameters to thousands of pore diameters. The resulting population of ganglia may assume several different shapes even in a homogeneous porous medium.

Once isolated, the NAPL ganglion can be mobilized only if viscous forces can be created hydrodynamically to squeeze the ganglion out of the pore body through the throat. The viscous forces should be sufficient in magnitude so that the NAPL is displaced by water at the downstream end of the ganglion and water is displaced by the NAPL at the upstream or leading end. The pressure difference required to mobilize the ganglion is proportional to the difference in the capillary pressures at these two ends. Note that, in addition to the viscous force, the density difference between the two liquids may also influence the mobilization. This may be of importance in the case of LNAPLs, where the blobs will be buoyant and will tend to rise vertically, thus aiding upward mobilization.

Mobilization of trapped ganglia is therefore dependent on a number of variables, including: (1) geometry of the pore network, (2) interfacial tension between the liquids, (3) density difference between the two liquids, (4) wettability of the

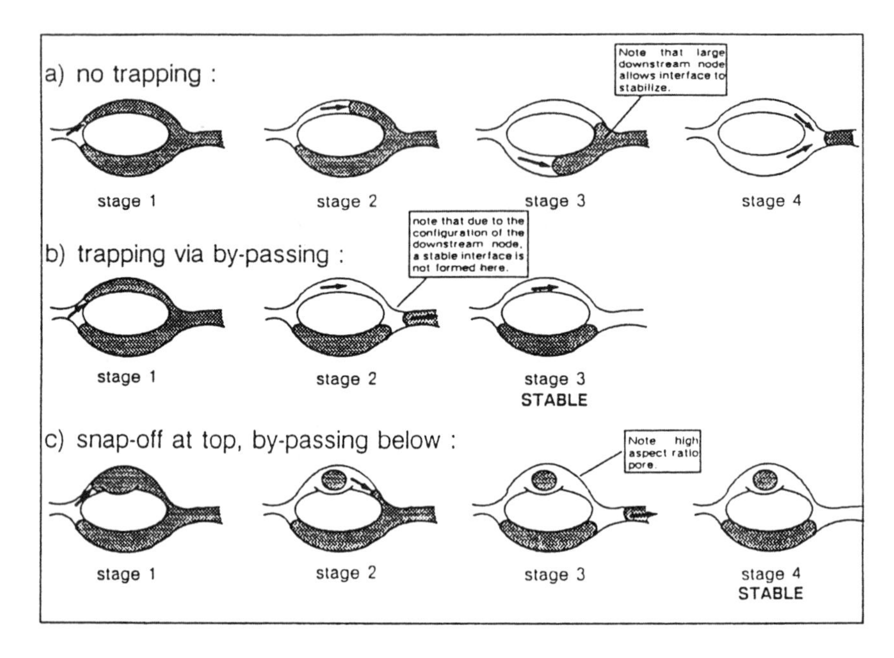

Figure 5.16 Conceptualization of NAPL trapping mechanisms using the pore doublet model (Chatzis et al., 1983).

medium by the two liquids, and (5) applied water-phase pressure gradient and its alignment with gravity. For soil–NAPL–water systems, these parameters are conveniently incorporated into two dimensionless numbers: capillary number, N_c, which is the ratio of viscous forces to capillary forces; and bond number, N_b, which is the ratio of gravity forces to capillary forces. Three different versions for capillary number N_c are used in the literature:

$$N_c^1 = \frac{V_w \mu_w}{\sigma_{nw}} \qquad \text{(Melrose and Brandner, 1974; Morrow and Songkran, 1981)} \qquad (5.24a)$$

$$N_c^2 = \frac{k \nabla P}{\sigma_{nw}} \qquad \text{(Ojeda et al., 1953; Larson et al., 1981)} \qquad (5.24b)$$

$$N_c^3 = \frac{KJ\mu_w}{\sigma_{nw}} \qquad \text{(Wilson and Conrad, 1984; Wilson et al., 1990)} \qquad (5.24c)$$

where the superscript is used to denote the version of the capillary number, K = water-saturated hydraulic conductivity of soil (L/T); ∇P = water phase pressure gradient expressed as (F/L^3); and J = hydraulic gradient in the water phase (including gravity), and all other terms are as defined earlier. All of these versions are equivalent in the sense that the numerator indicates a measure of the viscous force and the denominator indicates the capillary force. The bond number, N_b, also has been used in two different forms (Morrow and Songkran, 1981):

$$N_b^1 = \frac{\Delta\rho g R^2}{\sigma_{nw}} \tag{5.25a}$$

$$N_b^2 = \frac{\Delta\rho g k}{\sigma_{nw}} \tag{5.25b}$$

where $\Delta\rho$ = density difference between water and NAPL, g = acceleration due to gravity, and R = representative grain radius in the soil. Clearly, the intrinsic permeability of soil is related to the grain size, therefore the two equations are equivalent in nature. In random packings of uniform spheres, the relationship between R and k may be expressed as (Morrow and Songkran, 1981)

$$k = 0.00317R^2 \tag{5.26}$$

which, when substituted in Eq. (5.25), yields

$$N_b^2 = 0.00317N_b^1 \tag{5.27}$$

Most of the research on mobilization of NAPL ganglia was conducted in an attempt to correlate the mobilization with the capillary number. Although density differences influence mobilization, the effect of these forces were often considered to be negligible in comparison with viscous forces. Typical residual saturation versus capillary number curves for glass beads and sandstones containing NAPL ganglia are shown in Figure 5.17. The y axis represents the reduced residual saturation as a ratio of residual saturation at a given capillary number to the initial residual saturation. As seen in Figure 5.17, a minimum capillary number (indicated as N_c^*) is required to initiate mobilization of the ganglia from their initial state. Its magnitude is 2×10^{-5} for the sandstone curve and about 1×10^{-3} for the bead-pack. N_c^{**} represents the capillary number required to mobilize all of the entrapped ganglia; $N_c^{**} = 1.3 \times 10^{-3}$ for sandstones and 9×10^{-3} for bead pack. In order to achieve these capillary numbers, the water phase pressure gradient must be increased for a given soil–water–NAPL system.

The data from Figure 5.17 may be replotted as in Figure 5.18 for a better understanding of the range of hydraulic gradients required to mobilize the NAPL ganglia in a soil–NAPL–water system. Obviously, for NAPLs with smaller interfacial tensions, lower hydraulic gradients are needed to achieve N_c^{**}. As an example, consider the gradients required to initiate mobilization of NAPL ganglia with an interfacial tension of 10 dynes/cm. If the porous medium represents a fine gravel with $k = 10^{-5}$ cm^2, a gradient $J = 10^{-2}$ is required. To remove all

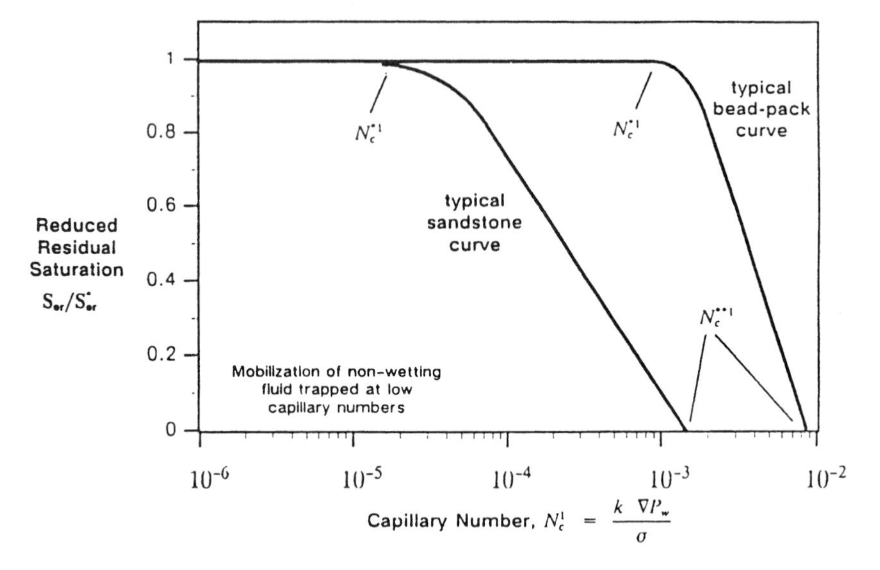

Figure 5.17 Residual saturation versus capillary number for sandstones and glass beads (Wilson et al., 1990).

of the ganglia from the system, a gradient of about 0.6 is needed (upper line in Fig. 5.18). Figure 5.19 shows the percentage recoveries as a function of the intrinsic permeability and the gradient, for a NAPL with σ_{nw} = 10 dynes/cm. It is clear from Figs. 5.18 and 5.19 that the gradients required to mobilize ganglia increase steeply as the interfacial tension is increased or the intrinsic permeability is decreased. This, of course, is a direct consequence of the capillarity principle.

Very often, the gradients required to mobilize residual ganglia are impractical to achieve in practice. Except in gravels and some coarse-grained soils, ganglia mobilization cannot be expected to occur with viscous forces alone. As an example, Wilson and Conrad (1984) estimated that a gradient of 0.1 is required to initiate PCE (density = 1.62 g/cm, and σ_{nw} = 47.5 dynes/cm) ganglia from a fine gravel with $k = 10^{-5}$ cm^2. To mobilize all of the ganglia, the necessary gradient would be 4.75. In comparison to this, the necessary gradient to initiate mobilization of the PCE ganglia would be 10 in a medium sand with $k = 10^{-7}$ cm^2. To remove all of the residual PCE ganglia in medium sands, the necessary gradient would jump up to 475, an impractical gradient to achieve at remediation sites. Based on this, one can conclude that viscous forces alone cannot mobilize ganglia at field sites, especially in systems with low permeability and/or high interfacial tension. An alternative measure is to increase capillary numbers by reducing interfacial tension. Detergents, known as surfactants, are known to ac-

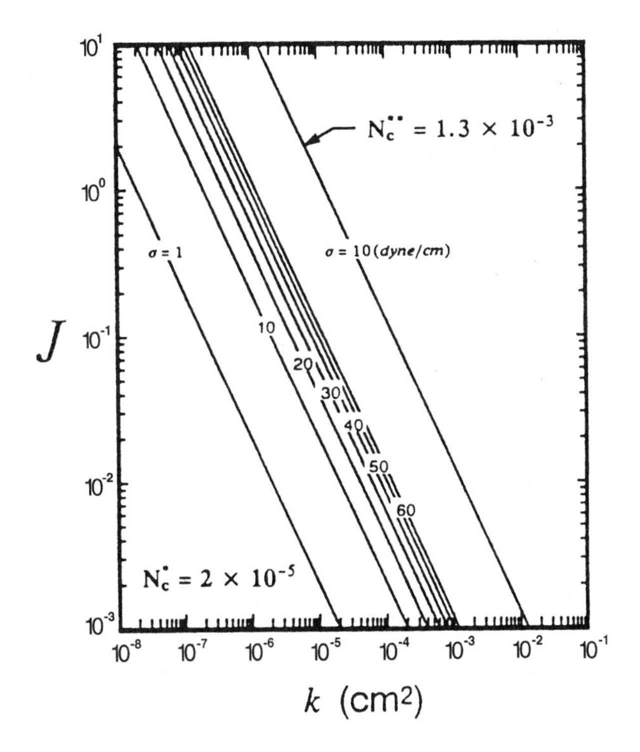

Figure 5.18 Hydraulic gradient required to initiate blob mobilization in porous media of various permeabilities, for organic liquids of various interfacial tensions (Wilson et al., 1990).

complish this to some extent. Development of innovative methods to clean up sites contaminated with residual NAPL ganglia is an area of intense research at present.

5.7 MASS TRANSFER PROCESSES

The sink-source terms in the governing equation for NAPLs transport [Eq. (5.19)] arise from a number of mass transfer processes occurring simultaneously among all the phases involved. As shown in Figure 5.20, the main processes pertinent to NAPLs are dissolution, sorption, and volatilization. Of these, dissolution deserves additional discussion in the context of NAPLs. It controls the movement of NAPLs to a great extent. For example, chlorinated solvents, which are only slightly soluble in water, penetrate deep into aquifers and remain as an immiscible

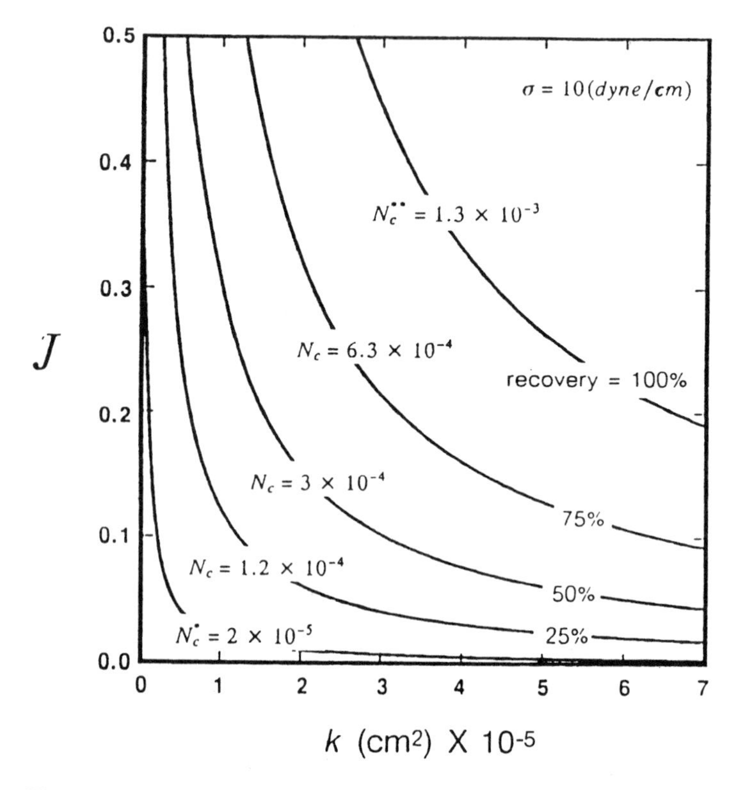

Figure 5.19 Recovery of residual saturation as a function of permeability and hydraulic gradient for an interfacial tension of 10 dynes/cm (Wilson et al., 1990).

phase causing very few reactions in the aqueous phase. On the other hand, compounds such as acetone, because of their relatively high solubility in water, dissolve quickly and participate in chemical and biological reactions in the water and gas phases. As a result, these compounds may not migrate to significant depths. Also, as the NAPL migrates through the unsaturated zone (see Fig. 5.9), the infiltrating water may dissolve some of its components and carry them down to the water table. The water with dissolved NAPL species may possess different properties of wettability and interfacial tension, thus altering the course of the incoming NAPL product. Perhaps the key aspects where dissolution becomes an important process are in the assessment of lifetimes of NAPL pools and the residual ganglia trapped in the saturated zone. Dissolution of the pool and the ganglia are strong functions of the water-phase velocities. Consider the dissolution times of TCE entrapped at a residual saturation of 20% in a soil with 35% porosity.

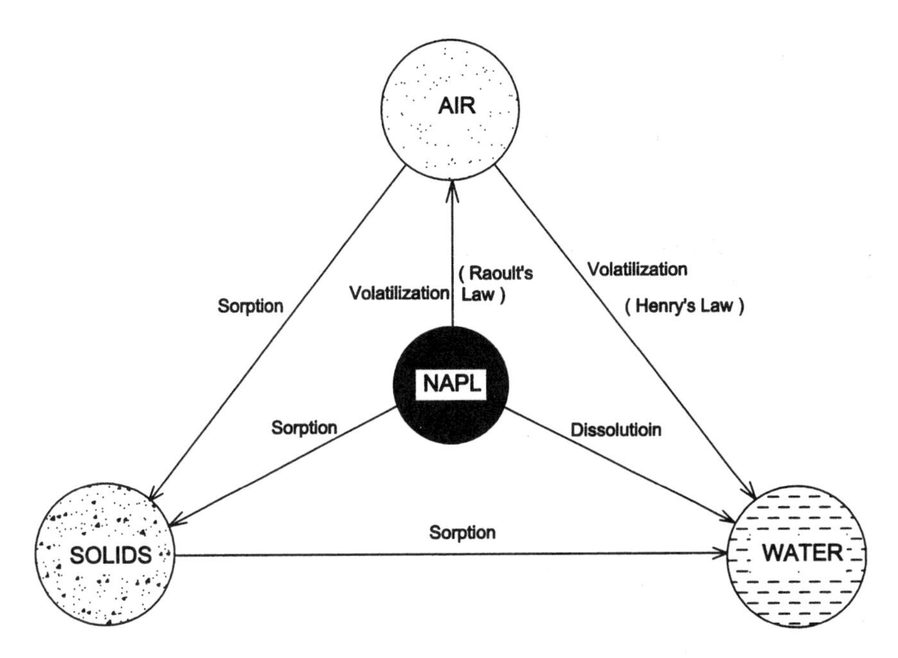

Figure 5.20 Mass transfer processes involved in partitioning of NAPLs into solid, water, and gas phases.

Palmer and Johnson (1989) estimate that it would take 15.4 years for the ganglia to be completely dissolved in a cubic meter of the soil when subjected to a continuous flow rate of 1.7 cm/day. NAPLs with lower solubility than TCE would require even longer periods, often up to several decades.

The aqueous solubilities of NAPL species govern to a great extent the time rate of pool and ganglia dissolution. As shown in Tables 5.1 and 5.2, the solubilities of NAPLs vary greatly. For NAPLs comprised of a mixture of several species, an *effective solubility* is often used. It is defined as the aqueous solubility of an organic constituent which is in chemical equilibrium with a NAPL mixture consisting of several such organic constituents. It can be estimated as a product of the mole fraction of the organic constituent in the NAPL mixture and its pure phase solubility.

In addition to the solubility of the NAPL in water, dissolution from NAPLs trapped in the form of ganglia is governed by the porosity of the medium, size and shape of the individual ganglia, and aqueous phase velocity. Several mathematical derivations exist in the literature correlating the flux from a dissolving spherical particle to all these parameters. In general, the dissolution flux J_d is expressed as

$$J_d = m(C_s - C_a) \tag{5.28}$$

where m = mass transfer coefficient, C_s = aqueous solubility of the NAPL, and C_a = average concentration in the aqueous phase in the vicinity of the dissolving particle. The difference in concentrations, $C_s - C_a$, is the driving force for dissolution flux. The mass transfer coefficient m may in turn be expressed for ganglia idealized as spheres, as (Wilson and Geankopolis, 1966)

$$m = 1.09 \frac{V_w}{n} \left(\frac{V_w a_n}{D} \right)^{-2/3} \tag{5.29}$$

where V_w = aqueous phase velocity, n = porosity, a_n = diameter of the dissolving ganglia, and D = molecular diffusivity of the NAPL in water. Equation (5.29) indicates that the dimensionless number known as the Peclet number, Pe (= $V_w a_n/D$) dominates the mass transfer coefficient. There is another class of equations for mass transfer coefficients, given in terms of dimensionless numbers. One such equation, given by Miller et al. (1990), relates the mass transfer coefficient in terms of Sherwood, Reynolds, and Schmidt numbers:

$$\mathrm{Sh} = \beta_0 \,\mathrm{Re}^{\beta_1}\, \theta_n^{\beta_2}\, \mathrm{Sc}^{1/2} \tag{5.30}$$

where $\beta_0 = 12 \pm 2$, $\beta_1 = 0.75 \pm 0.08$, $\beta_2 = 0.60 \pm 0.21$, θ_n = NAPL volumetric fraction, and *Sh*, *Re*, and *Sc* are the Sherwood, Reynolds, and Schmidt numbers, respectively, defined as

$$\mathrm{Sh} = \frac{m a_p^2}{D} \tag{5.31}$$

$$\mathrm{Re} = \frac{v_w \rho_w a_p}{\mu_w} \tag{5.32}$$

and

$$\mathrm{Sc} = \frac{\mu_w}{\rho_w D} \tag{5.33}$$

where a_p = diameter of the porous medium particle, v_w = mean pore velocity of the aqueous phase, and all other terms are as defined earlier.

Expressions such as (5.29) and (5.30) may be used to understand the nature of dissolution from ganglia and to predict the life span of ganglia under a given set of hydrogeological conditions. In general, dissolution of the ganglia begins at a threshold velocity, which is higher in magnitude for smaller ganglia. Once the Darcy velocity exceeds the threshold, the solubility of the ganglia is directly

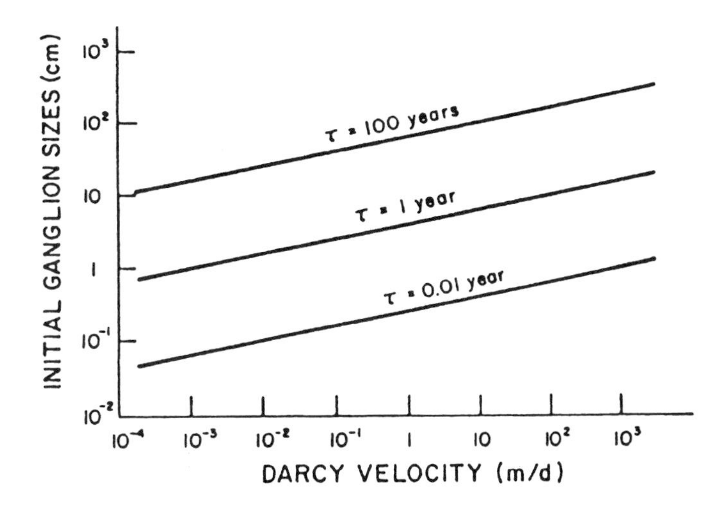

Figure 5.21 Predicted lifetime of a TCE ganglion (Hunt et al., 1988).

proportional to the velocity. Figure 5.21 shows the life spans in years of TCE ganglia of various initial sizes as a function of aqueous-phase velocity, estimated by Hunt et al. (1988). The dissolution flux was computed using Eqs. (5.28) and (5.29), where the average aqueous-phase concentrations C_a were assumed to be negligible compared to the solubility concentrations of the NAPL. Knowledge of the life spans of ganglia, such as those presented in Figure 5.21, is useful in designing remedial operations at NAPL-contaminated sites.

The dissolution from NAPLs in the form of pool or pancake has also been studied in the literature. Hunt et al. (1988) estimated the dissolution flux from NAPL pools using a steady-state advection-diffusion equation for transverse diffusion into a semi-infinite medium. The solution for dissolution fluxes were used by Johnson and Pankow (1992) to estimate the life spans of pools of various lengths assuming the pools have a rectangular configuration. The life span of a rectangular pool which maintains its areal dimensions constant during the dissolution process was expressed as

$$\tau_p = 2.43 \times 10^{-5} \rho_n \frac{\sqrt{L_p^3/DV_w}}{C_s} \tag{5.34}$$

where τ_p = life span of the pool in years, L_p = length of the pool, and all other terms are as defined earlier. Figure 5.22 shows how τ_p varies as a function of groundwater velocity for various values of L_p. Considering that groundwater velocities are typically in the range of 0.1–0.3 m/day, a 4-m-long pool may last

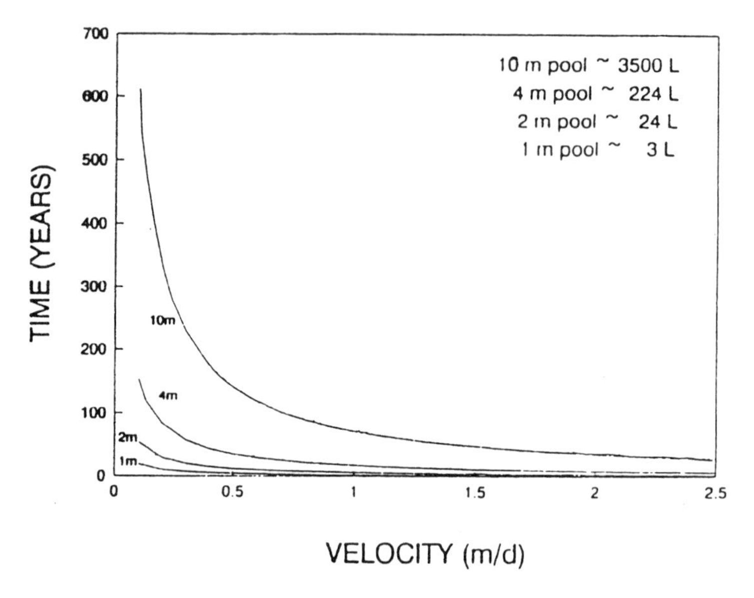

Figure 5.22 Pool dissolution time versus groundwater velocity for TCE for four pool lengths (Johnson and Pankow, 1992).

longer than 100 years. From a remediation standpoint, dissolution times cannot be reduced by more than a factor of 5 even if groundwater velocities are increased substantially above natural velocities.

Apart from the dissolution process, sorption, volatilization, hydrolysis, and biodegradation also play an important role in the transformation of NAPLs. The process of sorption has been one of the most studied in the literature, and the principles discussed in Chapter 4 are applicable to NAPLs. Volatilization process is of greater significance in the NAPLs than in the case of inorganic chemicals. Depending on its volatility, the NAPL may be partitioned into the gas phase, which then may be transported by molecular diffusion to the hitherto clean portions of the soils. When the LNAPL has left behind the residual ganglia (Fig. 5.8c), the volatilization becomes even more crucial because of the increased surface area of these ganglia. Volatilization from the NAPLs may be estimated using Henry's law and Raoult's law (described in Chapter 4) to partition between water (which consists of dissolved NAPLs) and air, and between NAPL and air, respectively. The process of hydrolysis is significant in the case of chlorinated hydrocarbons because they are not readily degradable. However, hydrolysis is controlled by dissolution process, since only the outer part of a NAPL ganglion is in touch with water. The organic molecules in the interior of the ganglion do not participate in the hydrolysis reaction. Biodegradation is a significant process in the case of LNAPLs such as benzene, toluene, and xylenes, which are known to be degraded by indigenous aerobic microorganisms. In general, in-situ biodegradation of LNAPLs is limited by the supply of dissolved oxygen. DNAPLs, on the other hand, are known to degrade only under anaerobic conditions or are resistant to degradation (Feenstra and Cherry, 1988).

REFERENCES

Abriola, L. M. (1988). Multiphase flow and transport models for organic chemicals: A review and assessment. Electric Power Research Institute, EA-5976, Palo Alto, CA.

Abriola, L. M., and Pinder, G. F. (1985a). A multiphase approach to the modeling of porous media contamination by organic compounds, 1. Equation development. *Water Resources Res.*, 21(1):11–18.

Abriola, L. M., and Pinder, G. F. (1985b). A multiphase approach to the modeling of porous media contamination by organic compounds, 2. Numerical simulation. *Water Resources Res.*, 21(1):19–26.

Anderson, W. G. (1986). Wettability literature survey, Part 1. Rock/oil/brine interactions, and the effects of core handling on wettability. *J. Petrol. Technol.*, Oct.: 1125–1149.

Chatzis, I., Morrow, N. R., and Lim, H. T. (1983). Magnitude and detailed structure of residual oil saturation. *SPE J.* 23(2):311–325.

Feenstra, S., and Cherry, J. A. (1987). Dense organic solvents in ground water: An introduction. In Dense Chlorinated Solvents in Ground Water, Institute for Ground Water Research, University of Waterloo, Progress Report 0863985, Waterloo, Ont.

Feenstra, S., and Cherry, J. A. (1988). Subsurface contamination by dense non-aqueous phase liquids (DNAPL) chemicals. Int. Groundwater Symp., Int. Assoc. Hydrogeol., Halifax, N.S., May 1–4, 1988.

Huling, S. G., and J. W. Weaver. (1991). Dense nonaqueous phase liquids. Ground Water Issue, EPA/540/4-91-002, Robert S. Kerr Environmental Research Laboratory, Ada, OK.

Hunt, J. R., Sitar, N., and Udell, K. S. (1988). Nonaqueous phase liquid transport and cleanup. 1. Analysis of mechanisms. *Water Resources Res.*, 24(8):1247–1258.

Johnson, R. L., and Pankow, J. F. (1992). Dissolution of dense chlorinated solvents into groundwater. 2. Source functions for pools of solvent. *Environ. Sci. Technol.*, 26(5): 896–901.

Kueper, B. H., and Find, E. O. (1988). An overview of immiscible fingering in porous media. *J. Contam. Hydrol.*, 2:95–110.

Larson, R. G., Davis, H. T., and Scriven, L. E. (1981). Displacement of residual nonwetting fluid from porous media. *Chem. Eng. Sci.*, 36:75–85.

Lenhard, R. J., and Parker, J. C. (1987). Measurement and prediction of saturation-pressure relationships in three phase porous media systems. *J. Contam. Hydrol.*, 1:407–424.

Lenhard, R. J., and Parker, J. C. (1988). Experimental validation of the theory of extending two phase saturation-pressure relations to three fluid phase systems for monotonic drainage paths. *Water Resources Res.*, 24:373–380.

Lenhard, R. J., and Parker, J. C. (1990). Estimation of free hydrocarbon volume from fluid levels in monitoring wells. *Ground Water*, 28:57–67.

Luckner, L., and Schestakow, W. M. (1991). *Migration Processes in the Soil and Groundwater Zone*. Lewis Publishers, Chelsea, MI.

Melrose, J. C., and Brandner, C. F. (1974). Role of capillary forces in determining microscopic displacement efficiency for oil recovery by waterflooding. *J. Can. Petrol. Technol.*, 13(42):54–62.

Mercer, J. W., and Cohen, R. M. (1990). A review of immiscible fluids in the subsurface: Properties, models, characterization and remediation. *J. Contam. Hydrol.*, 6:107–163.

Miller, C. T., Poirier-McNeill, M. M., and Mayer, A. S. (1990). Dissolution of trapped nonaqueous phase liquids: Mass transfer characteristics. *Water Resources Res.*, 26(11):2783–2796.

Morrow, N. R., and Songkran, B. (1981). Effect of viscous and buoyancy forces on nonwetting phase trapping in porous media. In D. O. Shah (ed.), *Surface Phenomena in Enhanced Oil Recovery*, pp. 387–411. Plenum, New York.

Ojeda, E., Preston, F., and Calhoun, J. C. (1953). Correlations of oil residuals following surfactant floods. *Producers Monthly*, 18(2):20.

Ostendorf, D. W., Richards, R. J., and Beck, F. P. (1993). LNAPL retention in sandy soil. *Ground Water*, 31(2):285–292.

Palmer, C. D., and Johnson, R. L. (1989). Physical processes controlling the transport of non-aqueous phase liquids in the subsurface. In Transport and Fate of Contaminants in the Subsurface, EPA/625/4-89/019, Chap. 3. U.S. Environmental Protection Agency, September.

Parker, J. C., and Lenhard, R. J. (1987). A model for hysteretic constitutive relations governing multiphase flow, I. Saturation pressure relations. *Water Resources Res., 23:2187–2196.*

Stone, H. L. (1973). Estimation of three-phase relative permeability and residual oil data. *J. Can. Petrol. Technol., 12(4):53–61.*

Van Dam, J. (1967). The migration of hydrocarbons in a water bearing stratum. In P. Hepple (ed.), *The Joint Problems of the Oil and Water Industries*, pp. 55–96. Inst. Petrol., London.

Van Genuchten, M. Th. (1980). A closed form equation for predicting the hydraulic conductivity of unsaturated soils. *Soil Sci. Soc. Am. J.*, 44:892–898.

Wilson, E. J., and Geankopolis, C. J. (1966). Liquid mass transfer at very low Reynolds numbers in packed beds. *Ind. Eng. Chem. Fundam.*, 5(1):9–14.

Wilson, J. L., and Conrad, S. H. (1984). Is physical displacement of residual hydrocarbons a realistic possibility in aquifer restoration? In *Proc., NWWA/API Conf. on Petroleum Hydrocarbons and Organic Chemicals in Ground Water—Prevention, Detection, and Restoration*, pp. 274–298. National Water Well Association, Worthington, OH.

Wilson, J. L., Conrad, S. H., Mason, W. R., Peplinski, W., and Hagan, E., (1990). Laboratory investigation of residual liquid organics from spills, leaks, and the disposal of hazardous wastes in groundwater, EPA/600/6-90/004. U.S. Environmental Protection Agency, Robert S. Kerr Environmental Research Laboratory, Ada, OK.

6
Site Characterization and Contaminant Release Mechanisms

6.1 SITE CONTAMINATION SCENARIOS

Contaminated sites for which remediation schemes are usually designed and implemented vary with respect to configuration, contaminant releasing unit, site hydrology, contaminant source relative to ground elevation, hydrogeological and geotechnical characteristics of the host media, chemical characteristics, and concentrations of contaminants. These factors are determinants of the hazards posed to human health and environment by any contaminated site. The most common site-specific sources of contaminants are

Solid waste management units
Land treatment units
Surface impoundments
Waste piles
Incinerators and other industrial installations
Tanks and other containers

A few of the site contamination scenarios that are associated with the sources listed above are illustrated in Figures. 6.1 and 6.2.

In Figure 6.1A, a landfill serves as the source of contaminants which have leaked through flaws in its liner into the surrounding media. In Figure 6.1B, the source is buried waste. No containment system was used. Consequently, waste constituents have migrated with little or no hindrance into the surrounding soil. The configuration shown in Figure 6.1B is representative of sources at relatively old waste sites initiated before the past 30–40 years, during which waste ''dumping'' in unlined systems was common. Figure 6.1C shows a waste pile that lies directly on an unprotected ground surface. Particles of waste have migrated into the ground. In general, the smaller the particle sizes of the waste relative to those

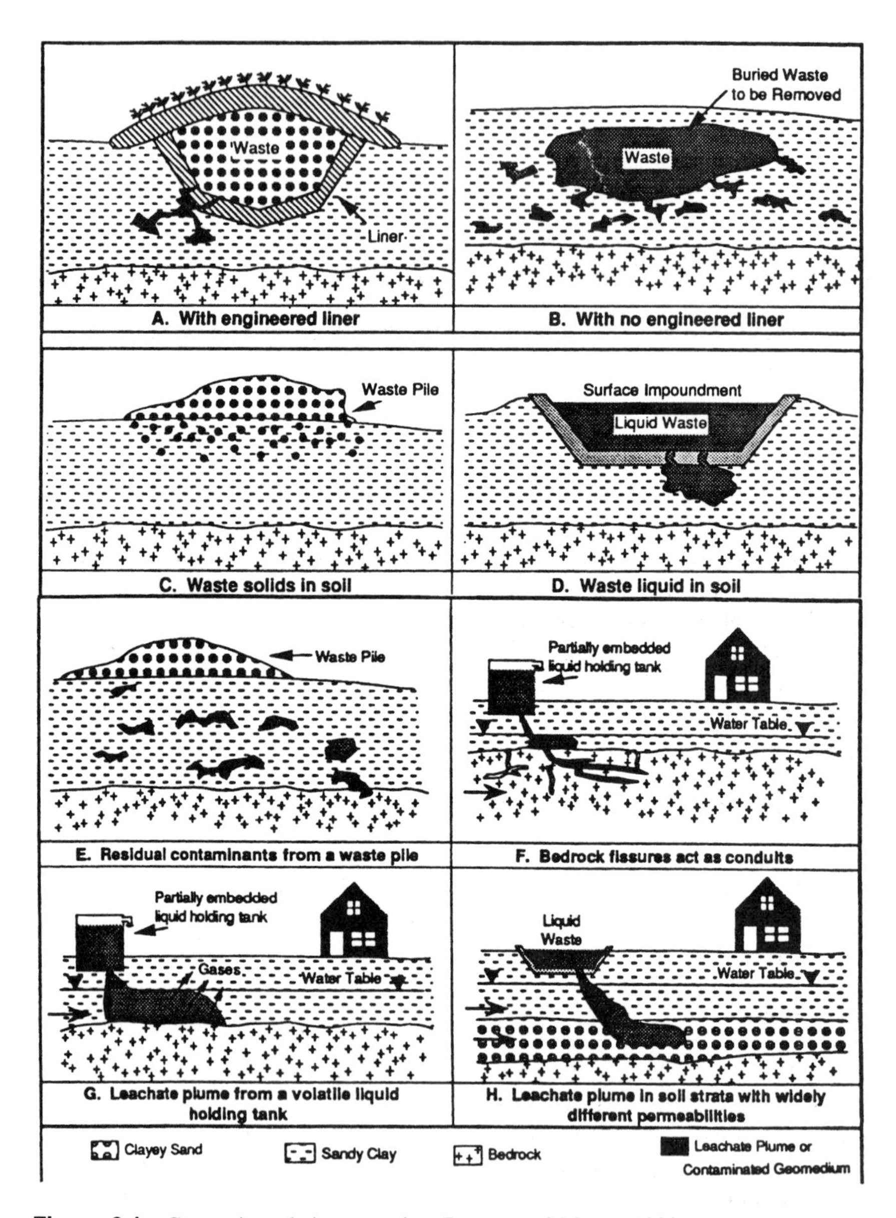

Figure 6.1 Contaminated site scenarios (Inyang and Myers, 1993).

of the ground on which it lies, the greater is the potential for significant migration of waste particles into the ground. Examples of situations where this scenario is common are stockpiling of mine tailings and excavated wastes, and above-ground heaping of industrial byproducts. Figure 6.1E illustrates a similar situation, but in this case the waste is soluble in a liquid transport medium such that dissolved constituents have leached downward and are partially captured by presumably fine-grained soil plugs that are randomly distributed within the underlying geomedium.

Figures 6.1D and 6.1H show lined surface impoundments which are frequently used in liquid waste treatment operations and sludge processing sites. In both illustrations, liquid-phase contaminants have been released. In Figure 6.1F, the travel of the contaminant away from the site is enhanced as it enters a relatively high permeability medium below the water table. Liquid contaminant release from partially embedded tanks is illustrated in Figures 6.1F and 6.1G. In the former, the contaminant is trapped in fractures. This scenario presents a formidable challenge with respect to the effectiveness of clean-up schemes. The liquid that has been released in Figure 6.1G is volatile, hence gases are emitted. Tank leakage scenarios are common at chemical processing plants, refineries, and petroleum dispensing installations. Figure 6.2 illustrates a site that has been re-

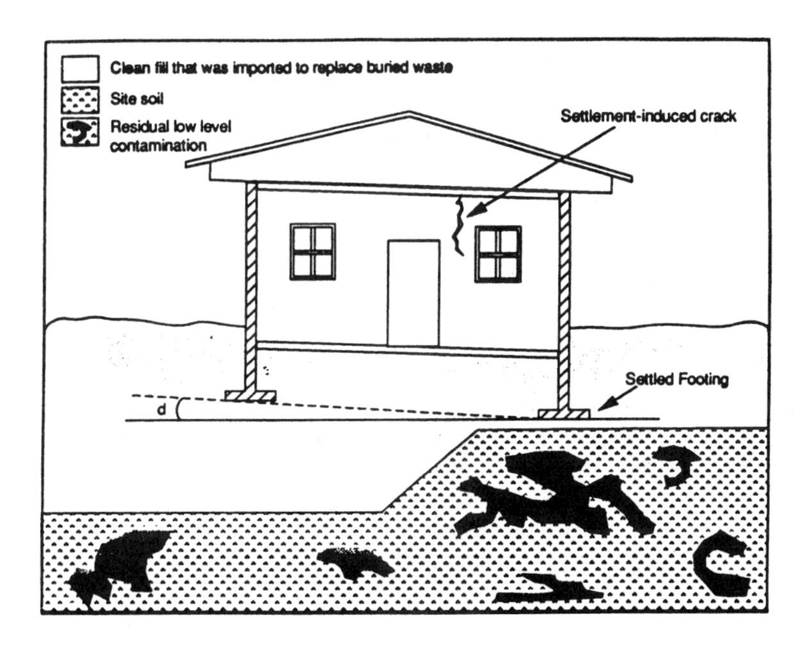

Figure 6.2 Contaminated site scenarios (Inyang and Myers, 1993).

claimed for building construction purposes. Residual contaminants are unevenly distributed in the subsurface. Land recycling "brownfields redevelopment" issues and associated technologies have been described by Yland and van Wachem (1988), Anderson and Hatayama (1988), and Blacklock (1987).

6.2 CHARACTERIZATION OF CONTAMINATED SITES

The objectives of site characterization are multifaceted. They may include one or more of the following:

Assessment of environmental/ecological settings
Determination of the concentrations and spatial distributions of contaminants
Assessment of environmental and human health risks (including exposure)
Determination of the treatability of sites and the most appropriate remediation techniques

Several basic questions usually need to be answered using data obtained during site characterization. Usually, the existence of data alone does not automatically imply that the objectives will be met. Analytical techniques need to be applied to acquired data to seek answers to the following categories of questions:

What are the contaminants?
In what physical forms are they?
What is the source of the contaminants?
Where are the contaminants and their boundaries?
Are they stationary or are they still moving?
What migration pathways are likely to be significant?
What and where are the potential receptors of the contaminants relative to the location of the release area?

It is advisable to develop a systematic approach to conducting site investigations to meet the intended objectives. For the site remediation objective which is frequently the focus of activities, several approaches which spell out the sequence of tasks as exemplified by Figure 6.3 (U.S. EPA, 1991) have been developed. The reader should note that the flowchart presented in Figure 6.3 focuses primarily on the site characterization tasks and relevant analyses needed to select a remediation technology for a contaminated site. The flowchart would be different if the objective were to determine locations of highest hazard at a site.

Some specific numerical ranges of parameters categorized in Table 6.1 have direct and measurable effects on the potential transport of contaminants from a source area. Rosenberg et al. (1990) have indexed the specific ranges for some site and waste characteristics into low, medium, high, and other relevant designa-

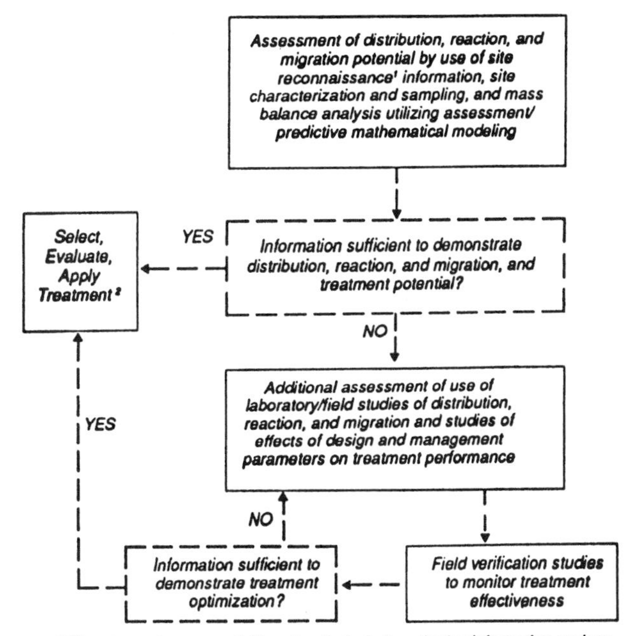

Figure 6.3 Flowchart for contaminated site characterization tasks targeted at remediation (U.S. EPA, 1991).

tions in terms of contaminant migration potential. These qualitative descriptions are presented for liquid and gas migration potentials in Tables 6.2 and 6.3, respectively.

6.2.1 Characterization techniques

Site characterization techniques vary in terms of the level of detail that can be provided about the nature of contamination at a site. Generally, a single technique is usually not adequate for all contaminants in all media. The principal types of media sampled at a site are

Surface and subsurface geomedia
Groundwater
Dusts
Surface water

Table 6.1 Categories of Site Characterization Parameters on Which Information Is Acquired[a]

Data category	Method of data acquisition	
	Lab	Field
Site history and land use pattern		
Facility type and design		
Distribution of population near site		
Proximity of drinking water and surface water resources to site		•
Present and past ownership		
Contaminant release history		•
Applicable regulations and regulatory history of site		
Geologic and hydrologic data		
Proximity to sensitive environments		•
Topography of area		•
Geologic setting of site		
Precipitation data		•
Groundwater depth and flow direction		•
Nature of vegetation		•
Type of soil overburden and bedrock	•	•
Geotechnical data		
Soil profile (thickness and classification)	•	•
Hydraulic conductivities of site soils	•	•
Dispersivities of site soils	•	•
Soil strength parameters	•	•
Chemistry of soils	•	
Waste data		
Water monitoring data	•	•
Size and configuration of contaminated area		•
Type and concentration of contaminants	•	•
Physical and chemical properties (viscosity, solubility, specific gravity, volatility, etc.)	•	•
Partition coefficients	•	
Hazard assessments (toxicity, ignitability, persistence, etc.)	•	

[a] Parameters for which directed measurement is not indicated are those on which information is usually obtained from appropriate agencies.

Table 6.2 Data Ranges and General Effects on Liquid Contaminant Migration from a Polluted Site

Factor	Units	Increasing migration potential →		
Release-specific parameters				
Time since last release	Months	Long (>12)	Medium (1–12)	Short (<1)
Site-specific parameters				
Hydraulic conductivity	cm-s	Low ($<10^{-6}$)	Medium (10^{-5}–10^{-3})	High (>30)
Soil porosity	%	Low (<19)	Medium (10–30)	High (>30)
Soil surface area	cm^2g	High (>50)	Medium (5–50)	Low (<5)
Liquid contaminant content	%	Low (<10)	Medium (10–30)	High (>30)
Soil temperature	°C	Low (<10)	Medium (10–20)	High (>20)
Rock fractures	—	Absent	—	Present
Water content[a]	%	High (>30)	Medium (10–30)	Low (<10)
Contaminant-specific parameters				
Liquid viscosity	CP	High (>20)	Medium (2–20)	Low (<2)
Liquid density	g/cm^3	Low (<1)	Medium (1–2)	High (>2)

[a] Although the referenced authors stated this, it is subject to debate.
Source: Rosenberg et al. (1990).

Chemical sensing and geophysical techniques are increasingly being used to delineate contaminated media without acquiring samples. In this section, focus is placed on techniques for contaminated zone delineation using soil sampling, groundwater sampling, and sensing techniques.

The stratigraphy of contaminated sites is the arrangement of beds of different soils and rocks at a site. It may play a significant role in the rates and directions of contaminant migration. Above the water table, it is usually convenient to sample soils and/or rocks for analyses. Analyses typically include contaminant concentration measurements and geotechnical tests to measure flow-related characteristics such as particle size distribution, hydraulic conductivity, and fracture intensity. Several surface and bore-hole geophysical techniques summarized in Table 6.4 and 6.5, respectively, can be used for hydrogeological char-

Table 6.3 Data Ranges and General Effects on Gas Migration from a Polluted Site

Factor	Units	Increasing migration potential →		
Site-specific parameters				
Air-filled porosity	%	Low (<10)	Medium (10–30)	High (>30)
Total porosity	%	Low (<10)	Medium (10–30)	High (>10)
Water content	%	High (>30)	Medium (10–30)	Low (<10)
Depth below surface	m	Deep (>10)	Medium (2–10)	High (<2)
Contaminant-specific parameters				
Liquid density	g/cm^3	Low (<50)	Medium (50–500)	High (>500)

Source: Rosenberg et al. (1990).

acterization of sites. The principles of some of these techniques are summarized below.

Ground-Penetrating Radar

Ground-penetrating radar (GPR) uses electromagnetic waves to penetrate the ground and delineate differences in pore-water content and quality, soil texture, and soil density. Usually, wave frequencies that range from 80 to 1000 MHz (broad band) are used. Upon contact with materials of different properties, a fraction of the wave energy is reflected back to an antenna located on the ground surface and the remaining fraction travels deeper into the ground, where it is reflected at contacts between beds of dissimilar properties. As illustrated in Figure 6.4, reflected electromagnetic pulses are received by recorders and converted to plots of amplitude versus wave travel time. Travel time is usually of the order of 10^{-9} s.

The depth of a reflecting bed in the subsurface can be calculated as

$$d = \frac{tv}{2} \tag{6.1}$$

where

d = depth to the reflector (L)
t = wave travel time in both directions (T)
v = wave velocity (L/T)

Table 6.4 Principal Surface Geophysical Techniques and Their Applications

Surface geophysical survey method	Applications	Advantages	Limitations
Seismic Refraction and Reflection Determines lithological changes in subsurface	Groundwater resource evaluations Geotechnical profiling Subsurface stratigraphic profiling including top of bedrock	Relatively easy accessibility High depth of penetration dependent on source of vibration Rapid areal coverage	Resolution can be obscured in layered sequences Susceptibility to noise from urban development Difficult penetration in cold weather (depending on instrumentation) Operation restricted during wet weather
Electrical resistivity Delineates subsurface resistivity contrasts due to lithology, groundwater, and changes in groundwater qualtiy	Depth to water table estimates Subsurface stratigraphic profiling Groundwater resource evaluations High ionic strength contaminated groundwater studies	Rapid areal coverage High depth of penetration possible (400–800 ft) High mobility Results can be approximated in the field	Susceptibility to natural and artificial electrical interference Limited use in wet weather Limited utility in urban areas Interpretation that assumes a layered subsurface Lateral heterogeneity not easily accounted for
Ground-penetrating radar Provides continuous visual profile of shallow subsurface objects, structure, and lithology	Locating buried objects Delineation of bedrock subsurface and structure Delineation of karst features Delineation of physical integrity of man-made earthen structures	Great areal coverage High vertical resolution in suitable terrain Visual picture of data	Limited depth of penetration (a meter or less in wet, clayey soils; up to 25 m in dry, sandy soils) Accessibility limited due to bulkiness of equipment and nature of survey Interpretation of data qualitative Limited use in wet weather
Magnetics Detects presence of buried metallic objects	Location of buried ferrous objects Detection of boundaries of landfills containing ferrous objects Location of iron-bearing rock	High mobility Data resolution possible in field Rapid areal coverage	Detection dependent on size and ferrous content of buried object Difficult data resolution in urban areas Limited use in wet weather Data interpretation complicated in areas of natural magnetic drift.

Source: Adapted from U.S. EPA (1991) and O'Brien and Gere Engineers (1988).

Table 6.5 Summary of Important Site Characterization Parameters and Bore-Hole Techniques for Their Measurement

Required information on the properties of rocks, fluid, wells, or the groundwater system	Widely available logging techniques that might be utilized
Lithology and stratigraphic correlation of aquifers and associated rocks	Electric, sonic, or caliper logs made in open holes; nuclear logs made in open or cased holes
Total porosity or bulk density or gamma-gamma	Calibrated sonic logs in open holes, calibrated neutron logs in open or cased holes
Effective porosity or true resistivity	Calibrated long-normal resistivity logs
Clay or shale content	Gamma logs
Permeability	No direct measurement by logging; may be related to porosity, injectivity, sonic amplitude
Secondary permeability—fractures, solution openings	Caliper, sonic, or bore-hole televiewer or television logs
Specific yield of unconfined aquifers	Calibrated neutron logs
Grain size	Possible relation to formation factor derived from electric logs
Location of water level or saturated zones	Electric, temperature, or fluid conductivity in open hole or inside casing; neutron or gamma-gamma logs in open hole or outside casing
Moisture content	Calibrated neutron logs
Infiltration	Time-interval neutron logs under special circumstances or radioactive tracers
Direction, velocity, and path of groundwater flow	Single-well tracer techniques—point dilution and single-well pulse; multiwell tracer techniques
Dispersion, dilution, and movement of waste	Fluid conductivity and temperature logs, gamma logs for some radioactive wastes, fluid sampler
Source and movement of water in a well	Injectivity profile; flowmeter or tracer logging during pumping or injection; temperature logs
Chemical and physical characteristics of water, including salinity, temperature, density, and viscosity	Calibrated fluid conductivity and temperature in the well; neutron chloride logging outside casing; multielectrode resistivity
Determining construction of existing wells, diameter and position of casing, perforations, screens	Gamma-gamma, caliper, collar, and perforation locator, bore-hole television
Guide to screen setting	All logs providing data on the lithology, water-bearing characteristics, and correlation and thickness of aquifers
Cementing	Caliper, temperature, gamma-gamma; acoustic for cement bond
Casing corrosion	Under some conditions, caliper or collar locator
Casing leaks and/or plugged screen	Tracer and flowmeter

Source: (Adapted from U.S. EPA (1991) and Keys and MacCary 1971).

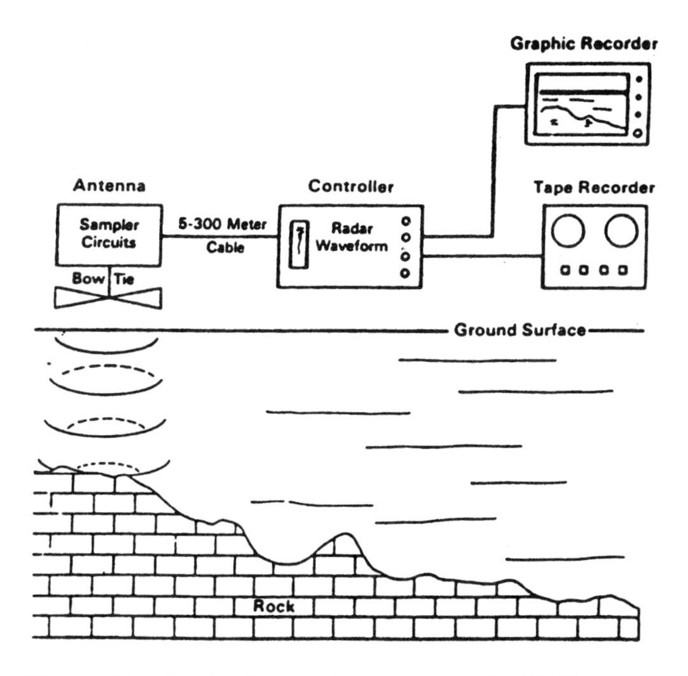

Figure 6.4 Sketch of a ground-penetrating radar (GPR) system (Evans, 1982).

considering that

$$v = \frac{C}{\varepsilon^{0.5}} \tag{6.2}$$

where

C = velocity of light in a vacuum (L/T)
ε = dielectric permittivity of the transport medium (unitless)

Then,

$$d = \frac{tC}{2\varepsilon^{0.5}} \tag{6.3}$$

Generally, soils that have high conductivities to electromagnetic waves dissipate radar energy quickly such that penetration depths are limited. Beres and Haeni (1991) attribute attenuation to spherical spreading losses, exponential losses due to the conversion of electromagnetic energy to thermal energy, and dielectric relaxation losses. In general, penetration depths up to 10 m are com-

Table 6.6 Approximate Values for Electromagnetic Parameters of Various Media

Material	Approximate conductivity σ(mho/m)	Approximate dielectric Constant, ε_r	Two-way travel time (ns/m)
Air	0	1	6.6
Fresh water	10^{-4}–3×10^{-2}	81	59
Fresh water ice	10^{-4}–10^{-2}	4	13
Permafrost	10^{-5}–10^{-2}	4–11	13–15
Granite	10^{-9}–10^{-3}	5.6–8	18.7
Dry sand	10^{-7}–10^{-3}	4–6	13–16
Sand, saturated (fresh water)	10^{-4}–10^{-2}	30	36
Silt, saturated (fresh water)	10^{-3}–10^{-2}	10	21
Clay, saturated (fresh water)	10^{-1}–1	8–25	18.6–23
Average "dirt"	10^{-4}–10^{-2}	16	23–30

Source: Benson et al. (1984).

mon. About 5–10% (by weight) of montmorillonite clay can reduce penetration depth to about 1 m (Walther et al., 1986). Table 6.6 shows typical values of dielectric constants, approximate conductivities, and two-way travel times for a variety of media. The conductivity of any geomedium is directly proportional to the concentration of dissolved salts in its pore solution. Operationally, the receiving antenna is usually towed in the field along preselected traverse lines at speeds up to 8 km/h. In Port Washington, New York, GPR was used at the Roslyn Beacon Hill Landfill Site to optimize monitoring well layout design (Kardos and Ennis, 1993).

Electromagnetic Resistivity

The electromagnetic resistivity method is also based on the electrical conductivity contrasts that usually exist between zones of different physical and chemical properties in the ground. It is not necessary to have electrodes embedded in the ground to transmit and receive current. Electromagnetic phenomena are generated in the media investigated. The method is an effective means of delineating contaminated zones beneath covered areas such as structural foundations and paved areas, because there is no requirement for the use of direct contact electrodes.

The principle of inducement of an electromagnetic field in geomedia for resistivity measurements is illustrated in Figure 6.5. A transmitter coil is held or fixed on or close to the ground surface. An alternating current is applied to the terminals of the coil to induce the flow of a current. This phenomenon generates an alternating magnetic field, which in turn causes electric currents to permeate the earth. Within the earth, a secondary magnetic field is induced. The secondary and primary magnetic fields are detected by a receiving coil placed near the transmitting coil. For a fixed intercoil spacing and operating frequency, the magnitude of the secondary magnetic field (or ratio of the secondary to primary magnetic fields) is directly proportional to the conductivity of the ground. Resistivity can be computed from conductivity data. In a geomedium, the solid particles act as insulators while soil moisture acts largely as a conductor of the electric current. Furthermore, the conductivity of the pore fluid is proportional to the concentration of ions within the fluid. Thus, these measurements can be used to delineate contaminated zones around waste containment barriers. If metallic objects are present near the surveyed site, the readings recorded may be negatively affected. The effective depth of penetration can be as high as 60 m. This method is often called the frequency domain electromagnetic (FDEM) method.

There are several variations of the electromagnetic induction method. The decay rate of magnetic field after the transmission is turned off can be measured. An eddy current flows through the ground at successively greater depths. The data obtained are interpreted to obtain resistivity variation with depth. This is called time domain electromagnetic (TDEM) survey. It has been applied by Hoekstra et al. (1992) to trace the migration of brines from an oilfield brine pit in southwestern Texas. Penetration depths of up to 500 m are estimated by Hoekstra et al. (1992). Another type of (EM) measurement involves the use of very low frequencies (15–25 kHz). In this method, ground contact is required for

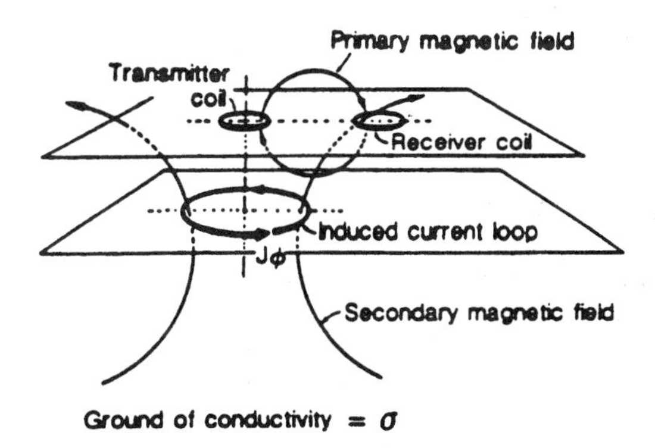

Figure 6.5 Induced electromagnetic fields (McNeil, 1982).

the potential electrodes. It is very suitable for investigating relatively shallow contaminant plumes (20–50 m).

Direct-Current Resistivity Method

Except for the means by which an electric current is generated, the direct-current (DC) resistivity method is similar to the electromagnetic resistivity method. Both are based on contrasts in the electrical conductivity or resistivity of geomedia with different textures, moisture contents, and contamination levels. In DC resistivity measurements, voltage is applied across a pair of electrodes that are embedded in the soil. Directly, this causes an electric current to flow through the soil. A pair of receiving electrodes is used to record the incoming voltage (from the soil). At least three different configurations of electrodes (arrays) exist, as discussed by Hempen and Hatheway, 1992, but the Wenner array illustrated in Figure 6.6 is commonly used in geoenvironmental site characterization. The depth of penetration of current is approximated by the interelectrode spacing (McNeil, 1982). The resistivity is computed as follows:

$$\rho = 2\pi a \left(\frac{V}{I} \right) \tag{6.4}$$

where

ρ = resistivity of the soil (ohm-m)
a = interelectrode spacing (m)

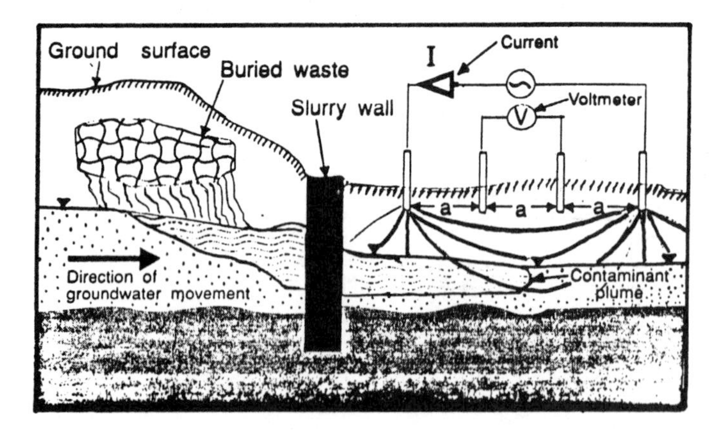

Figure 6.6 Schematic representation of the configuration of direct-current resistivity electrodes for contaminant plume remediation (Inyang, 1994).

I = current flow through outer electrodes (A)
V = voltage across inner electrodes (V)

The relationship between the physicochemical and textural characteristics of a medium and its resistivity is represented by Archie's law. It is given as

$$\rho = \frac{\rho_w}{\phi^{\mu} S^2} \tag{6.5}$$

where

ρ = resistivity of the geomedium
ρ_w = resistivity of water in the medium
ϕ = porosity of the medium
μ = cementation constant of the medium $\approx$ 1.3 for consolidated sands
S = degree of saturation of the medium with water

Figure 6.7 (Rhoades and Halvorson, 1977) shows the relationship between electrode configuration and the area monitored. The flowchart of resistivity measurements and data analysis steps is shown in Figure 6.8 (Nielson, 1991). Vertical resistivities of horizontally bedded media are represented by ρ_1, ρ_2, ρ_3, and ρ_4.

For any four-electrode configuration such as the Werner system, two categories of measurements can be conducted:

Sounding: The electrodes that transmit current are placed at increasing distances apart and measurements are made. Deeper parts of a geomedium are investigated using wider spacings.

Profiling: The entire array is moved from one location to another using the same electrode spacing. This provides an opportunity to assess lateral variations in electrical resistivity of media.

At contaminated sites, both lateral and vertical variabilities in geochemistry are often found. As shown in Figure 6.9, a geoelectrical soil profile that is conceptually stratified in terms of electrical resistivity may be considered for the purpose of estimating a composite value of resistivity. Dar Zarrouk parameters configured by Maillet (1947) as presented below are used for this purpose. Following the treatment by van Zijl (1978) in conjunction with Figure 6.9,

$$T = \sum_{i=1}^{n-1} \rho_i h_i \tag{6.6}$$

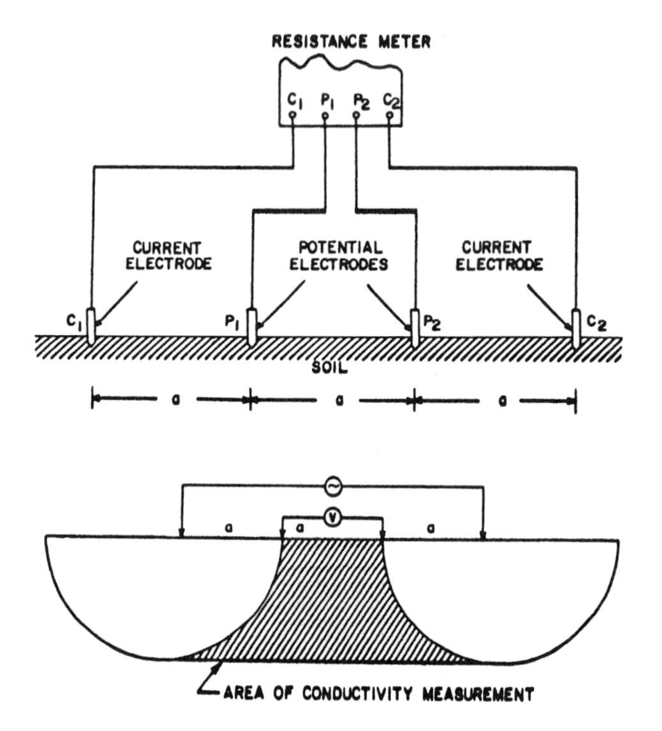

Figure 6.7 Relationship between electrode arrangement and area monitored (Rhoades and Halvorson, 1977).

where

T = total transverse resistance of all beds (ohm-m^2)
ρ_i = resistivity of bed i (ohm-m)
h_i = thickness of bed i (m)

For current that flows horizontally through any layer i,

$$\frac{1}{R_i} = \frac{1}{\rho_i} \cdot \frac{A}{L} \tag{6.7}$$

Considering that A, the cross-sectional area perpendicular to electrical flow, equals $(h_i)(I)$ when a unit width of the medium is considered, S_i, the longitudinal conductance, can be computed as

$$S_i = \frac{h_i}{\rho_i} \tag{6.8}$$

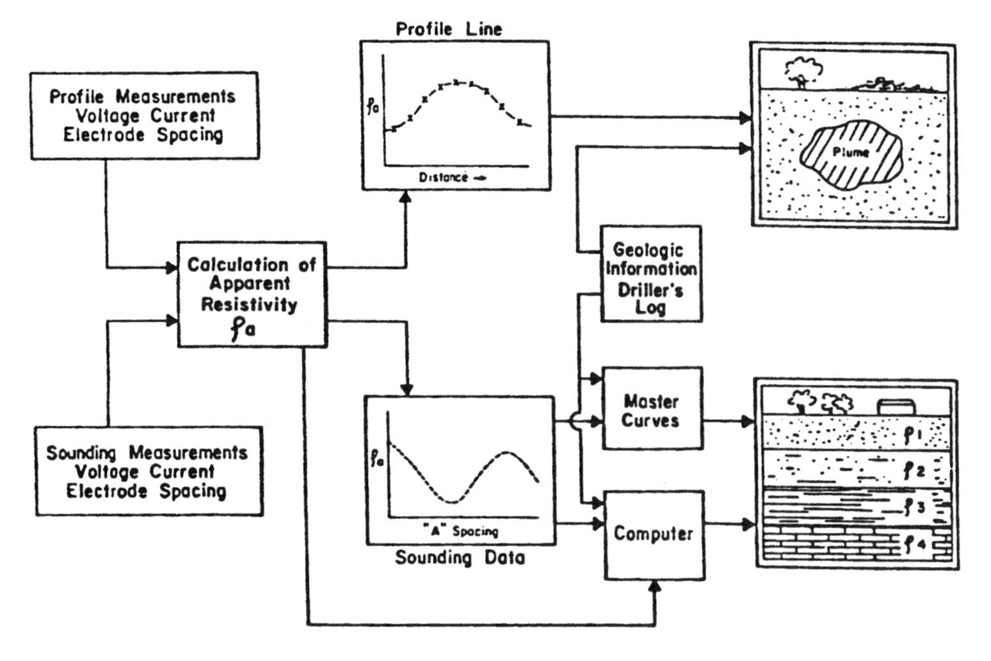

Figure 6.8 Flowchart of resistivity measurement and data analysis steps (Nielson, 1991).

and for all beds,

$$S = \sum_{i=1}^{n-1} \frac{h_i}{\rho_i} \tag{6.9}$$

where S = total longitudinal conductance (siemens).

The average transverse resistivity and average longitudinal resistivity can be computed using Eqs. (6.10) and (6.11), respectively:

$$\rho_a = \frac{T}{H} = \frac{\Sigma h\rho}{\Sigma h} \tag{6.10}$$

$$\rho_L = \frac{H}{S} = \frac{\Sigma h}{\Sigma h/p} \tag{6.11}$$

If a geomedium of very high vertical variability in electrical properties is evaluated, its anisotropy can be expressed as

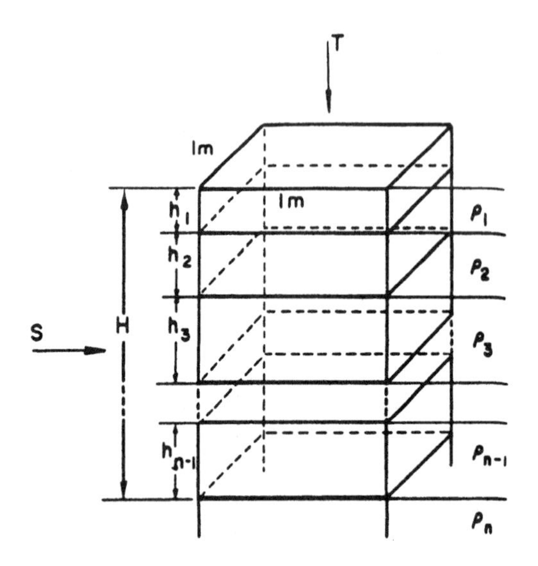

Figure 6.9 Schematic representation of a geoelectrical section for composite apparent resistivity analysis (van Zijl, 1978).

$$\lambda = \sqrt{\frac{\rho_a}{\rho_L}} = \sqrt{\frac{TS}{H^2}} \tag{6.12}$$

For practical purposes, it may be convenient to consider the entire section as having a resistivity value ρ_c that may be estimated as in Eq. (6.13):

$$\rho_c = \sqrt{\rho_a \rho_L} \tag{6.13}$$

Then

$$T = \lambda H \cdot \rho_c \tag{6.14}$$

$$S = \frac{\lambda H}{\rho_c} \tag{6.15}$$

and

$$\rho_c = \sqrt{\frac{T}{S}} = \lambda \rho_L \tag{6.16}$$

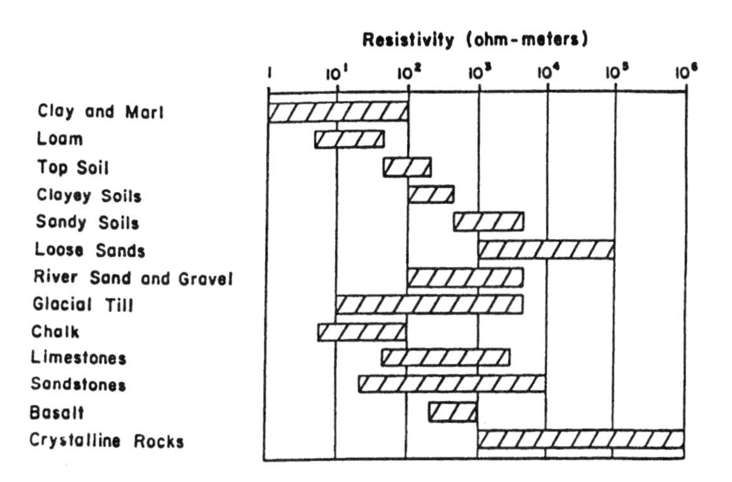

Figure 6.10 Typical ranges of resistivity data for various geomedia (Nielsen, 1991).

Figure 6.10 and Table 6.7 show typical ranges of resistivities of aqueous and geological media.

Seismic Methods

Seismic methods are useful for assessment of the structural characteristics of a contaminated site. They are not suitable for use in contaminated zone delineation because contaminants do not exhibit sufficient variations in their responses to seismic waves. Elastic waves generated by seismic sources travel faster in dense, tightly cemented and wet materials than in loose dry materials, hence seismic wave velocity is directly proportional to the density of the transport medium. This principle can be applied to the identification of potential migration pathways for contaminants in the subsurface. Equations (6.17), (6.18), and (6.19) show the relationships between seismic wave velocities and geological properties.

$$V_p = \left(\frac{k}{\rho}\right)^{0.5} \tag{6.17}$$

$$V_p = \left(\frac{4\mu/3}{\rho}\right)^{0.5} \tag{6.18}$$

$$V_s = \left(\frac{\mu}{\rho}\right)^{0.5} \tag{6.19}$$

Table 6.7 Typical Ranges of Resistivity Data for Various Aqueous and Geological Media

Earth material[a]	Resistivity (Ω-m)
Saline groundwater	0.01–1
Clay soil	1–30
Fresh groundwater	2–50
Calcareous shale (or chalk)	10–100
Sand (SP, moderately to highly saturated)	20–200
Shale (mudstone/claystone)	1–500
Shale (siltstone)	50–1,000
Limestone (low-density)	100–1,000
Volcanic flow rock (scoriaceous basalt)	300–1,000
Lodgement (dense, clayey, basal) till	50–5,000
Sandstone, uncemented	30–10,000
Ablation (dry, loose, chsionless) till	1,000–10,000
Fluvial sands and gravels (GW, unsaturated)	1,000–10,000
Loose, poorly sorted sand (SP, unsaturated)	1,000–100,000
Metamorphic rock	50–1,000,000
Crystalline igneous rock	100–1,000,000
Limestone (high-density)	1,000–1,000,000

[a] GW and SP are Unified Soil Classification terms for well-graded gravel and poorly graded sand, respectively.
Source: Hempen and Hatheway (1992).

where

V_p = compressional wave velocity (L/T)
V_s = shear wave velocity (L/T)
k = bulk modulus of the medium
μ = shear modulus of the medium
ρ = density of the medium (M/L^3)

Table 6.8 (Hempen and Hatheway, 1992) and Figure 6.11 (Nielsen, 1991) show typical seismic wave velocity ranges for geomedia. Surface refraction method has utility in the delineation of bedrock boundaries at contaminated sites. It is generally effective within a depth range of about 30 m. The operating principle of this method is that compressional waves are bent at the interface between media of significantly different physical characteristics. Among such characteristics are elasticity, density, and other parameters which can be correlated with them such as porosity and degree of cementation. Usually, a hammer hit or drop weight is used to generate waves in the case of shallow investigations at waste

Table 6.8 Typical Ranges of Compressional Wave Velocities for Various Media

Material	mps	fps
Air	330	1,100
Loam, dry	180–300	600–900
Loam, wet	300–750	1,000–2,500
Sand, dry	450–900	1,500–3,000
Gravel	600–800	2,000–2,600
Sand, cemented	850–1,500	2,800–5,000
Sand, loose saturated	1,500	5,000
Water (shallow)	1,450–1,600	4,800–5,100
Clayey soil, wet	900–1,800	3,000–6,000
Till, basal/lodgment	1,700–2,300	5,600–7,500
Rock, weathered sedimentary	600–3,000	2,000–10,000
Rock, weathered igneous and metamorphic	450–3,700	1,500–12,000
Shale	800–3,700	2,600–12,000
Sandstone	2,200–4,000	7,200–13,000
Basalt, fresh	2,600–4,300	8,500–14,000
Metamorphic rock	2,400–6,000	8,000–20,000
Steel	6,000	20,000
Dolostone and limestone, fresh	4,300–6,700	14,000–22,000
Granite, fresh	4,800–6,700	16,000–22,000

Source: Hempen and Hatheway (1992).

sites. Geophones which are placed as illustrated in Figure 6.12 (Benson et al., 1984) detect incoming refracted waves. This information is processed using the circuit shown in Figure 6.13 (Bensen et al., 1984).

Time–distance graphs are usually plotted and straight lines used to connect points using the analyzer's interpretation of subsurface structures. Changes in the slope of lines indicate differences in wave velocities and hence material characteristics of different layers of the subsurface. A critical assumption in the use of seismic refraction is that, within each stratum, density increases with depth. Generally, where denser strata underly less dense strata, interpretation is easier. Figure 6.14 illustrates the use of time–distance plots to analyze the geological stratification of subsurface media.

Electrochemical and Electrooptical Sensing Methods

Electrochemical and electrooptical sensing systems are the most recently developed approaches to detecting contaminants in the subsurface. They are based on

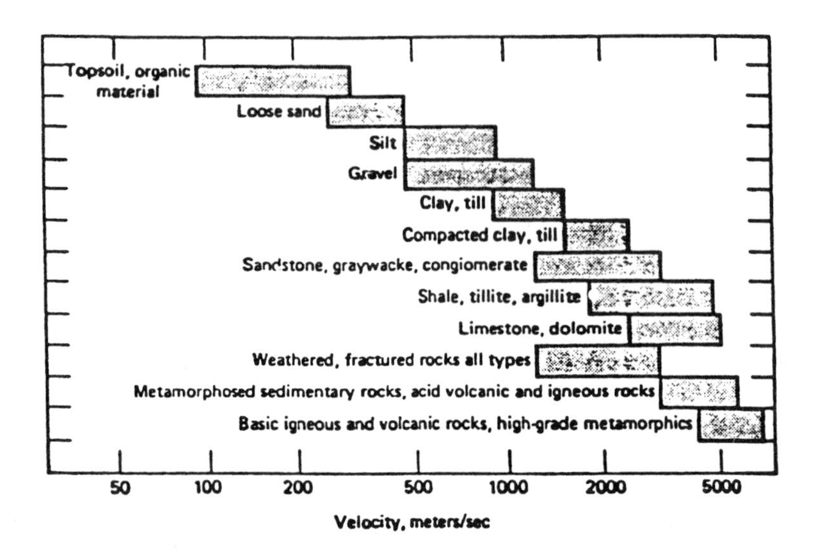

Figure 6.11 Typical ranges of compressional wave velocities for various geomedia (Nielsen, 1991).

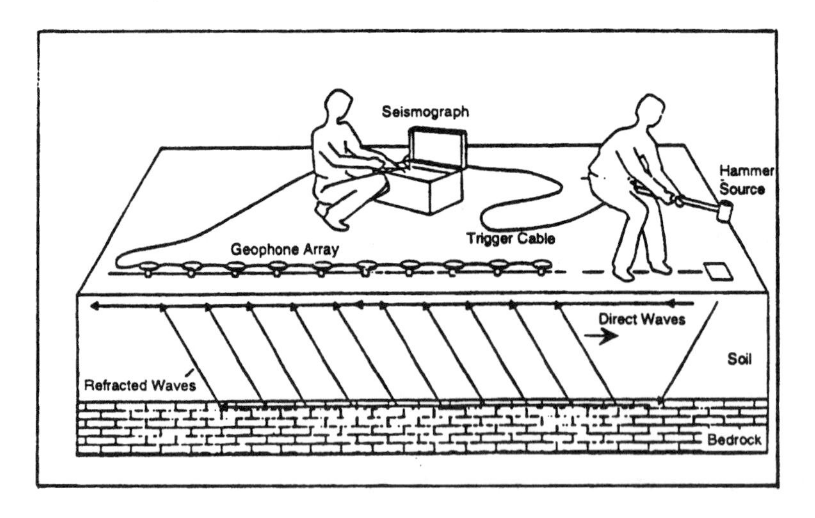

Figure 6.12 Layout of a 12-channel seismograph on a two-layered solid/rock subsurface site (Benson et al., 1984).

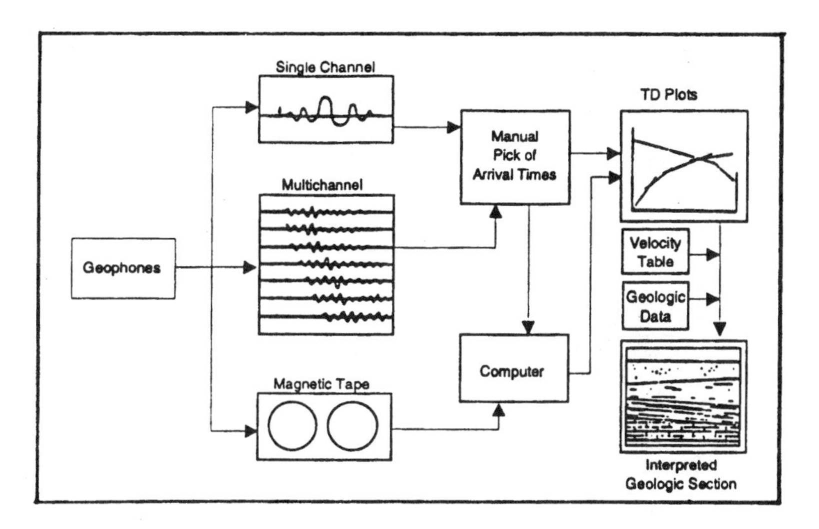

Figure 6.13 Flow diagram of the processing and interpretation of seismic refraction data (Benson et al., 1984).

the sensitivity of the detector to variations in the moisture content or concentrations of the target contaminants in the subsurface. Unlike geophysical and direct-current resistivity methods, the sensor has to establish contact with the contaminant or moisture in order to record a reading. Essentially, the fluid of concern must travel to sensing points before detection can occur. The closer the spacing among the sensors, the smaller is the probability that the target fluid will short-circuit the sensors. In terms of spatial coverage of a contaminated site, sensing systems can be categorized as follows.

Point sensor: The system comprises a stem which terminates in a single sensor.

Linear sensor: The system comprises a set of sensors that are arranged in single line, such as a cable.

Areal sensor: A network of sensors comprising point and/or linear systems.

Several principles are employed in the development of the sensing head, some of which are described briefly below.

Chemical paste: The sensor head is a paste capable of changing color when in contact with a contaminant. The sensor is retrieved from the well and evaluated using a colorimetric chart.

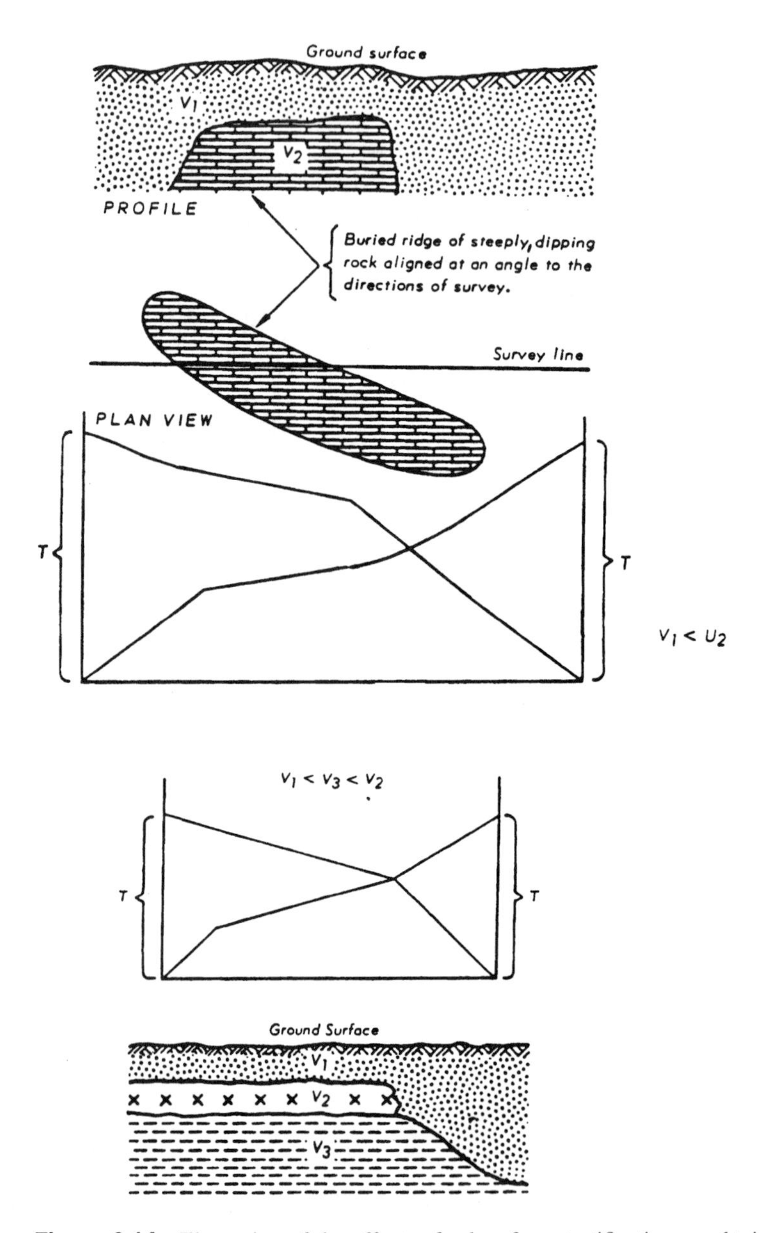

Figure 6.14 Illustration of the effects of subsurface stratification on obtained seismic velocity (v) plots for a buried dense ridge (top) and a dense ledge (bottom) (Henbest et al., 1971).

Permeable coating: The coating repels water and some contaminants, but is permeable by others which activate a signal upon entry into the sensing head.

Photo ionization: Molecules of gases emitted from contaminants are ionized when they absorb ultraviolet light. The ionization parameters are scaled against concentration. Inyang et al. (1992) have described the principle of photoionization and its application to hydrocarbon vapor detection. Generally,

$$R + h\nu \Rightarrow R^{+} + e^{-} \quad (6.20)$$

where

R = the species prior to ionization
$h\nu$ = photon with energy in excess of the species ionization potential
R^{+} = ionized species
e^{-} = released electron

Generally, photon energies of the order of 10.2 eV are used for compounds that may be present in petroleum-contaminated media.

Fiber-optic sensing: These sensors are based on the transmission of probe signals within the visible light regime by optical fibers through very long distances to embedded sensors. Wavelength-dependent optical attenuation of the probe beam or production of fluorescent emissions by the contaminant is measured and scaled as an index of the presence of the contaminant. Often, an organic dye is added to the sensor to promote fluorescence upon reaction with low concentrations of the target compound. Most of the relevant reactions are reversible, making it possible to sense new contaminant concentration levels. In general, the Beer-Lambert law can be applied to the optical absorbency of the detected compounds.

$$I_c = I_0 \exp[-a(\lambda,\mathrm{pH})Cd] \quad (6.21)$$

where

I_c = transmitted light intensity due to the presence of the contaminant
I_0 = light source intensity
a = molar absorptivity of the compound
λ = selected wavelength
d = optical path length
C = velocity of light

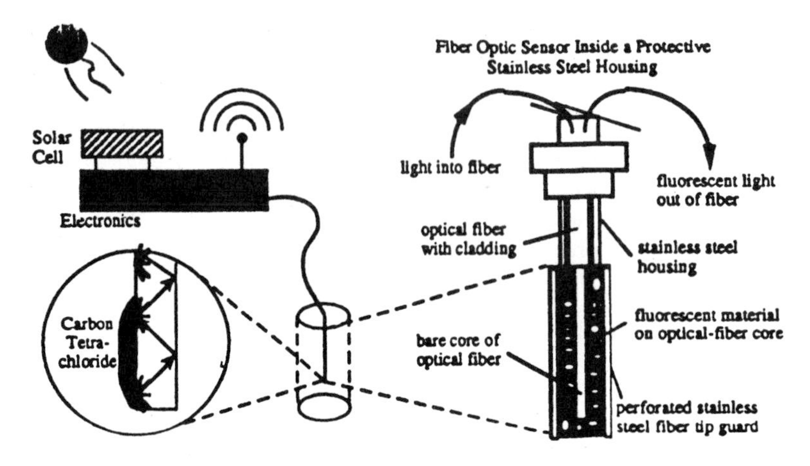

Figure 6.15 Schematic of reversible fiber-optic carbon tetrachloride sensor (U.S. DOE, 1994).

Figure 6.15 shows the details of a reversible fiber-optic sensor for carbon tetrachloride developed by Lawrence Livermore National Laboratory (U.S. DOE, 1994). Usually, fiber-optic sensors are designed for specific contaminants.

Traditional Soil and Groundwater Sampling Methods

Several agencies in various countries have developed protocols and test techniques for soil and sediment sample coring and groundwater well construction and sampling. These techniques, which are discussed in technical guidance documents, textbooks, and technical papers (U.S. EPA, 1991; Gibbons, 1994; Wilson, 1995; Boulding, 1994; Pungor, 1995), focus on sampling point selection, sample collection, sample transfer and preservation, and methods of instrument analysis.

6.3 GEOSTATISTICAL APPLICATIONS

Geomedia are characterized by variabilities at various spatial scales. The existence of contaminants at spatially and temporally variable concentrations further enhances the complexity and associated uncertainties of a contaminated site. The ground is opaque. Therefore, certain inferences have to be made about the size of the contaminated site and variabilities in the degree of contamination within the site through interpretation of acquired data. Inevitably, these analyses require interpolation of data because cost considerations place a constraint on the number of samples that can be obtained for testing. These interpolations and their representation through contouring of concentrations may be a requirement for the selection and design of potentially effective remediation schemes. Uncertainties about the nature of the contamination at a site can be minimized through the application of geostatistical techniques to sampling design and analysis.

6.3.1 Spatial Distribution of Monitoring Points

Monitoring points should be distributed such that contaminant concentrations and hydrogeological parameters can be estimated for the entire three-dimensional space using measurements at the selected points. It is desirable that monitoring network design be such that biases in the interface about hydrogeological and contamination scenarios at the site are minimal. Two important geostatistical parameters are critical with respect to monitoring point distribution.

Sampling pattern: This is the geometric configuration of monitoring points in three-dimensional space. Often, more emphasis is placed on the areal distribution of projections of monitoring points on the ground surface.

Sampling density: This is the number of sampling points per space unit. The space unit could be a unit area or a unit volume of the sampled geomedia.

Figure 6.16 shows various types of monitoring networks with descriptions of their utilities as described by the U.S. EPA (1991) and by Gilbert (1987). In general, samples may be taken along transects (lines) from within grids. Some general knowledge or informed suspicion about the hydrogeological conditions of a site is necessary for the selection of an effective monitoring network. The source of such information could be reconnaissance surveys and/or historical records as outlined in Table 6.1.

Sampling point density is another critical determinant of sampling efficiency. For a given network, sampling efficiency is directly proportional to sampling point density. Olea (1984) estimates that the efficiency of the network with a higher density relates to that of the lower density in consistence with Eq. (6.22):

$$\Delta E = E_h - E_L = \left(\frac{D_L}{D_h}\right)^{0.5} \tag{6.22}$$

where

ΔE = difference between sampling efficiencies of two networks
E_h = sampling efficiency of the higher-density network
E_L = sampling efficiency of the lower-density network
D_L = lower sampling point density (number/space unit)
D_h = higher sampling point density (number/space unit)

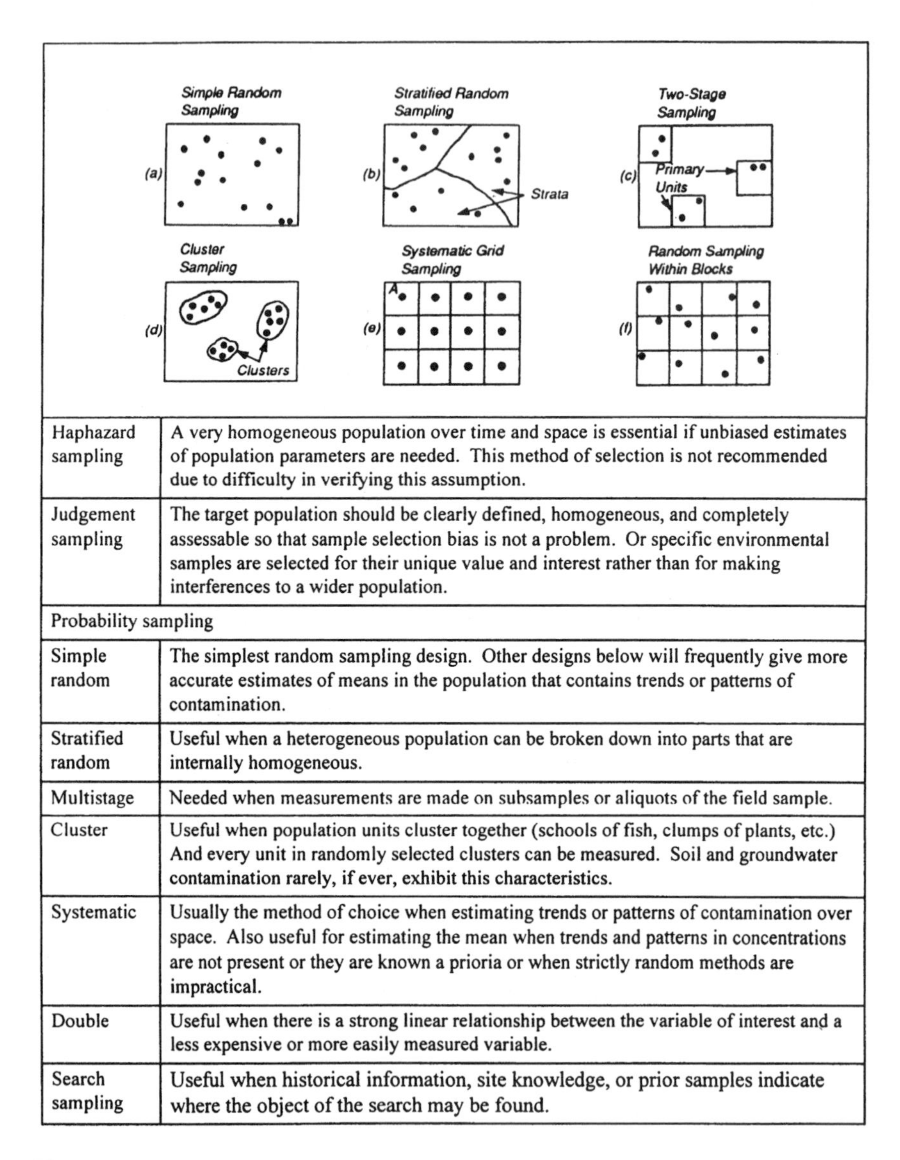

Haphazard sampling	A very homogeneous population over time and space is essential if unbiased estimates of population parameters are needed. This method of selection is not recommended due to difficulty in verifying this assumption.
Judgement sampling	The target population should be clearly defined, homogeneous, and completely assessable so that sample selection bias is not a problem. Or specific environmental samples are selected for their unique value and interest rather than for making interferences to a wider population.
Probability sampling	
Simple random	The simplest random sampling design. Other designs below will frequently give more accurate estimates of means in the population that contains trends or patterns of contamination.
Stratified random	Useful when a heterogeneous population can be broken down into parts that are internally homogeneous.
Multistage	Needed when measurements are made on subsamples or aliquots of the field sample.
Cluster	Useful when population units cluster together (schools of fish, clumps of plants, etc.) And every unit in randomly selected clusters can be measured. Soil and groundwater contamination rarely, if ever, exhibit this characteristics.
Systematic	Usually the method of choice when estimating trends or patterns of contamination over space. Also useful for estimating the mean when trends and patterns in concentrations are not present or they are known a prioria or when strictly random methods are impractical.
Double	Useful when there is a strong linear relationship between the variable of interest and a less expensive or more easily measured variable.
Search sampling	Useful when historical information, site knowledge, or prior samples indicate where the object of the search may be found.

Figure 6.16 Types of sampling patterns and their utilities. (Adapted with modifications from U.S. EPA, 1991; Gilbert, 1987.)

6.3.2 Contaminated Zone Contouring Using Measured Data

One of the most useful geostatistical techniques for contouring of spatially variable site characteristics data is kriging. This is a method of estimating the magnitudes of parameters for a network of points given measurements on points that do not necessarily coincide with the network points. In essence, kriging is a form of spatial interpolation which is useful for contouring of geostatistical data.

As represented in Figure 6.17, Z_0 is an unknown value of a parameter to be estimated from measured values Z_1, Z_2, Z_3. For the two-dimensional case, C_{1xy}, C_{2xy}, and C_{3xy} are designated as corresponding coordinates. Similarly, the coordinates of Z_0 can be designated as C_{0xy}. Each of the measured values contributes to Z_0^*, the estimated value of Z_0. The general kriging equation can be written as

$$Z_0^* = \sum_{i=1}^{n} \lambda_i Z_i \tag{6.23}$$

$$\sum_{i=1}^{n} \lambda_i \tag{6.24}$$

where

Z_0^* = estimated magnitude of unmeasured parameter at interpolation location C_{0xy}

Z_i = measured values of the parameter at locations that surround C_{0xy}: locations such as C_{1xy} and C_{2xy}

λ_i = weight of the kriging estimator

The weighting factors can be developed by direct correlation with distances of measurement points from C_{0xy}. Additional information on kriging techniques is provided by de Marsily (1986), Journel (1989), and Jury (1986).

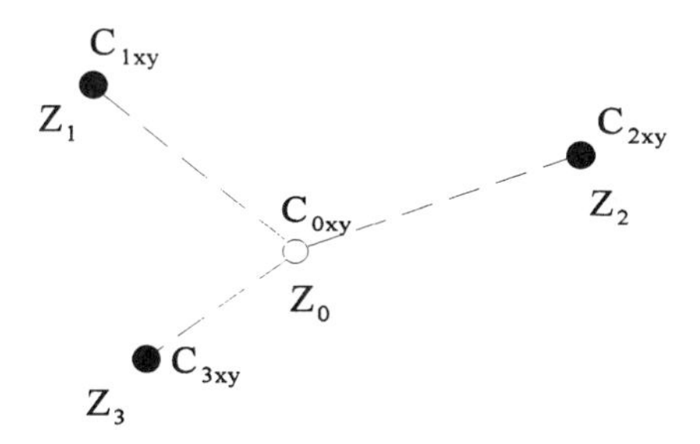

Figure 6.17 Illustration of the use of measured data for spatial interpolation.

6.4 CONTAMINANT RELEASE MECHANISMS: VAPORIZATION

Vaporization is the release of contaminants from a solid or liquid or mixed medium into the atmosphere. Vaporization usually involves a transition from solid or liquid into the vapor phase. It is driven by molecular transfer processes that are embodied in Henry's law for liquid–air interfaces. Henry's law states that for dilute solutions, the vapor pressure of any chemical is directly proportional to its concentration. Vapor pressure can be regarded as a fundamental characteristic of compounds (especially organics). It is the pressure exerted by the vapor of a compound when the vapor is in equilibrium with the liquid or solid phase of the compound. Equilibrium is defined here as a state in which the number of molecules that evaporate from the liquid or solid is equal to the number of molecules that return to these phases through condensation.

As discussed by Inyang et al. (1992), molecular structures of compounds influence the magnitude of vapor pressure and hence their rates of vaporization. Strong intermolecular forces in the liquid state produce low vapor pressures. Compounds that are very soluble tend to remain in the solution phase rather than in the vapor phase. Media that may be contaminated at hazardous waste sites fall into two general categories: free water bodies; and contaminated geomedia of various moisture contents. The vaporization processes of contaminants from these media are described as follows.

6.4.1 Vaporization from Open Water Bodies

The rate constant for vaporization of an organic chemical from water can be estimated as

$$k_v = \frac{1}{L}\left\{\left[\frac{1}{k_{oi}(D_{ci}/D_{oi})^m}\right] + \left[\frac{RT}{(10^6)Hk_{wg}(D_{cg}/D_{wg})^n}\right]\right\}^{-1} \tag{6.25}$$

where

k_v = rate constant (s^{-1})
L = mixing depth of water (cm)
k_{oi} = oxygen mass transfer coefficient in water (cm/s)
D_{ci}, D_{oi} = diffusion coefficient of the chemical and oxygen, respectively, in water (cm^2/s)
m = turbulence exponent of water, ranging from 0.7 to 1.0 and corresponding to turbulent rivers and still lakes, respectively

R = ideal gas constant (see below)
T = temperature (K)
H = Henry's law constant (atm-m^3/mol)
k_{wg} = mass transfer coefficient of water vapor in air (~0.58 cm/s)
D_{cg}, D_{wg} = diffusion coefficient of the chemical and water, respectively, in air (cm^2/s)
n = gas turbulence exponent, ranging from 0.5 to 1.0

Usually the quantity RT is combined such that it approximates 2.40×10^{-2} atm-m^3/mol. As discussed by the U.S. EPA (1988), D_{cg} can be estimated as follows if measured data are unavailable:

$$D_{cg} = 0.0067\, T^{1.5}(0.034 + M^{-1})^{0.5}\, M^{-0.17}[(\mathrm{M}/2.5d)^{0.33} + 1.81]^{-2} \quad (6.26)$$

where

M = molecular weight of the chemical compound (g/g-mol)
d = density of the liquid (g/cm^3)

Values of k_v computed through the use of Eq. (6.25) can be introduced into Eq. (6.27) to obtain the half-life of the vaporizing chemical compound. The half-life is the time it takes for half the original quantity of the substance to vaporize:

$$t_{0.5} = \frac{\ln 2}{k_v} = \frac{0.693}{k_v} \quad (6.27)$$

Equation (6.27) is applicable to a situation in which the vaporization process of the compound is of the first order. For a second-order process,

$$t_{0.5} = \frac{1}{k'_v C_0} \quad (6.28)$$

where

k'_v = second-order vaporization rate constant (L^3/MT)
C_0 = initial concentration of the vaporizing substance (M/L^3)

For impoundments and other quiescent water masses of fluids such as those illustrated in Figures 6.1D and 6.1H, the rate of emissions of contaminants into the atmosphere is obtainable from

$$E_r = k_c A_e C_L \quad (6.29)$$

where

E_r = rate of emissions from the surface of the liquid (M/T)
k_c = composite mass transfer coefficient (M/T)

A_e = exposed surface area of the liquid (L^2)
C_L = concentration of the vaporizing constituent in the liquid (M/L^3)

Essentially, k_c is a function of the mass transfer coefficients of the liquid and gas phases and Henry's law constant. Using specific units,

$$\frac{1}{k_c} = \frac{1}{k_L} + \frac{1}{k_G k_e} \tag{6.30}$$

where

k_c = composite mass transfer coefficient (m/s)
k_L = liquid-phase mass transfer coefficient (m/s)
k_G = gas-phase mass transfer coefficient (m/s)
k_e = equilibrium constant or partition coefficient that represents the ratio of the chemical's concentration in the gas phase to that in the liquid phase (dimensionless)

Here k_e is a function of Henry's law constant as follows:

$$k_e = \frac{H}{RT} \tag{6.31}$$

where

H = Henry's law constant (atm-m^3/mol)
R = universal gas constant = 8.21×10^{-5} atm-m^3/mol-K
T = temperature (K)

If vaporization occurs at the standard environmental temperature of 25°C, Eq. (6.31) can be written as

$$k_e = (40.9)H \tag{6.32}$$

Values of H for a variety of chemical compounds are obtainable from standard chemistry and chemical engineering texts. Generally, H can be estimated as the ratio of vapor pressure (expressed in atm) to aqueous solubility (expressed in mol/m^3) of the chemical concerned. Values of Henry's law constant for some organic compounds are presented in Table 6.9.

Huang (1982) has provided a comprehensive review of volatile emissions from waste management facilities.

6.4.2 Vaporization from Geomedia

The site contamination scenarios represented by Figures 6.1B, 6.1D, 6.1F, 6.1G, and 6.1H illustrate situations in which volatile contaminants can be emitted from geomedia. In contrast to open water bodies and impoundments, the presence of soil and rock particles affects both the vaporization and transport processes of chemical compounds into the open atmosphere. For example, as explained by Inyang et al. (1992), the solution phase is not spatially continuous in a moist soil, but is smeared on soil grains. Consequently, capillary forces may influence the vapor pressures and hence the vaporization rates of volatile contaminants. Barrow (1961) developed some numerical relationships for estimating reductions in vapor pressures of chemicals due to fluid surface curvature induced by the presence of soil particles. The reader is referred to that textbook.

For a soil contamination case in which gas adsorption by soil particles, depletion rate effects of the vaporizing substances, and internal gas generation processes exemplified by biodegradation are ignored, Fick's first law of steady-state diffusion can be applied to develop expressions for the emission rates of contaminants from buried or exposed waste. Emission is considered to occur at a horizontal plane coincident with the ground surface, and contaminant immediately above the ground surface is zero. With these largely conservative assumptions, Farmer et al. (1978) developed emission equations which were subsequently modified by Shen (1981) and Farino et al. (1983) in the following final form:

$$E_i = D_i C_{si} A_g (n^{4/3}) \frac{M_i}{d_e} \tag{6.33}$$

where

E_i = emission rate of contaminant i (M/T)
D_i = diffusion coefficient of contaminant i in air (L^2/T)
C_{si} = saturation vapor concentration of component i (M/L^3)
A_g = exposed area on the ground surface (L^2)
n = total porosity of the soil (dimensionless fraction)
M_i = mole fraction of the toxic contaminant i in the waste (mole/mole)
d_e = effective depth of the soil cover (L)

$$n = 1 - \left(\frac{\rho_b}{\rho_p}\right) \tag{6.34}$$

$$C_{si} = \frac{V_p M_i}{RT} \tag{6.35}$$

where

ρ_b = bulk density of soil (M/L^3), which varies generally between 1 and 2 g/cm^3
ρ_p = particle density of the soil (M/L^3), which is typically about 2.65 g/cm^3

Table 6.9 Values of Henry's Law Constants for Some Organic Compounds

Compound	Vapor pressure (atm × 10^{-3})	Temperature (K)	Solubility (g/m^3)	Temperature (K)	Molecular weight (g/mol)	H m^3-atm/mol × 10^{-3}	
						Calculated	Experimental
Acenaphthene	—	—	3.93	298	154.2	—	0.241
Benzene	125	298	1,780	298	78.1	5.48	5.55
Carbon tetrachloride	149	298	800	298	153.8	28.6	30.2
Chlorobenzene	15.5	298	472	298	112.6	3.70	3.93
1,2,4-Trichlorobenzene	0.383	298	30	—	181.5	2.32	1.42
Hexachlorobenzene	—	—	0.006	298	284.8	—	1.70
1,2-Dichloroethane	113	298	8,300	298	99.0	1.35	1.10
1,1,1-Trichloroethane	168	298	5,497	298	133.4	4.08	4.92
Hexachloroethane	—	—	50	295	236.7	—	9.83
1,1-Dichloroethane	3.8	298	5,500	293	99.0	5.54	5.45
Chloroform	260	298	9,600	298	119.2	3.23	3.39
1,2-Dichlorobenzene	1.97	298	145	298	147.0	2.00	1.94
1,3-Dichlorobenzene	2.48	298	123	298	147.0	2.96	2.63
1,4-Dichlorobenzene	—	—	79	298	147.0	—	2.72
1,1-Dichloroethylene	778	298	5,000	293	97.0	15.1	15.0
1,2-*trans*-Dichloroethylene	263	287	6,300	293	96.9	4.05	5.32
1,2-Dichloropropane	65.8	298	2,700	293	113.0	2.75	2.82

1,3-Dichloropropylene	32.9	293	2,700	298	111.0	1.35	3.55
Ethylbenzene	12.5	298	206	298	106.2	6.44	6.44
Methylene chloride	599[a]	298	16,700	298	84.9	3.04	3.19
Bromoform	7.37	298	3,130	298	252.8	0.595	0.532
Bromodichloromethane	—	—	—	—	163.8	—	2.12
Trichlorofluoromethane	833	298	1,100	298	137.4	104	58.3
Dibromochloromethane	65.8	293	—	—	168.8	—	0.783
Hexachlorobutadiene	0.197	293	2	293	260.8	25.7	10.3
Hexachlorocyclopentadiene	0.107	298	0.805	298	272.7	36.5	16.4
Nitrobenzene	0.374	298	2,000	298	123.1	0.023	0.024
4,6-Dinitro-*o*-cresol	—	—	—	—	198.1	—	0.0014
Phenol	—	—	67,000	298	94.1	—	0.0013
Acenaphthylene	—	—	3.93	298	152.2	—	0.114
Fluorene	—	—	1.98	298	116.2	—	0.117
Tetrachloroethylene	25.8[a]	298	150	298	165.8	28.5	28.7
Toluene	37.4	298	535	298	92.1	6.44	5.93
Trichloroethylene	97.8	298	1,100	298	131.5	11.7	11.7
Aldrin	—	—	0.2	293	364.9	—	0.496
Dieldrin	—	—	0.186	298	380.9	—	0.058

Source: Resummarized from Huang (1982).

Table 6.10 Diffusion Coefficients of Some Organic Compounds in Air

Compound	Formula	Molecular weight	Atomic diffusion volume	Diffusion coefficients (cm^2/s)		
				At 10°C	At 30°C	At 50°C
Acetaldehyde	C_2H_4O	44	46.40	0.11758	0.13249	0.14816
Acetic acid	$C_2H_4O_2$	60	51.88	0.10655	0.12007	0.13427
Acetone	C_3H_6O	58	66.86	0.09699	0.10930	0.12223
Aniline	C_6H_7N	93	118.55	0.07157	0.08065	0.09019
Benzene	C_6H_6	78	90.68	0.08195	0.09234	0.10327
Bromoethane	CH_3Br	95	57.44	0.09611	0.10830	0.12111
Bromoform	$CHBr_3$	118	53.48	0.09655	0.10880	0.12167
Carbon tetrachloride	CCl_4	154	94.50	0.07500	0.08451	0.09451
Chlorobenzene	C_6H_5Cl	113	128.40	0.06769	0.07627	0.08530
Chloroethane	C_2H_5Cl	65	62.40	0.09789	0.11031	0.12336
Chloroform	$CHCl_3$	120	76.89	0.08345	0.09404	0.10517
Chloromethane	CH_3Cl	51	57.94	0.10496	0.11827	0.13226
Cyclohexane	C_6H_{12}	84	122.76	0.07139	0.08045	0.08996
Dichloroethane	$C_2H_4Cl_2$	99	75.96	0.08557	0.09643	0.10784
Dichloroethylene	$C_2H_2Cl_2$	97	106.96	0.07442	0.08386	0.09377
Dichloropropalene	$C_3H_6Cl_2$	113	100.38	0.07519	0.08473	0.09475
Dimethylamine	C_2H_7N	45	52.55	0.11161	0.12577	0.14065
Ethanol	C_2H_6O	46	50.36	0.11297	0.12730	0.14236
Ethyl acetate	$C_4H_8O_2$	88	92.80	0.07991	0.09005	0.10070

Ethylamine	C_2H_7N	45	52.55	0.11161	0.12577	0.14065
Ethylbenzene	C_8H_{10}	116	151.80	0.06274	0.07070	0.07906
Fluorotoluene	C_7H_7F	110	154.36	0.06262	0.07056	0.07891
Heptane	C_7H_{16}	100	146.86	0.06467	0.07287	0.08149
Hexane	C_6H_{14}	86	126.72	0.07021	0.07912	0.08848
Isopropanol	C_3H_8O	60	37.82	0.12004	0.13526	0.15126
Methanol	CH_4O	32	29.90	0.14808	0.16686	0.18660
Methyl acetate	$C_3H_6O_2$	74	72.34	0.09054	0.10203	0.11410
Methyl chloride	CH_2Cl_2	85	59.46	0.09610	0.10830	0.12111
Methylethyl ketone	C_4H_8O	72	87.32	0.08417	0.09485	0.10607
PCB (1 Cl)	$C_{12}H_9Cl$	189	235.32	0.04944	0.05571	0.06230
Pentane	C_5H_{12}	72	106.26	0.07753	0.08737	0.09770
Phenol	C_6H_6O	84	96.16	0.07919	0.08924	0.09980
Styrene	C_8H_8	104	137.84	0.06620	0.07460	0.08343
Tetrachloroethane	$C_2H_2Cl_4$	168	1143.96	0.06858	0.07729	0.08643
Tetrachloroethylene	C_2Cl_4	166	111.00	0.06968	0.07852	0.08781
Toluene	C_7H_8	92	111.14	0.07367	0.08301	0.09283
Trichloroethane	$C_2H_3Cl_3$	133	97.44	0.07496	0.08447	0.09446
Trichloroethylene	C_2HCl_3	131	93.48	0.07638	0.08606	0.09625
Trichlorofluoromethane	CCl_3F	138	100.00	0.07391	0.08329	0.09314
Vinyl chloride	C_2H_3Cl	63	58.44	0.10094	0.11375	0.12720
Xylene	C_8H_{10}	106	131.60	0.06742	0.07597	0.08495

Source: U.S. EPA (1988).

V_p = vapor pressure of the chemical (mmHg)
M_i = molar weight of contaminant i (M/mole)
R = universal gas constant (62,361 mmHg-cm^3/mol-K)
T = absolute temperature (K)

The diffusion coefficient, D_i, of a contaminant in air can be estimated as

$$D_i = D'(M'/M_i)^{0.5} \tag{6.36}$$

where

D_i = diffusion coefficient of the contaminant (L^2/T)
D' = known diffusion coefficient of a specific compound (L^2/T)
M' = molecular weight of the compound that had D' (M)
M_i = molecular weight of the compound for which D_i is being estimated (M)

Values of D_i for selected organic compounds are presented in Table 6.10.

It should be noted that the presence of moisture in soil reduces the flow rate of emitted gases by increasing the tortuosity of flow paths.

6.5 CONTAMINANT RELEASE MECHANISMS: DUSTING

Dusting is the removal of fine soil particles by mechanical action. Usually, human activities and wind are the agents of dusting. Fine dust particles are considered to be pollutants due to their negative impacts on human health when inhaled and their hazard to plants. Both organic and inorganic contaminants may be present in dust, especially if it is generated from contaminated media such as those illustrated in Figures 6.1B, 6.1C, and 6.1E.

Cowherd et al. (1977) developed Eq. (6.37) for prediction of the average annual suspended particulate emissions due to wind action on flat or rolling terrain:

$$E = 3900 \frac{(e/110)(s/15)(f/25)V}{(P_i/50)^2} \tag{6.37}$$

where

E = emission rate of suspended particles smaller than 30 μm in Stokes diameter (kg/ha-yr)
e = surface erodibility or potential annual loss rate for a wide, open field with erodible dry aggregates (−20 mesh screen; kg/ha-yr)
s = surface silt content [particles smaller than 75 μm (%)]
f = percent of time that the wind velocity at 30 cm above the dust

source exceeds the nominal wind erosion threshold of 5.4 m/s (%)

V = fractional value that reflects the reducing effect of vegetation on wind erosion (1.0 for bare soil, and a fraction for vegetated media)

P_i = Thornthwaite's precipitation-evaporation Index for average moisture content of the source media

The value 3900 kg/ha-yr is a proportionality constant that is based on measurements in west Texas.

The revolving action of automobile tires on the surface of terrain also causes the emission of fine particulate matter. There exists an aerodynamic wake region behind the moving vehicle in which dust particles are thrown upward into the air. Dust particles that adhere to automobile tires can also be thrown up directly. Empirical relationships are commonly used to estimate emission rates for vehicular action. Equation (6.38) is an example.

$$E = 0.61(S/12)(V_h/48)(W/2.7)^{0.7}(n_w/4)^{0.5}\,[(365 - p)/365] \tag{6.38}$$

where

E = emission rate (kg/vehicle-km-h)
S = silt content of the surface (%)
V_h = mean vehicular speed (km/h)
W = mean vehicular weight (Mg)
n_w = mean number of wheels
p = number of days with precipitation $\geq$ 0.254 mm

6.6 CONTAMINANT RELEASE MECHANISMS: LEACHING

Leaching is the process by which chemical substances are removed from a matrix and transported to the boundaries of the releasing matrix. The removal process may be dominated by dissolution of the chemical substance by the leaching solution (leachant) or through the detachment of minute particles from the matrix. Thus the essential difference between leaching and dissolution is that the latter does not cover transport material to the boundary of the matrix while the former does. Dissolution can be considered as one of the leaching phenomena. The transport involved in leaching is within the source matrix boundaries and should be differentiated from the contaminant fate and transport processes that occur after the leached constituents exit the matrix. Establishment of the physical boundary for the two categories of processes differentiated above is important from the standpoint of contaminant emission rate computation.

The concentration of the target contaminant outside that boundary is an important determinant of the leaching rate if the leaching process is diffusion controlled. This is attributable to the fact that the diffusion rate of the contaminant is directly proportional to the concentration gradient from between both sides of the boundary.

Another important distinction is that between the leaching rate of contaminants from exposed or buried waste and the leachability/leaching rates of waste constituents from sampled matrices. Generally, the latter rarely include site-specific environmental parameters. Therefore, results obtained need to be integrated into a numerical predictive framework for use in site-specific field assessments. Waste leaching scenarios are too varied and associated phenomena are too complex to be amenable to a single generic numerical treatment. Nevertheless, some of the fundamental aspects are treated herein. Leaching of contaminants through a waste pile or buried mass of contaminated media fall into two principal categories: leaching that occurs predominantly through displacement of contaminated pore fluid by percolating (infiltrating) water; and leaching due to the transport of contaminant species through a relatively stagnant pore fluid of the source medium. The first case is more significant for media that have relatively high hydraulic conductivities ($>10^{-5}$ cm/s), whereas cemented and compacted fine-grained media are likely to leach bound contaminants through diffusion.

6.6.1 Leaching Through Contaminated Pore Fluid Displacement

The phenomenological aspects of contaminated pore fluid displacement involve the percolation (infiltration) of the leachant through a contaminated mass (cemented or discrete particulate). While the pore fluid moves downward at a velocity that depends on the rate of supply of the leachant and the hydraulic characteristics of the waste and surrounding media, the waste is stationary and supplies contaminant by wash-off and/or diffusion to the mobile leachant. Essentially, pore fluid displacement is by advection while the introduction of contaminants into it is a diffusive process, driven by the waste-pore concentration differential. For a particular location in the waste mass, the differential may change with time consistent with changes in pore fluid velocity and/or depletion of the contaminant in the waste matrix.

For a percolating liquid, the concentration of the contaminant at any location within the bed and time can be described by the following equation (Grant and Merrell, 1985)

$$\left(\frac{\delta C_L}{\delta t}\right)_z = -fV\left(\frac{\delta C_L}{\delta z}\right)_t - kS(C_L - C_{Le}) \tag{6.39}$$

where

$C_L(z,t)$ = bulk concentration of contaminant species i at depth z from the waste surface at time t (M/L^3)
C_{Le} = concentration of contaminant species i at the interface between waste particle and leachant at depth z from the waste surface at time t(M/L^3)
V = interstitial (pore fluid) velocity of the leachant in the waste (L/T)
f = fluid-filled interparticulate porosity of the waste (dimensionless fraction)
k = mass transfer coefficient of contaminant species i from the solid medium (1/T)
S = specific surface of the solid medium = surface area/volume (1/L)

The reader should note that the Darcy velocity $V_d = (f)(V)$ and that C_{Le} is not necessarily equal to the concentration of the contaminant in the interior of the waste particles, C_S. The change in C_S with time can be described by

$$\left(\frac{\delta C_S}{\delta t}\right)_z = kS(C_{Le} - C_L) \tag{6.40}$$

Equation (6.40) is amenable to the incorporation of such quantitative schemes as the shrinking core theory and diffusion models for internally fractured particles. Equations (6.39) and (6.40) are coupled partial differential equations. If C_s is assumed to be constant over a relatively long time interval, as in the case of slow leaching from low-permeability media, Eq. (6.40) can be ignored, such that Eq. (6.39) is rewritten as Eq. (6.41) with the boundary condition $C_L = 0$ at $z = 0$.

$$fV\left(\frac{\delta C_L(z)}{\delta z}\right) = kS(C_{Le} - C_L) \tag{6.41}$$

Integration of Eq. (6.41) results in a relationship for estimating the contaminant concentration at any depth z:

$$C_L(z) = C_{Le}(1 - e^{-kSz/fV}) \tag{6.42}$$

In the analysis of source terms for leachable wastes, it is often necessary to estimate the leach fraction, F_L, which is defined as the quantity of contaminant removed relative to the quantity of contaminant available in waste per unit time. Thus,

$$F_L = \frac{q}{Q} = \frac{C_L V_L}{C_S V_W} = \frac{I}{z_0} M_0(1 - e^{ks_{z_0}/fV}) \tag{6.43}$$

where

F_L = leach fraction per unit time (dimensionless)
q = quantity of waste leached during a given time interval (M)
Q = quantity of waste available for leaching at the beginning of the time interval considered (M)
V_L = volume of leachant = $I\, A_w$ (L^3)
V_w = volume of waste = $Z_0\, A_w$ (L^3)
I = infiltration during the time interval considered (L)
A_w = average cross-sectional area of the waste perpendicular to infiltration (L^2)
z_0 = thickness of the waste (L)
M_0 = partition coefficient = C_{Le}/C_s (dimensionless)

M_0 and k are characterized by uncertainties due to the numerous physicochemical parameters that influence their magnitudes. Noting that M_0 as incorporated into Eq. (6.43) is the reciprocal of the distribution coefficient that is usually obtained from batch sorption experiments, M_0 for a specific set of leachant pH and Eh conditions can be estimated as follows.

$$M_0 = \frac{1}{\rho k_d} \tag{6.44}$$

where

k_d = distribution coefficient of contaminant for the waste matrix
ρ = bulk density of the waste (M/L^3)

The mass transfer coefficient k is the function of interstitial flow turbulence and the particle size distribution of the waste. It can be obtained through measurement or estimation using correlative relationships found in chemical engineering texts. Equation (6.43) can be rearranged for use in estimating the rate of leaching (M/T) of the targeted contaminant from the undersurface of a buried waste mass:

$$q = F_L\, Q \tag{6.45}$$

$$q = \frac{I}{Z_0} M_0\, (1 - e^{-kS_{Z_0}/fV}) \cdot Q \tag{6.46}$$

In the case of Eq. (6.46), q has units of mass per unit time for contaminant release from the surface of fixed area provided I is expressed in units of distance per time.

6.6.2 Leaching Through Diffusion from Consolidated or Stabilized Waste Mass

In the case of consolidated or stabilized waste masses, advective flow is severely limited by negligible intergranular permeability. The buried waste can be considered to release the contaminant largely by diffusion through its external surfaces. A rough estimate of the emission rate of the contaminant from the waste can be made through an extension of the quantitative treatment of the diffusion of chemical species from a monolithic slab. Derivations that are based on Fick's second law of diffusion [presented as Eq. (6.47)] can be adapted for use in release quantification.

$$\frac{\delta C}{\delta t} = D_e\left(\frac{\delta^2 C}{\delta t^2}\right) \tag{6.47}$$

where

C = contaminant concentration in the pore space of a material (M/L^3)
t = time (T)
x = distance in the direction of diffusion (L)
D_e = Dn/τ = effective diffusion coefficient of the chemical species in the material (L^2/T)
n = porosity of the material (dimensionless fraction)
D = diffusion coefficient of the contaminant in a free solution (L^2/T)
τ = tortuosity factor of the material (dimensionless)

With simplifying assumptions, the most significant of which is that D_e will remain constant as diffusion progresses (time increases), the analytical solution developed by Crank (1975) can be adapted for use. The relevant expression is Eq. (6.48):

$$m = \frac{q}{A_w} = 2(C_0 - C_i)\left(\frac{D_e t}{\pi}\right)^{0.5} \tag{6.48}$$

where

m = total amount of contaminant leached per unit surface area of the waste monolith after a given time interval (M/L^2)
C_0 = initial concentration of the contaminant in the pore solution of the monolith (M/L^3)
C_i = concentration of the contaminant at the monolith/leachant interface (M/L$_3$)

q = total mass of contaminant leached after the given time (M)
A = area of the waste monolith through which leaching occurs (L^2)

If the leachant is renewed as in the case of buried waste that is submerged in flowing water, C_i is small relative to C_0. Then the assumption $C_i = 0$ is reasonable. Therefore,

$$m = 2C_0\left(\frac{D_e t}{\pi}\right)^{0.5} \tag{6.49}$$

Equation (6.49) can be used to estimate the mass of contaminant removed from a waste mass for a given duration, given pore fluid concentration, and material properties of the medium.

REFERENCES

Anderson, J. K., and Hatayama, H. K. (1988). Beneficial reuses of hazardous waste sites in California. In *Hazardous Materials Control Research Institute Monograph Series: Closure Consideration*, Odenton, MD, pp. 28–32.

Barrow, G. M. (1961). *Physical Chemistry*. McGraw-Hill, New York.

Benson, R. C., Glaccum, R. A., and Noel, M. R. (1984). Geophysical techniques for sensing buried wastes and waste migration, EPA/600/7-84/064. Office of Research and Development, U.S. Environmental Protection Agency, Washington, DC.

Beres, M., and Haeni, F. P. (1991). Application of ground penetrating radar methods in hydrogeological studies. *Ground Water*, 29:375–386.

Blacklock, J. R. (1987). Landfill stabilization for structural purposes. *Proc. Specialty Conf., Geotechnical Engineering Division*, pp. 275–293. American Society of Civil Engineers, Ann Arbor, MI.

Boulding, J. R. (1994). *Description and Sampling of Contaminated Soils: A Field Guide*. CRC Press, Boca Raton, FL.

Cowherd, C., Jr., Maxwell, C. M., and Nelson, D. W. (1977). Quantification of dust entrainment from paved roads, EPA-450/3-77-027. U.S. Environmental Protection Agency, Research Triangle Park, NC.

Crank, J. (1975). *The Mathematics of Diffusion*. Oxford University Press, London.

De Marsily, G. (1986). *Quantitative Hydrogeology*. Academic Press, San Diego, CA.

Evans, R. B. (1982). Currently available geophysical methods for use in hazardous waste site investigations. *Proc. Am. Chem. Soc. Symp. on Risk Assessment at Hazardous Waste Sites*, Las Vegas, pp. 94–115.

Farino, W., Spawn, P., Jasinski, M., and Murphy, B. (1983). Evaluation and selection of models for estimating air emissions from hazardous waste treatment, storage and disposal facilities. Draft Final Report Prepared by GCA Corporation, Bedford, MA, for the Office of Solid Waste, U.S. Environmental Protection Agency, Washington, DC.

Farmer, W. J., Yang, M. S., Letey, J., Spencer, W. F., and Roulier, M. H. (1978). Land disposal of hexa chlorobenzene waste: Controlling vapor movement in soils. *Proc. 4th Annual Symp. on Land Disposal*, San Antonio, TX.

Gibbons, R. D. (1994). *Statistical Methods for Groundwater Monitoring*. Wiley, New York.

Gilbert, R. O. (1987). *Statistical Methods for Environmental Pollution Monitoring*. Van Nostrand Reinhold, New York.

Grant, M. W., and Merrell, G. B. (1985). Models for the estimation of the leaching of radionuclides from low-level waste disposal facilities. *Proc. Symp. on Waste Management*, Tucson, AZ, vol. 1, pp. 189–196.

Hempen, G. L., and Hatheway, A. W. (1992). Geophysical methods for hazardous site characterization. Special publication No. 3, Association of Engineering Geologists, Sudbury, MA.

Henbest, O. J., Erinakes, D. C., and Hixson, D. H. (1971). Seismic and resistivity methods of geophysical exploration, Technical Release No. 44. Engineering Division, Soil Conservation Services, U.S. Department of Agriculture, Washington, DC.

Hoekstra, P., Lahti, R., Hild, J., Bates, C. R., and Phillips, D. (1992). Case histories of shallow time domain electromagnetics in environmental site assessment. *Ground Water Management and Research*, Fall: 110–117.

Huang, S. T. (1982). Toxic emissions from land disposal facilities. *Environ. Prog.*, 1(1): 3–9.

Inyang, H. I. (1994). Cost-effective post-construction integrity verification monitoring and testing techniques for subsurface barrier containment facilities, Technical Analysis Document. DuPont Corporate Remediation, Wilmington, DE.

Inyang, H. I., and Myers, V. B. (1993). Geotechnical systems for structures on contaminated sites, Technical Guidance Document EPA 530-R-93-002. Office of Solid Waste and Emergency Response, U. S. Environmental Protection Agency, Washington, DC.

Inyang, H. I., Von Ruden, J., and Giles, A. (1992). Hydrocarbon vaporization from soil into a controlled atmosphere. *Int. J. Environ. Issues Minerals. Energy Ind.*, 1(1):21–25.

Journel, A. G. (1989). Fundamentals of geostatistics in five lessons. Short Course in Geology, vol. 8. American Geophysical Union, Washington, DC.

Jury, W. A. (1986). Spatial variability of soil properties. In S. C. Hern and S. M. Melancon (eds.), *Vadose Zone Modeling of Organic Pollutants*, pp. 245–269. Lewis Publishers, Chelsea, MI.

Kardos, J. J., and Ennis, G. B. (1993). Subsurface interface radar as an investigative tool. *Public Works*, Sept.: 82–83.

Keys, W. S., and MacCary, L. M. (1971). Application of borehole geophysics to water resources investigations. U.S. Geological Survey Techniques of Water Resources Investigations TW1-2E1.

Maillet, R. (1947). The fundamental equations of electrical prospecting. *Geophysics*, 12: 529–556.

McNeil, J. D. (1982). Electromagnetic resistivity mapping of contaminant plumes, pp. 1–6. Proc. Natl. Conf. on Management of Uncontrolled Hazardous Waste Sites, Washington, DC.

Nielsen, D. M. (1991). Investigative techniques for determining locations for ground water monitoring wells. A slide presentation.

O'Brien and Gere Engineers, Inc. (1988). *Hazardous Waste Site Remediation*. Van Nostrand Reinhold, New York.

Olea, R. A. (1984). Systematic sampling of spatial functions. Series on Spatial Analysis, No. 7. Kansas Geological Survey, University of Kansas, Lawrence, KS.

Pungor, E. (1995). *A Practical Guide to Instrumental analysis*. CRC Press, Boca Raton, FL.

Rhoades, J. D., and Halvorson, A. D. (1977). Electrical conductivity methods for detecting and delineating saline seeps and measuring salinity in Northern Great Plains Soils. Report ARS W-42, U.S. Department of Agriculture, Washington, DC.

Rosenberg, M. S., Tafuri, A. N., and Goodman, I. (1990). The role of site investigation in the selection of corrective actions for leaking underground storage tanks. *Proc. Fifteenth Annual Research Symp.*, EPA/600/9-90/006, pp. 64–82. Risk Reduction Engineering Laboratory, U.S. Environmental Protection Agency, Cincinnati, OH.

Shen, T. (1981). Estimating hazardous air emissions from disposal sites. Pollution Eng., 13(8):31–34.

U. S. DOE. (1994). Fibre-optic sensors. Technology Catalogue DOE/EM-013P, pp. 43–47. Office of Environmental Management, Office of Technology Development, U.S. Department of Energy, Washington DC.

U.S. EPA. (1988). Superfund exposure assessment manual, EPA/540/1-88/001. Office of Remedial Response, U.S. Environmental Protection Agency, Washington, DC.

U.S. EPA. (1991). Site characterization for subsurface remediation. Seminar Publication, EPA/625/4-91/026. Office of Research and Development, U.S. Environmental Protection Agency, Washington, DC.

van Zijl, J. S. V. (1978). On the uses and abuses of the electrical resistivity method. *Bull. Assoc. Engi. Geol.*, XV(1):85–111.

Walther, E. G., Pitchford, A. M., and Olhoeft, G. R. (1986). A strategy for detecting subsurface organic contaminants. *Proc. NWWA/API Conf. on Petroleum Hydrocarbons and Organic Chemicals in Ground Water: Prevention, Detection and Restorations*, Houston, Texas, pp. 357–381.

Wilson, N. (1995). *Soil Water and Groundwater Sampling*. CRC Press, Boca Raton, FL.

Yland, M. W. F., and van Wachem, E. G. (1988). Soil covering system as remedial action in contaminated housing areas in the Netherlands. In K. Wolf, W. J. van den Brick, and F. J. Colon (eds.), Contaminated Soils, pp. 597–599. Kluwer Academic Press, Amsterdam.

7

Technical Basis for Treatment Technique Selection

7.1 IDENTIFICATION OF HAZARDOUS WASTES

Materials that are collectively referred to as "wastes" vary in form and physicochemical characteristics. The hazards posed by materials may not be derived from all measures of hazard but from one or a combination of a few components of hazard measures. The factors that are most commonly used to identify wastes are toxicity to living organisms, reactivity, corrosivity, and ignitability. Regulatory agencies have established minimum quantitative levels of each of these parameters for classification of materials as hazardous wastes. In some cases, parameters that are surrogates of one or more of these parameters are used by regulatory agencies. The reader should note that hazardous waste may be identified on the basis of just one of the parameters listed above.

7.1.1 Toxicity

A toxic substance is a material which produces detrimental effects on biological tissue and associated processes when organisms are exposed to concentrations above a certain level. The time period before the manifestation of toxic effects in exposed plants and animals varies widely. *Acute toxicity* is a description of an effect that occurs at or immediately after exposure, while *chronic toxicity* covers effects that may take longer intervals of time to manifest. The effects of a toxicant may be reversible or irreversible at one or more exposure levels. The *dose* is the quantity of the potentially toxic substance which is in contact with the living form. It is customary to quantify contaminant dose in units of mass of contaminant per unit mass of the receiving organism.

Dose–response curves and associated statistical analyses are used to determine the extent to which acute effects such as the death of organisms are attributable to toxic substance contacts at various doses. In general, the cumulative percent of organismal deaths is plotted on the vertical axis while the $\log_{10}$ dose is plotted on the horizontal axis. Plots are generally S-shaped and can be used to determine typical regulatory parameters such as LD_{90}, the dose at which 90% of the organisms died. Organisms that respond negatively to very low doses of the toxic substance are termed *hypersensitive*, while those that exhibit negative effects only to very high doses are described as being *hyposensitive* to the toxic substance.

Manahan (1990) has proposed a toxicity rating scale of 1.0–6.0 based on the minimum mass of a substance, per unit body mass of an organism, required to produce a toxic effect

1. Nontoxic	15 g/kg
2. Slightly toxic	5–15 g/kg
3. Moderately toxic	0.5–5 g/kg
4. Very toxic	50–500 mg/kg
5. Extremely toxic	5–50 mg/kg
6. Supertoxic	<5 mg/kg

Ultimately, the toxicity of a substance must be expressed in terms of effects which occur subsequent to the exposure of an organism to the substance. Two major phases are recognized.

Kinetic phase: During this phase, which involves transport processes, the substance is absorbed and passed through the organism's metabolic system. The substance may be detoxified or metabolized during biochemical processes, or it may be excreted in the same form as it entered the organism. Models which describe the absorption, transport, distribution, metabolism, binding, and excretion of substances that enter the body from an external source are described as *pharmacokinetic models*.

Dynamic phase: The toxicant and/or its derivatives that are not excreted interact with biological matter, such as tissue and cells of the host organism. Cellular and other processes may be affected negatively, leading to diseases.

In terms of the effects on organisms, toxic substances can be divided into three principal types.

Carcinogens: These substances are capable of affecting the rate of replication of cells. They form covalent bonds with large molecules such as DNA. The composite molecule then replicates spontaneously to create cancerous masses within the organism. Alkyl organic compounds are generally believed to be carcinogenic. Unfortunately, except for a few cases such as vinyl chloride, most carcinogens so-classified in regulatory programs have been identified through extrapolations from animal tests.

Teratogens: These chemical toxins cause mutations in primary cells such as germ cells. Relevant biochemical interactions can cause damages to human fetus, e.g., cutoff of the source of supply of vitamins.

Mutagens: These chemical toxins can induce undesirable changes in DNA. Such changes can be passed from generation to generation. As with teratogens, birth defects can result.

The interaction of a number of chemical toxins can produce composite toxic effects that are significantly different from those of the individual components. When the effect is simply a summation of those of the individual components, it can be described as being additive. Synergistic toxic effects are larger than the simple sum of the individual component effects. Some toxicities may also be partially or fully negated when combined.

Cancer risks form the criteria for most of the toxicity assessments of contaminants and site clean-up programs. As discussed by the National Research Council (1988), the combined effects of exposure to a variety of potential carcinogens on cancer risks can be estimated by assuming that effects are additive, at least at low doses. The generalized "multistage" model, in which a progressive transformation of a normal cell to a cancer cell occurs through a small number of stages, is discussed herein, following the method of Whittemore and Keller (1978).

$$T_{it} = [\lambda_{it}][P_{it}] \tag{7.1}$$

where

T_{it} = instantaneous probability that the organic cell concerned will be transformed from stage i to stage $(i + 1)$ at time t
P_{it} = probability that the organic cell is in stage i at time t
λ_{it} = rate of transition of the cell from state i to stage $(i + 1)$ within time t

If a number of carcinogenic agents are considered, the transition rate of the cell from stage i to stage $(i + 1)$ is given by

$$\lambda_{it} = \varepsilon_i + \sum_{j=1}^{m} [\gamma_{ij} x_{jt}] \tag{7.2}$$

where

ε_i = background transition rate for ith stage
γ_{ij} = unit exposure rate for jth agent on the ith stage
x_{jt} = exposure to the jth agent at time t

As explained by the National Research Council (1988), in Eq. (7.2) it is assumed that the chemical substances act independently on the host cells. Additional assumptions can be made to extend this treatment to use in estimating the age-specific death rate, A_t, due to a particular type of tumor in a specific organ. If the number of cells $[N]$ is assumed to be large and the time interval w between the final stage and growth of a cell into a tumor is constant, then

$$A_t = \frac{[N]\ \partial P_k(t - w)}{\partial t} \tag{7.3}$$

$$A_t = N[\lambda_{k-1}\ (t - w)][P_{k-1}\ (t - w)] \tag{7.4}$$

The parameter t is the time up to death; k is the final transformation state; and f_k is the probability that the cell is in the kth stage at time t.

In standard practice, the concentrations of specific contaminants above which the substance is considered toxic in media are specified in regulations. Test protocols such as the Toxicity Characteristics Leaching Procedure (TCLP) have been developed (Stults, 1993; U.S. EPA, 1989b; Lerose, 1995; Lee et al., 1994) for extracting chemical substances from potential toxic wastes. The toxic concentration levels specified by the U.S. EPA for a variety of chemical substances leached from waste samples are shown in Table 7.1.

7.1.2 Reactivity

Reactivity of a contaminant is the level of its tendency to interact chemically with other substances. This tendency is considered a hazard only when any of the following conditions is associated with the chemical interaction: violent or explosive reaction with water and/or other substances, generation of toxic gas, and explosive decomposition. Environmental conditions, which can be affected by temperature, pressure, and pH, have significant effects on reactivity. Substances react to establish thermodynamic equilibrium under an imposed environment. In some cases, the reaction may be spontaneous (without the additional supply of external energy).

The chemical structure of a compound is a significant factor with respect to its reactivity. For example, unsaturated bonds in an organic compound constitute one index of reactivity. Manahan (1990) explains that many organic compounds that contain a combination of nitrogen, carbon, and hydrogen tend to be very reactive. Examples of these compounds are azo dyes and triazenes. Inorganic compounds such as halogens of nitrogen, compounds with metal–nitrogen bonds, and halogens of oxides are very reactive.

7.1.3 Corrosivity

Corrosive contaminants degrade materials such as human tissue through chemical attack and removal of matter. In standard practice, the extent to which metals

Table 7.1 Concentration Levels of Chemical Substances Specified as Being Toxic in Leachates Produced Using the U.S. EPA TCLP Procedure

Arsenic	5.0
Barium	100.0
Benzene	0.5
Cadmium	1.0
Carbon tetrachloride	0.5
Chlordane	0.03
Chlorobenzene	100.0
Chloroform	6.0
Chromium	5.0
o-Cresol	200.0
m-Cresol	200.0
p-Cresol	200.0
Cresol	200.0
2,4-D	10.0
1,4-Dichlorobenzene	7.5
1,2-Dichloroethane	0.5
1,2-Dichloroethylene	0.7
2,4-Dinitrotoluene	0.13
Endrin	0.02
Heptachlor and its epoxide	0.008
Hexachlorobenzene	0.13
Hexachlorobutadiene	0.5
Hexachloroethane	3.0
Lead	5.0
Lindane	0.4
Mercury	0.2
Methoxychlor	10.0
Methyl ethyl ketone	200.0
Nitrobenzene	2.0
Pentachlorophenol	100.0
Pyridine	5.0
Selenium	1.0
Silver	5.0
Tetrachloroethylene	0.7
Trichloroethylene	0.5
2,4,5-Trichlorophenol	400.0
2,4,6-Trichlorophenol	2.0
2,4,5-TP (Silvex)	1.0
Vinyl chloride	0.2

can deteriorate under contact with the chemical substance is also used as an index of the corrosivity hazard of the chemical with respect to biological matter. Strong acids, strong bases, oxidants, and dehydrating agents are corrosive. For regulatory purposes in the United States, substances that have pH levels lower than 2.0 (highly acidic) or higher than 12.5 (highly basic) are considered hazardous because of high corrosivity. Substances are also identified as being hazardous if they can corrode steel at the rate of 6.35 mm/year. Examples of highly corrosive substances are nitric acid (HNO_3), sulfuric acid (H_2SO_4), sodium hydroxide ($NaOH$), and sodium chlorate ($NaClO_3$).

7.1.4 Ignitability

Ignitability is the ease with which a substance can burn. Other surrogate terms are flammability and combustibility. In general, a combustible substance will ignite less readily than a flammable substance. The temperature at which a mixture of a chemical's vapor and air ignite is called the *flash point* of the chemical substance. U.S. hazardous waste characterization rules classify contaminants as hazardous because of ignitability if any of the following situations holds:

The substance is a liquid with a flash point of less than 60°C.
The substance is a nonliquid that can cause fire and burn vigorously when ignited.
The substance is a compressed gas or an oxidizer.

Typically, for each flammable substance, there is a range of vapor/air volume ratios at which the substance can ignite. Below the lower ratio, called the lower flammability limit, the vapor content in the mix is insufficient to allow ignition. Above the upper flammability limit, the mixture is too depleted of air for ignition to occur. The flash points of some common compounds are methanol (12°C), toluene (4°C), and acetone (−20°C). The corresponding lower and upper flammability limits (expressed in percent volumetric content in air) are 6.0 and 37; 1.27 and 7.1; and 2.6 and 13, respectively.

Oxygen that is necessary for combustion can be most efficiently supplied by a group of substances called oxidizers. Halogens and their compounds are common oxidants. Examples of some oxidants are bromine (Br_2), chlorine (Cl_2), hydrogen peroxide (H_2O_2), nitrous oxide (N_2O), potassium permanganate ($KMnO_4$) and sodium dichromate ($Na_2Cr_2O_7$).

Pulverization may increase the flammability of combustible maaterials. For example, a dispersion of fine particles of liquid hydrocarbon in oxygen-rich air may have a flash point that is lower than that of the hydrocarbon liquid. The same is true for some fine particulate solids such as dispersed powder of coal, magnesium, and some polymers.

7.2 INTRODUCTION TO EXPOSURE ASSESSMENT

Exposure assessment is a qualitative and/or quantitative analysis of the magnitude, frequency, duration, and routes of contact of organisms with a chemical or physical agent. For human subjects, exposure assessment is commonly done as part of health-risk assessment of a contaminated site or for analysis of clean-up technologies. The U.S. EPA (1989b) has provided a detailed description of exposure assessment factors and the sequence of steps for their evaluation. The major steps are:

1. Characterization of the physical setting
2. Identification of potentially exposed populations
3. Identification of potential exposure pathways
4. Estimation of exposure concentrations
5. Estimation of chemical intakes

The interrelationships among these factors with respect to determination of the magnitude of contaminant concentrations taken by an exposed individual is illustrated in Figure 7.1.

Quantitative exposure assessments require two calculation steps: estimation of exposure concentrations; and estimation of pathway-specific intakes. It is usually convenient to use exposure magnitudes totaled over a given time period for

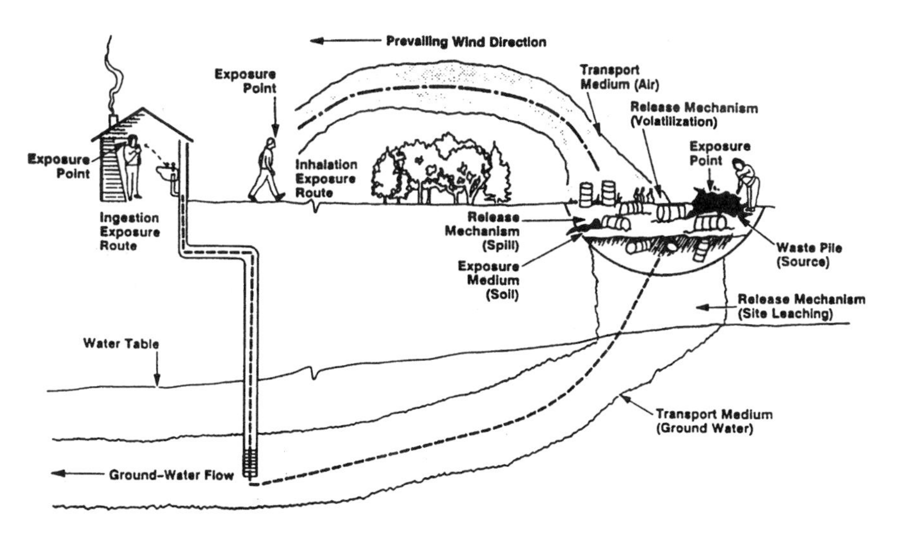

Figure 7.1 Pathways of human exposure to contaminants from a waste site (U.S. EPA, 1989b).

assessments. If the total exposure is normalized for time and body weight, then it is called *intake* and expressed in units of concentration per body weight-time (for example, mg of chemical/kg body weight-day). Three principal categories of parameters are included in the most basic intake equation as shown in Eq. (7.5). The principal categories are

A variable that describes the concentration of the chemical at the exposure point
Variables that describe the population with respect to human characteristics and opportunity for contact with the contaminant (e.g., contact rate, exposure frequency and duration, and body weight)
A variable that describes the duration of contact (time)

The most basic equation for quantification of intake is as follows:

$$I = C\left[\frac{(\mathrm{CR})(\mathrm{EF})(\mathrm{ED})}{(\mathrm{BW})}\right]\frac{1}{\mathrm{AT}} \tag{7.5}$$

where

I = intake = quantity of the chemical at the exchange interface of the body (M/M·T)
C = chemical concentration = the average concentration contacted over the duration of exposure (M/L^3)
CR = contact rate = quantity of contaminated medium contacted per unit time or event (L^3/T)
EF = exposure frequency (T/T)
ED = exposure duration (T)
BW = body weight = average body weight over the exposure period (M)
AT = averaging time = time period over which the exposure is averaged

More details on procedures for estimating each of the intake parameters are provided by the U.S. EPA (1989b). It should also be noted that Eq. (7.5) and its surrogates are typically intended for specific exposure pathways and land use types. Table 7.2 is a summary of the common exposure pathways for contaminants in various media relative to three land uses. Several physicochemical processes that can alter and/or transport contaminants from a waste source to exposure points are listed in Table 7.3. Some of these processes (leaching, dusting, and vaporization) have been discussed in Chapter 6. A variety of parameters may be used to compute the contact rate, CR. In the case of dermal contact with chemicals present in water,

$$\mathrm{CR} = (\mathrm{SA})(\mathrm{AF})(\mathrm{ABS}) \tag{7.6}$$

Table 7.2 Pathways to Human Exposure to Chemicals Present in Various Media with Respect to Three Land Uses[a]

Exposure medium/ exposure route	Residential population	Commercial/industrial population	Recreational population
Groundwater			
Ingestion	L	A	—
Dermal contact	L	A	—
Surface water			
Ingestion	L	A	L, C
Dermal contact	L	A	L, C
Sediment			
Incidental ingestion	C	A	C
Dermal contact	C	A	L, C
Air			
Inhalation of vapor phase chemicals			
Indoors	L	A	—
Outdoors	L	A	L
Inhalation of Particulates			
Indoors	L	A	—
Outdoors	L	A	L
Soil/dust			
Incidental ingestion	L, C	A	L, C
Dermal contact	L, C	A	L, C
Food			
Ingestion			
Fish and shellfish	L	—	L
Meat and game	L	—	L
Dairy	L, C	—	L
Eggs	L	—	L
Vegetables	L	—	L

[a] L = lifetime exposure; C = exposure in children may be significantly greater than in adults; A = exposure to adults (highest exposure is likely to occur during occupational activities); — = exposure of this population via this route is not likely to occur.
Source: U.S. EPA (1989b).

where

CR = contact rate (L^3/T or M/T)
SA = surface area of the skin available for contact (L^2/T)
AF = soil-to-skin adherence factor (M/L^2)
ABS = unitless absorption factor

Table 7.3 Principal Processes of Contaminant Alteration and Transport to Exposure Points and Human Intake Boundaries

Receiving medium	Release mechanism	Release source
Air	Volatilization	Surface wastes—lagoons, ponds, pits, spills
		Contaminated surface water
		Contaminated surface soil
		Contaminated wetlands
		Leaking drums
	Fugitive dust generation	Contaminated surface soil
		Waste piles
Surface water	Surface runoff	Contaminated surface soil
	Episodic overland flow	Lagoon overflow
		Spills, leaking containers
	Groundwater seepage	Contaminated groundwater
Groundwater	Leaching	Surface or buried wastes
		Contaminated soil
Soil	Leaching	Surface or buried wastes
	Surface runoff	Contaminated surface soil
	Episodic overland flow	Lagoon overflow
		Spills, leaking containers
	Fugitive dust generation/ deposition	Contaminated surface soil
		Waste piles
		Contaminated surface soil
	Tracking	
Sediment	Surface runoff	Surface wastes—lagoons, ponds, pits, spills
	Episodic overland flow	Contaminated surface soil
	Groundwater seepage	Contaminated groundwater
	Leaching	Surface or buried wastes
		Contaminated soil
Biota	Uptake (direct contact, ingestion, inhalation)	Contaminated soil, surface water, sediment, groundwater or air
		Other biota

Source: U.S. EPA (1989b).

The numerical magnitudes of AF and SA are of the order of 1.45–2.77 mg/cm^2 and <0.3 m^2 per specified time, respectively. ABS varies widely for different contaminants.

It is often necessary to develop and use numerical indices for assessing the factor of safety against damaging health effects of human exposure to contaminants. Thus, the risk of a damaging health effect (e.g., cancer) can be estimated by combining contaminant exposure equations with potency estimates of the contaminant with respect to the disease for which the estimate is made. Within regulatory frameworks, it is convenient to consider carcinogenic risks and noncarcinogenic risks separately in the risk characterization phase.

7.2.1 Carcinogenic Risks

Some chemical contaminants are carcinogens (cancer causing). The approach to quantifying cancer risk is to estimate the incremental probability that an individual will develop cancer over a lifetime due to exposure to the chemical. Two methods are commonly used for this estimate: the *linear low dose model* for risks that are below 0.01; and the *one-hit model* for exposure conditions in which the probable dose is high (>0.01). In the linear low-dose model, the carcinogenic risk is computed as follows:

$$R_c = I_c \cdot \mathrm{SF} \tag{7.7}$$

where

R_c = carcinogenic risk = probability of an individual developing cancer due to exposure to the contaminant (unitless fraction)
I_c = chronic intake averaged over 70 years (mg/kg-day)
SF = slope factor $(\text{mg/kg-day})^{-1}$

The chronic daily intake is computed using the relevant form of the exposure equation (7.5). Conceptually, the slope factor is the ratio of cancer frequency to the dose of the contaminant that is being evaluated. Generally it is the upper-bound (usually 95th percentile) confidence limit of the slope of the dose–response plot expressed in units of $(\text{mg/kg-day})^{-1}$. The magnitude of the slope factor can be considered as the risk per unit dose of the contaminant. An intrinsic assumption in the use of slope factors is that there is no threshold dose level below which cancer risk is zero.

It may also be found convenient to express the carcinogenic effects of a contaminant in terms of the risk per unit concentration of the contaminant in the medium of human contact. This is particularly useful in assessments of the potential health benefits of risk management technologies that may be effective in contaminant clean-up, containment, or other remedial actions. Unit risk can be estimated using Eq. (7.8):

$$\mathrm{UR} = \frac{\mathrm{SF} \cdot E_i}{70\ \mathrm{kg} \cdot 1000} \tag{7.8}$$

where

UR = unit risk for the particular medium of interest (fraction per μg/m^3)
E_i = consumption rate for water (2 liters/day) or inhalation rate for air (20 m^3/day)

The 70 kg used represents body weight and is included to cancel out the normalization for body weight in the definition of the slope factor. Also, the factor 1000 is used as the denominator to convert from milligrams [unit of the slope factor: (mg/kg-day)$^{-1}$] to μg [unit of unit risk: (μg/m^3)$^{-1}$ for air and (μg/L)$^{-1}$ for water].

Considering that cancer risk may be attributable to more than one contaminant through a number of pathways or exposure routes, composite low-dose cancer risk can be estimated using Eq. (7.9):

$$TR_c = \sum_{j=1}^{k} \sum_{i=1}^{n} (I \cdot SF)_{ij} \tag{7.9}$$

where

TR_c = total cancer risk = probability that an individual will develop cancer (unitless fraction)
I_{ij} = chronic daily intake of the ith contaminant through the jth exposure pathway (mg/kg-day)
SF_{ij} = slope factor for the ith contaminant contacted through the jth exposure pathway [(mg/kg-day)$^{-1}$]
n = total number of carcinogeneous contaminants
k = total number of exposure pathways

For the one-hit model, the risk of an exposed individual developing cancer due to a contaminant through a specific pathway can be computed using Eq. (7.10):

$$R_c = 1 - \exp(-I_c \cdot SF) \tag{7.10}$$

The parameters are as defined for Eq. (7.7). The corresponding relationship for an aggregate value of risk for all carcinogeneous contaminants for all exposure routes is

$$TR_c = \sum_{j=1}^{k} \sum_{i=1}^{n} [1 - \exp(-I_c \cdot SF)_{ij}] \tag{7.11}$$

All parameters are as defined for Eq. (7.9). In using Eqs. (7.9) and (7.11), strict additivity of cancer risks from identified carcinogens is assumed. Synergistic effects are neglected. Furthermore, the effects are summed equally for the carcinogens.

7.2.2 Noncarcinogenic Risks

In the case of noncarcinogens, the adverse effect of exposure to a contaminant is assumed not to be significant (or manifestive) unless the subject is exposed to doses above a critical level. The existence of the threshold dose is attributable to the body's protective physiological mechanism. The challenge is to determine the upper bound of the threshold dose for use as the body's tolerance limit. Variabilities in human susceptibilities to the same dose of a contaminant and uncertainties associated with extrapolations of toxicological data from one population to another necessitate continuous updating of data on critical doses.

Assessments of the risk of noncarcinogenic effects of contaminants are usually based on the reference dose, RfD, which is discussed in detail by the U.S. EPA (1986). As defined by the U.S. EPA (1989b), the chronic RfD is the daily exposure level for the human population (including subpopulations) that is likely to be without an appreciable risk of deleterious effects during a lifetime. Values of chronic RfD are meant to be protective of the effects of long-term exposure to a contaminant, whereas subchronic RfD addresses the effects of shorter-term exposures. RfD is currently used as a replacement for its precursors: the acceptable daily intake (ADI) and the acceptable intake for chronic exposure (AIC).

Considerable uncertainty (of the order of one magnitude or more) characterizes estimates of RfD. This parameter can be estimated from the results of dose–response experiments for noncarcinogens and/or through the use of relevant toxicological data. In dose–response experiments, it is useful to determine the response limits which can be used with appropriate uncertainty factors to estimate RfD as shown in Eq. (7.12):

$$\mathrm{RfD} = \frac{\mathrm{NOAEL}}{\left[\prod_{i=1}^{n} \mathrm{UF}_i\right]\mathrm{MF}} \tag{7.12}$$

where

RfD = reference dose (mg/kg·day)
NOAEL = no-observed-adverse-effect-level contaminant dose (mg/kg · day)
UF_i = uncertainty factor
MF = modifying factor ranging in magnitude from 1 to 10 to be applied by the evaluator to account for uncertainties in the database for the chemical when such uncertainties are not covered by any of the UFs

NOAEL, which can be computed from the results of dose–response experiments, is the exposure level at which no biologically or statistically significant increases in the severity or frequency of adverse effects occur in the exposed population relative to the control population. Other dose–response levels are, the no-observed-effect level (NOEL) and the lowest-observed-adverse-effect level (LOAEL). NOEL pertains to all effects (adverse or positive). Uncertainty factors are generally applied to address the following sources of uncertainty:

1. Variability in the general population which renders subpopulations such as children and the elderly more sensitive to contaminants
2. Extrapolation of dose–response data from experiments on animals to risk characterization for humans (interspecies variability between humans and other mammals)
3. Use of NOAEL derived from subchronic rather than from chronic data
4. Use of LOAEL rather than NOAEL

Reference dose data are used to develop a noncancer quotient for a contaminant for an exposure route as follows:

$$\mathrm{H} = \frac{I_c}{\mathrm{RfD}} \tag{7.13}$$

where

H = noncancer hazard quotient (unitless)

I_c and RfD are as previously defined. Their units as well as the period of coverage (chronic or subchronic) must be kept identical. Considering that at most polluted sites, several noncarcinogens within a medium may be taken in through one or more exposure pathways, the aggregate risk can be expressed using Eq. (7.14):

$$\mathrm{TH} = \sum_{j=1}^{k} \sum_{i=1}^{n} H_{ij} \tag{7.14}$$

$$\mathrm{TH} = \sum_{j=1}^{k} \sum_{i=1}^{n} \left(\frac{I_{ij}}{\mathrm{RfD}_{ij}} \right) \tag{7.15}$$

where

TH = total hazard index (unitless)
I_{ij} = exposure level (intake) for the *i*th contaminant through the *j*th exposure pathway
RfD_{ij} = reference dose for the *i*th contaminant for the *j*th pathway (mg/kg·day)
H_{ij} = hazard quotient for the *i*th contaminant and *j*th pathway

n = total number of contaminants that exhibit noncarcinogenic effects
k = total number of exposure pathways

From the configuration of Eq. (7.15), values of H and TH that are much less than 1.0 are more desirable than those that approach 1.0 or greater. Values greater than 1.0 can be used to justify risk management actions for the site concerned.

Within the context of geoenvironmental engineering, the purpose of exposure assessment, and the risk assessment which encompasses it, is to assess the need for implementation of an engineering control measure. Essentially, such measures may result in risk reduction through impact on the concentration of contaminants at the source and/or exposure points. The configuration of a contaminated site, the characteristics of the transport medium for released contaminants, climatic factors, and proximity of exposure points to human habitat are some of the critical factors that affect (directly or indirectly) the magnitudes of risks for which equations have been discussed in this chapter. The focus of this analysis has been human health risk. Ecological risk can also be assessed, although many more factors need to be considered. Some investigators have developed site vulnerability indices for use in scaling risks and hazards associated with contaminated sites and sites that may be sensitive to contamination. Examples of these indices are the DRASTIC Index (Aller et al., 1985), the Multi-attribute Utility Analysis Model (Call and Merkhofer, 1988), and a host of other models described by Hushon and Kornreich (1984), Shahane and Inman (1987), Mayernik and Fehrenkamp (1992), Merz (1988), and Mikroudis et al. (1986).

7.3 RISK-BASED ESTIMATION OF REQUIRED CLEAN-UP LEVELS

The primary objective of polluted-site remediation programs such as clean-up and containment is to reduce the concentrations of targeted contaminants at the source and/or exposure points such that risk can be reduced to an acceptable level. By specifying the maximum acceptable risk and rearranging risk estimation equations such as (7.5), (7.7), (7.9), (7.10), (7.11), (7.13), and (7.15), the maximum concentration of specific pollutants that are allowable in each medium can be back-calculated. This approach has utility with respect to the selection of remediation schemes (including technology) that could be effective in reducing contaminant concentrations to desirable levels.

Usually, it is advisable to specify the land use (e.g., commercial, residential, or agricultural) before applying the equations. The land use time primarily affects the default values or direct estimates of the magnitudes of some of the parameters of the risk equations used in back-calculations of the maximum concentrations of target contaminants. As discussed in the preceding sections, carcinogenic effects are treated separately from noncarcinogenic effects. Also, each contaminated medium (water, soil, or air) can exhibit risks that are attributable to one or more pathways. For example, for residential land use, risk from the water medium can be represented as follows, for the purpose of setting up equations for the back-calculation of the maximum acceptable concentration limits for contaminants:

Total risk from water = risk of water ingestion + risk of inhalation of volatiles from water

For carcinogenic effects, the maximum acceptable risk is usually specified as 10^{-6}, while a total hazard quotient of 1.0 is usually the upper limit for noncarcinogenic effects. The approach is to calculate the maximum concentration of a contaminant at the exposure point through consideration of all the pathways that can contribute to risk attributable to the contaminant.

7.3.1 Clean-up Level for Soil

For both residential and commercial/industrial land uses, the ingestion exposure pathway applies. Additionally, for the commercial/industrial pathway, other pathways are inhalation of volatiles from soil and inhalation of particulates from soil (by workers).

Residential

For residential land use, the total risk is attributable to the ingestion pathway. For carcinogenic effects, the relevant risk equation and its rearrangement to soil for maximum concentration are presented as Eqs. (7.16) and (7.17).

$$R_c = \mathrm{SF} \cdot I_c = \mathrm{SF}_o \left[\frac{(C \times 10^{-6}\ \mathrm{kg/mg})(\mathrm{EF})(\mathrm{IF})}{(\mathrm{AT} \times 365\ \mathrm{days/year})} \right] \tag{7.16}$$

$$C = \frac{(\mathrm{R}_c)(\mathrm{AT} \times 365\ \mathrm{days/year})}{(\mathrm{SF} \times 10^{-6}\ \mathrm{kg/mg})(\mathrm{EF})(\mathrm{IF})} \tag{7.17}$$

where

C = risk-based remediation level for the target contaminant (mg/kg of the contaminant in soil)
R_c = target excess individual cancer risk (usually specified as 10^{-6})
SF_o = oral cancer slope factor [$(\mathrm{mg/kg{\cdot}day})^{-1}$]
AT = averaging time (years); a default value of 70 years

EF = exposure frequency (days/year); a default value of 350 days/year
IF = age-adjusted ingestion factor (mg · yr/kg · day); default value of 114 mg · yr/kg · day [Note that ED of Eq. (7.5) is incorporated herein]

For noncarcinogenic effects, the noncancer hazard quotient presented as Eq. (7.13) can be rearranged as follows:

$$H = \frac{1}{\mathrm{RfD}_o}[I_c] = \frac{1}{\mathrm{RfD}_o}\left[\frac{\mathrm{C} \times 10^{-6}\ \mathrm{kg/mg)(EF)(IF)}}{(\mathrm{AT} \times 365\ \mathrm{days/year})}\right] \quad (7.18)$$

$$C = \frac{(\mathrm{RfD}_o)(H)(\mathrm{AT} \times 365\ \mathrm{days/year})}{(10^{-6}\ \mathrm{kg/mg})(\mathrm{EF})(\mathrm{IF})} \quad (7.19)$$

where

H = target hazard index (unitless); usually specified as <1.0
RfD_o = oral chronic reference dose (mg/kg · day)
AT = Averaging time (years); default value = 30 years

Other parameters and their corresponding default values are as stated for Eqs. (7.16) and (7.17).

Commercial/Industrial

For commercial or industrial use, the disturbance of soils by heavy equipment and traffic is assumed to increase the potential for emission of particulates and volatiles which could be inhaled by workers. Additional exposure pathways exemplified by dermal exposure may also be relevant at some sites. It is always necessary to identify the relevant exposure pathways for the particular site.

The concentration limit for a contaminant in soil at a site that is to be used for commercial/industrial purposes, based on health effects, can be computed from the following setup of the risk equation (only three pathways are used in the following example).

Risk = risk form intake due to soil ingestion by workers
+
risk from intake due to inhalation of volatiles by workers
+
risk from intake due to inhalation of particulates by workers

For carcinogenic effects, the two inhalation risks (for volatiles and particulates) can be combined such that the resulting total risk equation assumes the form of Eq. (7.20), which is the expanded form of Eq. (7.9).

$$\mathrm{TR}_c = \frac{(\mathrm{SF}_o)(C \times 10^{-6}\ \mathrm{kg/mg})(\mathrm{EF})(\mathrm{ED})(\mathrm{IR}_s)}{(\mathrm{BW})(\mathrm{AT} \times 365\ \mathrm{days/year})} + \frac{(\mathrm{SF}_i)(\mathrm{C})(\mathrm{EF})(\mathrm{ED})(\mathrm{IR}_s)(1/\mathrm{VF}) + (1/\mathrm{PEF})}{(\mathrm{BW})(\mathrm{AT} \times 365\ \mathrm{days/year})} \tag{7.20}$$

By rearrangement to produce an expression for the concentration C,

$$C = \frac{(\mathrm{TR}_c)(\mathrm{BW})(\mathrm{AT} \times 365\ \mathrm{days/year})}{(\mathrm{EF})(\mathrm{ED})\ [(\mathrm{SF}_o \times 10^{-6}\ \mathrm{kg/mg})(\mathrm{IR}_s) + (\mathrm{SF}_i)(\mathrm{IR}_a)(1/\mathrm{VF}) + (1/\mathrm{PEF})]} \tag{7.21}$$

where

C = chemical concentration in soil (mg/kg)
TR_c = target excess individual lifetime cancer risk (unitless; usually specified as 10^{-6})
SF_i = inhalation cancer slope factor for the contaminant $[(\mathrm{mg/kg} \cdot \mathrm{day})^{-1}]$
SF_o = oral cancer slope factor for the contaminant $[(\mathrm{mg/kg} \cdot \mathrm{day})^{-1}]$
BW = adult body weight (kg); default value = 70 kg
AT = averaging time (year); default value = 70 years
EF = exposure frequency (days/year); default value = 250 days/year
ED = exposure duration (years); default value = 25 years
IR_s = soil ingestion rate (mg/day); default value = 50 mg/day
IR_a = workday inhalation rate ($\mathrm{m^3/day}$); default value = 20 $\mathrm{m^3/day}$
VF = soil-to-air volatilization factor for the chemical ($\mathrm{m^3/kg}$)
PEF = particulate emission factor ($\mathrm{m^3/kg}$); default value = $4.63 \times 10^9\ \mathrm{m^3/kg}$

The emission factors, VF and PEF, can be estimated as follows:

$$\mathrm{VF} = \left[\frac{(\mathrm{LS})(V)(\mathrm{DH})}{A}\right]\left[\frac{[(3.14)\alpha\,(\mathrm{T})]^{0.5}}{(2)(D_{ei})(E)(K_{\mathrm{as}})(10^{-3}\ \mathrm{kg/g})}\right] \tag{7.22}$$

where

$$\alpha = \frac{(D_{ei})(E)}{E + (\rho_s)[(1 - \mathrm{E})/K_{\mathrm{as}}]}$$

VF = soil to air volatilization factor ($\mathrm{m^3/kg}$)

LS = length of the contaminated area (m)
V = wind speed in the mixing zone (m/s); default value = 2.25 m/s
DH = diffusion height (m); default value = 2 m
A = area of contamination (cm^2)
D_{ei} = effective diffusivity of the contaminant; default estimate = $D_i \times E^{0.33}$
E = true soil porosity (unitless); default value = 0.35
K_{as} = soil–air partition coefficient (g soil/cm^3 air); default value = $41(H/K_d)$
ρ_s = true soil density (particle density) (g/cm^3); default value = 2.65 g/cm^3
T = exposure interval(s); default value = 7.9×10^8 s
D_i = molecular diffusivity of the contaminant (cm^2/s)
H = Henry's law constant of the contaminant (atm · m^3/mol)
K_d = soil–water partition coefficient of the contaminant (cm^3/g) = $(K_{oc})(OC)$ for organic contaminants
K_{oc} = organic carbon partition coefficient of the contaminant (cm^3/g)
OC = organic carbon content of the soil (fraction); default value = 0.02

It should be noted that α does not represent any specific fundamental factor but is merely a collection of terms and has units of cm^2/s. The U.S. EPA (1991) also provides Eq. (7.23) for estimation of the particulate emission factor, PEF:

$$\text{PEF} = \left[\frac{(\text{LS})(V)(\text{DH})(3600\ \text{s/h})}{A}\right]\left\{\frac{1000\ \text{g/kg}}{(0.036)(1 - \text{G})(\mu_m/\mu_t)}[\text{F(X)}]\right\} \tag{7.23}$$

where

PEF = particulate emission factor (m^3/kg); default value = 4.63×10^9 m^3/kg
G = fraction of vegetative cover (unitless); default value = 0
μ_m = mean annual wind speed (m/s); default value = 4.5 m/s
μ_t = equivalent threshold value of wind speed at 10 m height (m/s); default value = 12.8 m/s
$F(X)$ = A function of the ratio μ_m/μ_t, unitless); default value = 0.0497

All other parameters are as defined for Eq. (7.22). The number 0.036 in Eq. (7.23) is the approximate magnitude of the respirable fraction of particulates in units of g/m^2-hr.

For noncarcinogenic effects relevant to the soil medium in commercial/

industrial land use, the setup of the risk equation for back-calculation of the target contaminant concentration is similar to the approach represented by Eq. (7.20). A total target hazard index value of 1.0 and relevant values of reference dose are used.

7.3.2 Clean-up Level for Groundwater

Following an approach similar to that used for soils in Section 7.3.1, the risk equation can be set up to cover as many exposure pathways as are found to be significant for a site. For residential land use, ingestion of water and inhalation of volatiles from household water are two common pathways. It may be necessary to reduce the exposure frequency, EF, for commercial land use calculations. Using the example of carcinogenic effects for a groundwater contaminant within a residential environment, the risk-based maximum concentration limit for a combination of ingestion and volatiles inhalation pathways is as follows:

$$\mathrm{TR}_c = \left[\frac{(\mathrm{SF}_o)(C)(\mathrm{IR}_w)(\mathrm{EF})(\mathrm{ED})}{(\mathrm{BW})(\mathrm{AT})(365\ \text{days/year})}\right] + \left[\frac{(\mathrm{SF}_i)(C)(\mathrm{K})(\mathrm{IR}_a)(\mathrm{EF})(\mathrm{ED})}{(\mathrm{BW})(\mathrm{AT})(365\ \text{days/year})}\right] \tag{7.24}$$

By rearrangement,

$$C = \frac{(\mathrm{TR}_c)(\mathrm{BW})(\mathrm{AT} \times 365\ \text{days/year})}{(\mathrm{EF})(\mathrm{ED})[(\mathrm{SF}_i)(K)(\mathrm{IR}_a) + (\mathrm{SF}_o)(\mathrm{IR}_w)]} \tag{7.25}$$

where

C = risk-based chemical concentration in water (mg/liter)
TR_c = target individual lifetime cancer risk (unitless), usually specified as 10^{-6}
SF_i = inhalation cancer slope factor for the contaminant $[(\text{mg/kg} \cdot \text{day})^{-1}]$
SF_o = oral cancer slope factor for the contaminant $[(\text{mg/kg}\cdot\text{day})^{-1}]$
BW = adult body weight (kg); default value = 70 kg
AT = averaging time (years); default value = 70 years
EF = exposure frequency (days/years); default value = 350 days/year
ED = exposure duration (years); default value = 30 years
IR_a = daily indoor inhalation rate (m^3/day); default value = 15 m^3/day
IR_w = daily water ingestion rate (liters/day); default value = 2 liters/day
K = volatilization factor (unitless); default value = 0.0005×1000 liters/m^3 as suggested by U.S. EPA (1991)

Essentially, the calculation of risk-based concentration levels for remedial action requires various categories of data: toxicological data, data on the characteristics of the contaminant and exposure media, land use data, as well as the cultural activities of the target population. Several uncertainties are associated with the data used in such calculations. Nevertheless, the risk-based approach provides the most rational way of justifying clean-up actions and evaluating technologies of knownoperational efficiencies with respect to the extent to which their application can reduce human health risk at an exposure point.

REFERENCES

Aller, L. Bennett, T., Lehr, J. H., and Petty, R. J. (1985). DRASTIC: A standardization system for evaluating groundwater pollution potential using hydrogeologic settings. Technical Report Under Cooperative Agreement CX-819715-01, Office of Research and Development, U.S. Environmental Protection Agency, Ada, OK.

Call, Hollis J., and Merkhofer, Miley W. (1988). A Multi-attribute Utility Analysis Model for Ranking Superfund Sites. *Proc. 9th Natl. Superfund Conf.*, Washington, DC, pp. 44–54.

Hushon, J. M., and Kornreich, M. R. (1984). Scoring systems for hazard assessment. In: J. Sacena (ed.), *Hazard Assessment of Chemicals, Vol. 3.*, pp. 63–109, Academic Press, Orlando, FL.

Lee, C. H., Wang, H. C., Lin, C. M., and Yang, C. C. (1994). A long-term leachability study of solidified wastes by the multiple toxicity characteristic leaching procedure. *J. Hazardous Materials*, 38:65–74.

Lerose, G. A. (1995). When to use the TCLP test method and when not to. *Virginia's Environment*, July.

Manahan, S. E. (1990). *Hazardous Waste Chemistry, Toxicology and Treatment.* Lewis Publishers, Chelsea, MI.

Mayernik, J. A., and Fehrenkamp, K. (1992). A new model for conducting quantitative ecological risk assessments at hazardous waste sites. *Proc. 13th Annual Natl. HMC/ Superfund Conf. and Exhibition*, Washington, DC, pp. 813–819.

Merz, E. R. (1988). Criteria for long-term hazard assessment of chemotoxic and radiotoxic waste disposal. *Proc. Int. Topical Meeting on Nuclear and Hazardous Waste Management*, Pasco, WA, pp. 316–318.

Mikroudis, G. K., Fang, H-Y, and Wilson, J. L. (1986). Development of Geotox expert system for assessment of hazardous waste sites. *Proc. Int. Symp. on Environmental Geotechnology*, Bethlehem, PA, Vol. 1, pp. 223–232.

National Research Council. (1988). Complex mixtures. A report of the Committee on Methods for the in Vivo Toxicity Testing of Complex Mixtures, Board on Environmental Studies and Toxicology, National Research Council, Washington, DC.

Shahane, A. N., and Inman, R. C. (1987). Review of available pesticide assessment rating systems. In: J. Saxena (ed.), *Hazard Assessment of Chemicals*, pp. 103–300, Hemisphere, Washington, DC.

Stults, R. G. (1993). Hazardous waste: Identification. *Hazmat World*, March: 55–60.

U. S. EPA. (1986). Guidelines for the health assessment of suspect developmental toxicants. 51 profiles. U.S. Environmental Protection Agency, in 52 *Federal Register*, 34028, September 24.

U. S. EPA. (1989a). Stabilization/solidification of CERCLA and RCRA wastes: Physical tests, chemical testing procedures, technology screening, and field activities, EPA/625/6-89/022. Office of Research and Development, U. S. Environmental Protection Agency, Washington, DC.

U. S. EPA. (1989b). Risk assessment guidance for Superfund: Volume 1, human health evaluation manual, part A, EPA/540/1-89/002. Office of Emergency and Remedial Response, U. S. Environmental Protection Agency, Washington, DC.

U. S. EPA. (1991). Risk assessment guidance for Superfund: Volume 1, human health evaluation manual, part B, development of risk-based preliminary remediation goals, EPA/540/R-23/003. Office of Research and Development, U. S. Environmental Protection Agency, Washington, DC.

Whittemore, A. and Keller, J. B. (1978). Quantitative theories of carcinogens *SIAM Rev*., 20:1–30.

8

Principles of Site and Geomaterial Treatment Techniques

The focus of this chapter is on the principles that underlie various treatment techniques for contaminated sites and excavated geomaterials. Several vendors have given a plethora of trade names to derivatives and combinations of technologies that are based on the basic treatment approaches and techniques described herein. This chapter focuses not on the technologies but on the general principles. Where necessary and possible, equations which describe the quantitative relationships among the parameters that define the effectiveness of techniques are provided.

8.1 TREATMENT APPROACHES

In Chapter 7, the characteristics which determine the categorization of wastes as "hazardous," were discussed. Furthermore, quantitative expressions for exposure assessment were found in that chapter to contain contaminant concentration. For off-site areas, exposure magnitudes are also influenced by the capacity of a contaminant to travel (mobility) from its source, through bounding media, to locations of concern. Treatment systems generally target the parameters that define risk (human and/or ecosystem risks). The major categories of treatment approaches aimed at reducing risks and exposure levels of contaminants are:

Modification of the chemical structure of the contaminant to reduce its toxicity
Modification of the chemical form of the contaminant to reduce its mobility (gas form is more mobile than liquid form, and liquid form is more mobile than solid form)
Change of contaminant phase to reduce volume
Modification of the characteristics of the transport media of the contaminant
Removal of the contaminant from the site

With respect to the direct treatment of waste to change the negative characteristics that render the waste "hazardous," the following basic treatment actions are employed.

Corrosive waste	pH adjustment
Ignitive waste	Oxidation or reduction
Reactive waste	Oxidation or reduction
Toxic waste	pH adjustment for metals, and oxidation, reduction, and pretreatment for organics

Usually, a "treatment train" is designed to treat contaminated media where the chemistry is complex and existing technologies are effective for only some contaminants in specific concentration ranges.

8.2 IN-SITU VERSUS EX-SITU TREATMENT

To the extent possible, in-situ treatment is preferable to ex-situ treatment because the former does not involve the expensive excavation and handling of materials. However, in-situ treatment is sometimes not possible because the treatment technique selected requires greater control of the processes than can be done in the subsurface.

Batch treatment systems are commonly used in ex-situ treatment. The contaminated media have the maximum possibility of contact with the energy or material used in treatment. Under in-situ conditions, that potential decreases immensely and is determined by several geotechnical/soil hydrological factors such as soil permeability, stratification, flow channeling, and intensity of the energy or treatment fluid pressure imposed on the soil. Table 8.1 shows ratings developed by a U.S. National Research Council Panel (Macdonald and Kavanaugh, 1994) for the cleanup of groundwater contaminated by various substances in a variety of hydrogeologic settings.

8.3 BASIS FOR TREATMENT TECHNOLOGY SELECTION

For any contaminated site or material, there is usually a number of treatment technologies that could be selected. The selection criteria for such technologies are summarized below, using information published by Ram et al. (1993) and LaGrega and Evans (1987).

Table 8.1 Ratings of the Relative Ease of Cleaning up Contaminated Groundwater

	Contaminant chemistry					
Hydrogeology	Mobile, dissolved (degrades/ volatilizes)	Mobile, dissolved	Strongly sorbed, dissolved (degrades/ volatilizes)	Strongly sorbed, dissolved	Separate phase LNAPL[a]	Separate phase DNAPL[b]
Homogeneous, single layer	1	1–2	2	2–3	2–3	3
Homogeneous, multiple layers	1	1–2	2	2–3	2–3	3
Heterogeneous, single layer	2	2	3	3	3	4
Heterogeneous, multiple layers	2	2	3	3	3	4
Fractured	3	3	3	3	4	4

The difficulty of cleanup is influenced by the hydrogeologic conditions and contaminant chemistry at a site. The NRC report classified the relative ease of cleanup as a function of these two conditions on a scale of 1–4, where 1 is the easiest and 4 the most difficult. The 1–4 scale used in this table should not be viewed as objective and fixed, but as a subjective, flexible method for evaluating sites. Other factors that influence ease of cleanup, such as the total contaminant mass at a site and the length of time since it was released, are not shown in this table.
[a] Light non-aqueous-phase liquid.
[b] Dense non-aqueous-phase liquid.
Source: Macdonald and Kavanaugh (1994).

1. *Applicability of the technology to the chemistry and concentrations of contaminants at the site*. This is the principal criterion for the selection of treatment technologies. Treatment system effectiveness may vary with changes in reaction conditions, contaminant concentrations, and contaminant compositions. Preliminary information about the contaminants present at a site may give an indication about the treatment technique that is likely to work. For example, contaminants with Henry's law constant greater than 0.1 are removable by air stripping technique (Ram et al., 1993).

2. *Possibility of effectiveness enhancement with other treatment technologies*. Some technologies may be selected because they can be combined with others in a "treatment train" that is more effective than each of the constituent technologies alone. Each technology may be effective only for a particular con-

taminant, in a specific concentration range, under a given set of environmental conditions. Outside these bounds, the ancillary technology in the treatment train would be designed to clean up the contaminant(s). Examples of treatment trains used in the Superfund Remedial Actions of the U.S. Environmental Protection Agency during the period 1982–1995 are illustrated in Figure 8.1.

3. *Site Characteristics*. Site characteristics play a very significant role

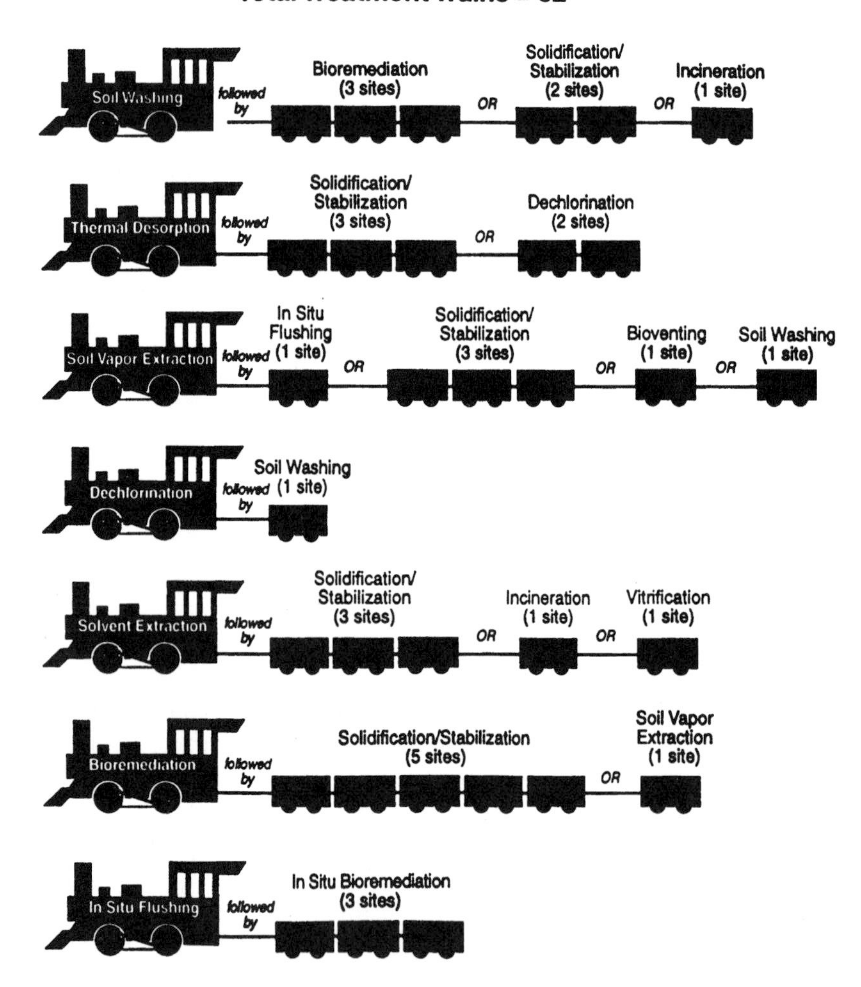

Figure 8.1 Treatment trains used in various superfund projects of the U.S. Environmental Protection Agency for the period 1982–1995 (U.S. EPA 1996).

in determining the potential effectiveness of a technology. For example, some treatment technologies such as soil venting and air sparging require that the soil formation be reasonably permeable to air for the removal of contaminants. Also, in clays that have negligible permeability, it may be difficult to deliver nutrients to microbes if in-situ bioremediation is selected as the treatment technology.

4. *Required treatment time*. Due to differences in the kinetics of the processes involved, different treatment technologies require different durations for completion. However, it should be noted that chemical and/or physical reaction time is only a fraction of the time required to implement a treatment technology. Time is also required for site preparation, equipment setup, and site safety programs. Time may be a limitation for some technologies if risk assessments and other regulatory considerations specify a time limit for the completion of cleanup.

5. *Regulatory acceptance of the technology*. Some technologies may be favored by regulations such that the decision maker has a limited number of treatment options from which to select. For example, if regulations specify that off-gas from a treatment technique must be treated, a technology such as air stripping that produces off-gas may fall out of contention because of the additional cost of off-gas treatment.

6. *Cost-effectiveness of the treatment technology*. In most cases, this is the overriding criterion. Most technologies could be effective if resources are unlimited for their implementation. Cost-effectiveness is a good measure of the returns on investment of resources on a technology at a contaminated site. It also relates to the duration of implementation, which affects cost.

Using various criteria, most of which are subcategories of the primary criteria described above, the U.S. EPA (1993) developed the screening table shown as Table 8.2 for selecting treatment technologies for various classes of contaminants. In order to address uncertainties about the potential effectiveness of specific technologies in treating contaminated media present at a site, treatment feasibility studies are usually conducted in two stages before full implementation in the field. The first stage is bench-scale feasibility studies, in which treatment process parameters are analyzed through testing of samples of the contaminated media in the laboratory. Based on the results obtained, a pilot-scale feasibility study is conducted to demonstrate the effectiveness of the technology in the field before its full-scale operation.

8.4 PUMP-AND-TREAT PRINCIPLES

Pump-and-treat systems involve the removal of contaminated groundwater through the use of extraction wells. The extracted groundwater is then treated above ground. There are two principal categories of pump-and-treat systems:

Table 8.2 Remediation Technologies Screening Table

NOTE: There are factors that may limit the applicability and effectiveness of any of the technologies and processes listed below. These factors are discussed in the *Remediation Technologies Screening Matrix Reference Guide*. This Matrix should always be used in conjunction with the *Reference Guide*, which contains additional information that can be useful in identifying potentially applicable technologies.

	Reference Guide Page Number	Status — (F)ull-scale or (P)ilot-scale	Contaminants/Pollutants Treated ª	Overall Cost	Capital (Cap) or O&M Intensive?	Commercial Availability	Typically Part of a Treatment Train? (excludes off-gas treatment)	Residuals Produced — (S)olid, (L)iquid, or (V)apor?	Minimum Contaminant Concentration Achievable	Addresses (T)oxicity, (M)obility, or (V)olume?	Long-Term Effectiveness/ Permanence?	Time to Complete Cleanup	System Reliability/ Maintainability	Awareness of Remediation Consulting Community	Regulatory/Permitting Acceptability	Community Acceptability
SOIL SEDIMENT AND SLUDGE																
***In Situ* Biological Processes**																
Biodegradation	21	F	3,4,5 1,2,6	○	O&M	■	No	None	○	T	Yes	▲	▲	○	▲	■
Bioventing	23	F	3,4,5 1,2,6	■	Neither	■	No	None	■	T	Yes	○	■	○	○	■
***In Situ* Physical/Chemical Processes**																
Soil Vapor Extraction (SVE)	25	F	1,3,5	■	O&M	■	No	L	○	V	Yes	○	■	■	■	■
Soil Flushing	27	P	1,3,7 2,4-6	I	O&M	■	No	L	▲	V	Yes	▲	○	○	▲	○
• Solidification/Stabilization	29	F	7 2,4,6	■	Cap	■	No	S	NA	M	I	■	■	○	○	○
Pneumatic Fracturing (enhancement)	31	P	1-7	■	Neither	▲	Yes	None	NA	M	Yes	NA	■	▲	I	I
***In Situ* Thermal Processes**																
Vitrification	33	P	7 1-6	▲	Both	▲	No	L	NA	M	Yes	■	▲	○	▲	▲
Thermally Enhanced SVE	35	F	2,4,6 1,3,5	○	Both	○	No	L	○	V	Yes	■	○	○	○	○
***Ex Situ* Biological Processes (assuming excavation)**																
Slurry Phase Biological Treatment	37	F	3,5 1,2,4,6	○	Both	○	No	None	○	T	Yes	○	○	○	■	○
Controlled Solid Phase Bio. Treatment	39	F	3,5 1,2,4,6	■	Neither	■	No	None	○	T	Yes	○	■	■	■	○
• Landfarming	41	F	3,5 1,2,4,6	■	Neither	■	No	None	○	T	Yes	▲	■	■	○	○
***Ex Situ* Physical/Chemical Processes (assuming excavation)**																
Soil Washing	43	F	2,4,5,7 1,3,6	○	Both	○	Yes	S,L	○	V	Yes	■	○	○	○	■
• Solidification/Stabilization	45	F	7 2,4,6	■	Cap	■	No	S	NA	M	I	■	■	■	○	○
Dehalogenation (Glycolate)	47	F	2,6 1	▲	Both	○	No	L	■	T	Yes	▲	▲	○	○	○
Dehalogenation (BCD)	49	F	2,6 1	I	I	▲	No	V	I	T	Yes	I	I	▲	I	I
Solvent Extraction (chemical extraction)	51	F	2,4,6 1,3,5	▲	Both	○	Yes	L	○	V	Yes	▲	○	○	○	○
Chemical Reduction/Oxidation	53	F	7 3-6	○	Neither	■	Yes	S	NA	T,M	I	■	■	○	○	○
Soil Vapor Extraction (SVE)	55	F	1,3	■	Neither	■	No	L	○	V	Yes	○	■	■	○	○
***Ex Situ* Thermal Processes (assuming excavation)**																
Low Temperature Thermal Desorption	57	F	1,3,5 2,4,6	■	Both	■	Yes	L	■	V	Yes	■	○	■	○	○
High Temperature Thermal Desorption	59	F	2,4,6 1,3,5	○	Both	■	Yes	L	■	V	Yes	■	○	○	○	○
Vitrification	61	F	7 1-6	▲	Both	○	No	L	NA	M	Yes	○	○	○	▲	▲
• Incineration	63	F	2,4,6 1,3,5	▲	Both	■	No	L,S	■	T	Yes	■	○	■	○	▲
Pyrolysis	65	P	2,4,6 1,3,5	▲	Both	▲	No	L,S	■	T	Yes	■	I	▲	○	▲

	Other Processes																
* Natural Attenuation	67	NA	3,4,5	1,2,6	■	Neither	■	No	None	I	T	Yes	▲	■	○	▲	▲
Excavation and Off-Site Disposal	71	NA		1-7	▲	Neither	■	No	NA	NA	M	No	■	■	■	▲	■
	GROUNDWATER																
	In Situ Biological Processes																
Oxygen Enhancement with H_2O_2	73	F	3,4,5	1,2,6	○	O&M	■	No	None	■	T	Yes	○	▲	■	○	■
Co-metabolic Processes	75	P	1,2	3-6	○	O&M	▲	No	None	■	T	Yes	○	▲	▲	I	I
Nitrate Enhancement	77	P	3,4,5	1,2,6	■	Neither	▲	No	None	■	T	Yes	○	○	▲	▲	○
Oxygen Enhancement with Air Sparging	79	F	3,4,5	1,2,6	■	Neither	■	No	None	■	T	Yes	○	■	○	○	■
	In Situ Physical/Chemical Processes																
* Slurry Walls (containment only)	81	F		1-7	■	Cap	■	NA	NA	NA	M	I	■	■	■	▲	○
Passive Treatment Walls	83	P	1,2,7	3,4,5	I	Cap	▲	No	S	I	T	I	▲	I	▲	I	I
Hot Water or Steam Flushing/Stripping	85	P	2,4,5	1,3	○	Cap	○	Yes	L,V	○	V	Yes	■	▲	▲	○	○
Hydrofracturing (enhancement)	87	P		1-7	○	Neither	I	Yes	None	NA	M	Yes	■	■	▲	■	○
Air Sparging	89	F	1,3,5		■	Neither	■	Yes	V	○	V	Yes	■	■	○	■	■
Directional Wells (enhancement)	91	F		1-7	I	Neither	▲	Yes	L,S	NA	NA	Yes	■	○	○	■	■
Dual Phase Extraction	93	F	1,3,5		○	O&M	■	Yes	L,V	○	V	Yes	○	○	■	○	■
Vacuum Vapor Extraction	95	P	1,2,5	3,4,6,7	○	Cap	▲	No	L,V	■	V	Yes	○	■	▲	○	■
* Free Product Recovery	97	F	4,5		■	Neither	■	No	L	NA	V	Yes	■	○	■	■	■
	Ex Situ Biological Processes (assuming pumping)																
Bioreactors	99	F	3,4,5	1,2,6	■	Cap	■	No	S	○	T	Yes	NA	○	○	■	○
	Ex Situ Physical/Chemical Processes (assuming pumping)																
* Air Stripping	101	F	1,3	2,4,5	■	O&M	■	No	L,V	■	V	Yes	NA	○	■	▲	○
* Carbon Adsorption (liquid phase)	103	F	2,4	1,5-7	▲	O&M	■	No	S	■	V	Yes	NA	■	■	■	■
UV Oxidation	105	F	1,2,6	3,5	○	Cap	■	No	None	■	T	Yes	NA	▲	○	○	○
	Other Processes																
* Natural Attenuation	107	NA	3,4,5	1,2,6	■	Neither	■	No	None	I	T	Yes	▲	■	○	▲	▲
	AIR EMISSIONS/OFF GAS TREATMENT																
* Carbon Adsorption (vapor phase)	111	F	1-6		■	Neither	■		S	■	V	Yes	NA	■	■	○	■
* Catalytic Oxidation (non-halogenated)	113	F	3,4,5		■	Neither	■		None	■	T	Yes	NA	■	■	■	■
Catalytic Oxidation (halogenated)	115	F	1,2		■	Neither	○	NA[b]	None	■	T	Yes	NA	○	▲	○	○
Biofiltration	117	F	3,5	1	■	Neither	○		None	■	T	Yes	NA	○	▲	I	I
* Thermal Oxidation	119	F	3,4,5		■	Neither	■		None	■	T	Yes	NA	■	■	■	○

[a] The listing of contaminant groups is intended as a general reference only. A technology may treat only selected compounds within the contaminant groups listed. Further investigation is necessary to determine applicability to specific contaminants.

[b] The definition of this factor is not applicable to these technologies. They are, by design, the final step in treatment processes.

* Conventional technologies/processes. Contaminant Codes: 1–Halogenated volatile organics; 2 -Halogenated semivolatile organics; 3–Non-halogenated volatile organics; 4–Non-halogenated semivolatile organics; 5–Fuel hydrocarbons; 6–Pesticides; 7–Inorganics.Target contaminants are listed first and in bold type. Rating Codes: ■, better; ○, average; ▲, worse; I, inadequate information; NA, not applicable.

Source: U.S. EPA (1993).

basic pump-and-treat systems and chemically (surfactant) enhanced pump-and-treat systems.

8.4.1 Basic Pump-and-Treat Principles

In the basic pump-and-treat system for contaminated site treatment, a series of recovery wells is installed to penetrate the medium containing contaminated groundwater and operated to remove the water to above-ground tanks for treatment. During pumping, the groundwater table is lowered. In an unconfined aquifer of relatively homogenous geomedia, a regular cone of depression is formed, the dimensions of which depend on hydraulic properties of the geomedia and the pumping rate. In most cases, the concentrations of contaminants in extracted groundwater decrease as pumping proceeds. This is often attributable to the inability of the contaminants to diffuse from inaccessible pores to accessible ones at a rate that can keep up with pumping. Also, the quantity of contaminants available for removal decreases as pumping proceeds.

An innovative development in pump-and-treat systems is pulsed pumping, in which resting phases are introduced into the pumping process, as illustrated in Figure 8.2, to allow sufficient time for contaminants to diffuse into high-permeability zones from which they can be more easily removed.

The Cooper and Jacob equation provides a means of relating operational parameters of an extraction well to aquifer characteristics for a case in which the regional groundwater gradient is less than 0.001. In traditional units, Eq. (8.1) can be used for relevant computations.

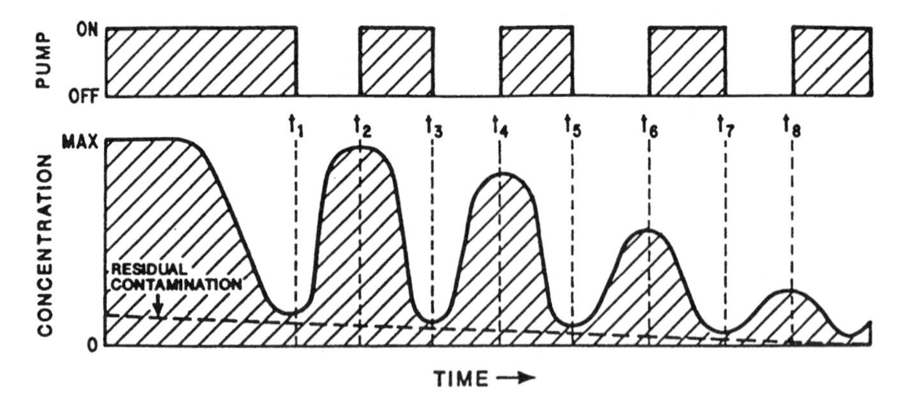

Figure 8.2 Pulsed pumping removal of residual contaminants from saturated media (U.S. EPA, 1989a).

$$Q = \left(\frac{sT}{264}\right)\left[\frac{1}{\log(0.3Tt/r_w^2S)}\right] \tag{8.1}$$

where

Q = pumping rate of a well (gal/min)
s = draw-down in the pumping well (ft)
T = aquifer transmissivity (gal/day per ft^2)
t = duration of pumping (days)
r_w = radius of well (ft)
S = aquifer storativity (percentage of connected pore volumes in the aquifer)

The desired draw-down and other parameters such as duration of pumping can be specified and pumping rate optimized to achieve the desired removal.

The effectiveness of pump-and-treat techniques depends on several physicochemical factors besides the adequacy of the design and efficiency of the pumping installation.

The presence of NAPLs. NAPLs are non-aqueous-phase liquids and are discussed in greater detail in chapter 5. They can be trapped in the pores of subsurface media by interfacial tension as illustrated in the top portion of Figure 8.3 such that they continue to be a source of contaminants with the advance of pumping. Examples of NAPLs are halogenated aliphatic hydrocarbons, halogenated benzenes, phthalate esters, and polychlorinated biphenyls. The transfer of the NAPLs to the surrounding aqueous phase is by diffusive liquid–liquid partitioning. At high groundwater velocities, the contaminant is not allowed to build up significantly in the aqueous phase before advection, as illustrated in Figure 8.3. At low groundwater velocities, the extent to which NAPLs transfer occurs is controlled by its solubility.

Sorption effects. As contaminants travel with water under imposed hydraulic gradients, a fraction of the contaminants is removed through sorption to subsurface media. As illustrated in Figure 8.4, initial desorption of contaminants from contaminated media may be relatively rapid. It will slow down as more tightly bound layers of the contaminant are reached.

Transport media hydraulic conductivity effects. High hydraulic conductivities in the regime of 10^{-5} cm/s, favor pump-and-treat operations. Heterogeneous media such as stratified layers of sand and clay may cause interlayer exchange of contaminants in directions that are not necessarily toward the extraction wells. Heterogeneity, even at a much smaller scale than thick stratified layers of fine–and coarse-grained media, is a large contributor to the ''tailing effect,'' in which contaminant concentration

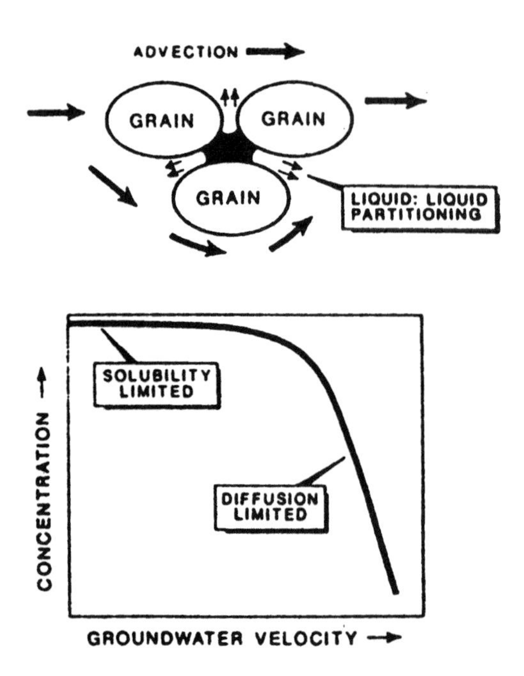

Figure 8.3 Schematic illustration of solubility and diffusion limitations to pump-and-treat systems (Keeley, 1989).

in the extracted moisture persists at low levels even after prolonged pumping.

The following mathematical expression can be used to estimate contaminant removal for a given rate and duration of pumping in a contaminated geologic medium. It should be noted that this expression is essentially a delay function of contaminant removal per unit of time and the goodness of fit of field data with this simple model can be affected negatively by transport media stratification and other heterogeneities.

$$R = \left(\frac{C_0 - C_t}{C_0}\right) = 1 - \exp\left[\left(\frac{Q}{K_d V}\right)t\right] \tag{8.2}$$

where

R = contaminant removal rate (M/T)
C_0 = initial concentration of contaminant (M/L^3)
C_t = concentration of contaminant at time t (M/L^3)
V = volume of contaminated groundwater (L^3)

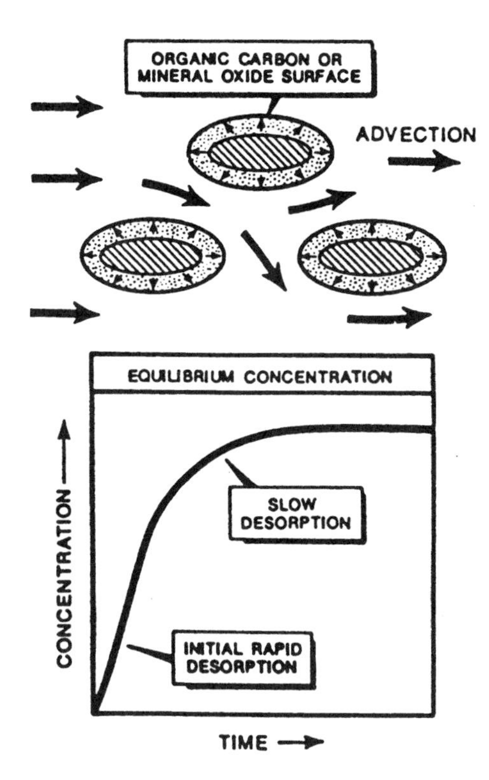

Figure 8.4 Schematic illustration of sorption limitations to pump-and-treat systems (Keeley, 1989).

Q = pumping rate of the groundwater extraction system (L^3/T)
K_d = distribution coefficient of the contaminant (M/L^3)

K_d can be obtained, typically through contaminant sorption measurements in the laboratory, for both organic and inorganic chemical compounds. For organic compounds, it can be estimated using Eq. (8.3):

$$K_d = K_{oc} f_{oc} \tag{8.3}$$

where

K_{oc} = partition coefficient of the organic compound (M/L^3)
f_{oc} = organic carbon fraction in the soil containing the contaminated groundwater (unitless fraction)

The organic compound partition coefficient, K_{oc}, can be estimated by correlation with other parameters such as the octanol–water partition coefficient and solubil-

Table 8.3 Characteristics of Chemical Compounds Including Their Solubilities in Water at 20°C and Octanol–Water Partition Coefficients (U.S. EPA 1990)

PESTICIDES												
Acrolein [2-Propenal]	107-02-8	PP	2.08E+05	H	2.69E+02	H	9.54E-05	X			8.13E-01	H
Aldicarb [Temik]	116-06-3		7.80E+03	E							5.00E+00	F
Aldrin	309-00-2	HPP	1.80E-01	A	6.00E-06	A	1.60E-05	A	9.60E+04	A	2.00E+05	A
Captan	133-06-2		5.00E-01	A	6.00E-05	A	4.75E-05	A	6.40E+03	B	2.24E+02	A
Carbaryl [Sevin]	63-25-2		4.00E+01	A	5.00E-03	A	3.31E-05	X	2.30E+02	G	2.29E+02	A
Carbofuran	1563-66-2		4.15E+02	G	2.00E-05	G	1.40E-08	X	2.94E+01	F	2.07E+02	F
Carbophenothion [Trithion]	786-19-6								4.66E+04	F		
Chlordane	57-74-9	HPP	5.60E-01	A	1.00E-05	A	9.63E-06	A	1.40E+05	A	2.09E+03	A
p-Chloroaniline [4-Chlorobenzenamine]	106-47-8	HSL	5.30E+03	L	2.00E-02	G	6.40E-07	X	5.61E+02	F	6.76E+01	H
Chlorobenzilate	510-15-6		2.19E+01	A	1.20E-06	A	2.34E-08	A	8.00E+02	B	3.24E+04	A
Chlorpyrifos [Dursban]	2921-88-2		3.00E-01	E	1.87E-05	J	2.87E-05	X	1.36E+04	E	6.60E+04	F
Crotoxyphos [Ciodrin]	7700-17-6		1.00E+03	E	1.40E-05	J	5.79E-09	X	7.48E+01	F		
Cyclophosphamide	50-18-0		1.31E+09	A					4.20E-02	B	6.03E-04	A
DDD	72-54-8	HPP	1.00E-01	A	1.89E-06	A	7.96E-06	A	7.70E+05	A	1.58E+06	A
DDE	72-55-9	HPP	4.00E-02	A	6.50E-06	A	6.80E-05	A	4.40E+06	A	1.00E+07	A
DDT	50-29-3	HPP	5.00E-03	A	5.50E-06	A	5.13E-04	A	2.43E+05	A	1.55E+06	A
Diazonin [Spectracide]	333-41-5		4.00E+01	E	1.40E-04	J	1.40E-06	X	8.50E+01	P	1.05E+03	F
1,2-Dibromo-3-chloropropane [DBCP]	96-12-8		1.00E+03	A	1.00E+00	A	3.11E-04	A	9.80E+01	B	1.95E+02	A
1,2-Dichloropropane	78-87-5	HPP	2.70E+03	A	4.20E+01	A	2.31E-03	A	5.10E+01	A	1.00E+02	A
1,3-Dichloropropene [Telone]	542-75-6	HPP	2.80E+03	A	2.50E+01	A	1.30E-01	A	4.80E+01	A	1.00E+02	A
Dichlorvos	62-73-7		1.00E+04	E	1.20E-02	J	3.50E-07	X			2.50E+01	E
Dieldrin	60-57-1	HPP	1.95E-01	A	1.78E-07	A	4.58E-07	A	1.70E+03	A	3.16E+03	A
Dimethoate	60-51-5		2.50E+04	A	2.50E-02	A	3.00E-07	X			5.10E-01	E
Dinoseb	88-85-7		5.00E+01	A	5.00E-05	G	3.16E-07	X	1.24E+02	E	1.98E+02	F
N,N-Diphenylamine	122-39-4		5.76E+01	A	3.80E-05	A	1.47E-07	A	4.70E+02	B	3.98E+03	A
Disulfoton	298-04-4		2.50E+01	E	1.80E-04	E	2.60E-06	X	1.60E+03	F		
alpha-Endosulfan	115-29-7	HPP	1.60E-01	H	1.00E-05	H	3.35E-05	X			3.55E+03	H
beta-Endosulfan	115-29-7	HPP	7.00E-02	H	1.00E-05	H	7.65E-05	X			4.17E+03	H
Endosulfan Sulfate	1031-07-8	HPP	1.60E-01	H							4.57E+03	H
Endrin	72-20-8	HPP	2.40E-02	E	2.00E-07	G	4.17E-06	X			2.18E+05	E

Endrin Aldehyde	7421-93-4	PP										
Endrin Ketone		HSL										
Ethion	563-12-2		2.00E+00	E	1.50E-06	J	3.79E-07	X	1.54E+04	E		
Ethylene Oxide	75-21-8		1.00E+06	A	1.31E+03	A	7.56E-05	A	2.20E+00	B	6.03E-01	A
Fenitrothion	122-14-5		3.00E+01	E	6.00E-06	J	7.30E-08	X			2.40E+03	E
Heptachlor	76-44-8	HPP	1.80E-01	A	3.00E-04	A	8.19E-04	A	1.20E-04	A	2.51E+04	A
Heptachlor Epoxide	1024-57-3	HPP	3.50E-01	A	3.00E-04	A	4.39E-04	A	2.20E+02	A	5.01E+02	A
alpha-Hexachlorocyclohexane	319-84-6	HPP	1.63E+00	A	2.50E-05	A	5.87E-06	A	3.80E+03	A	7.94E+03	A
beta-Hexachlorocyclohexane	319-85-7	HPP	2.40E-01	A	2.80E-07	A	4.47E-07	A	3.80E+03	A	7.94E+03	A
delta-Hexachlorocyclohexane	319-86-8	HPP	3.14E+01	A	1.70E-05	A	2.07E-07	A	6.60E+03	A	1.26E+04	A
gamma-Hexachlorocyclohexane [Lindane]	58-89-9	HPP	7.80E+00	A	1.60E-04	A	7.85E-06	A	1.08E+03	A	7.94E+03	A
Isophorone	78-59-1	HPP	1.20E+04	H	3.80E-01	H	5.75E-06	X			5.01E+01	H
Kepone	143-50-0		9.90E-03	A					5.50E+04	B	1.00E+02	A
Leptophos	21609-90-5		2.40E+00	E					9.30E+03	E	2.02E+06	E
Malathion	121-75-7		1.45E+02	A	4.00E-05	A	1.20E-07	X	1.80E+03	F	7.76E+02	A
Methoxychlor	72-43-5	HSL	3.00E-03	E					8.00E+04	E	4.75E+04	E
Methyl Parathion	298-00-0		6.00E+01	A	9.70E-06	A	5.59E-08	A	5.10E+03	F	8.13E+01	A
Mirex [Dechlorane]	2385-85-5		6.00E-01	C	3.00E-01	C	3.59E-01	X	2.40E+07	G	7.80E+06	D
Nitralin	4726-14-1		6.00E-01	E	9.30E-09	J	7.04E-09	X	9.60E+02	G		
Parathion	56-38-2		2.40E+01	G	3.78E-05	J	6.04E-07	X	1.07E+04	F	6.45E+03	F
Phenylurea [Phenylcarbamide]	64-10-8								7.63E+01	F	6.61E+00	M
Phorate [Thimet]	298-02-2		5.00E+01	E	8.40E-04	J	8.49E-11	X	3.26E+03	F		
Phosmet	732-11-6		2.50E+01	E	<1.0E-03	J					6.77E+02	E
Ronnel [Fenchlorphos]	299-84-3		6.00E+00	E	8.00E-04	J	5.64E-05	X			4.64E+04	E
Strychnine	57-24-9		1.56E+02	A							8.51E+01	M
2,3,7,8-Tetrachlorodibenzo-p-dioxin	1746-01-6		2.00E-04	A	1.70E-06	A	3.60E-03	A	3.30E+06	A	5.25E+06	A
Toxaphene	8001-35-2	HPP	5.00E-01	A	4.00E-01	A	4.36E-01	A	9.64E+02	A	2.00E+03	A
Trichlorfon [Chlorofos]	52-68-6		1.54E+05	A	7.80E-06	A	1.71E-11	A	6.10E+00	B	1.95E+02	A
HERBICIDES												
Alachlor	15972-60-8		2.42E+02	E					1.90E+02	E	4.34E+02	F
Ametryn	834-12-8		1.85E+02	E					3.88E+02	F		
Amitrole [Aminotriazole]	61-82-5		2.80E+05	A					4.40E+00	B	8.32E-03	A
Atrazine	1912-24-9		3.30E+01	G	1.40E-06	K	2.59E-13	X	1.63E+02	F	2.12E+02	F
Benfluralin [Benefin]	1861-40-1		<1.0E+00	E	3.89E-04	J			1.07E+04	E		
Bromocil	314-40-9		8.20E+02	P					7.20E+01	F	1.04E+02	F

Table 8.3 Continued

Cacodylic Acid	75-60-5		8.30E+05	A					2.40E+00	B	1.00E+00	A
Chloramben	133-90-4		7.00E+02	E	<7.0E-03	J			2.10E+01	E	1.30E+01	F
Chlorpropham	101-21-3		8.80E+01	E					8.16E+02	F	1.16E+03	F
Dalapon [2,2-Dichloropropanoic Acid]	75-99-0		5.02E+05	E							5.70E+00	F
Diallate	2303-16-4		1.40E+01	A	6.40E-03	A	1.65E-04	A	1.90E+03	G	5.37E+00	A
Dicamba	1918-00-9		4.50E+03	E	2.00E-05	G	1.30E-09	X	2.20E+00	F	3.00E+00	F
Dichlobenil [2,6-Dichlorobenzonitrile]	1194-65-6		1.80E+01	E	3.00E-06	J	3.77E-08	X	2.24E+02	F	7.87E+02	F
2,4-Dichlorophenoxyacetic Acid [2,4-D]	94-75-7		6.20E+02	A	4.00E-01	A	1.88E-04	A	1.96E+01	F	6.46E+02	A
Dipropetryne	47-51-7		1.60E+01	J	7.50E-07	J	1.53E-08	X	1.18E+03	F		
Diuron	330-54-1		4.20E+01	E	<3.1E-06	J			3.82E+02	F	6.50E+02	F
Fenuron	101-42-8		3.85E+03	E	<1.6E-04	K			4.22E+01	F	1.00E+01	E
Fluometuron	2164-17-2		9.00E+01	G					1.75E+02	G	2.20E+01	E
Linuron	330-55-2		7.50E+01	E	1.50E-05	J	6.56E-08	X	8.63E+02	F	1.54E+02	E
Methazole [Oxydiazol]	20354-26-1		1.50E+00	E					2.62E+03	E		
Metobromuron	3060-89-7		3.30E+02	E	3.00E-06	J	3.10E-09	X	2.71E+02	F		
Monuron	150-68-5		2.30E+02	E	5.00E-07	J	5.68E-10	X	1.83E+02	F	1.33E+02	F
Neburon	555-37-3		4.80E+00	E					3.11E+03	F		
Oxadiazon	19666-30-9		7.00E-01	E	<1.0E-06	J			3.24E+03	E		
Paraquat	4685-14-7		1.00E+06	E					1.55E+04	E	1.00E+00	F
Phenylmercuric Acetate [PMA]	62-38-4		1.67E+03	A								
Picloram	1918-02-1		4.30E+02	E	<6.2E-07	K			2.55E+01	F	2.00E+00	F
Prometryne	7287-19-6		4.80E+01	E	1.00E-06	J	6.62E-09	X	6.14E+02	F		
Propachlor	1918-16-7		5.80E+02	E					2.65E+02	E	5.60E+02	E
Propazine	139-40-2		8.60E+00	E	1.60E-07	K	5.63E-09	X	1.53E+02	F	7.85E+02	E
Silvex [Fenoprop]	93-72-1		1.40E+02	E					2.60E+03	E		
Simazine	122-34-9		3.50E+00	E	3.60E-08	K	2.73E-09	X	1.38E+02	F	8.80E+01	F
Terbacil	5902-51-2		7.10E+02	E					4.12E+01	F	7.80E+01	F
2,4,5-Trichlorophenoxyacetic Acid	93-76-5		2.38E+02	E					8.01E+01	F	4.00E+00	E
Triclopyr	55335-06-3		4.30E+02	E	1.26E-06	J	9.89E-10	X	2.70E+01	E	3.00E+00	E
Trifluralin	1582-09-8		6.00E-01	E	2.00E-04	G	1.47E-04	X	1.37E+04	E	2.20E+05	E
ALIPHATIC COMPOUNDS												
Acetonitrile [Methyl Cyanide]	75-05-8		infinite	A	7.40E+01	A	4.00E-06	A	2.20E+00	B	4.57E-01	A
Acrylonitrile [2-Propenenitrile]	107-13-1	PP	7.94E+04	A	1.00E+02	A	8.84E-05	A	8.50E-01	A	1.78E+00	A
Bis(2-chloroethoxy)methane	111-91-1	HPP	8.10E+04	I	<1.0E-01	I					1.82E+01	I
Bromodichloromethane [Dichlorobromometh]	75-27-4	HPP	4.40E+03	Q	5.00E+01	H	2.40E-03	Q	6.10E+01	Q	7.59E+01	I
Bromomethane [Methyl Bromide]	74-83-9	HPP	1.30E+04	G	1.40E+03	G	1.30E-02	G			1.26E+01	I
1,3-Butadiene	106-99-0		7.35E+02	A	1.84E-03	A	1.78E-01	A	1.20E+02	B	9.77E+01	A

Chloroethane [Ethyl Chloride]	75-00-3	HPP	5.74E+03	C	1.00E+03	C	6.15E-04	X	1.70E+01	C	3.50E+01	C
Chloroethene [Vinyl Chloride]	75-01-4	HPP	2.67E+03	A	2.66E+03	A	8.19E-02	A	5.70E+01	B	2.40E+01	A
Chloromethane [Methyl Chloride]	74-87-3	HPP	6.50E+03	A	4.31E+03	A	4.40E-02	A	3.50E+01	B	9.50E-01	A
Cyanogen [Ethanedinitrile]	460-19-5		2.50E+05	A								
Dibromochloromethane	124-48-1	HPP	4.00E+03	Q	1.50E+01	A	9.90E-04	Q	8.40E+01	Q	1.23E+02	A
Dichlorodifluoromethane [Freon 12]	75-71-8		2.80E+02	A	4.87E+03	A	2.97E+00	X	5.80E+01	A	1.45E+02	A
1,1-Dichloroethane [Ethylidine Chloride]	75-34-3	HPP	5.50E+03	A	1.82E+02	A	4.31E-03	A	3.00E+01	A	6.17E+01	A
1,2-Dichloroethane [Ethylene Dichloride]	107-06-2	HPP	8.52E+03	A	6.40E+01	A	9.78E-04	A	1.40E+01	A	3.02E+01	A
1,1-Dichloroethene [Vinylidine Chloride]	75-35-4	HPP	2.25E+03	A	6.00E+02	A	3.40E-02	A	6.50E+01	A	6.92E+01	A
1,2-Dichloroethene (cis)	540-59-0		3.50E+03	A	2.08E+02	A	7.58E-03	A	4.90E+01	B	5.01E+00	A
1,2-Dichloroethene (trans)	540-59-0	HPP	6.30E+03	A	3.24E+02	A	6.56E-03	A	5.90E+01	A	3.02E+00	A
Dichloromethane [Methylene Chloride]	75-09-2	HPP	2.00E+04	A	3.62E+02	A	2.03E-03	A	8.80E+00	A	2.00E+01	A
Ethylene Dibromide [EDB]	106-93-4		4.30E+03	A	1.17E+01	A	6.73E-04	A	4.40E+01	A	5.75E+01	A
Hexachlorobutadiene	87-68-3	HPP	1.50E-01	A	2.00E+00	A	4.57E+00	A	2.90E+04	A	6.02E+04	A

Hexachlorocyclopentadiene	77-47-4	HPP	2.10E+00	A	8.00E-02	A	1.37E-02	A	4.80E+03	A	1.10E+05	A
Hexachloroethane [Perchloroethane]	67-72-1	HPP	5.00E+01	A	4.00E-01	A	2.49E-03	A	2.00E+04	A	3.98E+04	A
Iodomethane [Methyl Iodide]	77-88-4		1.40E+04	A	4.00E+02	A	5.34E-03	A	2.30E+01	B	4.90E+01	A
Isoprene	78-79-5				4.00E+02	A						
Pentachloroethane [Pentalin]	76-01-7		3.70E+01	C	3.40E+00	C	2.44E-02	X	1.90E+03	D	7.76E+02	C
1,1,1,2-Tetrachloroethane	630-20-6		2.90E+03	A	5.00E+00	A	3.81E-04	A	5.40E+01	B		
1,1,2,2-Tetrachloroethane	79-34-5	HPP	2.90E+03	A	5.00E+00	A	3.81E-04	A	1.18E+02	A	2.45E+02	A
Tetrachloroethene [PERC]	127-18-4	HPP	1.50E+02	A	1.78E+01	A	2.59E-02	A	3.64E+02	A	3.98E+02	A
Tetrachloromethane [CarbonTetrachloride]	56-23-5	HPP	7.57E+02	A	9.00E+01	A	2.41E-02	A	4.39E+02	Q	4.37E+02	A
Tribromomethane [Bromoform]	75-25-2	HPP	3.01E+03	A	5.00E+00	A	5.52E-04	A	1.16E+02	A	2.51E+02	A
1,1,1-Trichloroethane [Methylchloroform]	71-55-6	HPP	1.50E+03	A	1.23E+02	A	1.44E-02	A	1.52E+02	A	3.16E+02	A
1,1,2-Trichloroethane [Vinyltrichloride]	79-00-5	HPP	4.50E+03	A	3.00E+01	A	1.17E-03	A	5.60E+01	A	2.95E+02	A
Trichloroethene [TCE]	79-01-6	HPP	1.10E+03	A	5.79E+01	A	9.10E-03	A	1.26E+02	A	2.40E+02	A
Trichlorofluoromethane [Freon 11]	75-69-4	PP	1.10E+03	A	6.67E+02	A	1.10E-01	Q	1.59E+02	A	3.39E+02	A
Trichloromethane [Chloroform]	67-66-3	HPP	8.20E+03	A	1.51E+02	A	2.873-03	A	4.70E+01	C	9.33E+01	A
1,1,2-Trichloro-1,2,2-trifluoroethane	76-13-1		1.00E+01	A	2.70E+02	A					1.00E+02	A
AROMATIC COMPOUNDS												
1,1-Biphenyl [Diphenyl]	92-52-4		7.50E+00	E	6.00E-02	G	1.50E-03	G			7.54E+03	E
Benzene	71-43-2	HPP	1.75E+03	A	9.52E+01	A	5.59E-03	A	8.30E+01	A	1.32E+02	A
Bromobenzene [Phenyl Bromide]	108-86-1		4.46E+02	E	4.14E+00	O	1.92E-03	X	1.50E+02	P	9.00E+02	E
Chlorobenzene	108-90-7	HPP	4.66E+02	A	1.17E+01	A	3.72E-03	A	3.30E+02	Q	6.92E+02	A
4-Chloro-m-cresol [Chlorocresol]	59-50-7	HPP	3.85E+03	C	5.00E-02	C	2.44E-06	X	4.90E+02	C	9.80E+02	C
2-Chlorophenol [o-Chlorophenol]	95-57-8	HPP	2.90E+04	C	1.80E+00	C	1.05E-05	X	4.00E+02	C	1.45E+02	C

Table 8.3 Continued

Chlorotoluene [Benzyl Chloride]	100-44-7		3.30E+03	A	1.00E+00	A	5.06E-05	A	5.00E+01	B	4.27E+02	A
m-Chlorotoluene	108-41-8		4.80E+01	D	4.60E+00	C	1.60E-02	X	1.20E+03	D	1.90E+03	C
o-Chlorotoluene	95-49-8		7.20E+01	C	2.70E+00	C	6.25E-03	X	1.60E+03	D	2.60E+03	C
p-Chlorotoluene	106-43-4		4.40E+01	D	4.50E+00	C	1.70E-02	X	1.20E+03	D	2.00E+03	C
Cresol (Technical) [Methylphenol]	1319-77-3		3.10E+04	A	2.40E-01	A	1.10E-06	A	5.00E+02	A	9.33E+01	A
o-Cresol [2-Methylphenol]	95-48-7	HSL	2.50E+04	J	2.43E-01	O	1.50E-06	X			8.91E+01	M
p-Cresol [4-Methylphenol]	106-44-5	HSL			1.14E-01	O					8.51E+01	M
Dibenzofuran		HSL									1.32E+04	M
1,2-Dichlorobenzene [o-Dichlorobenzene]	95-50-1	HPP	1.00E+02	A	1.00E+00	A	1.93E-03	A	1.70E+03	A	3.98E+03	A
1,3-Dichlorobenzene [m-Dichlorobenzene]	541-73-1	HPP	1.23E+02	A	2.28E+00	A	3.59E-03	A	1.70E+03	A	3.98E+03	A
1,4-Dichlorobenzene [p-Dichlorobenzene]	106-46-7	HPP	7.90E+01	A	1.18E+00	A	2.89E-03	A	1.70E+03	A	3.98E+03	A
2,4-Dichlorophenol	120-83-2	HPP	4.60E+03	A	5.90E-02	A	2.75E-06	A	3.80E+02	A	7.94E+02	A
Dichlorotoluene [Benzal Chloride]	98-87-3		2.50E+00	D	3.00E-01	C	2.54E-02	X	9.90E+03	D	1.60E+04	D
Diethylstilbestrol [DES]	56-53-1		9.60E-03	A					2.80E+01	B	2.88E+05	A
2,4-Dimethylphenol [as-m-Xylenol]	1300-71-6	HPP	4.20E+03	C	6.21E-02	H	2.38E-06	X	2.22E+02	C	2.63E+02	C
1,3-Dinitrobenzene	99-65-0		4.70E+02	A					1.50E+02	B	4.17E+01	A
4,6-Dinitro-o-cresol	534-52-1	HPP	2.90E+02	A	5.00E-02	A	4.49E-05	A	2.40E+02	A	5.01E+02	A
2,4-Dinitrophenol	51-28-5	HPP	5.60E+03	A	1.49E-05	A	6.45E-10	A	1.66E+01	A	3.16E+01	A
2,3-Dinitrotoluene	602-01-7		3.10E+03	A					5.30E+01	B	1.95E+02	A
2,4-Dinitrotoluene	121-14-2	HPP	2.40E+02	A	5.10E-03	A	5.09E-06	A	4.50E+01	A	1.00E+02	A
2,5-Dinitrotoluene	619-15-8		1.32E+03	A					8.40E+01	B	1.90E+02	A
2,6-Dinitrotoluene	606-20-2	HPP	1.32E+03	A	1.80E-02	A	3.27E-06	A	9.20E+01	A	1.00E+02	A
3,4-Dinitrotoluene	610-39-9		1.08E+03	A					9.40E+01	B	1.95E+02	A
Ethylbenzene [Phenylethane]	100-41-4	HPP	1.52E+02	A	7.00E+00	A	6.43E-03	A	1.10E+03	A	1.41E+03	A
Hexachlorobenzene [Perchlorobenzene]	118-74-1	HPP	6.00E-03	A	1.09E-05	A	6.81E-04	A	3.90E+03	A	1.70E+05	A
Hexachlorophene [Dermadex]	70-30-4		4.00E-03	A					9.10E+04	B	3.47E+07	A
Nitrobenzene	98-95-3	HPP	1.90E+03	A	1.50E-01	A	2.20E-05	G	3.60E+01	A	7.08E+01	A
2-Nitrophenol [o-Nitrophenol]	88-75-5	HPP	2.10E+03	H							5.75E+01	H
4-Nitrophenol [p-Nitrophenol]	100-07-7	HPP	1.60E+04	H							8.13E+01	H
m-Nitrotoluene [Methylnitrobenzene]	99-08-1		4.98E+02	G							2.92E+02	M
Pentachlorobenzene	608-93-5		1.35E-01	A	6.00E-03	C			1.30E+04	B	1.55E+05	A
Pentachloronitrobenzene [Quintozene]	82-68-8		7.11E-02	A	1.13E-04	A	6.18E-04	A	1.90E+04	B	2.82E+05	A
Pentachlorophenol	87-86-5	HPP	1.40E+01	A	1.10E-04	A	2.75E-06	A	5.30E+04	A	1.00E+05	A
Phenol	108-95-2	HPP	9.30E+04	A	3.41E-01	A	4.54E-07	A	1.42E+01	A	2.88E+01	A
Pyridine	110-86-1		1.00E+06	A	2.00E+01	A					4.57E+00	A
Styrene [Ethenylbenzene]	100-42-5	HSL	3.00E+02	R	4.50E+00	R	2.05E-03	X				
1,2,3,4-Tetrachlorobenzene	634-66-2		3.50E+00	C	4.00E-02	C			1.80E+04	D	2.88E+04	C

1,2,3,5-Tetrachlorobenzene			2.40E+00	C	7.00E-02	C			1.78E+04	D	2.88E+04	C
1,2,4,5-Tetrachlorobenzene	95-94-3		6.00E+00	A	5.40E-03	O			1.60E+03	B	4.68E+04	A
2,3,4,6-Tetrachlorophenol	58-90-2		7.00E+00	C	4.60E-03	C			9.80E+01	B	1.26E+04	A
Toluene [Methylbenzene]	108-88-3	HPP	5.35E+02	A	2.81E+01	A	6.37E-03	A	3.00E+02	A	5.37E+02	A
1,2,3-Trichlorobenzene	87-61-6		1.20E+01	C	2.10E-01	C	4.23E-03	X	7.40E+03	D	1.29E+04	C
1,2,4-Trichlorobenzene	120-82-1	HPP	3.00E+01	A	2.90E-01	A	2.31E-03	A	9.20E+03	A	2.00E+04	A
1,3,5-Trichlorobenzene	108-70-3		5.80E+00	C	5.80E-01	C	2.39E-02	X	6.20E+03	D	1.41E+04	C
2,4,5-Trichlorophenol	95-95-4	HSL	1.19E+03	A	1.00E+00	A	2.18E-04	A	8.90E+01	B	5.25E+03	A
2,4,6-Trichlorophenol	88-06-2	HPP	8.00E+02	A	1.20E-02	A	3.90E-06	A	2.00E+03	A	7.41E+03	A
1,2,4-Trimethylbenzene [Pseudocumene]	95-63-6		5.76E+01	G	2.03E+00	O	5.57E-03	X				
Xylene (mixed)	1330-20-7	HSL	1.98E+02	A	1.00E+01	A	7.04E-03	A	2.40E+02	B	1.83E+03	A
m-Xylene [1,3-Dimethylbenzene]	108-38-3		1.30E+02	A	1.00E+01	A	1.07E-02	X	9.82E+02	D	1.82E+03	A
o-Xylene [1,2-Dimethylbenzene]	95-47-6		1.75E+02	A	6.60E+00	G	5.10E-03	G	8.30E+02	D	8.91E+02	A
p-Xylene [1,4-Dimethylbenzene]	106-42-3		1.98E+02	A	1.00E+01	A	7.05E-03	X	8.70E+02	D	1.41E+03	A
POLYAROMATIC HYDROCARBONS												
Acenaphthylene	208-96-8	HPP	3.93E+00	A	2.90E-02	A	1.48E-03	A	2.50E+03	A	5.01E+03	A
Acenapthene	83-32-9	HPP	3.42E+00	A	1.55E-03	A	9.20E-05	A	4.60E+03	A	1.00E+04	A
Anthracene	120-12-7	HPP	4.50E-02	A	1.95E-04	A	1.02E-03	A	1.40E+04	A	2.82E+04	A
Benz(c)acridine	225-51-4		1.40E+01	A					1.00E+03	B	3.63E+04	A
Benzo(a)anthracene	56-55-3	HPP	5.70E-03	A	2.20E-08	A	1.16E-06	A	1.38E+06	A	3.98E+05	A
Benzo(a)pyrene	50-32-8	HPP	1.20E-03	A	5.60E-09	A	1.55E-06	A	5.50E+06	A	1.15E+06	A
Benzo(b)fluoranthene	205-99-2	HPP	1.40E-02	A	5.00E-07	A	1.19E-05	A	5.50E+05	A	1.15E+06	A
Benzo(ghi)perylene	191-24-2	HPP	7.00E-04	A	1.03E-10	A	5.34E-08	A	1.60E+06	A	3.24E+06	A
Benzo(k)fluoranthene	207-08-9	HPP	4.30E-03	A	5.10E-07	A	3.94E-05	A	5.50E+05	A	1.15E+06	A
2-Chloronapthalene	91-58-7	HPP	6.74E+00	I	1.70E-02	I	4.27E-04	X			1.32E+04	I
Chrysene	218-01-9	HPP	1.80E-03	A	6.30E-09	A	1.05E-06	A	2.00E+05	A	4.07E+05	A
1,2,7,8-Dibenzopyrene	189-55-9		1.01E-01	A					1.20E+03	B	4.17E+06	A
Dibenz(a,h)anthracene	53-70-3	HPP	5.00E-04	A	1.00E-10	A	7.33E-08	A	3.30E+06	A	6.31E+06	A
7,12-Dimethylbenz(a)anthracene	57-97-6		4.40E-03	A					4.76E+05	A	8.71E+06	A
Fluoranthene	206-44-0	HPP	2.06E-01	A	5.00E-06	A	6.46E-06	A	3.80E+04	A	7.94E+04	A
Fluorene [2,3-Benzidene]	86-73-7	HPP	1.69E+00	A	7.10E-04	A	6.42E-05	A	7.30E+03	A	1.58E+04	A
Indene	95-13-6										8.32E+02	M
Indeno(1,2,3-cd)pyrene	193-99-5	HPP	5.30E-04	A	1.00E-10	A	6.86E-08	A	1.60E+06	A	3.16E+06	A
2-Methylnapthalene	91-57-6	HSL	2.54E+01	E					8.50E+03	E	1.30E+04	E
Napthalene [Napthene]	91-20-3	HPP	3.17E+01	G	2.30E-01	G	1.15E-03	G	1.30E+03	C	2.76E+03	C

Table 8.3 Continued

1-Napthylamine	134-32-7		2.35E+03	A	6.50E-05	A	5.21E-09	A	6.10E+01	B	1.17E+02	A
2-Napthylamine	91-59-8		5.86E+02	A	2.56E-04	A	8.23E-08	A	1.30E+02	B	1.17E+02	A
Phenanthrene	85-01-8	HPP	1.00E+00	A	6.80E-04	A	1.59E-04	A	1.40E+04	A	2.88E+04	A
Pyrene	129-00-0	HPP	1.32E-01	A	2.50E-06	A	5.04E-06	A	3.80E+04	A	7.59E+04	A
Tetracene [Napthacene]	92-24-0		5.00E-04	E					6.50E+05	E	8.00E+05	E
AMINES AND AMIDES												
2-Acetylaminofluorene	53-96-3		6.50E+00	A					1.60E+03	B	1.91E+03	A
Acrylamide [2-Propenamide]	79-06-1		2.05E+06	G	7.00E-03	R	3.19E-10	X				
4-Aminobiphenyl [p-Biphenylamine]	92-67-1		8.42E+02	A	6.00E-05	A	1.59E-08	A	1.07E+02	B	6.03E+02	A
Aniline [Benzenamine]	62-53-3	HSL	3.66E+04	G	3.00E-01	G	1.00E-06	X			7.00E+00	E
Auramine	2465-27-2		2.10E+00	A					2.90E+03	B	1.45E+04	A
Benzidine [p-diaminodiphenyl]	92-87-5	HPP	4.00E+02	A	5.00E-04	A	3.03E-07	A	1.05E+01	A	2.00E+01	A
2,4-Diaminotoluene [Toluenediamine]	95-80-7		4.77E+04	A	3.80E-05	A	1.28E-10	A	1.20E+01	B	2.24E+00	A
3,3'-Dichlorobenzidine	91-94-1	HPP	4.00E+00	A	1.00E-05	A	8.33E-07	A	1.55E+03	A	3.16E+03	A
Diethanolamine	111-42-2		9.54E+05	G							3.72E-02	M
Diethylaniline [Benzenamine]	91-66-7		6.70E+02	E							9.00E+00	E
Diethylnitrosamine [Nitrosodiethylamine]	55-18-5				5.00E+00	A					3.02E+00	A
Dimethylamine	124-40-3		1.00E+06	A	1.52E+03	A	9.02E-05	A	4.35E+02	F	4.17E-01	A
Dimethylaminoazobenzene	60-11-7		1.36E+01	A	3.30E-07	A	7.19E-09	A	1.00E+03	B	5.25E+03	A
Dimethylnitrosamine	62-75-9	HPP	infinite	A	8.10E+00	A	7.90E-07	A	1.00E-01	A	2.09E-01	A
Diphenylnitrosamine	86-30-6	HPP									3.72E+02	I
Dipropylnitrosamine	621-64-7	PP	9.90E+03	A	4.00E-01	A	6.92E-06	A	1.50E+01	A	3.16E+01	A
Methylvinylnitrosamine	4549-40-0		7.60E+05	A	1.23E+01	A	1.83E-06	A	2.50E+00	B	5.89E-01	A
m-Nitroaniline [3-Nitroaniline]	99-09-2	HSL	8.90E+02	G							2.34E+01	M
o-Nitroaniline [2-Nitroaniline]	88-74-4	HSL	1.47E+04	T							6.17E+01	M
p-Nitroaniline [4-Nitroaniline]	100-01-6	HSL	7.30E+02	T							2.45E+01	M
N-Nitrosodi-n-propylamine	621-64-7	HSL										
Thioacetamide [Ethanethioamide]	62-55-5		1.63E+05	J							3.47E-01	A
o-Toluidine Hydrochloride	636-21-5		1.50E+04	A	1.00E-01	A	9.39E-07	A	2.20E+01	B	1.95E+01	A
o-Toluidine [2-Aminotoluene]	119-93-7		7.35E+01	A	<1.0E+00	R			4.10E+02	B	7.58E+02	A
Triethylamine	121-44-8		1.50E+04	G	7.00E+00	G	1.30E+05	G				
ETHERS AND ALCOHOLS												
Allyl Alcohol [Propenol]	107-18-6		5.10E+05	A	2.46E+01	A	3.69E-06	A	3.20E+00	B	6.03E-01	A
Anisole [Methoxybenzene]	100-66-3		1.52E+03	C	2.60E+00	C	2.43E-04	X	2.00E+01	C	1.29E+02	C

Benzyl Alcohol [Benzenemethanol]	100-51-6	HSL	8.00E+02	S	1.10E-01	S	1.95E-05	X			1.26E+01	M
Bis(2-chloroethyl)ether	111-44-4	HPP	1.02E+04	A	7.10E-01	A	1.31E-05	A	1.39E+01	A	3.16E+01	A
Bis(2-chloroisopropyl)ether	108-60-1	HPP	1.70E+03	A	8.50E-01	A	1.13E-04	A	6.10E+01	A	1.26E+02	A
Bis(chloromethyl)ether	542-88-1		2.20E+04	A	3.00E+01	A	2.06E-04	A	1.20E+00	A	2.40E+00	A
4-Bromophenyl Phenyl Ether	101-55-3	HPP			1.50E-03	I					1.91E+04	I
2-Chloroethyl Vinyl Ether	110-75-8	HPP	1.50E+04	M	2.67E+01	M	2.50E-04	Q			1.90E+01	I
Chloromethyl Methyl Ether	107-30-2										1.00E+00	A
4-Chlorophenyl Phenyl Ether	7005-72-3	HPP	3.30E+00	M	2.70E-03	I	2.19E-04	X			1.20E+04	M
Diphenylether [Phenyl Ether]	101-84-8		2.10E+01	R	2.13E-02	S	8.67E-09	X			1.62E+04	M
Ethanol	64-17-5		infinite	A	7.40E+02	A	4.48E-05	A	2.20E+00	B	4.79E-01	A
PHTHALATES												
Bis(2-ethylhexyl)phthalate	117-81-7	HPP	2.85E-01	C	2.00E-07	C	3.61E-07	X	5.90E+03	D	9.50E+03	C
Butylbenzyl Phthalate	85-68-7	HPP	4.22E+01	G							6.31E+04	M
Di-n-octyl Phthalate	117-84-0	HPP	3.00E+00	M							1.58E+09	I
Dibutyl Phthalate	84-74-2	HPP	1.30E+01	A	1.00E-05	A	2.82E-07	A	1.70E+05	A	3.98E+05	A
Diethyl Phthalate	84-66-2	HPP	8.96E+02	A	3.50E-03	A	1.14E-06	A	1.42E+02	A	3.16E+02	A
Dimethylphthalate	131-11-3	HPP	4.32E+03	M	<1.0E-02	M					1.32E+02	I
KETONES AND ALDEHYDES												
2-Butanone [Methyl Ethyl Ketone]	78-93-3	HSL	2.68E+05	A	7.75E+01	A	2.74E-05	A	4.50E+00	B	1.82E+00	A
2-Hexanone [Methyl Butyl Ketone]	591-78-6	HSL	1.40E+04	R	3.00E+10	R	2.82E-05	R				
4-Methyl-2-Pentanone [Isopropylacetone]	108-10-1	HSL	1.70E+04	S	2.00E+01	R	1.55E-04	X				
Acetone [2-Propanone]	67-64-1	HSL	infinite	A	2.70E+02	A	2.06E-05	A	2.20E+00	B	5.75E-01	A
Formaldehyde	50-00-0		4.00E+05	A	1.00E+01	A	9.87E-07	A	3.60E+00	B	1.00E+00	A
Glycidaldehyde	765-34-4		1.70E+08	A	1.97E+01	A	1.10E-08	A	1.00E-01	B	2.82E-02	A
Acrylic Acid [2-Propenoic Acid]	79-10-7		infinite	A	4.00E+00	A					1.35E+00	A
CARBOXYLIC ACIDS AND ESTERS												
Azaserine	115-02-6		1.36E+05	A					6.60E+00	B	8.32E-02	A
Benzoic Acid	65-85-0	HSL	2.70E+03	G							7.41E+01	M
Dimethyl Sulfate [DMS]	77-78-1		3.24E+05	A	6.80E-01	A	3.48E-07	A	4.10E+00	B	5.75E-02	A
Ethyl Methanesulfonate [EMS]	62-50-0		3.69E+05	A	2.06E-01	A	9.12E-08	A	3.80E+00	B	1.62E+00	A
Formic Acid	64-18-6		1.00E+06	A	4.00E+01	A					2.88E-01	A
Lasiocarpine	303-34-4		1.60E+03	A					7.60E+01	B	9.77E+00	A
Methyl Methacrylate	80-62-6		2.00E+01	A	3.70E+01	A	2.43E-01	A	8.40E+02	B	6.17E+00	A
Vinyl Acetate	108-05-4	HSL	2.00E+04	J								

Table 8.3 Continued

PCBs												
Aroclor 1016	12674-11-2	HPP	4.20E-01	H	4.00E-04	I					2.40E+04	H
Aroclor 1221	11104-28-2	HPP	1.50E+01	I	6.70E-03	I					1.23E+04	H
Aroclor 1232	11141-16-5	HPP	1.45E+00	I	4.06E-03	I					1.58E+03	I
Aroclor 1242	53469-21-9	HPP	2.40E-01	G	4.10E-04	G	5.60E-04	G			1.29E+04	I
Aroclor 1248	12672-29-6	HPP	5.40E-02	G	4.90E-04	G	3.50E-03	G			5.62E+05	I
Aroclor 1254	11097-69-1	HPP	1.20E-02	G	7.70E-05	G	2.70E-03	G	4.25E+04	E	1.07E+06	I
Aroclor 1260	11096-82-5	HPP	2.70E-03	G	4.10E-05	G	7.10E-03	G			1.38E+07	I
Polychlorinated Biphenyls [PCBs]	1336-36-3	HPP	3.10E-02	A	7.70E-05	A	1.07E-03	A	5.30E+05	A	1.10E+06	A
HETEROCYCLIC COMPOUNDS												
Dihydrosafrole	94-58-6		1.50E+03	A					7.80E+01	B	3.63E+02	A
1,4-Dioxane [1,4-Diethylene Dioxide]	123-91-1		4.31E+05	A	3.99E+01	A	1.07E-05	A	3.50E+00	B	1.02E+00	A
Epichlorohydrin	106-89-8		6.00E+04	A	1.57E+01	A	3.19E-05	A	1.00E+01	B	1.41E+00	A
Isosafrole	120-58-1		1.09E+03	A	1.60E-08	A	3.25E-12	A	9.30E+01	B	4.57E+02	A
N-Nitrosopiperidine	100-75-4		1.90E+06	A	1.40E-01	A	1.11E-08	A	1.50E+00	B	3.24E-01	A
N-Nitrosopyrrolidine	930-55-2		7.00E+06	A	1.10E-01	A	2.07E-09	A	8.00E-01	B	8.71E-02	A
Safrole	94-59-7		1.50E+03	A	9.10E-04	A	1.29E-07	A	7.80E+01	B	3.39E+02	A
Uracil Mustard	66-75-1		6.41E+02	A					1.20E+02	B	8.13E-02	A
HYDRAZINES												
1,2-Diethylhydrazine	1615-80-1		2.88E+07	A					3.00E-01	B	2.09E-02	A
1,1-Dimethylhydrazine	57-14-4		1.24E+08	A	1.57E+02	A	1.00E-07	A	2.00E-01	B	3.80E-03	A
1,2-Diphenylhydrazine [Hydrazobenzene]	122-66-7	PP	1.84E+03	A	2.60E-05	A	3.42E-09	A	4.18E+02	A	7.94E+02	A
Hydrazine	302-01-1		3.41E+08	A	1.40E+01	A	1.73E-09	A	1.00E-01	B	8.32E-04	A
MISCELLANEOUS ORGANIC COMPOUNDS												
Aziridine [Ethylenimine]	151-56-4		2.66E+06	A	2.55E+02	A	5.43E-06	A	1.30E+00	B	9.77E-02	A
Carbon Disulfide	75-15-0	HSL	2.94E+03	A	3.60E+02	A	1.23E-02	A	5.40E+01	B	1.00E+02	A
Diethyl Arsine	692-42-2		4.17E+02	A	3.50E+01	A	1.48E-02	A	1.60E+02	B	9.33E+02	A
Dimethylcarbamoyl Chloride	79-44-7		1.44E+07	A	1.95E+00	A	1.92E-08	A	5.00E-01	B	4.79E-02	A
Mercury and Compounds (Alkyl)	7439-97-6	PP										
Methylnitrosourea	684-93-5		6.89E+08	A					1.00E-01	B	1.54E-04	A
Mustard Gas [bis(2-chloroethyl)sulfide]	505-60-2		8.00E+02	A	1.70E-01	A	4.45E-05	A	1.10E+02	B	2.34E+01	A
Phenobarbital	50-06-6		1.00E+03	A					9.80E+01	B	6.46E-01	A
Propylenimine	75-55-8		9.44E+05	A	1.41E+02	A	1.12E-05	A	2.30E+00	B	3.31E-01	A
Tetraethyl Lead	78-00-2		8.00E-01	A	1.50E-01	A	7.97E-02	A	4.90E+03	B		
Thiourea [Thiocarbamide]	62-56-6		1.72E+06	A					1.60E+00	B	8.91E-03	A
Tris-BP [2,3-Dibromoipropanol phosphate]	126-72-7		1.20E+02	A					3.10E+02	B	1.32E+04	A

INORGANICS												
Ammonia	7664-41-7		5.30E+05	A	7.60E+03	A	3.21E-04	A	3.10E+00	B	1.00E+00	A
Antimony and Compounds	7440-36-0	PP			1.00E+00	A						
Arsenic and Compounds	7440-38-2	PP			0.00E+00	A						
Barium and Compounds	7440-39-3											
Beryllium and Compounds	7440-41-7	PP			0.00E+00	A						
Cadmium and Compounds	7740-43-9	PP			0.00E+00	A						
Chromium III and Compounds	7440-47-3	PP			0.00E+00	A						
Chromium VI and Compounds	7440-47-3	PP			0.00E+00	A						
Copper and Compounds	7440-50-8	PP			0.00E+00	A						
Cyanogen Chloride	506-77-4		2.50E+03	A	1.00E+03	A	3.24E-02	X			1.00E+00	A
Hydrogen Cyanide	74-90-8		infinite	A	6.20E+02	A					5.62E-01	A
Hydrogen Sulfide	7783-06-4		4.13E+03	A	1.52E+04	R	1.65E-01	R				
Lead and Compounds	7439-92-1	PP			0.00E+00	A						
Mercury and Compounds (Inorganic)	7439-97-6	PP	3.00E-02	G	2.00E-03	A	1.10E-02	G				
Nickel and Compounds	7440-02-0	PP			0.00E+00	A						
Potassium Cyanide	151-50-8		5.00E+05	A								
Selenium and Compounds	7782-49-2	PP			0.00E+00	A						
Silver and Compounds	7440-22-4	PP			0.00E+00	A						
Sodium Cyanide	143-33-9		8.20E+05	A								
Thallium Chloride	7791-12-0	PP	2.90E+03	A	0.00E+00	A						
Thallium Sulfate	7446-18-6	PP	2.00E+02	A	0.00E+00	A						
Thallium and Compounds	7440-28-0	PP			0.00E+00	A						
Zinc and Compounds	7440-66-6	PP			0.00E+00	A						

Notes: PP = Priority Pollutant; HSL = Hazardous Substance List Parameter; HPP = PP and HSL Parameters.

A = CONCAWE, 4/79, Protection of Groundwater from Oil Pollution.

B = Payne, J.R., and C.R. Phillips, 1985, Petroleum Spills in the Marine Environment, Lewis Publishers, Chelsea, MI.

C = National Institute for Petroleum and Energy Research, 1988, Personal communication.

D = Breuel, A., 1981, Oil Spill Cleanup and Protection Techniques for Shorelines and Marshlands, Noyes Data, N.J.

E = Cole-Parmer Co., 1989–1990, Equipment Catalog.

F = ASTM, 1985, Annual Book of ASTM Standards, Section 5, Petroleum Products, Lubricants, and Fossil Fuels, Philadelphia.

G = Chevron USA, Inc., 1988, Product Salesfax Digest, San Francisco.

H = Weast, R.C., (ed.), 1980–1981, CRC Handbook of Chemistry and Physics, 61st Edition, Cleveland.

I = Values calculated using ASTM viscosity-temperature charts for liquid petroleum products (ASTM D 341–77).

J = U.S. Coast Guard, 1979, CHRIS Hazardous Chemical Data.

K = Chevron USA, Inc., 1989, Personal Communication.

L = Hunt, J.R., N. Sitar, and K.S. Udell, 1988, Monaqueous Phase Liquid Transport and Cleanup 1. Analysis of Mechanisms, in Water Resources Research, Vol. 24, No. 8, pp. 1247–1258.

* = Values calculated based on: Absolute Viscosity (centipoise) = Kinematic Viscosity (centistokes) × Specific Gravity.

ity of organic compounds. The most common correlation equations (Yang and Parker, 1998) are as follows:
For nonpolar organic compounds:

$$\log K_{oc} = -0.621 \log S + 3.95 \tag{8.4}$$

$$\log K_{oc} = 0.72 \log K_{ow} + 0.49 \tag{8.5}$$

For hydrophobic organic compounds:

$$\log K_{oc} = \log K_{ow} - 0.21 \tag{8.6}$$

where

S = aqueous solubility of the organic contaminant (mg/L)
K_{ow} = octanol–water partition coefficient of the organic compound (unitless)

Values of S and K_{ow} for various chemical compounds are presented in Table 8.3 as compiled by the U.S. EPA (1990) from a variety of references. Typical ranges of f_{oc} values are 0.0002–0.01 for sand and 0.01–0.08 for topsoil (which generally contains clay) (Knox et al., 1993).

8.4.2 Chemical Enhancements Including Surfactants

Traditional pump-and-treat systems often lose effectiveness as the contaminant is removed until it reaches its residual saturation in the contaminated media. Residual saturation is the fluid saturation of the porous medium below which fluid drainage will not occur. As illustrated in Figure 8.5, the capillary attraction of the residual liquid becomes excessively high at residual saturation, relative to the imposed drainage forces. As discussed in Chapter 5, magnitude of residual

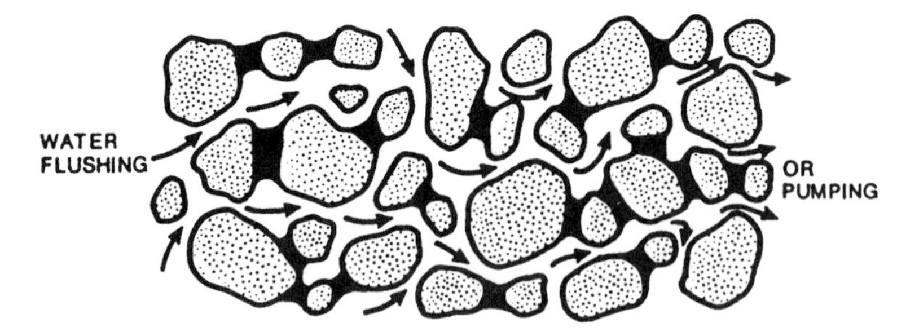

Figure 8.5 Trapped oily contaminant at residual saturation; flushing will not remove all of the trapped product because of capillary action (API, 1980).

saturation depends on the pore size distribution of the medium and physicochemical properties of the porous medium and liquid. Estimates are often imprecise because of the numerosity of determining factors. For the case of oil ganglia in water-filled soils, Hunt et al. (1988) used simple static force balance relations to develop equations for two parameters, bond number N_b and capillary number N_c, which define conditions for ganglia mobilization to occur in the vertical and horizontal directions, respectively:

$$N_b = \frac{(\rho_{cw} - \rho_w)gL_vD_p}{\sigma_{cw}} > 4 \tag{8.7}$$

$$N_b = \frac{(\mu_w v_w L_h D_p)}{K\sigma_{cw}} > 4 \tag{8.8}$$

where

N_b = bond number
N_c = capillary number
ρ_{cw} = contaminant (ganglia) − water density (M/L^3)
ρ_w = density of water (M/L^3)
g = gravitation acceleration (L/S^2)
L_v = length of ganglia mobilized in the vertical direction (L)
L_h = length of ganglia mobilized in the horizontal direction (L)
D_p = mean pore throat diameter of the porous medium (L)
σ_{cw} = contaminant–water interfacial tension (M g/T)
μ_w = dynamic viscosity of water (M/LT)
v_w = kinematic viscosity of water (L^2/T)
K = permeability of the porous medium (L^2/T)

Nonsatisfaction of Eqs. (8.7) and (8.8) implies that L_v and L_h are less than the lengths required for movement of ganglia to occur. Pressures or other parameters in the equations will need to change favorably before contaminant recovery can be increased.

For contaminants that are sorbed much more strongly at residual concentrations by physicochemical interactions with the geologic media than at higher concentrations, changes in the intensity of processes that control their distribution coefficients define how much of the contaminant can be made available for removal within the tailing phase of pump-and-treat systems and/or other systems that involve contaminant extraction.

Use of reactive agents, including surfactants, is intended to promote contaminant removal even at their residual saturations through one or a combination of the following phenomena.

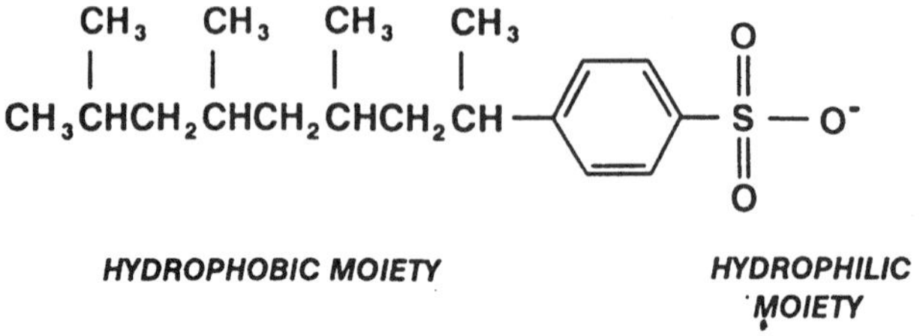

$$\begin{array}{l} \quad CH_3 \quad\; CH_3 \quad\; CH_3 \quad\; CH_3 \\ CH_3CHCH_2CHCH_2CHCH_2CH - C_6H_4 - SO_2 - O^- \end{array}$$

HYDROPHOBIC MOIETY

HYDROPHILIC MOIETY

Figure 8.6 Illustration of the configuration of a type of surfactant, alkylbenzene sulfonate (U.S. EPA, 1992a).

Surfactant Action

Surfactants (SURface-ACTive-ageNTs) are generally large amphiphilic molecules with a hydrophobic chain moiety (made up of hydrocarbon or fluorocarbon) and a hydrophilic head (made up of polar or ionic groups) as illustrated in Figure 8.6. Surfactants are amphiphilic in the sense that they can exist in one or more forms: as a dissolved phase in water; as part of a micelle with other molecules; and as an adsorbed phase between a solid and its surrounding liquid. The interaction of surfactants with contaminated media can reduce the contaminant–water interfacial tension, σ_{cw}, and enable the contaminant to be more available and mobile. With an appropriate type and concentration of surfactant, data in literature show that oil–water interfacial tension can be reduced from about 20 dynes/cm to about 10^{-4} dynes/cm.

Lowering of interfacial tension and dissolution of organic contaminants are largest when the surfactant attains its critical micelle concentration (CMC). A micelle, which is illustrated in Figure 8.7, is an aggregate formed (typically considered to be spherical in shape) by surfactant molecules such that nonpolar tails

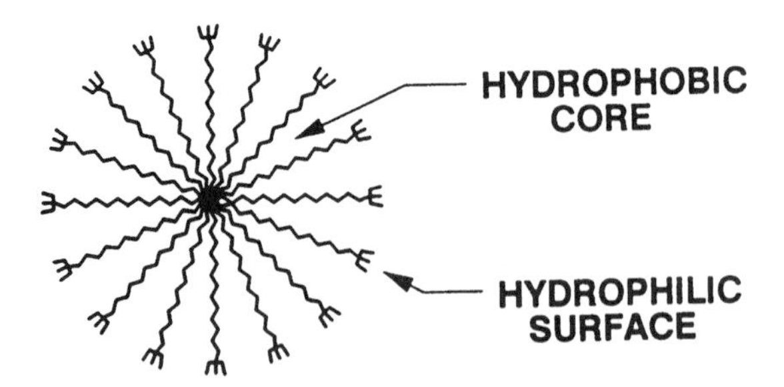

Figure 8.7 Aggregation of surfactant molecules into a micelle (U.S. EPA, 1992a).

occupy the interior of a sphere. Through a mechanism described as solubilization, nonpolar contaminants can dissolve in the micellar solution, thereby increasing the overall solubility of the contaminant. Although its use in predictions of solubility for surfactant–organic compound combinations are often imprecise, solubility increases of three orders of magnitude following the addition of as little as 5% by volume of an appropriate surfactant are possible.

There are more than 12,000 types of surfactants. However, they can be classified into four major categories based on chemical characteristics of their hydrophilic (water-linking) groups.

Anionic Surfactants In anionic surfactants the surface-active part of the dissolved surfactant is negatively charged. They are generally anions of alkali metal salts of fatty acid (salts), sulfates, sulfonates, and phosphates. They are the most common class of surfactants because of their relatively low cost and widespread use as wetting agents and detergents. Examples are

Sodium stearate (soap): $C_{17}H_{35}COO^{-}Na^{+}$
Sodium dodecyl sulfate: $C_{12}H_{25}OSO_3^{-}Na^{+}$

Cationic Surfactants In cationic surfactants the surface-active part carries a positive charge in solution. They are commonly amine salts, quartenary ammonium and pyridinium chemical compounds used primarily in surface sorption and coating applications. Examples are

Hexadecytlrimethylammonium bromide (CTAB): $C_{16}H_{33}N(CH_3)_3^{+}Br^{-}$
Ammonium salt of a long-chain amine: $RNH_3^{+}CL$

Nonionic Surfactants The surface-active parts nonionic surfactants do not carry ionic charge. They are generally polyoxyethylene and polyoxy propylene compounds, amine oxides, or alkanolamides. They are generally soluble in organic liquids and water and are somewhat expensive to produce. Examples are

Alkylpolyoxyethylene alcohols: $C_nH_{2n+1}(OCH_2CH_2)_xOH$
Monoglyceride of a fatty acid: $RCOOCH_2CHOHCH_2O$

Amphoteric or Zwitterionic Surfactants Amphoteric or Zwitterionic surfactants carry both positive and negative charges in their surface-active portions and are relatively uncommon. Examples are long-chain amino acids and sulfobetaine.

Co-solvent Action

The addition of water-miscible solvents to organic-contaminated soils can increase the solubility of organic compounds. Nkedi-Kizza et al. (1985) report that the adsorption coefficient of anthracene in methanol–water mixtures in soil decreased by four orders of magnitude for an increase of methanol fraction from 0 to 1. Apparently, the anthracene became more soluble, and this was reflected in less partition to the solid phase in soil.

Oxidation–Reduction

Reactive materials can be introduced into soils that contain contaminants at or below residual saturation to influence pH-Eh conditions and transform the contaminants into more mobile forms. For example, reduction of iron from Fe^{3+} to Fe^{2+} results in increased mobility. Other examples are oxidation of Se^{4+} to Se^{6+}, and reduction of Cr^{4+} to Cr^{3+}.

Dissolution

Some contaminants which may be released slowly over a relatively long time interval may be bound into chemical compounds which are not adequately soluble

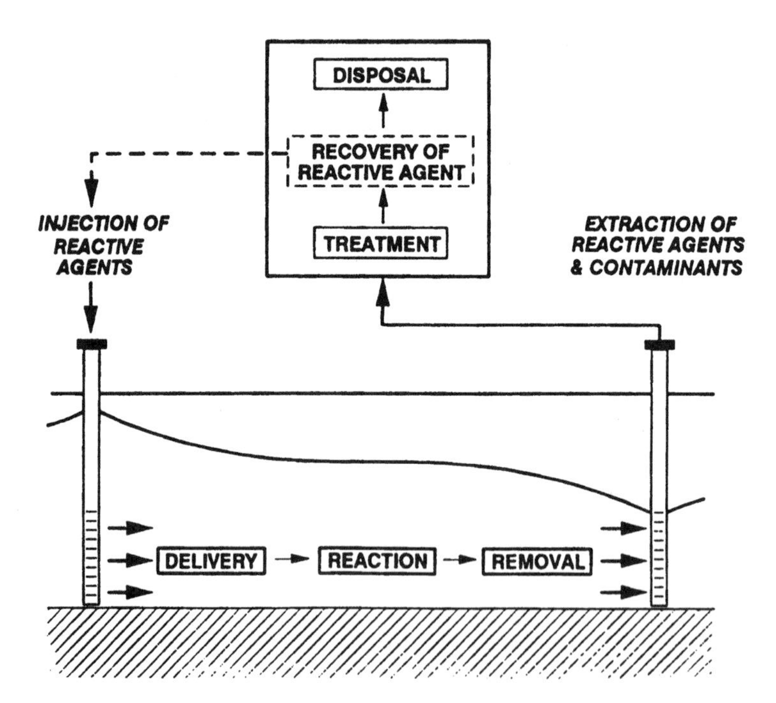

Figure 8.8 Schematic illustration of the arrangement of injection extraction, treatment, and disposal network for reactants used in enhancement of pump-and-treat systems (U.S. EPA, 1992a).

Table 8.4 Advantages and Disadvantages of Injection Systems in Reactant-Enhanced Pump-and-Treat Systems

	Advantages	Disadvantages
Fluid injection mode		
Continuous	Fluid distributed over wide area	Greater potential for clogging of screens
	Less maintenance of pumping schedules	Greater pumping costs
Slug	Less volume of water	Fluid distributed over very small volume of aquifer
	Less potential clogging of wells	
Pulsed	Can be developed around working schedules	Requires more design to ensure injection and off periods are balanced relative to natural groundwater flow
Active agent injection mode		
Continuous	Maintain concentration in high permeability zones allowing for diffusion into low-permeability zones	
Slug	Requires less mass of active agent	May not allow sufficient time for diffusion into low-permeability lenses
		Concentration decreases with time/distance, which can reduce effectiveness of the active agent
Pulsed	Less total mass of active agent	Requires greater maintenance/control
	Can be planned around work schedules	Requires more analysis to ensure that injection and off periods are of sufficient length
	Allows for sufficient time for diffusion	

Source: U.S. EPA (1992a).

under prevailing conditions. Particularly for soils that are contaminated by inorganic substances, modification of pH through the addition of substances such as lime and other materials that provide H^+ for competition with metal ions for sorption sites can make contaminants more available for extraction.

In using reactive agents to enhance the effectiveness of pump-and-treat systems, several operational factors need to be assessed. Among these factors are the feasibility of recovering and reusing the reactive agent, the possibility of interference of the reactive agent with the treatment processes for the extracted water, the environmental and human toxicity of the reactive agent, and the optimal method for introduction of the agent into the contaminated soil. A schematic illustration of the arrangement of injection and extraction points, and their links to treatment and disposal systems for both extracted water and reactive agents, is provided in Figure 8.8. Fluids which carry the reactive agent and the reactive agent itself may be injected using one of three optional modes: continuous, slug, and pulsed injection modes. The advantages and disadvantages of each of these modes are summarized in Table 8.4 (after U.S. EPA, 1992a)

8.5 IN-SITU SOIL FLUSHING

In-situ soil flushing is the injection of flushing solution to permeate soils to release bound contaminants and move them with the infiltrating solution toward extraction systems. As illustrated in Figure 8.9, the contaminated media to be treated are usually located in the vadose zone. The flushing solution may be neutral or may contain surface-active agents, some of which were discussed in Section 8.4. This treatment technique is usually designed for inorganic contaminants (metals) such as Pb, Cr, As, and Cd. In this process, the flushing solution infiltrates the

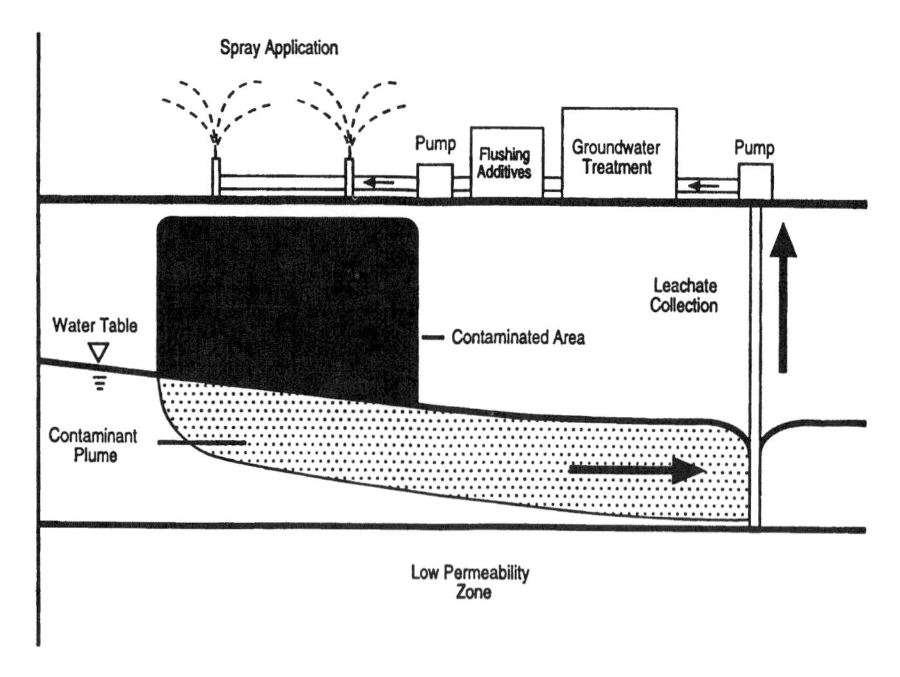

Figure 8.9 Illustration of a typical soil flushing system (U.S. EPA, 1997).

treatment zone and selectively extracts and displaces the target metals through the vadose zone into groundwater. Through pumping, the contaminated water is extracted and subjected to a treatment process in which the leached metal is precipitated for disposal. In some cases, it may be necessary to inject a stabilizing solution through residual media to fix the remaining substances.

Examples of substances that are commonly used in flushing solutions are HCL, EDTA, HNO_3, and $CaCl_2$. Leaching models such as those described in Chapter 6 (Section 6.6) can be used to estimate the rate of displacement of contaminants downward by percolating liquids. Detailed models for describing soil flushing have been published by Kayano and Wilson (1993) and Gabr et al. (1996). In general, soil flushing is effective in relatively permeable soils (silts and sands).

8.6 VOLATILIZATION AND AIR PRESSURIZATION PRINCIPLES

Extraction of contaminants from polluted media requires that the contaminants be made more available for transport out of the polluted media. The mobility of contaminants toward extraction points increases as they are transferred from the solid phase of soil systems to the liquid phase and the gaseous phase, in that order. The objective of remediation schemes that are based on volatilization and air pressurization techniques is to induce the transfer of contaminants into soil atmosphere from which they can be more easily extracted than from the original media.

8.6.1 In-Situ Soil Venting/Soil Vapor Extraction Principles

After the removal of free product from contaminated soils, the soil is unsaturated with contaminated liquid but the contaminant is still present in the residual liquid. In soil venting/soil vapor extraction, airflow through the contaminated zone is induced through the application of a vacuum pressure, typically, through a well located in the center of the contaminated zone. Air inlet wells are then installed at the boundary of the zone. Application of vacuum pressure draws in air through the inlet wells and the contaminants that transfer from the soil and residual liquids to the air are drawn out of the ground.

Efforts (Lingineni and Dhir, 1992; Johnson et al., 1990) have been made to develop predictive models for the rate of contaminant removed for a given set of contaminant properties, soil characteristics, vacuum pressures, and other operational features of soil venting systems. For the Johnson et al. (1990) analy-

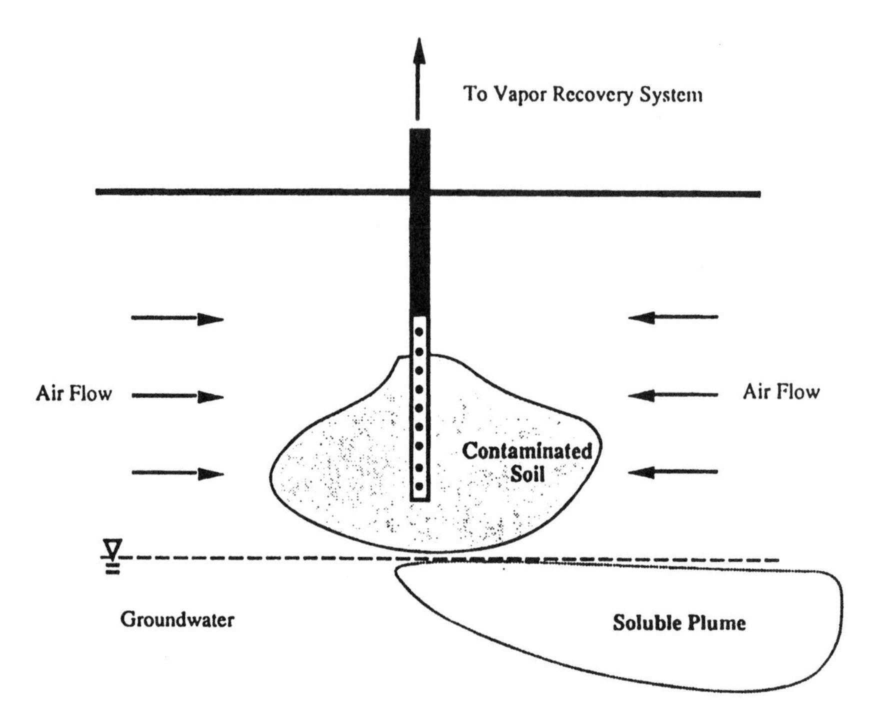

Figure 8.10 Schematic illustration of a soil venting system (Johnson et al., 1990).

sis, the configuration shown in Figure 8.10 is used. The governing equations for vapor flow through a confined porous medium of thickness m are as follows:

$$u = -\frac{K}{\mu}(\nabla P + \rho g) \tag{8.9}$$

$$\frac{\partial(\varepsilon \rho m)}{\partial t} = -\nabla(\rho m u) \tag{8.10}$$

where

ρ = vapor density (M/L^3)
g = gravity vector (L/T^2)
ε = vapor-filled void fraction of soil (fraction)
u = darcian vapor velocity (L/T)
K = soil permeability (L^2/T)
μ = vapor viscosity (M/LT)
P = vapor-phase pressure (M/L^3)

∇ = gradient operator (1/T)
m = stratum thickness (L)

Equations (8.9) and (8.10) are Darcy's law and the continuity equation, respectively. In Eq. (8.10), ρg can be neglected because vapor flow is considered to be primarily in a direction that is orthogonal to gravity. Also, it will be assumed that the soil permeability is affected by neither liquid removal nor by airflow. In order to establish the relationship between induced pressure and the density of the vapor, it is assumed herein that the vapor behaves as an ideal gas. Thus,

$$\rho = \rho_{atm}\left[\frac{P}{P_{atm}}\right] \tag{8.11}$$

where

ρ_{atm} = vapor density at reference pressure P_{atm} (M/L^3)

Equations (8.9) and (8.11) can be substituted into Eq. (8.10), resulting in the following equation:

$$\left(\frac{2\varepsilon\mu}{K}\right)\frac{\partial P}{\partial t} = \nabla^2 P \tag{8.12}$$

Under conditions of steady state, radial flow with the boundary conditions $P = P_w$, $r = R_w$, and $P = P_{atm}$, $r = R_i$, Eq. (8.12) can be solved as follows for a homogeneous porous medium:

$$P^2(r) - P_w^2 = \left(P_{atm}^2 - P_w^2\right)\left[\frac{\ln(r/R_w)}{\ln(r_i/R_w)}\right] \tag{8.13}$$

where

r = nominal radius of the vacuum well (L)
P_w = pressure at the vaccum well of R_w (M/LT)
R_w = radius of the vaccum well (L)
R_i = radius of influence of the well (radius to where pressure is equal to P_{atm}) (L)
P_{atm} = atmospheric pressure (M/LT)

Then, $U(r)$, the radial Darcian velocity distribution and Q, the volumetric flow rate, can be calculated using the following equations:

$$u(r) = -\frac{K}{2\mu}\left(\frac{P_w(r/n(R_w/R_i))[1 - (P_{atm}/P_w)^2]}{\{1 + [1 - (P_{atm}/P_w)^2]\}\{[\ln(r/R_w)]/[\ln(R_w/R_i)]\}^{1/2}}\right) \tag{8.14}$$

$$Q = 2\pi R_w u(R_w) H = H\left(\frac{\pi K}{\mu}\right)\left\{\frac{P_w[1 - (P_{atm}/P_w)^2]}{\ln(R_w/R_i)}\right\} \tag{8.15}$$

where

Q = volumetric vapor flow rate (L^3/T)
H = screened length of the vacuum well through the vadose zone (L)

Considering that the soil system may be layered, an appropriate expression for such layered soils is given as Eq. (8.16)

$$Q = \sum_{i=1}^{n} Q_i \tag{8.16}$$

where

Q_i = volumetric vapor flow rate through each homogeneous layer i (L^3/T)
n = number of homogeneous soil layers through which the screened segment of the vacuum well passes

Additional analyses for quantification of vapor extraction rates from soils are provided by Bass (1993) and Gierke et al (1995).

The total mass of contaminant removed can be estimated as follows:

$$M = QC_a t \tag{8.17}$$

where

M = cumulative mass of contaminant removed (M)
C_a = average concentration of contaminant in air (M/L^3)
t = time (T)

The average concentration C_a depends on the transfer of the contaminant from solid and liquid phases into vapor phase in the soil pores and the interaction of the vapor with the porous transport medium during its extraction. The relationship shown in Figure 8.11 is often observed between concentration and operation time. The volatility of contaminants and the permeability of the soils are the primary determinants of the effectiveness of soil vapor extraction systems (SVE). High permeability and high volatility enhance the effectiveness of SVE. In general, contaminants with vapor pressures greater than 0.5 mm Hg are considered to be suitable for remediation using SVE (U.S. EPA, 1995). The Henry's law constant is also a good indicator of volatility. It expresses the tendency of dissolved contaminants to partition between the vapor and dissolved (in liquid) phases. Contaminants that have Henry's law constant values in excess of 100 are

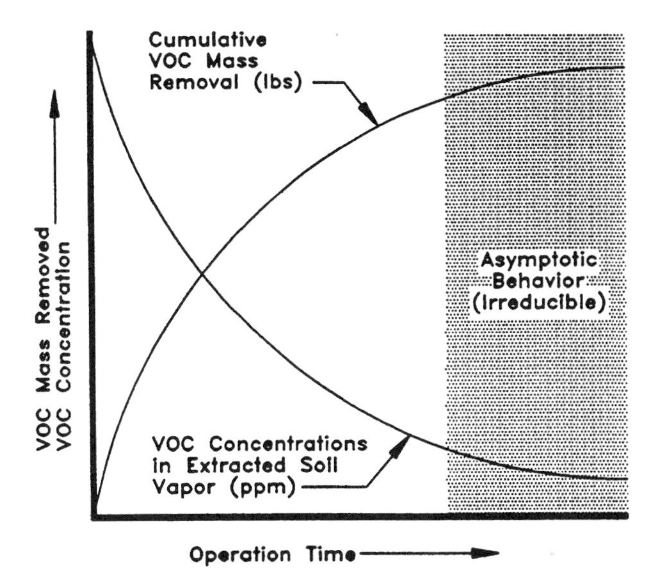

Figure 8.11 Relationship between contaminant concentration and extraction time in soil vapor extraction systems (U.S. EPA, 1995).

candidates for removal by SVE. Vapor pressures and Henry's Law constant values are provided for common constituents of petroleum in Tables 8.5 and 8.6, respectively.

8.6.2 In-Situ Steam Extraction Principles

In-situ steam extraction is an augmentation of the SVE process described in Section 8.6.1 through the injection of steam. The introduction of steam results in the extraction of semivolatiles which would normally not be removed at lower temperatures and drier conditions. This technology is useful for treating media contaminated with chlorinated hydrocarbons, aromatics, alcohols, and high-boiling-point chlorinated aromatics. Figure 8.12 shows a typical configuration of an in situ steam extraction system.

8.6.3 In-Situ Air Sparging Principles

Unlike the SVE process, in which air is injected into the vadose zone (consisting of contaminated, unsaturated media), in air sparging, air is injected into the contaminated soil below the water table. The injected air bubbles contact dissolved and absorbed-phase contaminants and even NAPLs such that contaminants are

Table 8.5 Vapor Pressure of Common Petroleum Constituents

Constituent	Vapor pressure (mmHg at 20°C)
Methyl *t*-butyl ether	245
Benzene	76
Toluene	22
Ethylene dibromide	11
Ethylbenzene	7
Xylenes	6
Naphthalene	0.5
Tetraethyl lead	0.2

Source: U.S. EPA (1995).

Table 8.6 Henry's Law Constants of Common Petroleum Constituents

Constituent	Henry's law constant (atm)
Tetraethyl lead	4700
Ethylbenzene	359
Xylenes	266
Benzene	230
Toluene	217
Naphthalene	72
Ethylene dibromide	34
Methyl *t*-butyl ether	27

Source: U.S. EPA (1995).

transferred into them as they travel into the vadose zone. The oxygen introduced by injected air may increase the rate of biodegradation of in-situ contaminants. An air sparging system is illustrated schematically in Figure 8.12. As can be seen, air is introduced through the sparger well and contaminant-laden bubbles travel into the vadose zone, where they are removed by extraction wells. A more elaborate illustration is shown in Figure 8.13.

The maximum concentration of contaminants in the air phase can be estimated as follows from the ideal gas law:

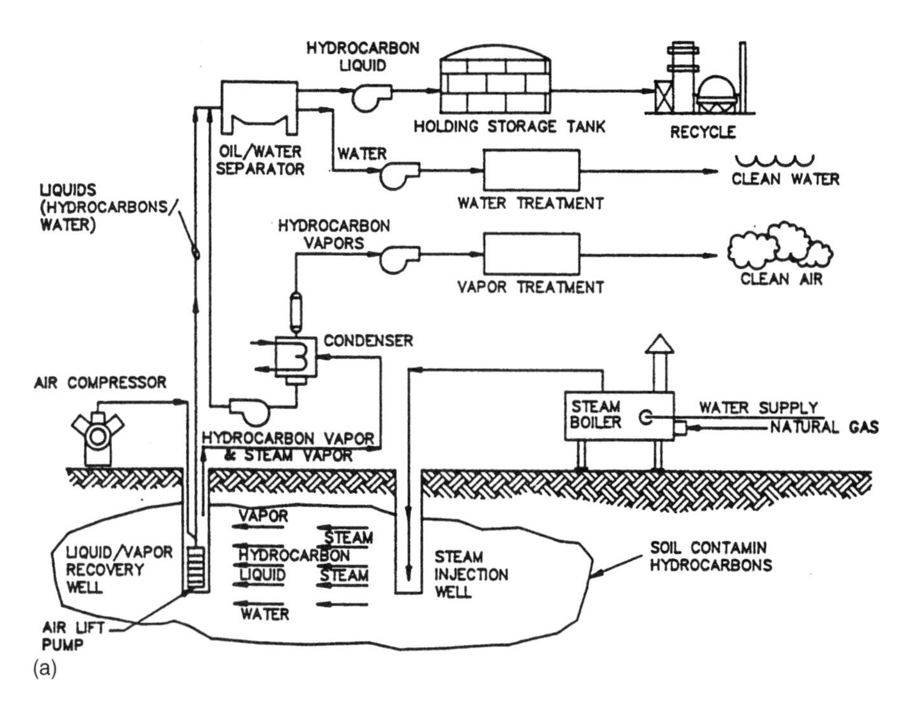

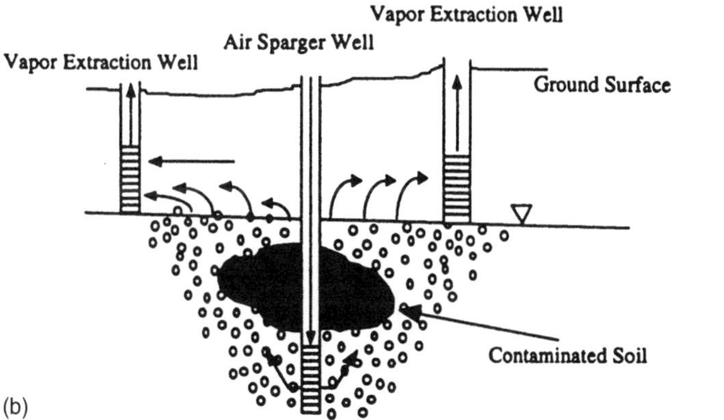

Figure 8.12 (a) Configuration of an in-situ steam extraction system (U.S. EPA, 1994a). (b) Schematic illustration of an air sparging system (U.S. EPA, 1996).

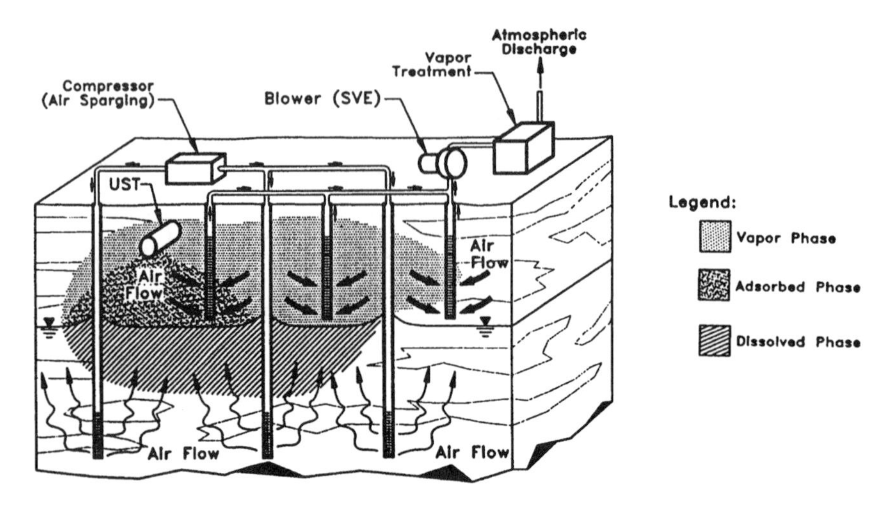

Figure 8.13 Schematic illustration of surface and subsurface components of an air sparging system (U.S. EPA, 1995).

$$C_m = \sum \frac{X_i P_i M_i}{RT} \tag{8.18}$$

where

C_m = maximum vapor concentration in the vapor phase (mg/L)
X_i = mole fraction of component i in the liquid phase of the contaminated media
P_i = pure-component vapor pressure of component i at temperature T (atm)
M_i = molecular weight of component i (mg/mole)
R = gas constant = 0.0821 atm/mole-K
T = absolute temperature (K)

Air sparging is most effective for volatile contaminants in unstratified soils where groundwater contains little or no dissolved iron. Dissolved iron tends to oxidize in the presence of introduced oxygen and precipitate, thereby reducing the permeability of the soil to fluids. Air sparging is not recommended when the concentration of Fe^{2+} in groundwater exceeds 20 mg/L.

8.7 IN-SITU VITRIFICATION PRINCIPLES

Vitrification is the process by which materials (which may include waste-containing materials) are converted into glass or substances that behave like glass.

Considering that glass is a rigid, noncrystalline substance with very limited porosity, contaminants that are fused into glass are held tightly so that they may not leach out significantly.

In hazardous waste remediation practice, vitrification is usually induced through heat application in situ, or the processing and heating of excavated wastes. The latter enables greater control of the vitrification process but has the disadvantage of greater human and environmental exposure risks if radioactive and dispersive contaminants are involved. In this section, the focus is on the principles of in-situ vitrification. A vitrification process that is based on heat application may destroy organic contaminants through pyrolysis or combustion, and fuse inorganic metals, including radioactive elements into the structure of glass. Glass is generally durable and has the capacity to incorporate a wide variety of substances in a reasonably high range of concentrations. However, glass formation requires that its component elements be available in appropriate forms and proportions. This is not always the case with contaminated media. Thus, some additives may need to be added to deficient media to improve glass formation. During vitrification, both organic and inorganic substances may vaporize as off-gas, which often requires control or treatment.

Glass is composed of oxides of silicon (primary constituent), boron, aluminum, and alkali and alkaline earth elements. The primary glass types that are used in waste vitrification processes are sodium silicates, borosilicates, and aluminosilicates. The following raw materials primarily supply the elements that are used in industrial glass formation. The mix proportions can be optimized to produce glass of specific physicochemical characteristics.

Sand: SiO_2
Limestone: $CaCO_3$
Dolomite: $CaMg(CO_3)_2$
Soda ash: Na_2CO_3, $KALSi_3O_8$

The roles of various oxides in vitrification and performance of vitrified materials is summarized in Table 8.7. The typical proportions of oxides in soda lime glass (used for making window glass), commercial borosilicate glass, and in-situ vitrified (ISV) glass are summarized in Table 8.8.

In contaminated media vitrification processes, the application of heat in the temperature range of 1000–2000°C melts materials into a liquid. Upon cooling, the liquid solidifies into a largely noncrystalline (amorphous) mass. An illustration of the in-situ heat application process in which electrodes are used is shown in Figure 8.14. The glass mass comprises a three-dimensional network of silicon–oxygen tetrahedra which lack crystallinity because of the irregularity and randomness of the silicon–oxygen bonds of the network. This structure is illustrated in Figure 8.15. Each silicon atom is bonded to four oxygen atoms which occupy the corners of the silicon–oxygen tetrahedra. Some or all of the oxygen atoms

Table 8.7 Effects of Waste-Glass Components on Processing and Performance of Vitrified Materials

	Processing	Product Performance
Frit Components		
SiO_2	Increases viscosity greatly; reduces waste solubility	Increases durability
B_2O_3	Reduces viscosity; increases waste solubility	Increases durability in low amounts, reduces in large amounts
Na_2O	Reduces viscosity and resistivity; increases waste solubility	Reduces durability
Li_2O	Same as Na_2O, but greater effect; increases tendency to devitrify	Reduces durability, but less than Na_2O
K_2O	Same as Na_2O; decreases tendency to devitrify	Reduces durability more than Na_2O
CaO	Increases then reduces viscosity and waste solubility	Increases then reduces durability
MgO	Is same as CaO; reduces tendency to vitrify	Is same as CaO, but more likely to decrease durability
TiO_2	Reduces viscosity slightly; increases then reduces waste solubility; increases tendency to devitrify	Increases durability
ZrO_2, La_2O_3	Reduces waste solubility	Increases durability greatly
Waste Components		
Al_2O_3	Increases viscosity and has tendency to devitrify	Increases durability
Fe_2O_3	Reduces viscosity; is hard to dissolve	Increases durability
U_3O_8	Reduces tendency to devitrify	Reduces durability
NiO	Is hard to dissolve; increases tendency to devitrify	Reduces durability
MnO	Is hard to dissolve	Increases durability
Zeolite	Is slow to dissolve; produces foam	Increases durability
Sulfate	Is an antifoam, melting aid; increases corrosion of processing equipment	Too much causes foam or formation of soluble second phase

Source: Adapted from U.S. EPA (1992), originally developed by Plodinec et al. (1982).

Table 8.8 Typical Ranges of Oxide Compositions in Soda-Lime Glass, Borosilicate Glass, and in-Situ Vitrified (ISV) Glass

Oxide	Typical soda-lime glass (wt%)	SRS borosilicate benchmark glass (wt%)	Sample ISV glass (wt%)
SiO_2	65–75	48.95	71.20
Al_2O_3	1–2	3.67	13.50
Na_2O	12–16	16.71	1.55
K_2O	0.1–3	0.04	2.47
MgO	0.1–5	1.66	1.87
CaO	6–12	1.13	3.58
B_2O_3	—	11.12	—
Fe_2O_3	—	8.08	4.63
FeO	—	0.89	—
La_2O_3	—	0.41	—
Li_2O	—	4.28	—
MnO	—	1.34	0.11
NiO	—	0.61	0.12
TiO_2	—	0.71	0.76
ZrO_2	—	0.41	0.07
SrO	—	—	0.02
BaO	—	—	0.10

Source: (Compiled by U.S. EPA, 1992).

may be shared by neighboring tetrahedra to form the glass network. If all the oxygen atoms were shared, a silicon/oxygen ratio of 1/2 would result from the satisfaction of charge neutrality because silicon has a valency of $+4$ and oxygen has a valency of -2. Some metals can replace some of the silicon atoms in the interior of tetrahedra (network-forming elements) or interrupt the linkages among neighboring tetrahedra (modifying elements).

Several energy-generation processes can be used to generate heat for vitrification of wastes. The primary types of heat processing are joule heating, plasma heating, and microwave heating. The first two types are technologies adapted from the metalworks and glassmaking industries, while microwave is an enhancement of technologies from a variety of materials and incineration industries.

Joule heating is the most common type of in-situ vitrification process. In this process, an electric current applied through electrodes embedded in the ground flows through the contaminated media. The internal resistance of the media to current flow causes a reduction in power, such that heat is transferred to

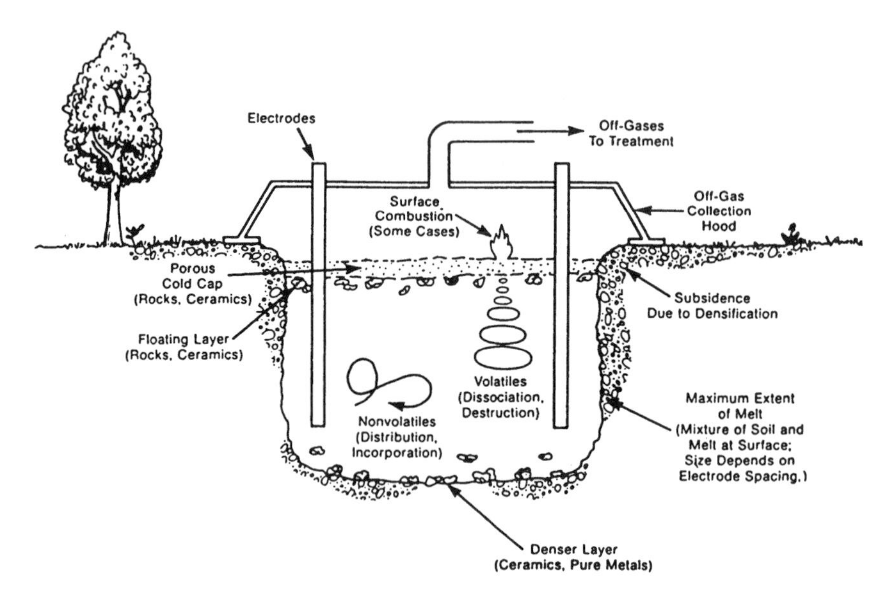

Figure 8.14 Schematic application of in-situ vitrification process in which electrodes are used for heat application (U.S. EPA, 1992).

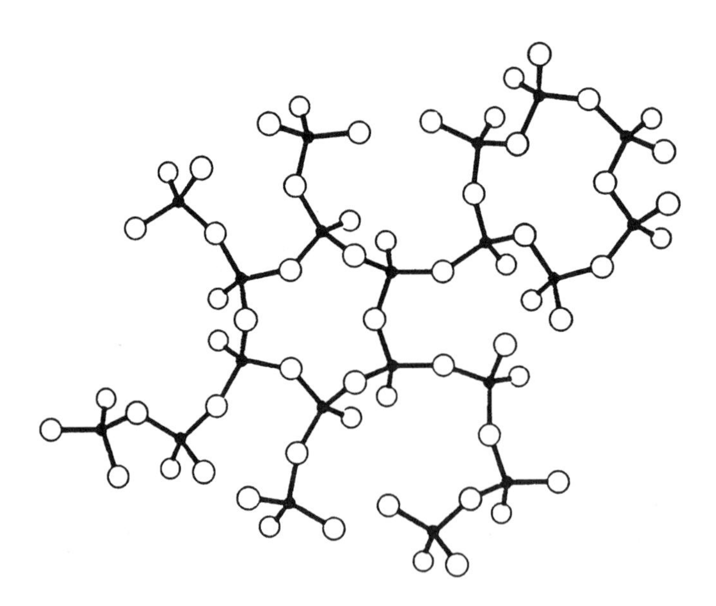

Figure 8.15 Example of a silicate glass network structure (McLelland and Strand, 1984).

● = silicon

○ = oxygen

the media. Joule's law provides a quantitative method for estimating the power that is dissipated in the media:

$$P = I^2R \tag{8.19}$$

$$R = \frac{V}{I} \tag{8.20}$$

where

P = dissipated power (watts)
I = current flow through the media (amperes)
R = resistance of the material (ohms)
V = voltage across the material (volts)

Generally, soils are very resistant because of the low electrical conductivity of SiO_2, its primary constituent. However, soils are less resistant at higher temperatures. Initially, high voltage in the regime of 2000–4000 volts is applied to compensate for the soil's high resistance at low temperatures. In addition, flaked graphite and glass frit are placed between electrodes to reduce the initial conductivity of soils to initiate soil heating. These enhancers are subsequently destroyed at higher temperatures and the heat in the system reaches levels at which soil is heated more efficiently. In this later stage high voltage is no longer necessary and can be reduced by as much as 10-fold.

In plasma heating (which is currently mostly ex-situ), heat is produced through conversion of a gas to plasma under energy supplied by an electric arc generated by direct current (DC) or alternating current (AC) sources. Essentially, a plasma is an ionized gas, typically produced from materials such as oxygen, nitrogen, air, noble gases, or their mixtures. The process is used because ionized materials are good conductors of electricity. A column of the gas(es) is heated in the equipment and heat is produced and transferred to the soil through convection, radiation, and electrical resistance.

In microwave heating, which is currently an ex-situ process, the contaminated media absorb electromagnetic radiation in the effective microwave range (3000–30,000 MHz). The radiation is directed into the contaminated media through waveguides. Soils are dielectric materials, implying that they are not effective conductors of heat. Dielectrics exhibit some polarization under an electric field. Alternation of the field causes distortion of the molecules of dielectric materials (Orfeuil, 1987), resulting in heating of the material. This is the principle that underlies the use of microwave for heating of soils to vitrified masses.

The vitrification processes described in the preceding sections result in the following vitrification treatment mechanisms for contaminants:

Table 8.9 Approximate Ranges of Solubility of Elements in Silicate Glasses

Less than 0.1 wt%:	Ag, Ar, Au, Br, H, He, Hg, I, Kr, N, Ne, Pd, Pt, Rh, Rn, Ru, Xe
Between 1 and 3 wt%:	As, C, Cl, Cr, S, Sb, Se, Sn, Tc, Te
Between 3 and 5 wt%:	Bi, Co, Cu, Mn, Mo, Ni, Ti
Between 5 and 15 wt%:	Ce, F, Gd, La, Nd, Pr, Th, B, Ge
Between 15 and 25 wt%:	Al, B, Ba, Ca, Cs, Fe, Fr, K, Li, Mg, Na, Ra, Rb, Sr, U, Zn
Greater than 25 wt%:	P, Pb, Si

Source: Volf (1984).

Covalent bonding of metal (contaminant) ions to oxygen atoms of tetrahedra through replacement of silicon
Ionic bonding of some inorganics to oxygen atoms outside tetrahedra such that the traditional glass network is interrupted
Encapsulation of both organics and inorganics by surrounding material without their being a part of the chemical structure of the glass
Destruction of organic compounds by heat

Deterioration processes such as matrix dissolution of glass, and interdiffusion of elements can result in the release of fractions of initially bound contaminants from vitrified waste masses. These processes are after collectively termed leaching and are exacerbated by the presence of aqueous media of suitable chemistry and temperature. Volf (1984) developed approximate ranges of solubility of elements in silicate glasses. This information is provided in Table 8.9. Metal concentrations in Toxicity Characteristics Leaching Procedure (TCLP) leachates from vitrified soils at Idaho National Engineering Laboratory are shown in Table 8.10. Table 8.11 shows metal retention data compiled by Hansen (1991) for various pilot and engineering-scale contaminated soil vitrification projects. The destruction and removal efficiencies for organics are provided in Table 8.12.

8.8 IN-SITU CHEMICAL TREATMENT IN REACTIVE WALLS

Reactive walls are subsurface walls that allow contaminated water to travel through granular media (which comprise the wall) while the contaminants are removed by physicochemical and/or biological treatment processes. Reactive walls have a variety of configurations, depending on the contaminant hydrogeology of the site and the design efficiency of the treatment process selected. The

Table 8.10 TCLP Extract Metal Concentrations in Leachate from Idaho National Engineering Laboratory Vitrified Soils

Metal	Maximum allowable leachate concentration (mg/liter)	Contaminated soil concentration (mg/kg)	Vitrified product leachate concentration (mg/liter)
Arsenic	5.0	200	<0.168
Barium	100.0	200	0.229
Cadmium	1.0	200	0.0098
Chromium	5.0	200	0.0178
Lead	5.0	2000	0.636
Mercury	0.2	200	<0.0001
Selenium	1.0	200	0.098
Silver	5.0	200	<0.023

Source: U.S. EPA (1994b).

most basic design involves funneling groundwater through one or more gates into one or multiple reaction zones. The system is designed such that the exit concentration C_e of the target contaminant is adequately reduced, relative to the entry concentration, C_0. Sampling points are usually placed in both the entry and exit zones of the wall as well as along the flow axis within the wall, to monitor contaminant concentrations in space and time. Schematic illustrations of reactive walls are shown in Figs. 8.16 and 8.17. Reactive walls can be placed in series or in parallel to achieve contaminated water treatment efficiencies or provide a ''treatment train'' in which each wall targets a different contaminant within the polluted groundwater.

Essentially, the wall is an in-situ reactor that can operate using one or more treatment process to transform and/or remove contaminants. The prevailing processes depend on the type of packing material and additives used in the wall. Commonly, earthen media such as fine sands and silts are mixed with reactive substances or coated with the latter such that adequate permeability can be maintained while reactions that occur at estimated rates are sustained. The choice of reactive media and other design elements is made to ensure that contaminated water residence time within the treatment zone is long enough for significant reactions to occur. With respect to typical dimensions, one system described by O'Hannesin and Gillham (1993) has a length of 5.5 m, a thickness of 1.6 m, and was located at a depth of 2.2m below the groundsurface and 1.0 m below the water table. Examples of reactive substances that can be used in reactive walls are zero-valent iron (iron filings), limestone, organic carbon, and microbial consortia.

Table 8.11 Metal Retention Efficiencies of in Situ Vitrification Systems Recorded in Pilot and Engineering Projects

Class	Metal	Retention efficiency (%)[a]	Scale[b]
Volatile	Mercury (Hg)	0	Engineering
Semivolatile	Arsenic (As)	70–85	Engineering
	Cadmium (Cd)	67–75	Pilot
	Cesium (Cs)	99–99.9	Pilot
	Lead (Pb)	90–99	Pilot
	Ruthenium (Ru)	99.8	Pilot
	Antimony (Sb)	96.7–99.9	Pilot
	Tellerium (Te)	50–99	Pilot
Nonvolatile	Americium (Am)	99.99	Pilot
	Barium (Ba)	99.9	Engineering
	Cerium (Ce)	98.9–99.9	Pilot
	Cobalt (Co)	98.7–99.8	Pilot
	Copper (Cu)	90–99	Engineering
	Chromium (Cr)	99.9	Engineering
	Lanthanum (La)	98.9–99.98	Pilot
	Molybdenum (Mo)	99.9–99.999	Pilot
	Neodymium (Nd)	99–99.98	Pilot
	Nickel (Ni)	99.9	Engineering
	Plutonium (Pu)	99.99	Pilot
	Radium (Ra)	99.9	Engineering
	Strontium (Sr)	99.9–99.998	Pilot
	Thorium (Th)	99.99	Engineering
	Uranium (Th)	99.99	Engineering
	Zinc (Zn)	90–99	Engineering

[a] Percentage of original amount remaining in the melt.
[b] Engineering-scale tests involve a melt depth of 1–2 ft. Pilot-scale tests involve a melt depth of 3–7 ft.
Source: Hansen (1991).

Unlike in slurry walls and other barrier systems, fluid flow through a permeable reactive wall is dominated by advection. If it is assumed that sorption of the contaminants onto the wall media is negligible, and that plug-flow reaction with dispersion is the dominant operational mode of the system, Eq. (8.21), developed by Tchobanoglous and Schroeder (1987), can be used to estimate the effluent concentration, C_0, of the contaminant.

$$\frac{C_e}{C_0} = \frac{2aP_e}{(1 + a)^2 \exp(0.5aP_e) - (1 + a)^2 \ \exp(-0.5aP_e)} \tag{8.21}$$

Table 8.12 Organics Destruction and Removal Efficiencies (DRE) Recorded for Contaminated Media Vitrification Systems (HWC, 1990)

Contaminant	Initial concentration (ppb)	Percent destruction	Total DRE (including off-gas removal)
Aldrin	113	>97	>99.99
Chlordane	535,000	99.95	>99.999
DDD, DDE, DDT	21–240,000	99.9–99.99	>99.999
Dieldrin	24,000	98–99.9	>99.99
Dioxins	>47,000	99.9–99.99	>99.9999
Fuel Oils	230–11,000	>99	>99.999
Furans	>9,400	99.9–99.99	>99.9999
Glycol	NA	>90	>99.99
Heptachlor	61	98.7	>99.99
MEK	NA	>99	>99.999
PCBs	19,400,000	99.9–99.99	>99.9999
Pentachlorophenol	>4,000,000	99.995	>99.99999
Toluene	203,000	99.996	>99.99999
Trichloroethane	106,000	99.995	>99.99999
Xylenes	3,533,000	99.998	>99.99999

Source: Hazardous Waste Consultant (HWC) (1990).

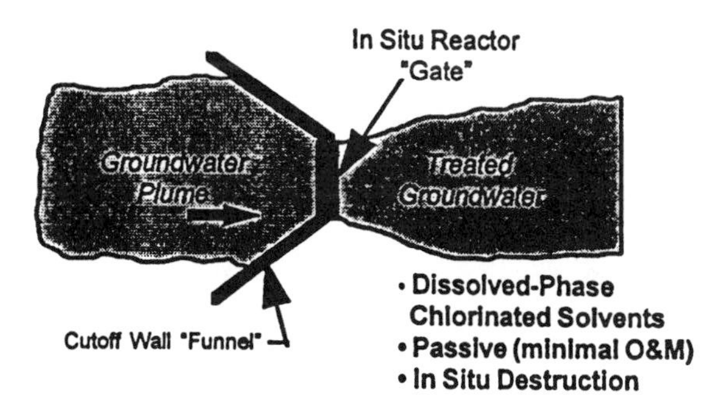

Figure 8.16 Funnel-and-gate configuration of a reactive wall for in-situ treatment of contaminated groundwater.

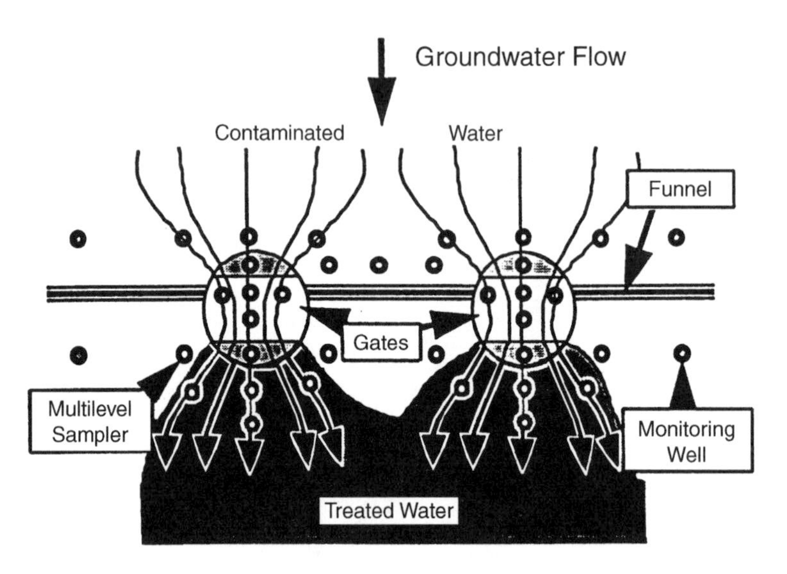

Figure 8.17 Array of monitoring points for assessing the effectiveness of a double-gate reactive wall treatment system.

$$a = (1 + 4\lambda_t P_e)^{0.5} \tag{8.22}$$

$$P_e = \frac{vL}{D} \tag{8.23}$$

where

C_e = exit concentration of the contaminant (M/L^3)
C_0 = entrance concentration of the contaminant (M/L^3)
P_e = Peclet number (unitless)
L = length of the treatment zone (L)
v = contaminated water flow velocity (L/T)
D = dispersion coefficient of the contaminant through the wall (L^2/T)
λ_t = decay rate constant of the target contaminant (I/T)

Most commonly, reactive walls in full-scale and bench-scale projects apply contaminant transformation processes (oxidation and reduction for organics) and precipitation of inorganic compounds through modification of Eh and pH conditions. Iron metal and iron compounds have been used to degrade hydrocarbons and precipitate some metals. In particular, dehalogenation of chlorinated solvents has been accomplished through the use of reduced forms of iron such as Fe^0,

Fe^{2+}, and FeS in aqueous systems. Iron minerals have high specific surface areas and are known to catalyze redox reactions in natural soils. The principal mechanisms by which iron reduces chlorohydrocarbons are illustrated in Figs. 8.18 and 8.19. The most widely recognized mechanism is that chlorinated hydrocarbons receive electrons from Fe^0 to dechlorinate. Sivavec and Horney (1995) postulate that surface-bound Fe^{2+} can serve as a mediator in the process of transfer of

Direct reduction at the metal surface

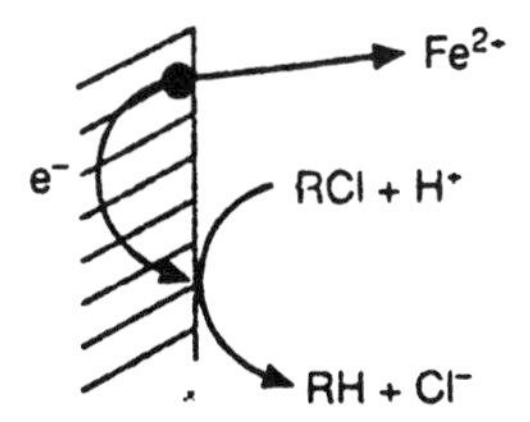

Reduction by ferrous iron

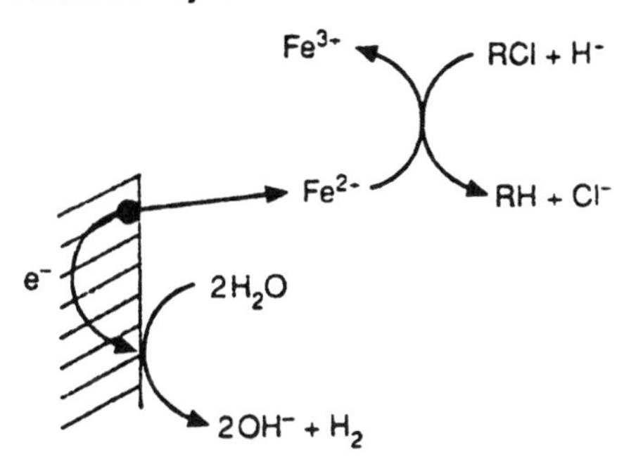

Reduction by hydrogen with catalysis

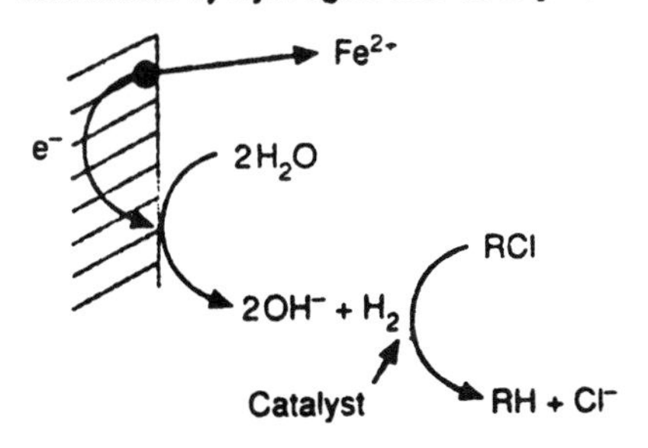

Figure 8.18 Principal mechanisms through which chlorinated hydrocarbons are reduced by iron (Wilson, 1995).

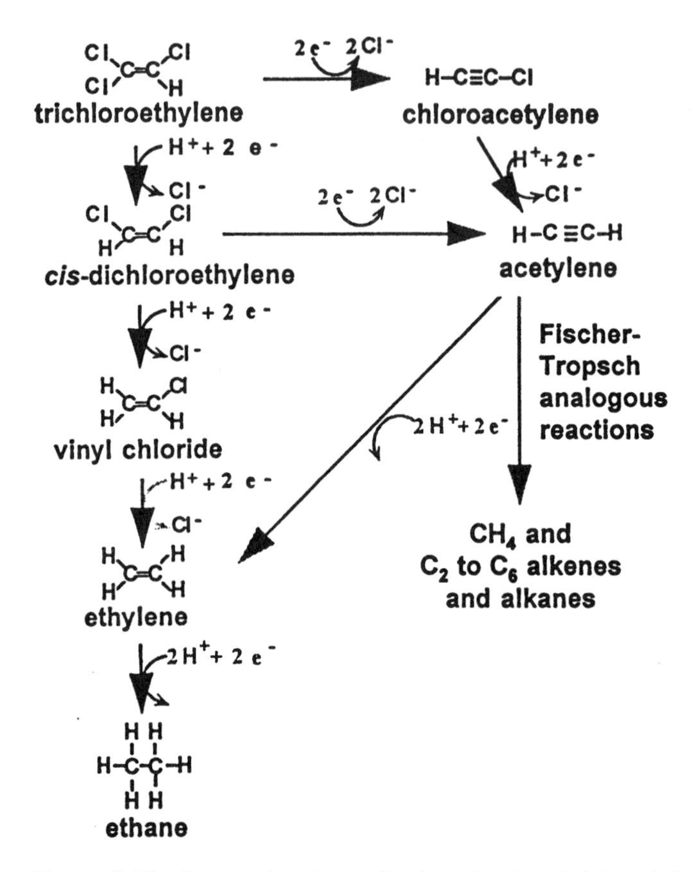

Figure 8.19 Suggested pathways for the reduction of chloroethylenes by zero-valent iron (courtesy of undated USEPA information sheet).

electrons from Fe^0 to chlorinate hydrocarbon. Reduced iron (Fe^{2+}) has been known (Heijman et al., 1995) to serve as an electron-transfer mediator in the reduction of compounds by iron-reducing bacteria. Dehalogenation of compounds occurs through direct electron transfer from Fe^0 to chlorinated hydrocarbon. Reduced iron, Fe^0, can be represented as follows:

$$Fe^0 + RX + H^+ \rightarrow Fe^{2+} + RH + X^- \tag{8.24}$$

In Eq. (8.24), RX is the adsorbed species, RH is the dehalogenated compound, and X^- is the chloride released.

The overall reaction represented by Eq. (8.24) actually occurs as the successor of two reactions as illustrated below for chlorinated organics.

$$2Fe^0 + O_2 + 2H_2O \rightleftharpoons 2Fe^{2+} + 4OH^- \quad (8.25)$$

$$Fe^0 + 2H_2O \rightleftharpoons Fe^{2+} + H_2 + 2OH^- \quad (8.26)$$

$$Fe^0 + RCL + H^+ \rightarrow Fe^{2+} + RH + X^- \quad (8.27)$$

Redox reactions that involve transformation of reducible organic compounds by iron are generally fast, with reaction constants that are either first-order or pseudo-first-order. Laboratory bench scale experiments show that the reaction constants correlate highly with the specific surface area of the iron materials and the quantity of iron present per unit reaction volume. Table 8.13 provides data on degradation rates of some chlorinated hydrocarbons by iron metal.

$$K = (M \pm X_m)S + (C \pm X_c)$$

where

K = first-order reaction rate constant for the transformation of organic compounds by iron-bearing materials (I/T)
M = central value of the correlation constant = slope of K versus S line (L/T)

Table 8.13 Degradation Rates of Some Chlorinated Hydrocarbons by Iron

	Half-life (days)	Halogenated products
Trichloromethane	6.5	Dichloromethane (low concentration)
Tetrachloromethane	5.4	Trichloromethane, dichloromethane
Monochloroethene	14.9	None detected
1,1-Dichloroethene	55	None detected
cis-1,2-Dichloroethene	37	None detected
trans-1,2-Dichloroethene	6.7	None detected
1,1,1-Trichloroethane	5.5	1,1-Dichloroethane
1,1,2-Trichloroethane	7.8	Trace of monochloroethene
Trichloroethene	7.1	cis-1,2-Dichloroethene
1,1,2,2-Tetrachloroethane	10.2	cis-1,2-Dichloroethene, trans-1,2-dichlorethene
1,1,1,2-Tetrachloroethane	5.2	1,1-Dichlorethene
Tetrachloroethene	13.9	Trichloroethene, cis-1,2-dichloroethene
Hexachloroethane	6.1	Tetrachloroethene, trichloroethene, cis-1,2-dichloroethene, 1,2-dichloroethane
Tetrabromoethene	6.7	Tribromoethene, 1,2-dibromoethene

Source: Developed by Wilson (1995), using U.S. EPA information.

X_m = numerical bounds of the range of the correlation constant about its average
S = total surface area of iron-bearing material per unit reaction volume (L^2/L^3)
C = correlation constant representing intercept of the trend line on the K axis (I/T)
X_c = upper or lower limit of C (I/T)

For chloroethenes (chlorinated ethenes) reduced by iron metal (Fe^0) in batch experiments conducted (Sivavec and Horney, 1995) at 25°C, K values are summarized in Table 8.14. The magnitudes of the correlation parameters are

$$M = 1.27 \times 10^{-4}\ \text{m/h}$$
$$X_m = 2.40 \times 10^{-5}\ \text{m/h}$$
$$C = 8.45 \times 10^{-3}\ \text{h}^{-1}$$
$$X_c = 3.67 \times 10^{-3}\ \text{h}^{-1}$$

For dechlorination of TCE (a chloroethene) by iron sulfide (FeS), Sivavec et al. (1995) obtained the following values:

$$M = 8.68 \times 10^{-4}\ \text{m/h}$$
X_m = not applicable
$$C = 7.16 \times 10^{-4}\ \text{h}^{-1}$$
X_c = not applicable

Table 8.14 Reductive Dechlorination Rates of Aqueous Chlorinated Ethenes Determined in Batch Experiments at 25°C (Milli-Q water buffered with 40 mg/liter $CaCO_3$, initial pH = 7) (25.0 g VWR untreated iron filings: 122 mg/liter aq PCE, TCE, DCE, or VC)

Chloroethene	Initial conc. (mg/liter)	Iron surface area/volume (m^2/liter)	Rate constant (h^{-1})	$t_{1/2}$ (h)	r^2
PCE	20.0	254	0.0549	13	0.990
TCE	19.2	254	0.0746	9	0.963
cis-DCE	22.2	254	0.0037	187	0.837
trans-DCE	22.9	254	0.0257	27	0.992
1,1-DCE	21.6	254	0.0047	147	0.615
VC	23.6	254	0.0039	177	0.910

Source: Sivavec and Horney (1995).

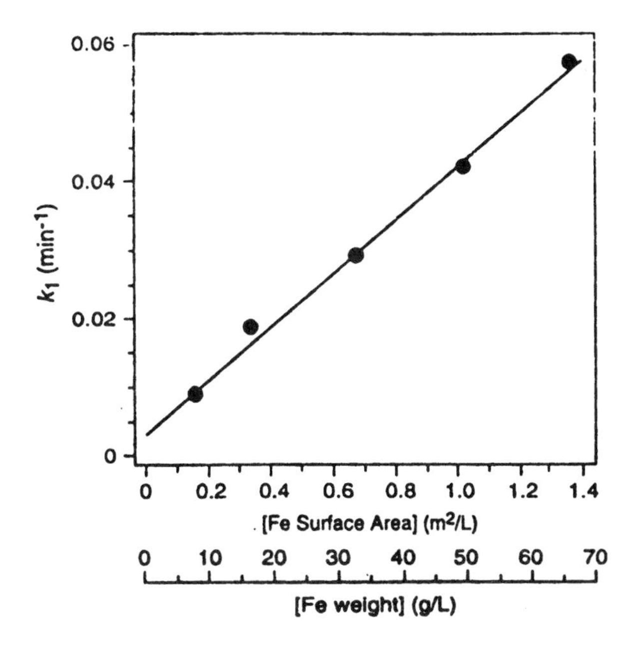

Figure 8.20 Effects of zero-valent iron metal surface area concentration on pseudo-first-order reaction rate constant for nitrobenzene reduction (Agrawal and Tratnyek, 1996).

For reduction of nitrobenzene by Fe^0, Agrawal and Tratnyek (1996) obtained the following data. The trend line is shown in Figure 8.20, noting that for the following data, K is in 1/min, and S is in M^2/liter.

$$M = 3.90 \times 10^{-3} \text{ m/min}$$
$$X_m = 2.00 \times 10^{-3} \text{ m/min}$$
$$C = 3.00 \times 10^{-3} \text{ min}^{-1}$$
$$X_c = 1.00 \times 10^{-3} \text{ min}^{-1}$$

Studies by Chuang et al. (1995) have indicated that Fe^0 can promote the dechlorination of polychlorinated biphenyls (PCBs) without a solvent at temperatures in the regime of 400–500°C. However, the reader should note that greater control of experimental conditions can be maintained in the laboratory than within the subsurface environment in which reactive walls are designed to operate.

8.9 SOLIDIFICATION/STABILIZATION (EX-SITU) PRINCIPLES

Solidification and stabilization are waste treatment processes that are intended primarily to reduce the mobility of waste constituents. Solidification is the change in waste physical state, from a nonsolid to a solid. Stabilization is the modification of the waste such that it is converted to a less soluble, mobile or toxic form. Thus, solidification/stabilization of waste materials, which is usually performed ex-situ following adaptations of the treatment scheme illustrated in Figure 8.21, has the following advantages.

1. Improvement in waste handling due to possible reduction in volume
2. Decrease in the surface area of waste available for leaching of contaminants
3. Reduction in the solubility of waste
4. Transformation of contaminants in waste from toxic to nontoxic forms.

The reader should note that regardless of the possibility of occurrence of advantages 3 and 4, solidification/stabilization (S/S) is considered to be primarily a physical treatment technique for wastes.

In the following paragraphs, the primary types of stabilization and their basic principles are described.

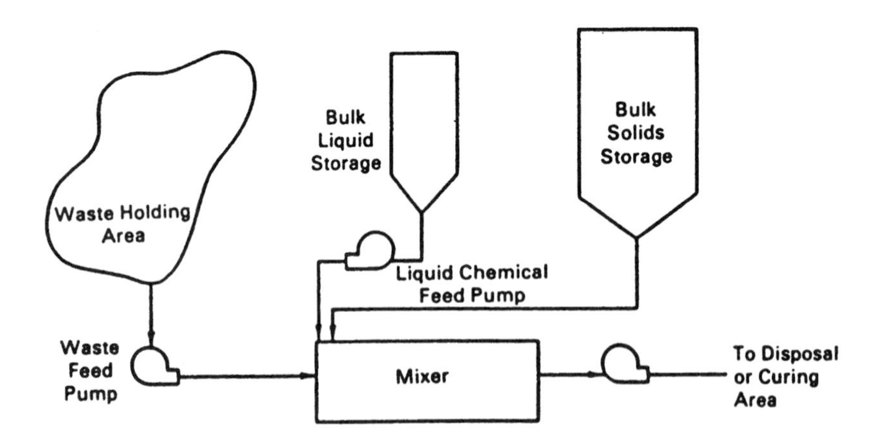

Figure 8.21 Schematic illustration of the solidification/stabilization process for contaminated media (courtesy of U.S. EPA).

8.9.1 Types of Stabilization

Cement-Based Stabilization/Solidification

In cement-based stabilization/solidification, portland cement is the most common stabilization agent used. It is mixed with the waste in predetermined proportions and water is added. Usually, hydroxides of metals are formed. These hydroxides are less soluble than the initial waste forms, enabling S/S to be used widely in the treatment of soils contaminated by metals. Upon curing, the waste–portland cement mass solidifies into a monolith.

Portland cement comprises particles in the size range 1–50 μm. Mineralogically, the composition (by weight) is as follows:

C_3S, tricalcium silicate, 45–60%
C_2S, dicalcium silicate, 15–30%
C_3A, tricalcium aluminate, 6–12%
$C_4AF \approx 8\%$ tetracalcium aluminoferrite, ~8%

The following oxides occur in portland cement:

$C = CaO$
$S = S1O_2$
$A = AL_2O_3$
$F = Fe_2O_3$
$S = SO_3$
$H = H_2O$

Table 8.15 provides data on the composition and characteristics of primary portland cement compounds.

The basic hydration reactions that ultimately result in the stabilization of waste within the solidified mass of waste–cement are as follows:

$$C_3A + 6H \rightarrow C_3AH_6 \tag{8.29}$$

Evolving 207 cal/g of heat

$$C_3A + 3CH + 32H \rightarrow C_6AS_3H_{32} \tag{8.30}$$

Evolving 347 cal/g of heat

$$2C_3S + 6H \rightarrow \underset{\text{(Tobermorite gel)}}{C_3S_2H_3} + 3CH \tag{8.31}$$

Evolving 120 cal/g of heat

$$2C_2S + 4H \rightarrow C_3S_2H_3 + CH \tag{8.32}$$

Evolving 62 cal/g of heat

Table 8.15 Composition and Characteristics of Primary Compounds in Portland Cement

Approximate composition	$3CaO{\cdot}SiO_2$	$\beta 2CaO{\cdot}SiO_2$	$3CaO{\cdot}Al_2O_3$	$4CaO{\cdot}Al_2O_3Fe_2O_3$
Abbreviated formula	C_3S	βC_2S	C_3A	C_4AF
Common name	Alite	Belite	—	Ferrite phase, Fss
Principal impurities	MgO, Al_2O_3, Fe_2O_3	MgO, Al_2O_3, Fe_2O_3	SiO_2, MgO, alkalies	SiO_2, MgO
Common crystalline form	Monoclinic	Monoclinic	Cubic, orthorhombic	Orthorhombic
Proportion present (%)				
Range	35–65	10–40	0–15	5–15
Average in ordinary cement	50	25	8	8
Rate of reaction with water	Medium	Slow	Fast	Medium
Contribution to strength				
Early age	Good	Low	Good	Good
Ultimate	Good	High	Medium	Medium
Heat of hydration	Medium	Low	High	Medium
Typical (cal/g)	120	60	320	100

$$C + H \rightarrow CH \tag{8.33}$$

Evolving 279 cal/g of heat

Wastes containing the following constituents have been stabilized using portland cement: Cd^{2+}, Cr(IV), Cu^{2+}, Pb^{2+}, Ni^{2+}, Zn^{2+}, PCBs, oils, sludges, and plastics.

Pozzolanic Stabilization/Solidification

A pozzolan is a material that is not cementitious on its own but that forms cementitious compounds when combined with lime (CaO) and water at ordinary temperatures. The most common pozzolans are fly ash from coal-fired power plants, kiln dust from lime or cement kilns, and blast furnace slack. Pozzolanic reactions produce compounds that are similar to those produced through the hydration of portland cement. In pozzolanic reactions, however, lime is consumed rather than produced. Also, pozzolanic reactions are slower than portland cement hydration reactions. Oil sludges, plating sludges containing heavy metals, waste acids, and creosote, have been stabilized with pozzolans.

Thermoplastic Stabilization/Solidification

Thermoplastics are substances that exhibit plastic behavior at elevated temperatures. Upon cooling to normal environmental temperatures, they become more solid (tougher). In contaminated media treatment applications, they are used in microencapsulations in which the thermoplastic occupies pore spaces among waste/soil particles. Examples of commonly used thermoplastics are asphalt (bitumen) and polyethylene. Microencapsulation, which is essentially a physical treatment technique, can result in waste volume reduction and easier handling, elimination of free liquids from waste, lower permeability, and increased strength of the waste matrix. The disadvantages are that it has a high air pollution potential, high sensitivity to organics and salt content, and requires the use of expensive equipment.

Macroencapsulation

Macroencapsulation is the enclosure of contaminated media or waste within overpack materials and containers. The waste or contaminated media may be microencapsulated before macroencapsulation. Usually, drums of encapsulated wastes are produced in this process. Examples of wastes that are usually considered for macroencapsulation are incinerator ash, polychlorinated biphenyls (PCBs), and dioxins.

8.9.2 Performance Analysis for Stabilized/Solidified Wastes

Stabilized/solidified materials usually need to be analyzed with respect to their durability and the associated potential for release of contaminants into environmental media. This is born of the concern that water (whether through precipitation or groundwater) and air will eventually have the opportunity to contact the stabilized/solidified media and thus promote the release of materials through processes of leaching. Although stabilized media strength and durability are sometimes assessed, contaminant leachability is the primary issue in performance evaluation of S/S wastes. It should be noted that the leachability discussion provided in this section also applies to other treated wastes.

Leachability is the release of contaminants from waste matrix into the aqueous phase and the exit of these contaminants from the bounding surfaces of the waste mass. Several direct and indirect factors influence contaminant leachability from a stabilized matrix. These factors are discussed below.

Surface Area of the Waste

Typically, contaminant leachability increases with the specific surface of particulate media. The specific surface is the ratio of total available surface area to the waste mass volume or weight. The specific surface increases from that of a monolith, through that of a granular mass, to the highest level in fine powders, for the same external volume or weight of a material.

For a regularly shaped monolith, the specific surface is easily computed because internal porosity is not taken into account in the estimation of its surface area. For a mass of granular materials or fine powders that comprises various size fractions, internal and intergranular porosity and particle size distribution have to be taken into consideration. The volumetric specific surface, S_f of such materials can be computed as follows for a particulate medium that comprises idealized spherical particles of various size fractions:

$$S_f = \frac{A_s}{V} \tag{8.34}$$

S_f = volumetric specific surface (1/L)
A_s = surface area of particles within the waste mass (L^2)
V = external volume of the waste mass (L^3)

It should be noted that V is not the summation of the volumes of individual particles of the waste mass. If this is done, the incorrect assumption is that the particulate medium, consisting of various particle sizes, has a porosity of zero. The volumetric specific surface, S_f can be further computed as follows:

$$S_f = \frac{\sum_{i=1}^{m} 4\pi r_i^2 N_i}{\sum_{i=1}^{m} [1.333\pi r i^3 N_i/(1 - n)]} \tag{8.35}$$

$$S_f = 3(1 - n)\sum_{i=1}^{m}\left(\frac{f_i}{r_i}\right) \tag{8.36}$$

$$S_f = \frac{3(1 - n)}{r_a} \tag{8.37}$$

where

r_i = representative radius (possibly average) of the ith particle fraction of the medium (L)
N_i = number of particles in the ith fraction of the medium (unitless number)
n = intergranular porosity of the medium (unitless fraction)
f_i = volume fraction of the ith particle size in the medium (unitless fraction)
r_a = harmonic mean radius of all particles in the medium (L)

Agitation Technique and Equipment

In laboratory tests, agitation is necessary to accelerate the attainment of equilibrium between solid samples and leachant. Contaminant leachability tends to be directly proportional to the intensity of agitation due to particle breakage (and hence increased specific surface) and leachant renewal at the waste solid–leachant interface.

Characteristics of the Leachant

The pH and Eh of the leachant affect contaminant leachability because of their controls on the solubility of compounds within the waste matrix. Some minerals and contaminants are more stable within the waste matrix at specific ranges of leachant pH and Eh, while they are less stable at other ranges. In the field, these parameters may fluctuate with time.

Leachant/Waste Volumetric Ratio

The higher the leachant/waste volumetric ratio, the higher is the total mass of contaminant released for a given time interval, although recorded concentrations of the leached contaminant may be less than those for lower leachant/waste volumetric ratios within the same time envelope. Field ratios are difficult to simulate in the laboratory because of the intermittent manner in which the leachant is supplied by rainfall and other sources over an extended time period.

Duration of Contact

The longer the duration of contact of the leachant with the waste, the greater is the opportunity for contaminant removal. Some waste constituents can only be removed when associated mineral phases have dissolved. Dissolution of such minerals may require a minimum duration under prevailing environmental conditions.

Temperature Conditions

The solubilities of most substances tend to be directly proportional to temperature, although it should be noted that solubility is not the sole determinant of the magnitude of leaching. Upon dissolution, the contaminant species must find their way (through diffusion or other transport processes) to the boundary through which they pass to establish leaching rates. The media through which this transport occurs (which may be the waste or stabilized waste itself) have various levels of permeability, dispersion, and diffusion constants with respect to the traveling contaminant species.

Number of Elutions of Leachate

Within a given time interval, the greater the number of elutions, the larger is the total quantity of contaminants leached (with the total available quantities of the contaminants in the waste being the upper-bound values). For most contaminants, after reaching a maximum, the concentration of the contaminant in the leachate decreases for each subsequent elution. This often follows a negative exponential pattern of concentration decrease. Essentially, as the pore fluid (leachant) receives more and more contaminant by diffusion from the waste matrix, the concentration gradient between the waste matrix and the leachant decreases, resulting in reduced transfer rate. If the contaminant-laden pore fluid is purged (eluted) and its concentration measured, the value obtained is usually less than that of subsequent elutions because for the latter, less and less of the contaminant is available in the matrix for leaching.

Upon rising to a maximum value, M_0, the concentration of a target contaminant in the leachate usually follows equation (8.38).

$$M_t = M_0 \exp(-kt) \tag{8.38}$$

where

M_t = contaminant concentration in leachant measured at time or number of elutions t (M/L^3)
M_0 = contaminant concentration in leachant at reference time or number of elutions, t_0 (M/L^3)
k = empirical leaching constant obtainable from laboratory tests (unitless number)
t = time (T) or number of elutions

8.9.3 Tests for Leachability Assessment

Several empirical tests that exhibit different levels of simulative capacity for waste leachability in the field have been developed to evaluate stabilized, solidified wastes, wastes treated using other methods, and untreated wastes (often for classification and treatment process selection purposes). The results obtained from laboratory tests for S/S wastes using a specific type of test depend on the following two sets of factors:

1. The simulative capacity of the test variables to field conditions, in terms of the factors discussed in the previous section.
2. The effectiveness of the S/S technique

The most common leachability assessment tests are described briefly below.

Toxicity Characteristics Leaching Procedure (TCLP) Test

The TCLP is a batch test conducted on samples that are crushed to a maximum particle size of 9.5 mm. Liquids are usually separated from solids before use of the solids in the test. The leaching solution is added to the waste in a zero headspace extractor (ZHE) at a liquid:solid ratio of 20:1. The sample is then agitated in a rotary tumbler at 30 rpm for 18 h. The leaching solution is filtered out and combined with the free liquid that had been separated earlier. Contaminant concentrations are then measured. The TCLP test is mostly suitable for waste classification. Unless obtained concentrations are modified, the results lack the simulative capacity needed for extension of its utility to contaminant source-term and transport assessments for field scenarios.

Extraction Procedure Toxicity (EP TOX) Test

The EP TOX test is similar to the TCLP test except that the pH of the leachant is periodically adjusted up to a specified maximum acid addition. It should be noted that for the TCLP test, the acid is added only once: at the onset of the test. EP TOX tests tend to extract lesser concentrations of metals from media than TCLP tests because on a total basis the conditions (of the EP TOX tests) are less aggressive, as shown in Table 8.16.

Monofill Waste Extraction Procedure (MWEP)

The MWEP test, which was formerly called the Solid Waste Leach Test, involves multiple extractions of leachate from a monolith or crushed mass of waste using distilled/deionized water. Usually, the sample is tested as a monolith if it passes a structural integrity test (for strength); otherwise, it is crushed until all particles

Table 8.16 Comparison of TCLP and EP TOX Test Parameters

Experimental parameter	TCLP	EP TOX
Filter size, μm	0.6–0.8	0.45
Filter pressure, psi	50	75
Leaching solution	Acetate buffered solution (pH ≈3 or 5)	Acetic acid (pH ≈5)
Period of extraction, h	18	24
Liquid : solid ratio	20 : 1	16 : 1

Source: U.S. EPA (1989b).

pass a 9.5-mm sieve before use in batch testing. Leaching conditions are considered to be mild: a liquid:solid ratio of 10:1, operated for 18 h per extraction, using nonacidic leachant.

Multiple Extraction Procedure (MEP)

In the MEP test, contaminants are sequentially extracted from a material using a synthetic acid rain solution. This test is useful for addressing the effects of leachant pH changes on contaminant leachability from waste samples. Usually, acetic acid solution is used for the first extraction, followed with a concentrated H_2SO_4/HNO_3 solution (proportion of 60%/40% by weight) that is diluted to a pH of 3.0. In total, nine extractions are performed per test.

Materials Characterization Center Static Leach Test (MCC-IP)

The MCC-IP test was developed primarily for high-level radioactive waste monolith testing. Monoliths are used because grinding of samples containing radionuclides to granular mass and batch testing of ground mass may pose excessive levels of exposure risk. Water is used as the leachant in a quantity that provides a leachant volume-to-monolith surface area in the range of 10–200 cm.

American Nuclear Society (ANS-16.1) Leach Test

In field situations, the leachant is commonly not static but flows through pores or fractures in the waste mass, generating a dynamic leaching situation in which the leachant is continuously renewed. The ANS-16.1 test is a quasi-dynamic leaching test in the sense that leachant renewal is accomplished by replacing the leachant/leachate in the leaching chamber at specific time intervals. Usually, each waste sample is a monolithic waste cylinder with a length/diameter ratio between 0.2 and 5.0. The surface of the cylinder is cleaned by rinsing before leaching in demineralized water. The leachant is introduced into the leaching chamber illustrated in Figure 8.22, at a volume-to-monolith surface area ratio of 10 cm.

The leachate/leachant is replaced at the following time periods: 2 h, 7 h, 24 h, 48 h, 72 h, 4 days, 5 days, 7 days, 14 days, 28 days, 43 days, and 90 days. Contaminant concentration measured at these time intervals are plotted as illustrated schematically in Figure 8.23. The effective diffusion coefficient of the target contaminant from the monolith can be computed as follows.

$$D_e = b^2\left(\frac{\pi}{4}\right)\left(\frac{V^2}{S^2}\right) \tag{8.39}$$

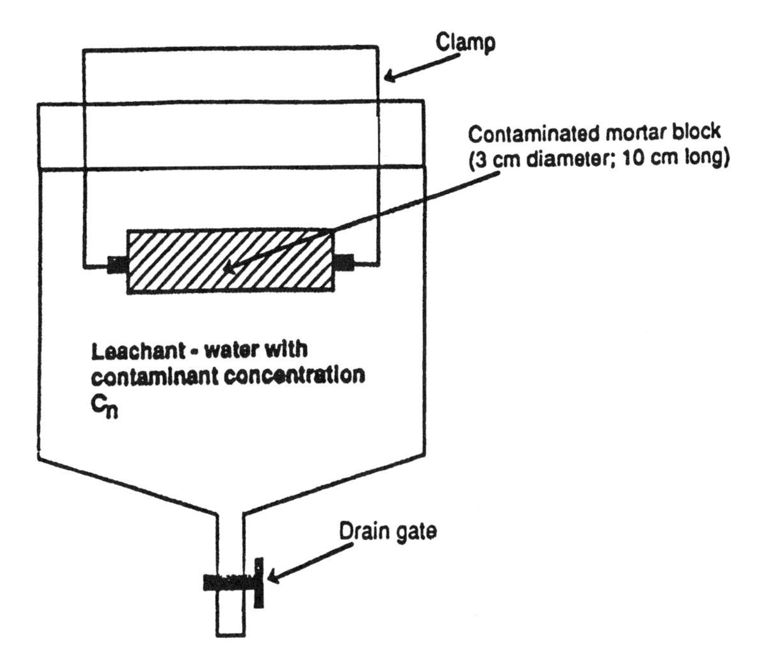

Figure 8.22 Test setup for measurement of the effective diffusion coefficient of the doped containment.

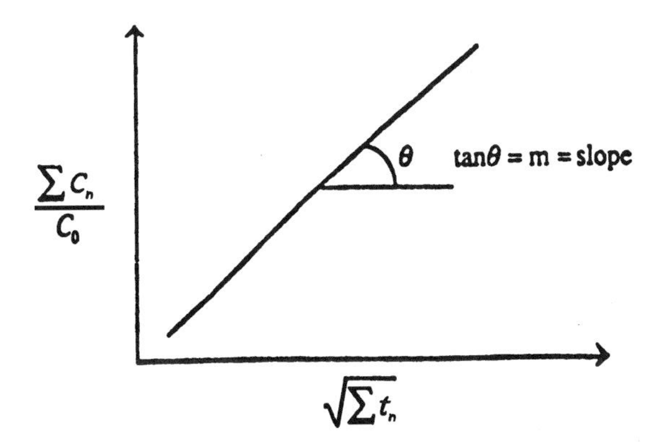

Figure 8.23 Method for determining the σ value for calculating the effective diffusion coefficient.

where

D_e = effective diffusion coefficient of the target contaminant through the waste matrix (L^2/T)
b = slope of the plot illustrated in Figure 8.23 (number/T)
t_n = resident time of each leachant volume (T)
V = volume of the monolith of waste (L^3)
S = surface area of the monolith of waste (L^2)
C_n = concentration of the contaminant in the leachate drained after time of residence t_n in the leaching chamber (M/L^3)
C_o = concentration of the contaminant within the waste (M/L^3)

The American Nuclear Society further defines a Leachability Index as follows:

$$L_w = \frac{1}{7}\sum_{n=1}^{7} \log\left(\frac{\beta}{D_e}\right) \tag{8.40}$$

where

L_w = leachability index, which ranges in magnitude from 5 for rapid diffusion to 15 for very slow diffusion
β = a test-specific constant that approximates 1 cm^2/s

The reader should note that in Eq. (8.40), $n = 7$ is used, meaning that the number of times the leachant/leachate is purged is 7. This number, n, can be modified to suit other purging times. A modification of this test is the Dynamic Leach Test (DLT) developed by Environment Canada and the Alberta Environmental Center. In this test, the leachant/leachate is renewed more frequently than in the ANS 16.1 test. Because of the increased frequency, the test is considered to be a dynamic leaching test. Renewal frequency is selected on the basis of data obtained from batch extraction tests conducted on the same material.

Column Leaching Tests (CLT)

Column leaching tests are usually conducted using granular materials. The material tested is packed into a cylindrical column and the leachant is introduced, usually at the bottom to avoid flow channeling (if the leachant is introduced at the top on permeable materials). The concentration of contaminants in the leachate is monitored at specific time intervals. This is the test that is most commonly used by researchers to simulate the leaching of contaminants from heaps or buried masses of discrete particulate media. It is a dynamic test.

8.9.4 Theoretical Analysis of Contaminant Leaching Rates

From a theoretical standpoint, two types of leaching can be identified based on the stabilized waste form:

1. Leaching of discrete particulate media
2. Leaching of monoliths of waste

For particulate media leaching, percolation leaching equations have been presented in Chapter 6. In addition, the release of substances from individual particles of the medium can be further analyzed using particle size reduction–leaching time functions such as those applied in the "shrinking core model." The model is described in standard mineral processing textbooks and is not presented here.

For monoliths, the release of the target contaminant into an adjacent leachant external to the monolith from locations x in the monolith measured from the monolith-leachant boundary can be described using the following form of the one-dimensional advection-dispersion equation:

$$D_e\left(\frac{\partial^2 C}{\partial x^2}\right) - V_v\left(\frac{\partial C}{\partial x}\right) - \left[1 + \left(\frac{1-f}{f}\right)\rho_t K_d\right]\left(\frac{\partial C}{\partial x}\right) = 0 \tag{8.41}$$

where

D_e = as defined for Eq. 8.39
C = concentration of the contaminant in the waste (M/L^3)
x = distance from the waste/leachant boundary into the waste monolith (L)
V_v = advective flow velocity of the contaminant from the monolith to the leachant (L/T)
f = porosity of the matrix (dimensionless fraction)
ρ_t = density of the matrix (M/L^3)
K_d = distribution coefficient of the contaminant for the matrix solid matter–pore fluid combination (L^3/M)
t = time (T)

Other versions of Eq. (8.41) may contain moisture content terms, put in to replace porosity terms. Furthermore, the tightness of pores in stabilized materials, which implies that both porosity and permeability magnitudes are very low, makes advective flow negligible. The primary mechanism for contaminant removal from the stabilized mass is then diffusion. Thus,

$$\frac{\partial C}{\partial t} = D_e\left(\frac{\partial^2 C}{\partial x^2}\right) \tag{8.42}$$

Equation (8.42) is commonly known as Fick's second law of diffusion. Following the analytical solution developed by Crank (1975) and with the simplifying assumption that D_e will remain constant as leaching (and hence time) increases, Eq. (8.43) results for use in estimating the quantity of contaminant leached from a monolith.

$$M_t = 2(C_0 - C_i)\left(\frac{D_e t}{\pi}\right)^{0.5} \tag{8.43}$$

where

M_t = amount of contaminant leached per unit surface area of a monolith after a given time interval (M/L^2)
C_0 = initial concentration of the contaminant in the pore solution of the monolith (M/L^3)
C_i = concentration of the contaminant at the monolith/leachant interface (M/L^3)

Other parameters are as defined earlier.

The concentration C_i can be assumed to be the same as the concentration of the contaminant in the leachant/leachate. In reality, C_i is greater than the bulk leachant/leachate concentration of the target contaminant. Equilibration can be attained through mixing or through other forms of agitation of the reaction chamber. When the leachant is renewed, C_i is initially zero in magnitude.

Within the monolith itself, the contaminant partitions between the matrix and the liquid in the pore spaces. A partition coefficient or mass transfer coefficient can be used to describe this partition. Hence,

$$C_0 = K_p C_d \tag{8.44}$$

$$C_0 = K_p\left(\frac{m_d}{v_d}\right) \tag{8.45}$$

where

K_p = mass transfer (or partition coefficient) of the target contaminant between the pore solution and the matrix (dimensionless fraction)
C_d = mix design concentration of the contaminant per bulk volume of the stabilized mass (M/L^3)
m_d = mass of the contaminant in the stabilized material (M)
v_d = design volume of the stabilized material containing M_d(L^3)

If K_p is assumed to have its maximum possible value of 1.0 (the target contaminant gets completely into the pore fluid) and $C_i = 0$, then Eq. (8.44) can be introduced into Eq. (8.43) resulting in the following equation:

$$M_t = 2\left(\frac{m_d}{v_d}\right)\left(\frac{D_e t}{\pi}\right)^{0.5} \tag{8.46}$$

Considering that M_t is expressed on a per-unit-surface-area basis, it needs to be multiplied by the total external surface area of the monolith as follows, to obtain the total quantity removed from the monolith:

$$M_{tT} = (M_t)(S) = 2S\left(\frac{m_d}{v_d}\right)\left(\frac{D_e t}{\pi}\right)^{0.5} \tag{8.47}$$

where

M_{tT} = mass of the contaminant that leaches through the entire external surface of the monolith during a given time interval (M)
S = total external surface area of the monolith (L^2)

Equation (8.47) can then be rearranged as follows, noting that to use Eq. (8.48), v_d must be the volume of the monolith that has the total surface area S.

$$\frac{M_{tT}}{m_d} = 2\left(\frac{S}{v_d}\right)\left(\frac{D_e t}{\pi}\right)^{0.5} \tag{8.48}$$

Essentially, Eq. (8.48) is an expression for estimating the quantity of the contaminant that can be removed from a monolith after a given duration of leaching expressed as a fraction of the quantity that was originally put into the material.

8.10 EX-SITU CHEMICAL TREATMENT PRINCIPLES

Chemical treatment is the destruction or transformation of contaminants from hazardous to less hazardous forms through direct or catalyzed chemical reactions. Catalysis reactions are more effective in the aqueous phase. Consequently, low treatment efficiencies are usually obtained for contaminants that are strongly sorbed to solid matrix components in soils. Secondly, the chemical reactions that are necessary for decontamination involve stoichiometric combinations of the reactants, in the presence of impurities (soil particles) which interfere with, and influence, the effectiveness of the reactions. Most chemical treatment processes were developed for treatment of aqueous streams of contaminants. Contact between reactants is more efficient in such streams than in contaminated soils. Therefore, chemical treatment of soil media that have high solids concentrations may require slurrying and aggressive mixing to produce adequate decontamination levels. Difficulties associated with delivery of reactants to contaminated materials in situ in proportions that satisfy stoichiometric requirements, control of reaction conditions, and mixing, constrain chemical treatment to being primarily an ex situ treatment process. However, recent developments have involved deep soil mixing (with the reaction agents) and/or installation of subsurface walls of reaction agents such as iron filings and limestone through which contaminated water can pass through to attain treatment. The most common chemical treatment methods are pH adjustment, oxidation/reduction, dehalogenation, hydrolysis, and supercritical water oxidation. The principles that underlie these techniques are discussed in this section. The reader should note that dehalogenation has been discussed in section 8.8.

8.10.1 Chemical Oxidation–Reduction Principles

Both chemical oxidation and reduction occur in the same process for different reactants. Usually, the oxidation state of one compound is increased (the oxidized compound) while it is decreased for another compound (the reduced compound). The oxygen supplier is called the oxidizing agent, and the compound that accepts oxygen is called the reducing agent. Chemical oxidation can also be defined as the loss of electrons by the oxidized compound, while reduction is the gain of electrons by the reduced compound.

The most common classes of contaminants treated using oxidation are phenols, chlorophenols, cyanides, some pesticides, and halogenated aliphatics. The process can be used in treating slurried contaminated soils that contain oxidizable contaminants. Oxidation rates can be improved through the use of catalysts, radiation, and electrolytic treatment. Some of the most commonly used oxidizing agents are discussed briefly below.

Hydrogen Peroxide (H_2O_2)

Hydrogen peroxide is a colorless, weakly acidic liquid. The reactions through which hydrogen peroxide can oxidize waste can be described as follows.

$$H_2O_2 + R \rightarrow RO + H_2O \qquad (8.49)$$

Catalysts such as Fe^{3+} and Fe^{2+} can cause the decomposition of $H_2\,O_2$ to radicals of perhydroxyl and hydroxyl, which are very effective oxidants.

$$Fe^{2+} + H_2O_2 \rightarrow Fe^{2+} + OH^- + OH \qquad (8.50)$$

$$Fe^{3+} + H_2O_2 \rightarrow Fe^{2+} + H^+ + HO_2 \qquad (8.51)$$

The perhydroxyl and hydroxyl are HO_2 and OH, respectively. A number of oxidation reactions referred to as "Fenton's reactions" (Dorfman and Adams, 1973) can be induced through the use of these two oxidants. In high concentrations, some organics and inorganics may produce heat in reactions with H_2O_2.

Ozone (O_3)

Ozone is a blue, excessively reactive gas which is capable of oxidizing a wide variety of compounds. When introduced into aqueous solutions that contain impurities such as soil particles and contaminants, ozone decomposes quickly in direct proportion to the content of impurities in the water. Masschelein (1982) identifies the following pathways for the oxidation of organic compounds by ozone, each of which results in different reaction products.

1. Direct oxidation
2. Oxidation by hydroxyl formed by the decomposition of ozone in water
3. Oxidation that results from ozone–solute interactions

The U.S. EPA (1986a) estimates that some compounds that have low reaction rates with ozone will react 100–1000 times faster if ultraviolet radiation or ultrasonic energy is applied. High pH favors an increase in the reactions that involve the hydroxyl free radical, which are normally slower than other reactions under lower pH conditions. Table 18.17 shows data on the ozonation of various organic compounds in water.

Hypochlorites

Hypochlorites are formed by the reaction of chlorine with an alkali. Hyprochlorites such as calcium hypochlorites ($CaOCl_2$) and sodium hypochlorite (NaOCl) are strong oxidizing agents and are used primarily as a safe method of delivering chlorine. The addition of chlorine to contaminated media causes reactions that fall into three classes of mechanisms:

1. Addition of chloride ion to the molecular structure of the compound
2. Replacement of molecular units of the compound by chlorine
3. Oxidation of the compound by chlorine

The first two mechanisms are generally called chlorination. If organic compounds are the target contaminants, chlorinated organic compounds, which are generally toxic, may be formed under suitable reaction conditions. This is common in strong solutions at low pH. Oxidation is favored in slightly acidic solutions and is the desirable reaction in use of hypochlorites for treatment of organic compounds.

Reduction is also a viable technique for decontamination of polluted soil media. Examples of reducing agents are compounds of iron (as previously discussed), zinc, sodium, and aluminum. Waste streams containing chromium, mer-

Table 8.17 Data on the Ozonation of Some Organic Compounds in Water

Compound(s)	Initial conc. (mg/liter)	Ozone dose (mg/liter)	Final Conc. (mg/liter)	Percent reduction
Methanol	2000	—	160	92
Ethanol	1000	—	90	91
Isoamyl alcohol	1000	—	80	92
Glycerine	1000	—	0	100
Hydrazine	100	—	0	100
Carbon Disulfide	100	—	0	100
Hydrogen Sulfide	10	—	0	100
Phenol	100	—	0	100
o-Cresol	100	—	0	100
Hydroquinone	100	—	0	100
Salicyclic acid	100	—	0	100
Gasoline	1000	—	0	100
Benzene	500	—	0	100
Toluene	500	—	0	100
Xylene	500	—	0	100
Acetone	100	—	30	70
Petroleum	10	4.5	0.2–0.3	97–98
Gasoline	50	1.29	1	98
		5.1	0.1	99.8
Benzene	200	20	5	97.5
Diethylbenzene	125–100	150–10	5–12.5	88–96
2,2,4-dinitrophenol	50	100	0.35	99.3
	3	17	0.05	98.3
		14	0	100
DDT	0.5	13.8	0.25	50
Malathion	10	3.5	2	80
		9.8	1	90
		2.6	0	100
Methylparathion	10	4.5	0.5	95
		9.5	0.1	99
Trichloromethyl parathion	10	8–10	0.07	99.3
	0.5–1	3.5–4.5	0	100
Dinitro-orthocresol	10	5–6	0	100

Source: U.S. EPA (1986).

cury, chlorinated organics, and unsaturated organics can be treated by the reduction method. For example, hexavalent chromium can be reduced to the less toxic and readily precipitable trivalent chromium. Both redox and oxidative treatment processes often require a subsequent process for removal of reaction products.

8.10.2 Supercritical Water Oxidation

At temperatures and pressures near its critical point, the characteristics of water make it an excellent medium for oxidation of compounds. At the critical point of water, which is 374°C and 221 atm, hydrogen bonds of the water assume the characteristics of a polar solvent. At this stage, oxygen and almost all hydrocarbons become readily soluble in water while the solubility of inorganic compounds becomes negligible. This implies that organic compounds can be oxidized while inorganics can be precipitated by supercritical water.

The process described by the U.S. EPA (1987) involves mixing of aqueous solutions or slurries of the target contaminated media with oxygen at pressures greater than 218 atm and temperatures in the regime of 500°C. This is accomplished in either an above-ground pressure vessel or a buried reactor vessel. This treatment technique requires high energy input. Emissions and residues include gases (N_2 and CO_2), inorganic precipitates, and aqueous solutions of organics and some salts. Oxidation and precipitation rates are typically very high.

8.10.3 Hydrolysis Principles

Hydrolysis is a complexation reaction in which hydroxyl (OH^-) liquid substitutes for a detached group in a molecule such that the compound becomes more stable. Water is usually the source of the (OH^-) liquid, and the displacement reaction can be represented as follows:

$$RX + H_2O \rightarrow ROH + HX \tag{8.52}$$

where R is the organic moiety and X is the chemical group that is cleaved off in the reaction. Even under natural conditions, hydrolysis of chemical compounds does occur. For decontamination of soils, the purpose is to induce environmental conditions that can increase its rate of occurrence. Organic compounds hydrolyze naturally in rates that range from a few seconds to millions of years. The rates are generally of first-order or pseudo-first-order form. The organic compounds that are susceptible to hydrolysis are listed below with representative reaction equations using information mostly adapted from U.S. EPA (1986).

$$\text{Alkyl halides:} \quad RX + H_2O \rightarrow ROH + HX \tag{8.53}$$

$$\text{Esters:} \quad R_1C(O)OR_2 + H_2O \rightarrow R_1C(O)OH + R_2OH \tag{8.54}$$

Amides: $RC(O)NR_1R_2 + H_2O \rightarrow RC(O)OH + R_1R_2NH$ (8.55)

Carbamates: $ROC(O)NR_1R_2 + H_2O \rightarrow ROH + HNR_1R_2 + CO_2$ (8.56)

Phosphoric and phosphoric acid esters:

$$ROPR_1R_2 + H_2O \rightarrow HOPR_1R_2 + ROH \quad (8.57)$$

As in the previous reaction equation, R represents the organic moiety, which may be of one type in the parent organic compound. In some cases some moieties within the same compound can be hydrolyzable while others may not be hydrolyzable. Also, the hydrolysis of various moities or part of the compound may be contingent upon different conditions of pH, temperature, etc.

Generally, the hydrolysis rate of a compound is regarded as the sum of neutral, acid-catalyzed, and base-catalyzed process constants, as follows:

$$-\frac{\partial(RX)}{\partial t} = k_N(RX) + k_A(RX)(H^+) + k_B(RX)(OH^-) \quad (8.58)$$

where

RX = the concentration of the organic contaminant
t = time
k_N = neutral hydrolysis reaction constant (for pH=7.0)
k_A = acid-catalyzed hydrolysis reaction constant (for pH < 7.0)
k_B = base-catalyzed hydrolysis reaction constant (for pH > 7.0)

$$k_h = k_N + k_A(H^+) + k_B(OH^-) \quad (8.59)$$

In Eq. (8.59), k_h is the overall hydrolysis reaction rate constant. Equation (8.58) can then be rewritten as Eq. 8.60:

$$-\frac{\partial(RX)}{\partial t} = k_h(RX) \quad (8.60)$$

The U.S. EPA (1986) has compiled data on hydrolysis rates of various organic compounds. These data are presented in Tables 8.18–8.21. The reader should note that these hydrolysis rates are at temperatures at or near 25°C. Application of hydrolysis as a treatment technique to contaminated media to achieve rates higher than those at normal temperature and pH conditions may require alteration of these conditions. Generally, hydrolysis intensifies with increase in temperature. Information provided on pH dependence tabulated in Tables 8.18–8.21 can also be used to increase rates.

Table 8.18 Data on the Hydrolysis of Halogenated Ethers, Epoxides, and Alcohols

Compound	Data source	k_A	k_N	k_B	$t_{1/2}$ at pH 5	6	7	8	9	10	11
Chloromethylmethylether	1	—	—	—	—	—	0.007s	—	—	—	—
bia(Chloromethyl)ether	1	—	—	—	—	—	38s	—	—	—	—
2-Chloroethanol	1	—	—	—	—	—	21y	—	—	—	—
1-Chloro-2-propanol	1	—	—	—	—	—	2 y	—	—	—	—
2-Chloroethylvinylether	2	—	—	—	—	—	0.48 y	—	—	—	—
3-Chloro-,1,2 epoxy, 2-methyl propane	3	1.84E-3	5E-7	—	16d	16d	16d	16d	16d	16d	16d
Alpha-epichlorohydria	3	8.0E-4	9.8E-7	—	8.2d	8.2d	8.2d	8.2d	8.2d	8.2d	8.2d
Epibromohydrin	3	6.1E-4	5E-7	—	16d	16d	16d	16d	16d	16d	16d

Note: All values reported for 25°C. Rate constants in s^{-1}; a = second; d = day; y = year.
Data sources: 1. Radding et al. (1977); 2. Versar, Inc. 1979; 3. Mabey and Mill (1978).
Source: U.S. EPA (1986).

Table 8.19 Data on the Hydrolysis of carbamates

Compound	Date source	k_A	k_N	k_B	$t_{1/2}$ at pH 5	6	7	8	9	10	11
$CH_3OC(O)N(H)C_6H_5$	1	—	—	5.5E-5	4×10^5y	40,000y	4,000y	400y	40y	4y	146d
$C_2H_5O(CO)N(CH_3)C_6H_5$	1	—	—	5.0E-6	4.4×10^6yr	4.4×10^5y	44,000y	4,400y	440yr	44y	4.4y
$C_6H_5O(CO)N(H)C_6H_5$	1	—	—	5.42E1	150d	15d	1.5d	3.6h	21m	2m	13sec
$C_6H_5OCCO)N(CH_3)C_6H_5$	1	—	—	4.2E-5	5.2×10^5y	52,000y	5200y	520y	52y	5.2y	191d
p-$CH_3OC_6H_4(O)N(H)$-C_6H_5	1	—	—	2.5E1	320d	32d	3.2d	7.7h	46m	4.6m	28sec
m-$ClC_6H_4OC(O)N(H)$- C_6H_5	1	—	—	1.8E3	4.5d	11h	1.1h	6.4m	39sec	3.9sec	0.4sec
p-$NO_2C_6H_4OC(O)N(H)$-C_6H_5	1	—	—	2.7E5	43m	4.3m	26sec	2.6sec	0.3sec	0.03sec	0.003sec
p-$NO_2C_6H_4OC(O)N$-$(CH_3)C_6H_5$	1	—	—	8.0E-4	27,500y	2750y	275y	27.5y	2.7y	100d	10d
1-$C_{10}H9OC(O)N(H)CH_3$	1	—	—	9.4EO	2.3y	85d	8.5d	20th	2h	1.2m	74sec
1-$C_{10}H9OC(O)N(CH_3)_2$	1	—	1.8E-11	—	1200y	1200y	1200y	1200y	1200y	1200y	1200y
$(C_2H5)_2NCH_2CH_2OC$-$(O)N(H)C_6H_5$	1	—	—	2.6E-5	8.5×10^5y	85,000y	8400y	850y	85y	8.5y	310d
$(C_2H_5)_2NCH_2CH_2OC$-$(O)N(H)C_6H_3(CH_3)_3$	1	—	—	9.4E-7	2.3×10^7y	2.3×10^6y	2.4×10^5y	23,000y	2300y	230y	23y
$(CH_3)_3NC_6H4OC(O)N$-$(H)CH_3$	1	—	—	6.7E-1	33y	3.3y	120d	12d	1.2d	2.9h	1.7m
$(CH)_3)_3NC_6H4OC(O)N$-$(CH_3)_2$	1	—	—	2.8E-4	89,500y	7850y	785y	78.5y	7.9y	268d	29d
$ClCH_2CH_2OC(O)N(H)$-C_6H_5	1	—	—	1.6E-3	14,000y	1400y	140y	14y	1.4y	50d	5d
$Cl_2CHCH_2OC(O)NHC_6H_5$	1	—	—	5.0E-2	440y	44y	4.4y	160d	16d	1.6d	3.9h
$CCl_3CH_2OC(O)NHC_6H_5$	1	—	—	3.2E-1	69y	6.9y	252d	25d	2.5d	6h	36m
$CF_3CH_2OC(O)NHC_6H_5$	1	—	—	1.0E-1	220y	22y	2.2y	80d	8d	20h	2h
Ethyl carbamate	2	—	—	—	—	—	11,000y	—	—	—	—
$C_2H_5OC(O)NHCH_3$	2	—	—	—	—	—	38,000y	—	—	—	—
$C_2H_5OC(O)N(CH_3)_2$	2	—	—	—	—	—	39,000y	—	—	—	—

Note: All values at 25°C. Rate constants in s^{-1} m = minute; h = hour; d = day; y = year.
Data sources: 1. Mabey and Mill (1978); 2. Radding et al. (1977).
Source: U.S. EPA (1986).

8.11 IN-SITU NATURAL ATTENUATION PRINCIPLES

Natural attenuation, which is also called intrinsic or passive remediation, is a technique in which natural processes in soil systems are relied upon to degrade and dissipate contaminants. Contaminant dissipation may occur through elimination or transformation. If conditions in a soil system favor the occurrence of natural attenuation, the implication is that the necessity to clean up the soil system to very low residual contaminant concentrations is minimized.

Soil systems contain natural materials such as microbes, aqueous iron, iron minerals, reduced sulfur, and soil organic matter. The interaction of these materials with contaminants can result in natural attenuation. The distinction between natural attenuation processes and associated processes that result in the transfer of contaminants from the aqueous phase to the solid phase (sorption and precipitation) has been controversial. Both sets of processes are interrelated, but natural attenuation implies a decrease in the concentrations of the target contaminants within the total soil system and not a transfer from one phase to another as in the case of sorption. Both natural attenuation and phase-transfer processes result in a decrease in contaminant concentration in the aqueous phase.

Figure 8.24 illustrates contaminant plume migration from a source that is induced by groundwater movement through the source zone. The plume travels at a velocity V_{min} from the source zone. Natural attenuation occurs within the plume zone and will occur in the intervening zone between the plume and the point of compliance (POC). Undoubtedly, the concentration of contaminants that will be obtained through sample retrieval from the POC and analytical measurements is expected to be lower than those for preceding points in this box model. This will be true if the plume travels at a rate faster than the rate of contaminant replenishment from the source. The reduction in contaminant concentration at the POC relative to all preceding points (in this homogeneous, isotropic, and continuous geomedia model) is attributable to the following factors.

1. Sorption of contaminants by soil materials with which the contaminants come in contact
2. Precipitation of contaminants along the path of travel if favored by prevailing groundwater chemistry and pH-Eh conditions
3. Destruction and transformation of contaminants by microbes and soil minerals

Natural attenuation is generally considered to be primarily factors 2 and 3. The U.S. EPA (1995) conveniently describes natural attenuation processes of organic contaminants as outlined in Table 8.22. For biological processes, aerobic biodegradation rates are commonly higher than anaerobic rates. Typically, the plume that undergoes natural attenuation comprises an inner anaerobic core and an outer aerobic core as shown in Figure 8.25. Aerobic biodegradation of the

Table 8.20 Data on the Hydrolysis of Esters

Compound	Data source	k_A	k_N	k_B	$t_{1/2}$ at pH 5	6	7	8	9	10	11
Ethyl acetate	1	1.1E-4	1.5E-10	1.1E-1	16y	16y	2.0y	0.2y	7.3d	0.73d	0.073d
Isopropyl acetate	1	6.0E-5	—	2.6E-2	35y	68y	8.4y	308d	31d	3.1d	0.31d
Butyl acetate	1	1.3E-4	—	1.5E-3	1.6y	78y	140y	1.5y	015y	5.5d	0.5d
Vinyl acetate	1	1.4E-4	1.1E-7	1.0E1	73d	67d	7.3d	0.8d	1.9h	12m	1.2m
Allyl acetate	1	—	—	7.3E-1	30y	3.0y	110d	11d	1.1d	2.6h	0.26h
Benzyl acetate	1	1.1E-4	—	2.0E-1	17y	10.4y	1.1y	40d	4.0d	9.6h	1h
O-Acetyl phenol	1	7.8E-5	6.6E-8	1.4E0	119d	100d	38d	5.5d	0.57d	1h	5m
2,4-Dinitrophenyl acetate	1	—	1.1E-5	9.4E1	17h	16h	9.4h	1.8h	0.2h	1.2m	1 + 1m
$ClCH_2C(O)^0CH_3$	1	8.5E-5	2.1E-7	1.4E2	23d	5.0d	14h	1.4h	8.3m	50s	5s
$Cl_2CHC(O)OCH_3$	1	2.3E-4	1.5E-5	2.8E3	11h	4.5h	38m	4.1m	255	2.5s	0.2s
$Cl_2CHC(O)OC_6H_5$	1	—	1.8E-3	1.3E4	6.4m	6.1m	3.7m	47s	4.7s	0.5s	0.05s
$F_2CHC(O)OC_2H_5$	1	—	5.7E-5	4.5E3	3.3h	1.9h	23m	2.6m	16s	1.6s	0.2s
$Cl_3CC(O)OCH_3$	1	—	7.7E-4	—	15m	15m	15m	15m	15m	15m	15m
$F_3CC(O)OC_2H_5$	1	—	3.2E-3	—	3.6m	3.6m	3.6m	3.6m	3.6m	3.6m	3.6m
$F_3CC(O)OC(CH_3)_3$	1	—	1.3E-3	—	8.9m	8.9m	8.9m	8.9m	8.9m	8.9m	8.9m
$CH_3SCH_2C(O)OC_2H_5$	1	—	—	9.2E-1	24y	2.4y	87d	8.7d	0.87d	2.1h	13m
$CH_3S(O)CHC(O)C_2H_5$	1	—	—	1.3E1	1.7y	62d	6.2d	0.62d	1.5h	9m	1m

$(CH_3)_2SCH_2C(O)\text{-}OC_2H_5$	1	—	—	2.0E2	40d	96h	9.6h	57m	5.7m	35s	3.5s
$C_2H_5C(O)OC_2H_5$	1	3.3E-5	—	8.7E-2	52.3y	24.4y	2.5y	91d	9.1d	0.9d	2.2h
$C_3H_7C(O)OC_2H_5$	1	1.8E-5	—	3.8E-2	100y	55y	5.8y	212d	21d	2.1d	5h
$(CH_3)_2CHC(O)OC_2H_5$	1	—	—	2.3E-2	960y	96y	9.6y	350d	35d	3.5d	0.35d
$CH_2CHC(O)OC_2H_5$	1	1.2E-6	—	7.8E-2	244y	35y	3.5y	128d	13d	1.3d	0.13d
trans-$CH_3CHCHC(O)\text{-}OC_2H_5$	1	6.3E-7	—	1.3E-2	1140y	160y	17y	1.7y	62d	6.2d	0.62d
$CHCC(O)OC_2H_5$	1	—	—	4.68E0	4.7y	170d	17d	1.7d	0.17d	24m	2.4m
$C_6H_5C(O)OCH_3$	1	4.0E-7	—	1.9E-3	3720y	1160y	118y	11.8y	1.2y	0.12y	4.4d
$C_6H_5C(O)OC_2H_5$	1	—	—	3.0E-2	730y	73y	7.3y	0.73y	27d	2.7d	0.27d
$C_6H_5C(O)OCH(CH_3)_2$	1	—	—	6.2E-3	3500y	350y	35y	3.5y	128d	12.8d	1.3d
$C_6H_5C(O)OCH_2C_6H_5$	1	—	—	8.0E-3	2700y	270y	27y	2.7y	99d	9.9d	1.0d
p-$NO_2\text{-}C_6H4C(O)OCH_3$	1	4.3E-7	—	7.4E-2	282y	30y	3.0y	0.3y	11d	1.1d	2.6h
p-$NO_2\text{-}C_6H4C(O)OCH_3$	1			6.4E-1	34y	3.4y	0.34y	12.4d	1.2d	2.6h	0.26h
p-$NO_2\text{-}C_6H4C(O)\text{-}OC_2H_5$	1	1.4E-7	—	2.4E-01	92y	9.2y	0.92y	34d	3.4d	0.34d	0.8h
1-$C_5H4NC(O)OC_2H_5$	1	—	—	5.4E-1	41y	4.1y	0.41y	15d	1.5d	3.6h	0.36h
o-$C_6H_4[C(O)OC_2H_5]2$	1	—	—	1.0E-2	2200y	220y	22y	2.2y	80.3d	8.0d	0.8d
o-$C_6H_4[C(O)OCH_2\text{-}C_6H_5]2$	1	—	—	1.7E-2	1300y	130y	13y	1.3y	47.5d	4.8d	0.48d
p-$C_6H_4(CCO)OCH_3]_2$	1	—	—	2.5E-1	880y	88y	0.88y	32d	3.2d	7.7h	0.77h
p-$C_6H_4[C(O)OC_2H_5]2$	1	—	—	6.9E-2	320y	32y	3.2y	117d	11.7d	1.2d	2.6h

Note: All values reported for 25°C. Rate constants in s^{-1}; s = second; m = minute; h = hour; d = day; y = year.
Data source: 1. Mabey and Mill (1978).
Source: U.S. EPA (1986).

Table 8.21 Data on the Hydrolysis of Miscellaneous Compounds Including Some Pesticides

Compound	Data source	k_A	k_N	k_B	$t_{1/2}$ at pH						
					5	6	7	8	9	10	11
beta-Propiolactone	1	—	3.3E-3	—	3.5m	3.5m	3.5m	3.5m	3.5m	3.5m	3.5m
$CH_2CH_2S(O_2)$	1	—	2.15E-5	—	8.9h	8.9h	8.9h	8.9h	8.9h	8.9h	8.9h
Dimethyl sulfate	1	—	1.66E-4	1.48E-2	1.2h	1.2h	1.2h	1.2h	1.2h	1.1h	1.1h
Bis(chloromethyl) ether	1	—	2.8E-2	—	25sec	25sec	25sec	25sec	25sec	25sec	25sec
Phenyldimethyltriaxine	1	—	2.75E-5	—	7h	7h	7h	7h	7h	7h	7h
Benzoyl chloride	1	—	4.2E-2	—	16sec	16sec	16sec	16sec	16sec	16sec	16sec
$(CH_3)_2NCO$	1	—	2.5E-3	—	4m	4m	4m	4m	4m	4m	4m
$CH_3OC(O)$	1	—	5.64E-4	—	20m	20m	20m	20m	20m	20m	20m
Methoxychlor	1	—	2.99E-8	3.64E-4	270d	270d	270d	267d	241d	121d	20d
Captan	1	—	1.87E-5	5.7E2	10h	8h	3h	20m	2m	12sec	1sec
Atraxine	1	3.9E-5	7.6E-5	—	2.5h	2.5h	2.5h	2.5h	2.5h	2.5h	2.5h
Malathion	2	4.8E-5	7.7E-9	5.5E0	1.6h	128d	14d	1.5d	3.5h	21m	2m
Parathion	2	—	4.5E-8	2.3E-2	178d	177d	170d	118d	29d	3.4d	8.4h
Paraoxon	2	—	4.1E-8	1.3E-1	195d	190d	149d	47d	6d	15h	1.5h

Diazinon	2	2.1E-2	4.3E-8	5.3E-3	32d	125d	176d	165d	14d	14d	1.5d
Diazoxon	2	6.4E-1	2.8E-7	7.6E-6	1.2d	9d	23d	28d	29d	29d	29d
Chlopyrifos	2	—	1.E-7	1E-1	80d	79d	73d	40d	7d	19h	2h
Sevin	2	—	—	7.7E0	2.9h	104d	10d	1d	2.5h	15m	1.5m
Sevin	2	—	—	3.4E0	6.5y	236d	24d	2.4d	5.7h	34m	3.4m
Baygon	2	—	—	4.6E-1	48y	4.8y	174d	17d	1.7d	4.2h	25m
Pyrolam	2	—	—	1.1E-2	2000y	200y	20y	2y	73d	7.3d	18h
Dimetilan	2	—	—	5.7E-5	3.9×10^5y	39,000y	3900y	390y	39y	3.9y	141d
P-Nitrophenyl-N-methyl carbamate	2	—	4E-5	3.0E3	4.5h	2.8h	34m	3.8m	23sec	2.3sec	0.23sec
2,4-D,m-butoxyethylester	2	2.0E-5	2.0E-5	3.02E1	9.6h	9.5h	8.4h	4.4h	36m	4m	23sec
Methoxychlor	2	—	2.8E-8	2.8E-4	286d	286d	286d	286d	252d	122d	20d
DDT	2	—	1.9E-9	9.9E-3	12y	11y	7.6y	1.9y	79d	8d	19h
2,4-D,methylester	2	—	—	1.7E1	1.3yr	47d	4.7d	11hr	1.1hr	6.8min	41sec

Note: All values for 25 ± 5°C. Rate constants in s^{-1}; m = minute; h = hour; d = day; y = year.
Data sources: 1. Mabey and Hill (1978); 2. Harris (1982).
Source: U.S. EPA (1986).

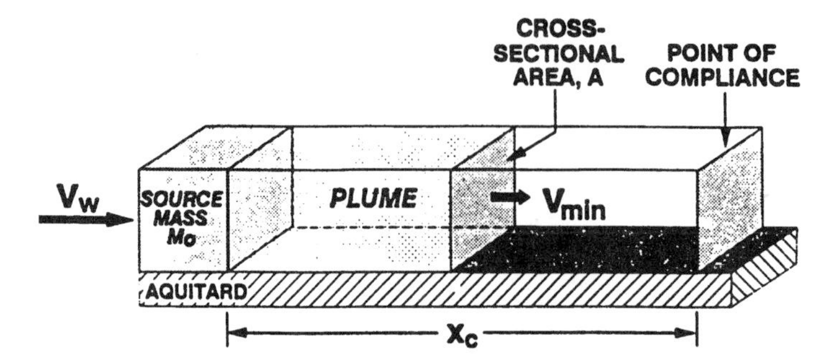

Figure 8.24 Illustration of contaminant plume movement within a box model: V_w = groundwater velocity through the contaminant (of mass M_0) source zone; V_{min} = plume velocity. (Adapted from U.S. EPA, 1994b.)

outer core is enhanced by its greater access to infiltrating oxygen than the inner core. From the analysis developed by Kerfoot (1994), the oxidation of a hydrocarbon (an organic compound class) can be described as follows:

$$C_xH_y + \left[\frac{(x + y)}{4}\right]O_2 = xCO_2 + \left(\frac{y}{2H_2O}\right) \tag{8.61}$$

C_xH_y is the empirical formula of this generic hydrocarbon. Considering that carbon has an atomic mass of 12 amu and hydrogen has an atomic mass of 1 amu, each mole of C_xH_y has a mass of $(12x + y)$ gram, and is degraded by $[x + (y/4)]$ moles or $(32x + 8y)$ gram of molecular oxygen to produce 1 mole or $(44x)$ gram of CO_2.

For hydrocarbons with the following x and y values, Kerfoot (1994) provides the approximations indicated below for rates of oxygen consumption, hydrocarbon degradation, and CO_2 production.

For $x \leq y \leq 2x$ and $4 \leq x \leq 12$:

$$\frac{\partial(C_xH_y)}{\partial(O_2)} = 0.31 \tag{8.62}$$

$$\frac{\partial(C_xH_y)}{\partial(CO_2)} = -0.31 \tag{8.63}$$

Inorganic contaminants such as metals can also be naturally attenuated in soil systems. Electron donors such as Fe^{2+}, ferrous iron minerals, organic matter, and reduced sulfur, which are present in soils in various proportions, can reduce

Table 8.22 The Most Prevalent Natural Attenuation Mechanisms

Mechanism	Description	Potential For BTEX attenuation
Biological		
Aerobic	Microbes utilize oxygen as an electron acceptor to convert contaminant to CO_2, water, and biomass.	Most significant attenuation mechanism if sufficient oxygen is present. Soil air (O_2) $\geq$2%. Groundwater D.O. $\geq$ 1–2 mg/liter.
Anaerobic Denitrification Sulfate-reducing Methanogenic Fe-reducing	Alternative electron acceptors (e.g., $NO^-{}_3$, $SO^4{}_{2-}$, Fe^{3+}, CO_2) are utilized by microbes to degrade contaminants.	Rates are typically much slower than for aerobic biodegradation: toluene is the only component of BTEX that has been shown to consistently degrade.
Hypoxic	Secondary electron acceptor required at low oxygen content for biodegradation of contaminants.	Has not been demonstrated in the field for BTEX.
Physical		
Volatilization	Contaminants are removed from groundwater by volatilization to the vapor phase in the unsaturated zone.	Normally minor contribution relative to biodegradation. More significant for shallow or highly fluctuating water table.
Dispersion	Mechanical mixing and molecular diffusion processes reduce concetrations.	Decreases concentrations, but does not result in a net loss of mass.
Sorption	Contaminants partition between the aqueous phase and the soil matrix. Sorption is controlled by the organic carbon content of the soil, soil mineralogy, and grain size.	Sorption retards plume migration, but does not permanently remove BTEX from soil or groundwater as desorption may occur.

Source: U.S. EPA (1995).

contaminants such as Cr(VI). The reaction equation for the reduction of Cr(VI) by Fe^{2+} is as follows:

$$3Fe^{2+} + HCrO^{4-} + 7H^{+} \rightarrow Cr^{3+} + 3Fe^{3} + 4H_2O \quad (8.64)$$

Eary and Rai (1988) estimate that if sufficient dissolved oxygen gains access to the reactants, the reaction may reach completion in less than 5 min.

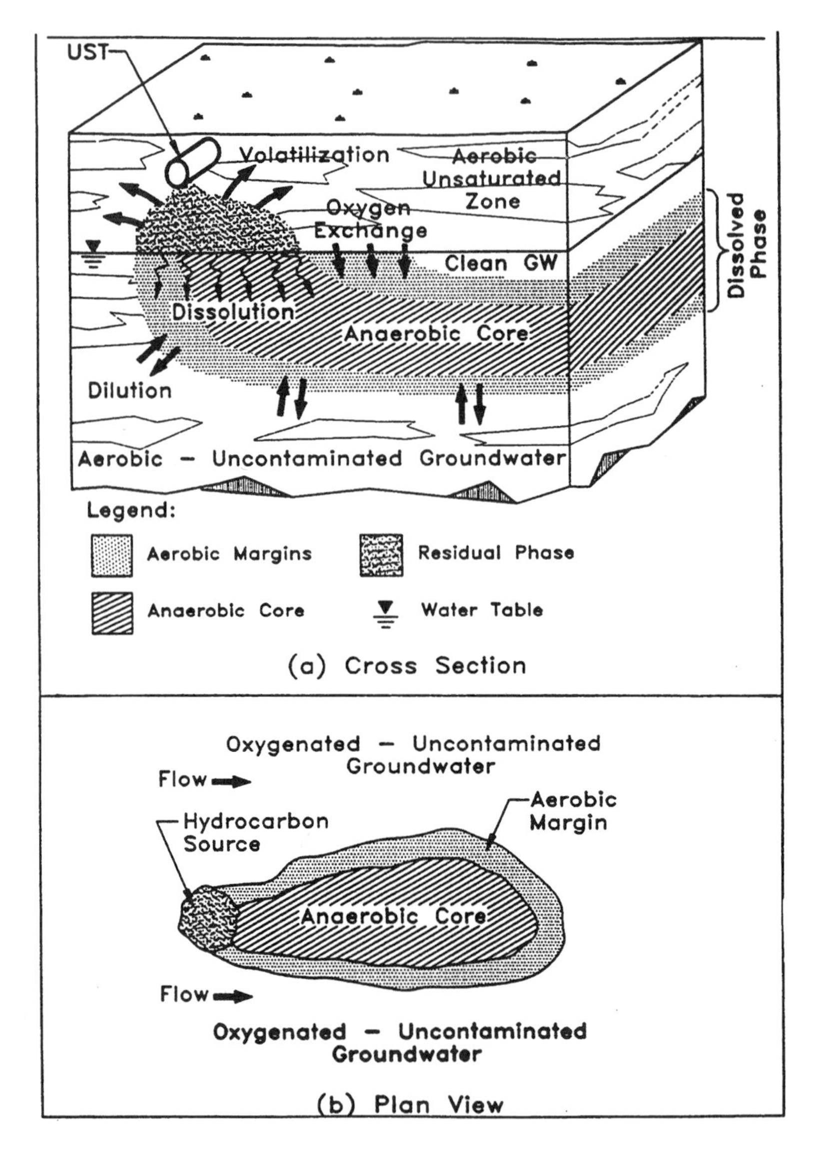

Figure 8.25 Illustration of the effects of oxygen access on biodegradation of a contaminant plume (U.S. EPA, 1995).

Many rock-forming minerals which are subsequently weathered to soil minerals contain iron. Among these minerals are olivines, micas, smectites, and pyroxenes. Iron-based minerals such as ilmenite, magnetite, and pyrite are common soil minerals that may reduce oxidants such as hexavalent chromium.

Organic matter is also known to reduce metallic radicals such as chromates. An example of this reaction is presented below.

$$2Cr_2O_7^{2+} + 3C^0 + 16H^+ \rightarrow 4Cr^{3+} + 3CO_2 + 8H_2O \quad (8.65)$$

In summary, natural attenuation is adequate only when contaminant concentrations are low enough to be dissipated by natural processes in soils. Excessive concentration of contaminants requires the application of active remediation techniques.

8.12 IN-SITU PHYTOREMEDIATION PRINCIPLES

Phytoremediation is the use of plants to remove contaminants from soil systems, either in situ or in batch reaction systems. In-situ phytoremediation involves the direct removal or transformation of contaminants in soil systems by growing plants, and is the focus of this section. This technique can be used to remediate soil systems contaminated with metals, chlorinated solvents, crude oil, pesticides, and explosives such as trinitrotoluene (TNT).

In phytoremediation, one or a combination of plants is selected and cultivated on a contaminated site. During their growth, the buried portions (roots) remove contaminants from soils through one or more of the following processes: direct absorption; transformation by microbes in the root zone; and stabilization through phase-change processes (example: precipitation) and reduction of soil permeability. Fractions of the mass of the contaminants that are absorbed by the roots of the plants are translocated to other parts (stem and leaves) of the plants and accumulate for use in metabolic processes. Some of the absorbed contaminants are volatilized, mostly through leaves into the atmosphere. The accumulated contaminants can be harvested along with the host plants and processed for disposal in reduced volume or for recycling.

Phytoremediation is useful at sites where targeted contaminants exist primarily within the depth of the root zone in concentrations that are low to moderate. Considering that reliance is on plant growth and efficient interaction with soil contaminants, only plants that can tolerate the soil environmental conditions, including contamination levels and micro and macroclimates, can be used effectively. Such plants must also have adequately rapid growth rates and root densities that are high enough to provide adequate surface area for contact with contaminated soil media. Most plants that show high potential as phytoremediators grow in tropical areas such as sub-Saharan Africa, Central and South America, and the Caribbean Islands. Moffat (1995) explains that contaminant accumulation in these plants is a natural defense against plant-eating insects and microbial pathogens that abound in the tropics. The different techniques applied in in-situ phytoremediation are described briefly below.

8.12.1 Phytoextraction

Phytoextraction is also referred to as phytoaccumulation. It is the uptake of contaminants (usually metals) by plants and subsequent distribution of the contaminants to plant stems and leaves, which can be harvested for concentration and recycling or disposal of the metals. The harvested plant is usually composted or incinerated. Incineration to ash reduces the volume of the material to only about 10% of the original. The contaminated site can be replanted with the selected plants until contaminant concentrations in surficial soils are reduced to desired levels. Plants which have high capacities to accumulate metals are called hyperaccumulators. Examples are presented in Table 8.23 and illustrated for nickel uptake in Figure 8.26.

Hundreds of plants in various locations around the world are known to absorb high quantities of metals such as zinc, copper, and nickel. The U.S. EPA (1996b) recognizes hyperaccumulation as an ecophysiological adaptation of plants to stress and as an indicator of resistance to metals that are usually found where such hyperaccumulators thrive. Baker and Brooks (1989) define a hyperaccumulator as a plant that contains more that 0.1% Ni, Co, Cu, or Cr, or 1.0% Zn and Mn, in its leaves on a dry-weight basis. Lack of sufficient biomass in

Table 8.23 Examples of Hyperaccumulators of Metals

Metal	Plant species	Percentage of Metal in Dry Weight of Leaves (%)	Native Location
Zn	Thlaspi calaminare	<3	Germany
	Viola species	1	Europe
Cu	Aeolanthus biformifolius	1	Zaire
Ni	Phyllanthus serpentinus	3.8	New Caledonia
	Alyssum bertoloni and 50 other species of Alyssum	>3	Southern Europe and Turkey
	Sebertia acuminata	25 (in latex)	New Caledonia
	Stackhousia tryonii	4.1	Australia
Pb	Brassuca juncea	<3.5	India
Co	Haumaniastrum robertii	1	Zaire

Source: U.S. EPA, 1996b.

Figure 8.26 Illustration of nickel uptake through the process of phytoremediation (U.S. EPA, 1996b).

some vegetation types such as some grasses hinder their use in phytoextraction. The U.S. DOE (1994) identifies the following traits as ideal for plants as hyperaccumulators:

1. High accumulation even at low concentrations of the contaminant in soil
2. Ability to accumulate high levels of the target contaminants
3. Ability to accumulate a variety of metals
4. Fast growth and high biomass production
5. High resistance to diseases and pests

Although lead (Pb) is the most common heavy metal contaminant in surficial soils, it has not been possible to identify plant species that can accumulate Pb adequately under natural conditions in the field. The bioavailability of Pb is low. Soil matter has high sorptive capacity for Pb. Thus, even where total concentrations are high in soil systems, the fraction present in the aqueous phase is usually low. It is this fraction that can be taken up by plants. To enhance the bioavailability and hence the phytoextraction of low yield metals, chelators are usually added to soils, as chelators improve metal bioavailability in soil and their translation from the roots to other parts of plants. Tables 8.24 and 8.25 present summaries of research data (Huang et al., 1997) from investigations of the effects of five chelators on the accumulation of Pb in corn and pea plants.

8.12.2 Phytodegradation

Phytodegradation is the process by which enzymes produced by plant roots break down contaminants in soils to less hazardous forms. This process applies primar-

Table 8.24 Effects of Adding EDTA to Pb-Contaminated Soil[a] with Total Soil Pb mg/kg on Pb Concentration in Xylem Sap and Pb Accumulation in Shoots[b] of 21-day-old Corn Grown in Contaminated Soil

EDTA added to soil (g kg^{-1})	Pb concentration in xylem sap (mg/liter)	Pb translocation to shoots (μg/plant/day)
0.0	0.15 ± 0.02	0.83 ± 0.11
0.5	6.93 ± 0.08	28.00 ± 0.42
1.0	21.30 ± 2.05	99.80 ± 5.52
ANOVA P > F	0.001	0.001

[a] EDTA treatment was initiated by adding the appropriate EDTA solution to the soil surface for each pot. Shoots were cut 1 cm above root–shoot junction 24 h after applying the EDTA. Immediately following the harvest, xylem sap was collected for 8 h for each treatment.
[b] Values are mean ± SE (n = 3).
Source: Huang et al. (1997).

Table 8.25 Relative Efficiency of Five Synthetic Chelates[a] in Enhancing Pb Accumulation in Shoots of Corn and Pea Plants Grown in Pb-Contaminated Soil with a Total Pb of 2500 mg/kg

	Total Pb accumulation in shoots[b]	
Chelate used	Corn (μg/plant)	Pea (μg/plant)
Control[c]	42 ± 7	81 ± 4
EDDHA	112 ± 13	276 ± 62
DTPA	300 ± 70	1930 ± 205
EGTA	909 ± 53	2080 ± 207
HEDTA	1110 ± 52	5670 ± 280
EDTA	2410 ± 140	8960 ± 620
ANOVA P > F	0.001	0.001

[a] Each chelate was used at a rate of 0.5 g/kg of soil and was mixed with the Pb-contaminated soil before transplanting corn and pea plants (14-day-old). The plants were harvested 1 week after transplanting to the contaminated soil.
[b] Values are mean ± SE (n = 3).
[c] Control denotes the plants grown in the Pb-contaminated soil without added chelate.
Source: Huang et al. (1997).

ily to organic compounds. The organic compounds are converted into simpler molecules which can be absorbed by the plant for growth. For example, poplar trees have been found (Newman et al., 1997) to be capable of taking up trichloroethylene (TCE) from contaminated soils and converting it inside the plants to metabolites such as trichloroethanol, trichloroacetic acid, and dichloracetic acid. Several plant enzymes, including nitroreductase, are also capable of breaking down TNT (2,4,6-trinitrotoluene). In general, herbicides, chlorinated solvents, and ammunition wastes can be remediated by phytodegradation.

8.12.3 Rhizosphere Biodegradation

In rhizosphere biodegradation, microorganisms such as yeasts, fungi, and bacteria that exist in the root zone degrade organic compounds present in soil. The compounds provide additional nutrients for the microbes to supplement the sugars, acids and alcohols secreted by plant roots.

8.12.4 Phytovolatilization

Phytovolatilization is the process by which plants are used to extract contaminants from soils and release them to the atmosphere as gases. It is estimated (U.S. EPA, 1996b) that poplar trees volatilize about 90% of the TCE quantity, that they take up from soils. Metals can be produced in their volatile forms by metabolic activities in plants and volatilized into the atmosphere. For example, introduction of the bacterial gene called mercury reductase into some mustard plants can reduce uptaken mercury to its gaseous form Hg(0), in which it can be volatilized by the plants.

8.12.5 Phytostabilization

Phytostabilization is the process by which plants immobilize contaminants in soils through sorption, complexation, precipitation, and changes in soil structure. Unlike most of the other phytoremediation processes, phytostabilization involves processes that are external to the internal structure of the plants. In the DuPont phytostabilization process, soil amendments such as alkalizing agents, mineral oxides, organic matter, and biosolids are added to soils which are then planted with phytostabilizers. The fibrous roots of these plants hold the soil particles together and promote the processes mentioned above, thereby reducing the leaching potential of contaminants that are present in the soil.

8.13 IN-SITU BIOREMEDIATION PRINCIPLES

Bioremediation techniques are very diverse, and include landfarming, bioreclamation, and even composting. However, the focus of this section is on bioreclamation, which is herein defined as the addition of nutrients and oxygen to subsurface contaminated media to enhance the biodegradation of contaminants by naturally occurring microbes and/or microbes introduced into the subsurface. While some naturally occurring microorganisms are capable of degrading organic compounds, others can be genetically engineered to target specific contaminants for degradation. Figure 8.27 shows the configuration of a bioreclamation system in the field. The contaminant present in a soil mass in the vadose zone is flushed down into groundwater using aerated water containing nutrients. This water is also introduced upstream. The contaminants leached from the waste mass are biodegraded in the plume especially at the aeration zone shown in the figure. Down gradient, the treated water is extracted for re-aeration, reinnoculation with nutrients/and possibly microbes, and recycling into the ground. Several factors influence the effectiveness of in-situ bioreclamation systems. The process factors are discussed briefly below.

8.13.1 Contaminant Intrinsic Biodegradability

It is believed that the intrinsic biodegradability of a chemical compound is significantly related to its chemical structure. Typically, laboratory measurements are conducted to obtain data on surrogate parameters that reflect the effects of chemical structure and form on biodegradation. A common parameter is the ratio of the 5-day biochemical oxygen demand (BOD_5) to the chemical oxygen demand (COD). In general, the following designations are made:

$$\frac{BOD_5}{COD} \geq 0.01 \qquad \text{biodegradable}$$

$$\frac{BOD_5}{COD} < 0.01 \qquad \text{nonbiodegradable}$$

Another empirical measure of contaminant biodegradability is the refractory index (RI), defined as the ratio of the ultimate biochemical oxygen demand (BOD_u) to the ultimate oxygen demand (UOD).

$RI > 0.5$ generally biodegradable
$RI \leq 0.5$ generally nonbiodegradable

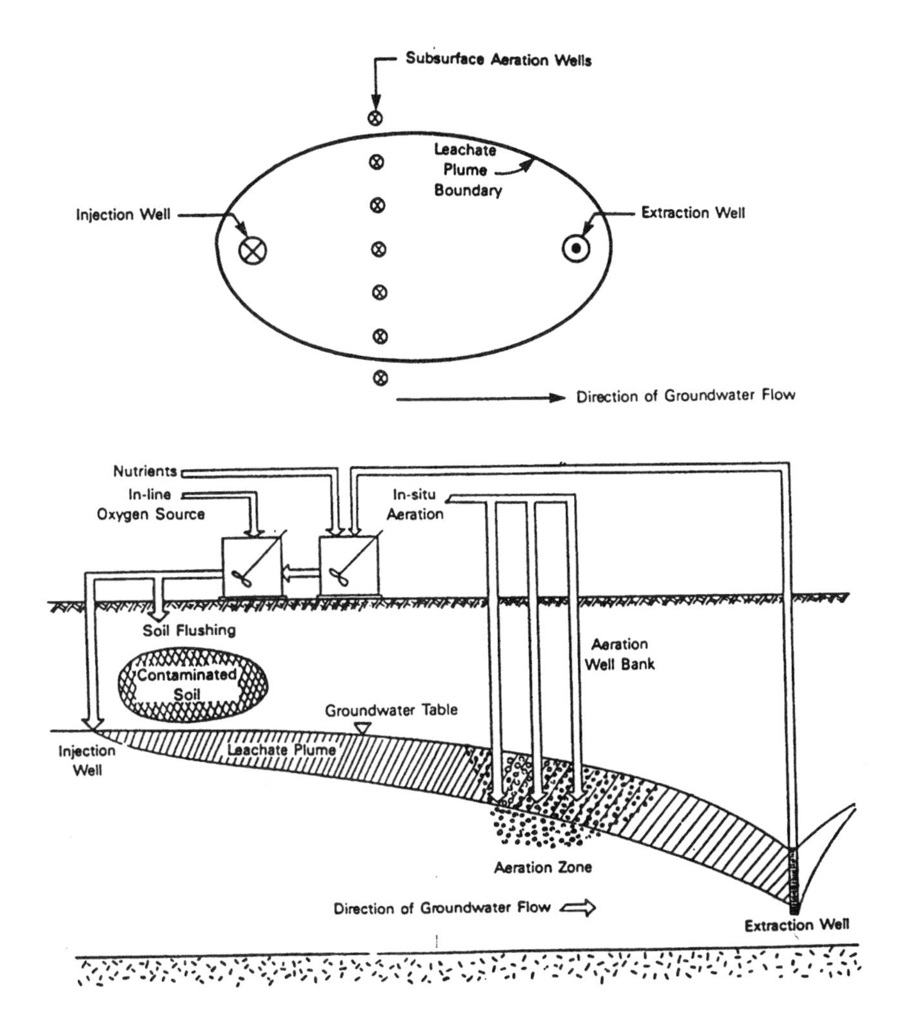

Figure 8.27 Schematic illustration of contaminated groundwater bioreclamation (U.S. EPA, 1986).

Tables 8.26 and 8.27 provide data on both the BOD_5/COD ratios and RI of some contaminants. The reader should note that these classifications are somewhat generic. Compounds with low ratios may still be biodegraded through use of specifically engineered microbes, provision of enough oxygen, and implementation of other process optimization measures.

Waste masses usually contain a combination of biodegradable compounds,

Table 8.26 BOD_5/COD Ratios for Various Organic Compounds[a]

Compound	Ratio	Compound	Ratio
Relatively Nondegradable		*p*-Xylene	<0.11
Butane	0	Urea	0.11
Butylene	0	Toluene	<0.12
Carbon tetrachloride	0	Potassium cyanide	0.12
Chloroform	0	Isopropyl acetate	<0.13
1,4-Dioxane	0	Amyl acetate	0.13–0.34
Ethane	0	Chlorobenzene	0.15
Heptane	0	Jet fuels (various)	0.15
Hexane	0	Kerosene	0.15
Isobutane	0	Range oil	0.15
Isobutylene	0	Glycerine	<0.16
Liquefied natural gas	0	Adiponitrile	0.17
Liquefied petroleum gas	0	Furfural	0.17–0.46
Methane	0	2-Ethyl-3-propylacrolein	<0.19
Methyl bromide	0	Methylethylpyridine	<0.20
Methyl chloride	0	Vinyl acetate	<0.20
Monochlorodifluoromethane	0		
Nitrobenzene	0	Naphthalene (molten)	<0.20
Propane	0	Dibutyl phthalate	0.20
Propylene	0	Hexanol	0.20
Propylene oxide	0	Soybean oil	0.20
Tetrachloroethylene	0	Paraformaldehyde	0.20
Tetrahydronaphthalene	0	*n*-Propyl alcohol	0.20–0.63
l-Pentene	<0.002	Methyl methacrylate	<0.24
Ethylene dichloride	0.002	Acrylic acid	0.26
l-Octene	>0.003	Sodium alkyl sulfates	0.30
Morpholine	<0.004	Triethylene glycol	0.31
Ethylenediaminetetracetic acid	0.005	Acetic acid	0.31–0.37
Triethanolamine	<0.006	Acetic anhydride	>0.32
o-Xylene	<0.008	Ethylenediamine	<0.35
m-Xylene	<0.008	Formaldehyde solution	0.35
Ethylbenzene	<0.009	Ethyl acetate	<0.36

Moderately degradable		Octanol	0.37
Ethyl ether	0.012	Sorbitol	<0.38
Sodium alkylbenxenesulfonates	0.017	Benzene	<0.39
Monoisopropanoalmine	<0.02	*n*-Butyl alcohol	0.42–0.74
Gas oil (cracked)	0.02	Propionaldehyde	<0.43
Gasolines (various)	0.02	*n*-Butyraldehyde	<0.43
Mineral spirits	0.02	Ethyleneimine	0.46
Cyclohexanol	0.03	Monoethanolamine	0.46
Acrylonitrile	0.031	Pyridine	0.46–0.58
Monanol	>0.033	Dimethylformamide	0.48
Undecanol	<0.04	Dextrose solution	0.50
Methylethylpyridine	0.04–0.75	Corn syrup	0.50
l-Hexane	<0.044	Maleic anhydride	>0.51
Methyl isobutylketone	<0.044	Propionic acid	0.52
Diethanolamine	<0.049	Acetone	0.55
Formic acid	0.05	Aniline	0.56
Styrene	>0.06	Isopropyl alcohol	0.56
Heptanol	<0.07	*n*-Anyl alcohol	0.57
sec-Butyl acetate	0.07–0.23	Isoamyl alcohol	0.57
n-Butyl acetate	0.07–0.24	Cresols	0.57–0.68
Methyl alcohol	0.07–0.73	Crotomaldehyde	<0.58
Acetonitrile	0.079	Phthalic anhydride	0.58
Ethylene glycol	0.081	Benzaldehyde	0.62
Ethylene glycol monothyl ether	<0.09	Isobutyl alcohol	0.63
Sodium cyanide	<0.09	2,4-Dichlorophenol	0.78
Linear alcohols (12–15 carbons)	>0.09	Tallow	0.80
Allyl alcohol	0.091	Phenol	0.81
Dodecanol	0.097	Benzoic acid	0.84
Relatively Degradable		Carbolic acid	0.84
Valeraldehyde	<0.10	Methyl ethyl ketone	0.88
n-Decyl alcohol	>0.10	Benzoyl chloride	0.94
		Hydrazine	1.0
		Oxalic acid	1.1

[a] BOD_5 values were not measured under the same conditions for all chemicals.

Source: Lyman et al (1982).

Table 8.27 Refractive Indices of Some Organic Compounds

Compound	RI
High degradability	
Biphenyl	1.14
Antifreeze	1.12
Sevin	1.0
d-Glutamic acid	1.00
d-Glucose	0.93
l-Valine	0.93
Acetone	0.93, 0.71
Phenol	0.87
Sodium butyrate	0.84
l-Aspartic acid	0.81
Sodium propionate	0.80
Propylene glycol	0.78, 0.52
Ethylene glycol	0.76
Medium-high degradability	
Potato starch	0.72, 0.64
l-Arginine	0.65
Acetic acid	0.61
Aniline	0.58
Soluble starch	0.54
l-Histidine	0.52
l-Lysine	0.52
Hydroquinone	0.41
Low degradability	
Benzene	0.23
Gasoline	0.21
Adenine	0.14, 0.12
Vinyl chloride	0
Carboxymethyl cellulose	0
Humics	0
DDT with carrier	0
p-Chlorophenol	0
Dichlorophenol	0
Bipyridine	0
Chloroform	0
Cyanuric acid	0

Source: Data from Lyman et al. (1982)

some of which may be preferentially biodegraded before others. The following general preference patterns (SCS, 1979; and U.S.EPA, 1996b) can be made:

Nonaromatics are preferentially biodegraded relative to aromatics.
Unsaturated bonds in substances increase biodegradation potential.
Straight-chain compounds can be biodegraded preferentially over branched-chain compounds and others with more complex structures such as polymers.
Soluble compounds are more easily biodegraded than insoluble compounds.
Halogenated compounds tend to be more biorefractory than nonhalogenated compounds.
The presence of some functional groups in organic compounds enhances their biodegradability. For example, alcohols, aldehydes, acids, esters, amides, and amino acids are generally biodegradable.

8.13.2 Oxygen Requirements

Oxygen is required to degrade organic compounds at fast rates. However, anaerobic biodegradation is also possible. The theoretical quantity of oxygen required to degrade organic compounds can be estimated through stoichiometry using equations presented in Section 8.11. Oxygen can be supplied to microorganisms during bioreclamation through any of the following methods: aeration, oxygenation, or use of compounds that can supply oxygen molecules. About 10 mg/liter of dissolved oxygen at 15°C is the maximum concentration that can be attained through injection of air. Sawyer and McCarty (1967) developed the following expression for estimating the quantity of dissolved oxygen in water:

$$C_0 = fPH \tag{8.66}$$

where

C_0 = concentration of oxygen in water (mg/L)
f = volume fraction of oxygen in air (0.21)
P = air pressure (atm)
H = Henry's law constant for oxygen = 43.8 (mg/L) • atm at 68°F (20°C)

More oxygen can be transferred into solution through oxygenation systems than through aeration systems.

8.13.3 Groundwater Temperature

Generally, the bulk of soil bacteria are classified as mesophiles that grow well within a temperature range of about 15–45°C. Additional classes of bacteria and their preferred temperatures are psychrophiles (below 20°C) and thermophiles (45–65°C). Specific bacteria within these classes thrive in narrower temperature ranges within the wider ranges stated above. Generally, every 10°C increase in temperature doubles biotransformation rate.

8.13.4 Soil pH Conditions

A soil pH range of about 6–8 generally produces the greatest growth rate of microorganisms, although most bacteria favor neutral pH conditions. Substances such as lime or acids can be introduced into soil systems to optimize bacterial growth.

8.13.5 Soil Nutrients

Microorganisms that degrade contaminants in soil need nutrients which are not normally available completely in the target contaminants. These nutrients are required in specific proportions as matter for the synthesis of new cells during bacterial growth. In their order of abundance, the typical composition of bacterial cells (on a dry-weight basis) are carbon (50%), oxygen (20%), nitrogen (14%), hydrogen (8%), phosphorus (3%), sulfur (1%), potassium (1%), sodium (1%), and calcium, magnesium, chlorine, iron, and other trace elements in percentages that are each less than 1%. The absence of one or more of these compounds may be the limiting step in the growth of microorganisms and hence the effectiveness of the bioreclamation process.

Aerobic bioreclamation is more commonly employed than anaerobic bioreclamation in the in-situ bioremediation of contaminated subsurface media. This is because, at target sites, high concentration of organics generally require oxygen. A very brief quantitative description of the aerobic biodegradation kinetics in the aqueous environment based on Monod kinetics (Lyman et al., 1982) is presented herein.

$$\frac{\partial C}{\partial t} = \frac{k_{max} C_m C}{Y(C_k + C)} \qquad (8.67)$$

where

$\partial(C)/\partial t$ = rate of disappearance of substrate
k_{max} = maximum growth rate of microorganisms (number/T)
C_m = concentration of microorganisms (number/L)
C = concentration of substrate (M/L^3)
C_k = concentration of the substrate supporting half of the maximum growth rate, approximates 0.1–10 mg/liter for most substances (M/L^3)

Y = yield coefficient = $-\partial C_m / \partial C$, approximates 0.5 for nondilute systems.

Two conditions can be stated and used to develop approximate estimates of final contaminant concentrations for known growth rates and initial contaminant concentrations. For $C \gg C_k$, which applies in the initial stages of bioreclamation, $(C_k + C)$ can be assumed to be equal to C and eq. (8.67) can be manipulated to arrive at

$$C_f = C_i e^{-k_{max} t} \tag{8.68}$$

C_f = final concentration of contaminant (M/L^3)
C_i = initial concentration of contaminant (M/L^3)
t = time (T)
k_{max} = maximum growth rate of the microorganisms using first-order rate constant (number/T)
e = base of natural logarithms

For $C < C_k$, $(C_k + C)$ approximates C_k. Then Eq. (8.67) becomes

$$-\frac{\partial C}{\partial t} = \frac{k_{max} C_m C}{Y C_k} \tag{8.69}$$

The growth rate k_{max} can be estimated using methods described in standard microbiology textbooks.

8.14 OTHER TECHNIQUES

Several other techniques, most of which have been used in ex-situ treatment of contaminated media, have been developed. Among these techniques are electrokinetic remediation, which involves the application of electrical energy to generate the movement of charged components of contaminants in desired directions, thermal treatment, incinceration, and solvent extraction.

REFERENCES

Agrawal, A., and Tratnyek, P. G. 1996. Reduction of nitro aromatic compounds by zero-valent iron metal. *Environmental Science and Technology*, Vol. 30, pp. 153–160.

API. 1980. *Underground Spill Cleanup Manual*, API Publication 1628. American Petroleum Institute, Washington, DC.

Baker, A. I. M., and Brooks, R. R. 1989. Terrestrial higher plants which hyperaccumulate metallic elements: a review of their distribution, ecology, and phytochemistry. *Biorecovery*, Vol. 1, pp. 81–126.

Bass, D. H. 1993. Estimation of effective cleanup radius for soil vapor extraction systems. *Journal of Soi Contamination*, Vol. 2, No. 2, pp. 191–202.

Chaung, F., Larson, R. A., and Wessman, M. S. 1995. Zero-valent iron-promoted dechlorination of polychlorinated biphenyls. *Environmental Science and Technology*, Vol. 29, No. 9, pp. 2460–2463.

Crank, J. 1975. *The Mathematics of Diffusion*. Oxford University Press, London, England.

Dorfman, L. M., and Adams, G. E. 1973. Reactivity of hydroxyl radical in aqueous solution. U.S. Department of Commerce, National Bureau of Standards, Washington, DC.

Eary, L. E., and Rai, D. 1988. Chromate removal from aqueous wastes by reduction with ferrous iron. Environmental Science and Technology, Vol. 22, No. 8, pp. 972–977.

Gabr, M. A., Wang, J., and Bowders, J. J. 1996. Model for efficiency of soil flushing using PVD-enhanced system. *Journal of Geotechnical Engineering*, Vol. 122, No. 11, pp. 914–919.

Gierke, J. S., Wang, C., West, O. R., and Siegrist, R. L. 1995. In situ mixed region vapor stripping in low-permeability media. 3. Modeling of field tests. *Environmental Science and Technology*, Vol. 29, No. 9, pp. 2208–2216.

Hansen, J. E. 1991. Treatment of heavy metal contaminated soils by in situ vitrification. Presented at the 4th. Annual Hazardous Materials Management Conference, Rosemont, IL, April 3–5.

Harris, J. C. 1982. Rate of hydrolysis. In: *Handbook of Chemical Property Estimation (Lyman, W. J., Reehl, W. F., and Rosenblatt, O. H., eds.). McGraw-Hill, New York, chap. 7.*

Heijman, C. G., Grieder, E., Hollinger, C., and Schwarzenbach, R. P. 1995. Reduction of nitroaromatic compounds coupled to microbial iron reduction in laboratory aquifer columns. *Environmental Science and Technology*, Vol. 29, pp. 775–783.

Huang, J. W., Chen, J., Berti, W. R., and Cunningham, S. D. 1997. Phytoremediation of lead-contaminated soils: role of synthetic chelates in lead phytoextraction. *Environmental Science and Technology*, Vol. 31, pp. 800–805.

Hunt, J., Sitar, N. and Udell, K. 1988. Nonaqueous phase liquid transport and cleanup: I. Analysis of mechanisms. *Water Resources Research*, Vol. 24, No. 8, pp. 1247–1258.

HWC. 1990. In situ vitrification update. *Hazardous Waste Consultant*, September/October, pp. 1–18.

Johnson, P. C., Kemblowski, M. W., and Colthart, J. D. 1990. Quantitative analysis for the cleanup of hydrocarbon contaminated soils by in-situ soil venting. *Groundwater*, Vol. 28, No. 3, pp. 413–429.

Kayano, S., and Wilson, D. J. 1993. Migration of pollutants in groundwater. VI. Flushing of DNAPL droplets/ganglia. *Environmental Monitoring and Assessment*, Vol. 25, pp. 193–212.

Keeley, J. F. 1989. Performance evaluations of pump-and-treat remediations, EPA Superfund Groundwater Issue Document EPA/540/4-89/005. Office of Solid/Waste and Emergency Response, U.S. Environmental Protection Agency, Washington, DC.

Kerfoot, H. B. 1994. In situ determination of the rate of unassisted degradation of saturated-zone hydrocarbon contamination. *Journal of Air and Waste Management Association*, Vol. 44, pp. 877–880.

Knox, R. C., Sabatini, D. A., and Canter, L. W. 1993. *Subsurface Transport and Fate Processes*. Lewis Publishers, Ann Arbor, MI.

LaGrega, M. D., and Evans, J. C. 1987. Overview of existing technologies for hazardous waste site clean-up. *Proceedings of the International Symposium on Environmental Geotechnology*, Allentown, PA, pp. 30–46.

Lingineni, S., and Dhir, V. K. 1992. Modeling of soil venting processes to remediate unsaturated soils. *Journal of Environmental Engineering*, Vol. 118, No. 1, pp. 135–151.

Lyman, W. J., Rechl, W. F., and Rosenblatt, O. H. 1982. *Handbook of Chemical Property Estimation Methods*. McGraw-Hill, New York.

Mabey, W., and Mill, T. 1978. Critical review of hydrolysis of organic compounds in water environmental conditions. *Journal of Physical Chemistry Reference Data*, Vol. 7, No. 2, pp. 383–415.

Macdonald, J. A., and Kavanaugh, M. C. 1994. Restoring contaminated groundwater: an achievable goal? *Environmental Science and Technology*, Vol. 28, No. 8, pp. 362A–368A.

Masschelein, W. 1982. *Ozonation Manual for Water and Wastewater Treatment*. John, Wiley, New York.

McLellan, G. W., and Strand, E. B. 1984. *Glass Engineering Handbook*, 3rd ed. McGraw-Hill, New York.

Moffat, A.S. 1995. Plants proving their worth in toxic metal cleanup. *Science*, Vol. 269, pp. 302–303.

Newman, L. A., Strand, S. E., Choe, N., Duffy, J., Ekuan, G., Ruszaj, M., Shurtleff, B. B., Wilmoth, J., Heilman, P., and Gordon, M. P. 1997. Uptake and biotransformation of trichloroethylene by hybrid poplars. *Environmental Science and Technology*, Vol. 31, pp. 1062–1067.

Nkedi-Kizza, P. Rao, P. S. C., and Hornsby, A. G. 1985. Influence of organic cosolvents on sorption of hydrophobic organic chemicals by soils. *Environmental Science and Technology*, Vol. 19, pp. 975–979.

O'Hannesin, S. F., and Gillham, R. W. 1993. In situ degradation of halogenated organics by permeable reaction wall. *Groundwater Currents*, EPA/542/N-93/003, March. Office of Solid Waste and Emergency Response, U.S. Environmental Protection Agency, Washington, DC, pp. 1–2.

Orfeuil, M. 1987. *Electric Process Heating*. Battelle Press, Columbus, Richland, WA.

Plodinec, M. J., Wicks, G. G., and Bibler, N. E. 1982. An assessment of Savannah River borosilicate glass in repository environment, DP-1629. Savannah River Laboratory, Aiken, SC.

Radding, S. B., Liu, D. H., Johnson, H. L., and Mill, T. 1977. Review of the environmental fate of selected chemicals, EPA 560/5-77-033. Office of Research and Development, U.S. Environmental Protection Agency, Washington, DC.

Ram, N. M., Bass, D. H., Falotico, R., and Leahy, M. 1993. A decision framework for selecting remediation technologies at hydrocarbon-contaminated sites. *Journal of Soil Contamination* Vol. 2, No. 2, pp. 167–189.

Sawyer, C. N., and McCarty, P. L. 1967. *Chemistry for Sanitary Engineers*. McGraw-Hill, New York.

SCS. 1979. Selected biodegradation techniques for treatment and/or ultimate disposal of organic materials, EPA-60/2-79-006. Soil Conservation Service, U.S. Environmental Protection Agency, Cincinnati, OH.

Sivavec, T. M., and Horney, D. P. 1995. Reductive dechlorination of chlorinated ethenes by iron metal. Extended abstract presented at the American Chemical Society Meeting, Anaheim, CA, April 2–7.

Sivavec, T. M., Horney, D. P., and Baghel, S. S. 1995. Reductive dechlorination of chlorinated ethenes by iron metal and iron sulfide minerals. Extended abstract presented at the Industrial and Environmental Chemistry Special Symposium of the American Chemical Society, Atlanta, Georgia GA, September 17–20.

Tchobanoglous, G., and Schroeder, E. D. 1987. *Water Quality*, Addison-Wesley, Reading, MA, p. 286.

U.S. DOE. 1994. Summary report of a workshop on phytoremediation research needs, DOE/EM-0224. U.S. Department of Energy, Washington, DC.

U.S. EPA. 1986. Systems to accelerate in situ stabilization of waste deposits, EPA/540/2-86/002. Hazardous Waste Engineering Research Laboratory, U.S. Environmental Protection Agency, Cincinnati, Ohio, OH.

U.S. EPA. 1987. A compendium of technologies used in the treatment of hazardous wastes, EPA/625/8-87/014. Center for Environmental Research Information, U.S. Environmental Protection Agency, Cincinnati, OH.

U.S. EPA. 1989a. Performance evaluations of pump-and-treat remediations. Groundwater Issue, EPA/540/4-89/005. Office of Research and Development, U.S. Environmental Protection Agency, Washington, DC.

U.S. EPA. 1989b. Stabilization/solidification of CERCLA and RCRA wastes: physical tests, chemical testing procedures, technology screening, and field activities, EPA/625/6-89/022. Office of Research and Development, U.S. Environmental Protection Agency, Washington, DC.

U.S. EPA. 1990. Basics of pump-and-treat groundwater remediation technology, EPA/600/8-90/003. Robert S. Kerr Environmental Research Laboratory, U.S. Environmental Protection Agency, Ada, OK.

U.S. EPA. 1992. Vitrification technologies for treatment of hazardous and radioactive waste, EPA/625/R-92/002. Office of Research and Development, U.S. Environmental Protection Agency, Washington, DC.

U.S. EPA. 1992a. Chemical enhancements to pump-and-treat remediation (Palmer, C. D., and Fish, W.), EPA/540/S-92/001. Office of Research and Development, U.S. Environmental Protection Agency, Washington, DC.

U.S. EPA. 1993. Remediation technologies screening matrix and reference guide, EPA-542-B-93-005. Office of Solid Waste and Emergency Response, U.S. Environmental Protection Agency, Washington, DC.

U.S. EPA. 1994a. Superfund innovative technology evaluation program: technology profiles, EPA/540/R-94/526. Office of Research and Development, U.S. Environmental Protection Agency, Washington, DC.

U.S. EPA. 1994b. In situ vitrification treatment. *Engineering Bulletin*, EPA/540/S-94/504. Office of Research and Development, U.S. Environmental Protection Agency, Cincinnati, OH.

U.S. EPA. 1994c. Natural attenuation of hexavalent chromium in groundwater and soils. *Ground Water Issue*, EPA/540/S-94/505. Office of Research and Development, U.S. Environmental Protection Agency, Washington, DC.

U.S. EPA. 1995. How to evaluate alternative cleanup technologies for underground storage tank sites, EPA/510-B-95-007. Office of Solid Waste and Emergency Response, U.S. Environmental Protection Agency, Washington, DC.

U.S. EPA. 1996a. Innovative treatment technologies: annual status report. Office of Solid Waste and Emergency Response, EPA-542-R-96-010, U.S. Environmental Protection Agency, Washington, DC.

U.S. EPA. 1996b. A citizen's guide to phytoremediation, EPA 542-F-96-014. Office of Solid Waste and Emergency Response, U.S. Environmental Protection Agency, Washington, DC.

U.S. EPA. 1997. Recent developments for in situ treatment of metal contaminated soils, EPA-542-R-97-004. Office of Solid Waste and Emergency Response, U.S. Environmental Protection Agency, Washington, DC.

Versar, Inc. 1979. Water-related environmental fate of 129 priority pollutants, Vols. I and II, EPA/440-4-029a and b. U.S. Environmental Protection Agency, Washington, DC.

Volf, M. B. 1984. *Chemical Approach to Glass*. Elsevier, New York.

Wilson, E. K. 1995. Zero-valent metals provide possible solution to groundwater problems. *Chemical and Engineering News*, July 3, pp. 19–22.

Yang, Y. J., and Parker, R. A. 1998. Pump-and-treat systems. *Chemical Engineering*, February, pp. 129–133.

9
Containment System Implementation

9.1 ESSENTIALS OF WASTE CONTAINMENT

In Chapter 6, various mechanisms and scenarios of contaminant generation from wastes were analyzed. This chapter focuses on approaches/techniques for containing such contaminants in the subsurface. Waste containment is one of the four general approaches to waste management. The other approaches are waste treatment (in situ or ex situ), natural attenuation, and waste excavation for disposal or recycling.

The contaminants that need to be contained can be present in the solid, liquid, or gaseous forms. In the subsurface, the liquid and gaseous-phase contaminants are usually the focus of control efforts because of the relatively high mobility of contaminants in these two phases. In this part of this text, the focus is primarily on the liquid-phase contaminants, which generally migrate as a contaminant plume in the subsurface. Wastes in the solid form could give rise to such plumes through leaching processes described in Chapter 6. Waste containment system components that can influence the rate of generation and migration of leachates from waste solids are also analyzed. The volatile components of wastes tend to migrate upward as permitted by the hydrogeological characteristics of the site concerned. Both active and passive collection schemes for vapor-phase contaminants have been successfully implemented.

The implementation of effective schemes for waste containment requires information on the size, shape, and directions of movement of the waste materials and/or contaminants plume. For solid and liquid wastes that are stored in landfills and impoundments/tanks, respectively, the size and shape can be inferred from the implemented design dimensions of the system. With respect to contaminant

Table 9.1 Principal Physicochemical and Biochemical Factors that May Influence Contaminant Mobility in the Subsurface

---

Physical processes

Dispersion—Causes dilution of wastes and smearing of the plume front. The dispersive capacity of a porous or fractured medium is directly dependent on the groundwater velocity and the heterogeneity of the aquifer materials, and is inversely proportional to the porosity.

Filtration—Favors reduction in amounts of substances associated with colloidal or larger-sized particles (e.g., sediments and microbes). Most effective in clay-rich materials, least effective in gravels or fractured or cavernous rock.

Gas movement—Requires unsaturated conditions and high porosity if not dissolved in groundwater. Where gas movement can occur, favors aerobic breakdown of organic substances and increased rates of decomposition. Constituents mobile under oxidized conditions (e.g., chromium) will then predominate. Restriction of gas movement by impermeable, unsaturated materials or by saturated materials can produce an anaerobic state and reduced rates of organic decay. This will mobilize substances that are soluble under anaerobic conditions (e.g., iron, manganese)

Geochemical processes

Complexation and ionic strength—Complexes and ion pairs most often form by combination of ions including one or more multivalent ions and increase in amount with increased amounts of ions involved. Ionic strength is a measure of the total ionic species dissolved in groundwater. Both ionic strength and complexation increase the total amounts of species in solution that would otherwise be limited by processes such as oxidation, precipitation, or sorption.

Acid–base reactions—Most constituents increase in solubility and thus in mobility with decreasing pH. In organic-rich waters, lower pH (4–6) is associated with high values of carbonic acid and often also of organic acids. These will be most abundant in moisture saturated soils and rock.

Oxidation–reduction—Many elements can exist in more than one oxidation state. Contaminants will often be oxidized or only partially reduced in unsaturated soils and groundwater recharge areas, but will become reduced under saturated conditions when excess organic matter is present. Mobility depends on the element and pH involved: chromium is most mobile under oxidizing conditions, whereas iron and manganese are most mobile under those reduced conditions in which dissolved oxygen and hydrogen sulfide are absent.

Precipitation–dissolution—The abundance of anions such as carbonate, phosphate, silicate, hydroxide, or sulfide may lead to precipitation, especially of multivalent cations, as insoluble compounds. Dilution or a change in oxygen content, where precipitation has resulted from oxidation or reduction, may return such constituents to solution.

Sorption–desorption—Ion exchange can withhold, usually temporarily, cations and to a lesser extent anions, on the surfaces of clays or other colloidal-sized materials. Amounts of sorbed metal cations will increase with increasing pH. Molecular species may be weakly retained on colloidal size materials by physical sorption. The much stronger binding forces caused by chemical action result in the formation of surface compounds involving metal ions and mineral grains. Sorbed species may return to solution when more dilute solutions come in contact with the colloidal material, depending on the nature of the sorption bond and sorption of organic chemicals by chemical interactions such as bonding and polar attraction.

Biochemical processes

Decay and respiration—Microorganisms can break down insoluble fats, carbohydrates, and proteins, and in so doing release their constituents as solutes or particulates to subsurface waters.

Cell synthesis—N, C, K, and P, and some minor elements are required for growth or organisms, and can thus be retarded in their movement away from a waste disposal site because they are temporarily incorporated within microbial cells.

Source: Jackson (1980).

plumes in the subsurface, there are uncertainties associated with the dimensions and shape of the plume to be contained. Such uncertainties are due to their physicochemical characteristics and interactions among waste constituents, site hydrogeological parameters, and climatic factors. Table 9.1 shows a summary of the primary physicochemical and biochemical factors that can influence the mobility of aqueous phase contaminants in the subsurface (Jackson, 1980). Aquifer characteristics such as hydraulic conductivity and its spatial variability, proximity and fracture intensity of the bedrock, and soil stratification influence the size and shape of the contaminant plume as well. As shown in Table 9.2 (Freeze and Cherry, 1979) various geomedia have widely different hydraulic conductivities. Assuming that other factors are constant, the rate of migration (hence, plume size) is proportional to hydraulic conductivity. Figures 9.1–9.3 (U.S. EPA, 1985) illustrate the effects of subsurface hydrogeological conditions on plume shape.

In Figure 9.1, the contaminant is miscible with water and moves in a homogeneous medium. In this relatively simple case, the plume is displaced laterally by the transverse-flowing groundwater. The plume migration speed is essentially the composite speed of movement of the various contaminants that constitute the plume. If the geomedium is clayey, physicochemical and biological processes will act to retard the migration of contaminants. For media that are dominated by silts and sands, interactions are mostly in the physical realm: contaminant

Table 9.2 Typical Ranges of Hydraulic Conductivity Values of Geomedia

Geologic material	Hydraulic conductivity	
	cm/s	gal/day/ft^2
Gravel	10^{-1}–10^{2}	10^{4}–10^{7}
Sand, well sorted	10^{-4}–1	10–10^{5}
Silty sand	10^{-5}–10^{-1}	1–10^{4}
Silt	10^{-7}–10^{-3}	10^{-2}–10^{2}
Clay, unweathered	10^{-10}–10^{-7}	10^{-5}–10^{-2}
Glacial till	10^{-10}–10^{-4}	10^{-5}–10
Carbonate rocks	10^{-7}–1	10^{-3}–10^{5}
Sandstones	10^{-8}–10^{-4}	10^{-3}–10
Shales	10^{-11}–10^{-7}	10^{-6}–10^{-2}
Crystalline rocks		
Highly fractured	10^{-6}–1	10^{-1}–10^{5}
Relative unfractured	10^{-12}–10^{-8}	10^{-7}–10^{-3}

1 gal/day/ft^2 = 1.74×10^{-6} ft/day
1 gal/day/ft^2 = 4.72×10^{-5} cm/s
Source: Freeze and Cherry (1979).

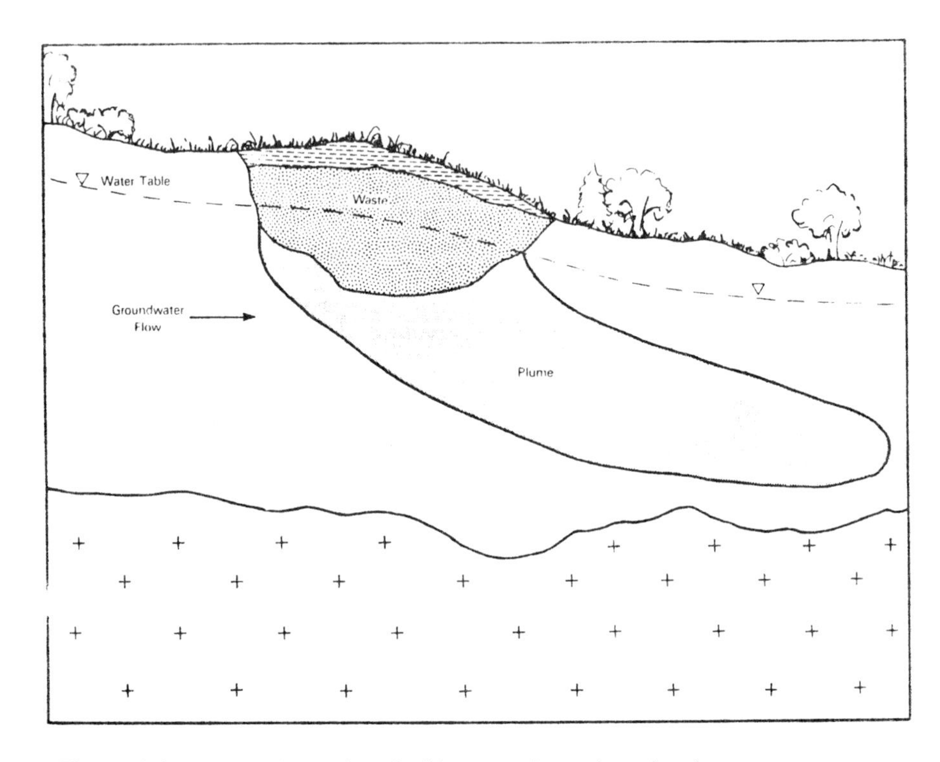

Figure 9.1 Typical shape of a miscible contaminant plume in a homogeneous porous geomedium (U.S. EPA, 1985).

retardation is minimal. Figure 9.2 shows the distortion that stratification can impose on a contaminant plume. Contaminant migration rate is greater in more permeable media such as gravel and sand than in the clay layer. As a result, the plume displays fingering. Fragmentation of bedrock through dissolution or fissuring as shown in Figure 9.3 can result in the entrainment of contaminants such that the plume appears to be stabilized in the overlying homogeneous medium.

Under normal temperature and pressure conditions, the density of water is approximately 1 g/cm^3. Contaminants which are released to groundwater exhibit a wide range of densities and aqueous solubilities. Figure 9.4 (U.S. EPA, 1985) shows the effects of density differences of insoluble contaminants on plume shape. Contaminants that are denser than water tend to sink and accumulate on relatively impermeable media from which they can migrate under gravitational and other imposed forces. The contaminants that are lighter than water float around the water table. Table 9.3 shows the solubilities and

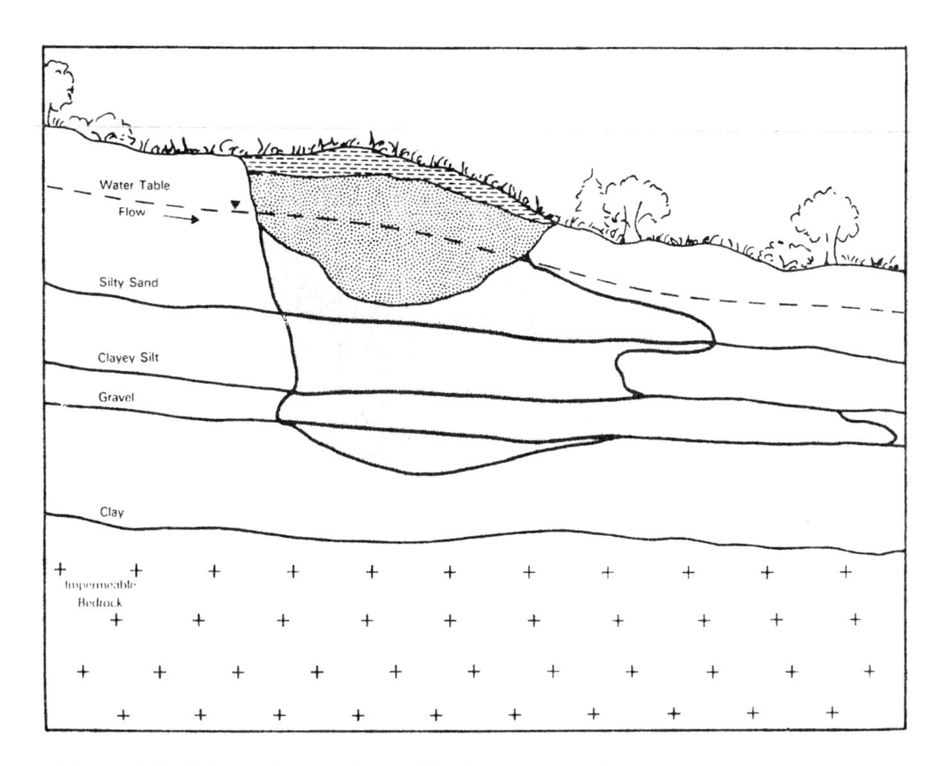

Figure 9.2 Effects of geomedia stratification on contaminant plume. (U.S. EPA, 1985).

densities of some common contaminants. Knowledge about the distribution of various contaminants within a plume is important in the design of containment schemes.

The primary functions of waste containment systems fall into the following categories:

- Minimization of the intrusion of moisture, which can generate and mobilize leachate
- Minimization of the transport of waste constituents into the surrounding environment
- Isolation of wastes such that the potential for contact by humans and other animals is minimized

More specific design functions of various types of containment systems are discussed in Chapter 10.

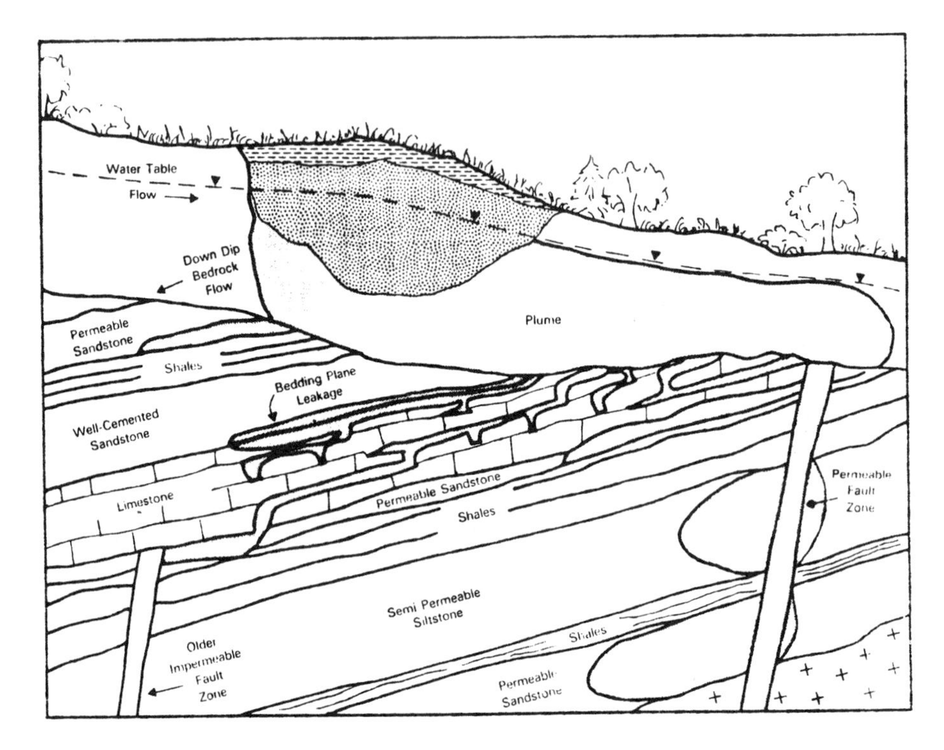

Figure 9.3 Effects of secondary porosity on contaminant plume movement (U.S. EPA, 1985).

9.2 HYDRAULIC AND PHYSICAL CONTAINMENT

Waste constituents and moisture can be contained through the installation of barriers or pore fluid extraction through gravity-induced drainage or pumping. The configurations of physical containment systems are presented and discussed in Chapter 10. Also, numerical techniques for assessing their structural stability and moisture/contaminant transport through their components are discussed in Chapter 11.

Physical containment involves the installation of barriers to attenuate contaminants, and/or change their direction and/or rate of transport in the subsurface. Attenuation can be accomplished through the use of sorptive, low-permeability materials, or permeable, but reactive materials. In both cases, the objective is to reduce output concentration, with respect to the input concentration of the targeted contaminants. In the case of permeable barriers, the flow rate of moisture

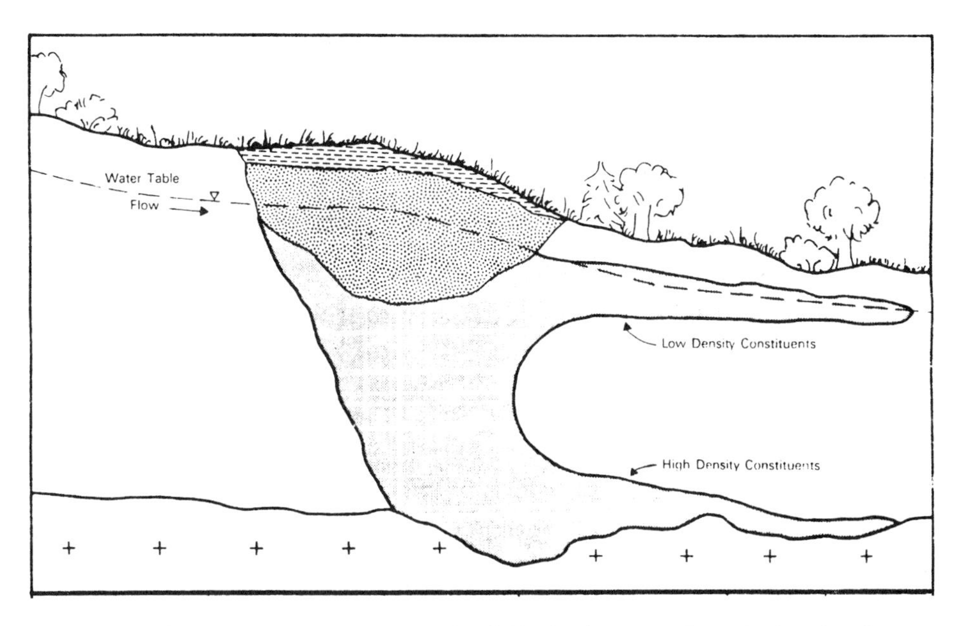

Figure 9.4 Typical migration pattern of variable-density contaminants in the subsurface (U.S. EPA, 1985).

Table 9.3 Categorization of Some Common Contaminants on the Basis of Aqueous Solubility and Density

Constituents	Low-density constituents	Moderate-density constituents	High-density constituents
Water soluble	Acetone	Acetic acid	Chloroform
	Acrylonitrile	Aniline	Halogenated ethanes
	Ammonia	Most metal salts	Halogenated phenols
	Benzene	Phenol	Nitrobenzene
	Cyclohexane	*t*-Butyl acetate	Carbon tetrachloride
Water insoluble	Ethylbenzene	Butyl cellosolve	Chlorobenzene
	Gasoline and some oils	Hexanol ester-alcohol	PCBs
	Toluene		Most pesticides
	Vinyl chloride		Most polyhalogenated benzenes and phenols
	Xylene		Most polycyclic aromatic hydrocarbons

Source: U.S. EPA (1985).

Table 9.4 Contaminant Plume Characteristics that Should Be Considered in the Selection of Leachate Migration Control Techniques

Plume characteristic	Leachate migration control technologies			
	Groundwater pumping	Subsurface drains	Low-permeability barriers	In-situ treatment
Viscosity	Highly viscous leachate may clog screens, pipes, and pumps thus reducing well efficiency	Highly viscous leachate may clog envelope materials, resulting in drain failure	Little or no effect	Effect depends on the delivery system used
Solubility in groundwater	Pumping systems can be designed for both soluble and insoluble contaminants	Drainage systems can be designed for both soluble and insoluble contaminants, although drains tend to function more effectively when contaminants are soluble in groundwater	Little or no effect	Effect depends on the treatment technology used
Reactiveness, explosiveness, corrosiveness, volatility	Care must be taken during installation and the operation of pumps and treatment systems; some contaminants may degrade well materials	Care must be taken during installation and the operation of pumps and treatment systems; some contaminants may degrade drain materials	Care must be taken during installation; some contaminants may degrade wall materials	Care must be taken in implementing in-situ treatment activities. The ability of contaminants to enter into chemical and biochemical reactions, however, is a requirement of in-situ treatment technologies.
Toxicity	Little or no effect except for treatment system	Little or no effect except for treatment system	Little or no effect	Effect depends on the treatment technique used; some highly toxic contaminants cannot be bioreclaimed

Flow direction	Little or no effect because the system will readjust flow direction	Little or no effect because the system will readjust flow directions	Very important for placing noncircumferential walls especially if flow directions change seasonally	Little or no effect
Flow rate	Important in system design and operation but generally does not affect selection or performance	High flow rates generally preclude the use of subsurface drains	Can be very important for noncircumferential walls especially downgradient walls	Can be very important depending on the treatment system used
Volume and extent	Little or no effect	Generally not practical for very large plumes	Generally not practical for isolating very large plumes	Generally not practical for very large plumes
Contaminant types	Can affect the selection of system materials and components; some contaminants can clog systems	Can affect the selection of system materials and components; some contaminants can clog systems	Can affect wall placement and performance if contaminants degrade wall materials	Generally the primary consideration in selecting a treatment system
Concentrations	Little or no effect if the system is designed and operated properly	High concentrations of some contaminants will clog drainage systems	High concentrations of some contaminants can degrade wall materials	High contaminant concentrations cannot be treated effectively by some treatment systems
Density	Will influence the placement of screens in individual wells	Can reduce system effectiveness if drain is not situated properly within the aquifer	Very important for hanging walls	May be a factor in the design of the delivery system

Source: U.S. EPA (1985).

is relatively high. The contaminant extraction approach involves the removal of moisture which may be allowed to drain into constructed ditches, or pumping of water from the subsurface. If extraction points are upstream of the waste, the objective may be to isolate the waste from groundwater to minimize the rate of leachate generation and migration. Plume drainage requires a treatment step and is often done to reduce the size of the leachate plume or flush out contaminants to acceptable residual concentrations.

Several factors such as those summarized by the U.S. EPA (1985) in Table 9.4 control the relative effectiveness of plume drainage and barriers as techniques for containing wastes in the subsurface.

9.3 CONTAINMENT EFFECTS ON SOURCE TERMS

Contaminant migration control as discussed in the preceding section is one of the primary objectives of waste containment schemes. In Chapter 4 of this text, numerical formulations for contaminant transport through advection and dispersion-diffusion have been presented in detail. Within the context of waste containment, contaminant migration control implies a reduction in the magnitude of the source terms for contaminant transport. Such an effect can be direct or indirect.

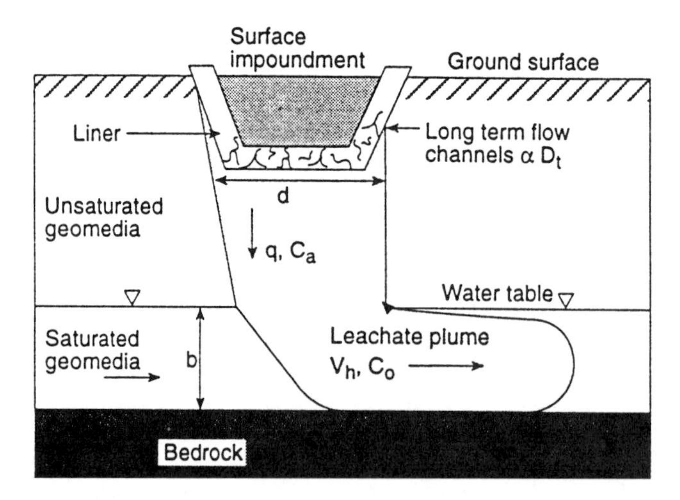

Figure 9.5 Illustration of the influence of containment on contaminant source term for fate and transport modeling.

Table 9.5 Factors that Influence Leachate Generation from Buried or Contained Waste

Factor	Effect
Precipitation	
Amount per year	The greater the amount of precipitation, the greater the volume of leachate generated.
Highest 24-h rainfall	No real effect, but the greater the number of short-duration, high-volume rainstorms, the less likely that precipitation will infiltrate to waste, and the more likely that most will run off as surface flow.
Mean annual temperature	The higher the mean annual temperature, the greater the amount of precipitation that will evaporate and not infiltrate into waste, reducing the volume of leachate generation.
Degree of insolation	Increasing insolation will increase evaporation, reducing the quantity of leachate produced.
Surface slope	Increasing slope at a disposal site will promote surface runoff, reducing the volume of precipitation infiltrating into the waste and thus reducing the volume of leachate generated.
Cover permeability	Low-permeability cover materials promote runoff, reducing infiltration and the volume of leachate generated.
Vegetation	Heavy vegetative cover of grasses will increase transpiration. Heavy vegetative cover of grasses can also impede runoff, thus increasing infiltration. Heavy vegetative cover of trees, bushes, and shrubs may increase infiltration by opening up channels of flow into waste because of deep root penetration.
Subsurface characteristics	
Water content of waste	Increased water content within waste will decrease time until leachate generation begins.
Volume of water flow through waste	Increased water flow through waste will increase the volume of leachate generated.
Rate of flow through wastes	Increased flow rates through waste will decrease time until start of leachate generation, but may decrease the concentration of contaminates present.

Source: U.S. EPA (1985).

An indirect effect is the minimization of infiltration rate of moisture, which in turn reduces the leachate production rate. A direct effect is the containment of the leachate, which reduces in its concentration at the clean side of the barrier and possibly reduces the fluid flow rate through the barrier.

An example of the direct effect is illustrated for a lined surface impoundment in Figure 9.5. The source term concentration, C_o, shown in the figure, is nominally related to the other parameters as follows:

$$C_o = \frac{C_a q d}{V_h b} \tag{9.1}$$

where

C_o = source term concentration per unit width of the plume, for contaminant fate and transport modeling (M/L^2)
q = contaminant flow rate per unit width in the unsaturated zone (L^2/T)
C_a = contaminant concentration in the vadose zone (M/L^3)
d = width of the satureate zone (L)
V_h = plume velocity (L/T)
b = plume thickness (L)

The liner influences both q and C_a, which are partial determinants of C_o.

Covers are the most common means of controlling source terms indirectly. In this regard the essential function is minimization of the infiltration rate of moisture. The factors that influence the effectiveness of the infiltration control approach are presented in Table 9.5 (after U.S. EPA, 1985). Barrier systems are not perfect and tend to deteriorate as time progresses, especially the buried components. Also, contaminant transport through nondegraded low permeability barriers is slow but certain. Consequently, contaminant transport source terms are time-variable, as discussed and illustrated in Chapter 12.

9.4 CONTAINMENT SITE SELECTION TECHNIQUES

Site factors are important in waste containment because of the following reasons.

Hydrological and hydrogeological factors which influence the performance of a containment system vary from one site to another.

Some sites are more sensitive ecologically than others. Thus, for the same level of intrinsic performance of a containment system, the potential for damage of a sensitive environment may be site-dependent.

For some waste containment scenarios, the site is already fixed. The option selection issue centers around alternative techniques for containment. For other scenarios, the waste management site has to be selected on the basis of engineering factors, socioeconomic factors, environmental factors, and direct cost. Several methods (Mustalish and Costanzo, 1991; Bain and Hanse, 1981; Charnpratheep et al., 1997; Anandalingam and Westfall, 1989; Lemme et al., 1990; Vita, 1987; and Aller et al., 1985) that incorporate one or more of the categories of factors mentioned above have been developed and used for facility site selection. The comprehensive site selection process illustrated in Figure 9.6 was applied by Armour (1986) to screen 150 candidate sites progressively to 8 and 6, before selecting a single site for a waste management facility in Alberta, Canada.

Assessment of technical factors in site selection for waste management usually involves an analysis of the vulnerability and/or risk-mitigation capacities of the alternative sites. For the case of a location near Lake Poinsette, South

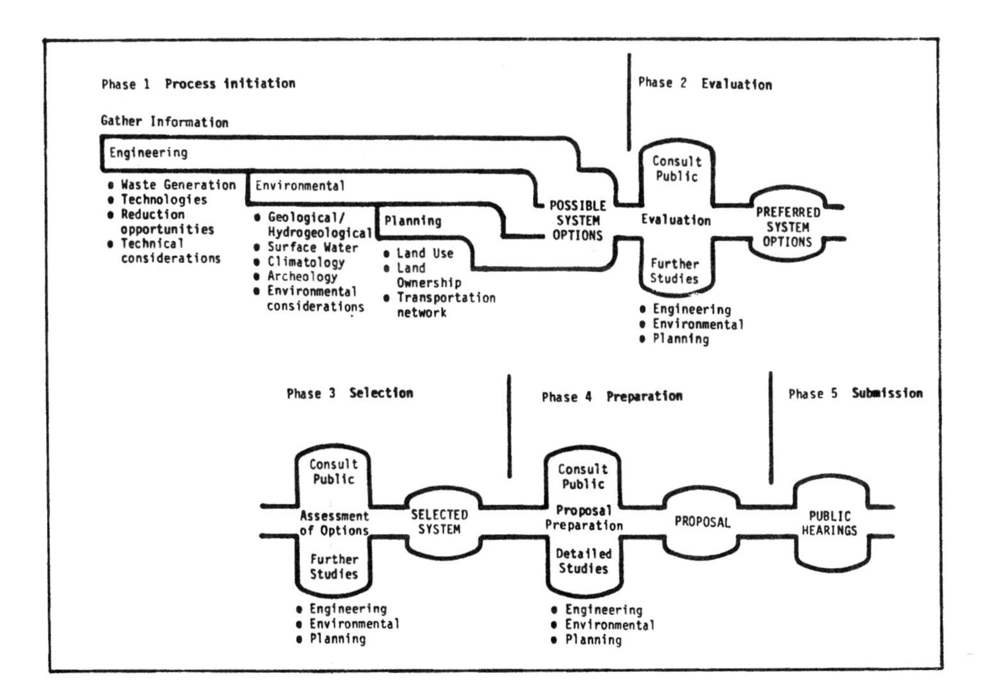

Figure 9.6 Example of a comprehensive site selection process for waste management facilities (Armour, 1986).

Dakota, USA, Lemme et al. (1990) developed the aquifer water vulnerability index shown in Eq. (9.2) for ranking potential sites.

$$V_i = 2.0 - [0.2666(S_o)(S_t)] + \sum_{d=0}^{60} [0.0067(\mathrm{L})] \tag{9.2}$$

where

V_i = vulnerability index (unitless)
S_o = surface organic matter (% by weight)
S_t = surface soil profile thickness (ft)
L = soil permeability at depth d (in./hr)

For the region that was evaluated, the range of V_i values was 0–10. Values of 0–2.0, 2.1–6.0, 6.1–8.0, and above 8.0 were designated as very low vulnerability, low vulnerability, medium vulnerability, and high vulnerability, respectively. Figure 9.7 shows the spatial distribution of vulnerability indices computed using Eq. (9.2).

Perhaps the most widely recognized site ranking method in waste management facility site selection is the DRASTIC method developed by Aller et al. (1985). This method is based on the justifiable assumption that each hydrogeological setting presents a set of physical characteristics which affect the pollution potential of groundwater. The factors that are included in DRASTIC are mappable and do not generally require detailed investigations for input data. The acronym ''DRASTIC'' relates to these factors as follows:

D = depth of water
R = (net) recharge
A = aquifer media
S = soil media
T = topography (slope)
I = impact of the vadose zone
C = conductivity (hydraulic) of the aquifer

The numerical formulation for applying the DRASTIC method is represented by Eq. (9.3):

$$\mathrm{PP} = D_R D_W + R_R R_W + A_R A_W + S_R S_W + T_R T_W + I_R I_W + C_R C_W \tag{9.3}$$

where

PP = pollution potential
R = subscript representing the rating on the parameter indicated
W = subscript representing the weight on the parameter indicated

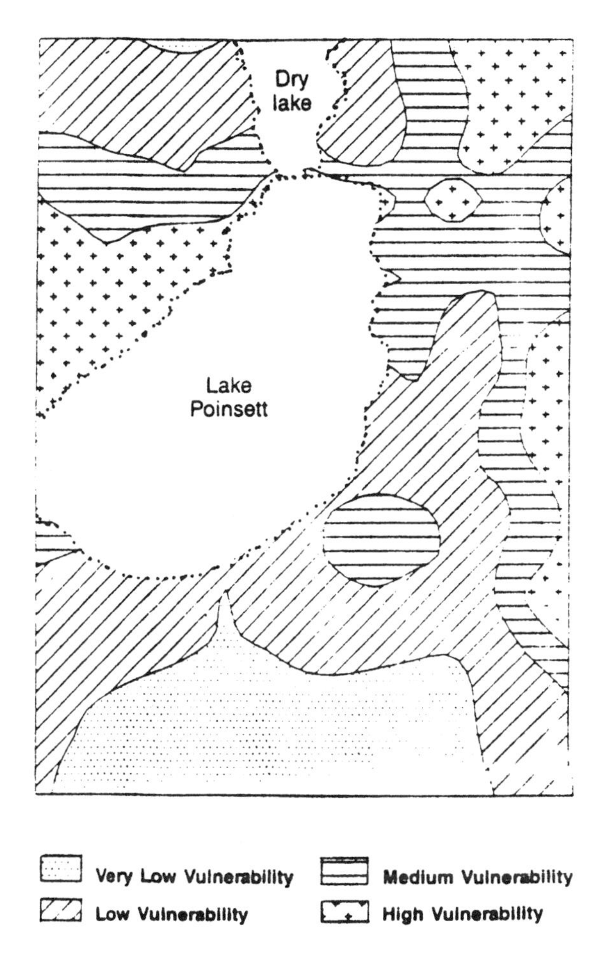

Figure 9.7 Aquifer vulnerability map of the Lake Poinsett Region of South Dakota, USA (Lemme et al. 1990).

Three categories of parameters (two direct and one indirect) determine the magnitude of PP. These categories are weight, ranges, and ratings. DRASTIC factors are weighted on a scale of 1–5 in direct proportion to the significance of each factor, relative to others. These weights are constant and are shown in Table 9.6. The magnitude of each factor has been divided into ranges or characteristic magnitudes and is supplied on each factor for various media. Each of the factors

Table 9.6 Weights Assigned to DRASTIC Factors

Feature	Weight
Depth to water table	5
Net recharge	4
Aquifer media	3
Soil media	2
Topography	1
Impact of the vadose zone	5
Hydraulic conductivity of the aquifer	3

Note: The weights do not change. *Source*: Aller et al. (1985).

D, R, S, T, and *C* is assigned one specific value per range within the possible magnitude spectrum. For *A* and *I*, opportunity is allowed for the evaluator to choose a typical value or a refined value on the basis of more precise information about the sites evaluated. Rating schemes for various DRASTIC factors are presented in Tables 9.7–9.13. Table 9.14 illustrates the application of the DRASTIC method to a site for which data on the parameters are presented. It should be noted that rating schemes such as DRASTIC are useful only for screening-level analysis of sites. More detailed analyses require more specific site data and details on the design of the facility to be sited.

More elaborate site selection models have been developed. Among these models are the fuzzy set/analytic hierarchy method (Charnpratheep et al., 1997)

Table 9.7 DRASTIC Ranges and Ratings for Depth to Groundwater

Depth to water (ft)	
Range	Rating
0–5	10
5–10	9
15–30	7
30–50	5
50–75	3
75–100	2
100+	1
Weight: 5	Agricultural weight: 5

Source: Aller et al. (1985).

Table 9.8 DRASTIC Ranges and Ratings for Net Recharge

Net recharge (in.)	
Range	**Rating**
0–2	1
2–4	3
4–7	6
7–10	8
10+	9
Weight: 4	Agricultural weight: 4

Source: Aller et al., (1985).

and the multiattribute/fuzzy set method (Anandalingam and Westfall, 1989). Most site selection models are adaptable to the application of Geographic Information Systems (GIS). Ecological factors that are important inputs to ecosystem-scale risk assessment, as discussed by Wright et al. (1993) and Mayernik and Fehrenkamp (1992), play a significant role, together with environmental justice factors (English et al., 1993), in facility site selection.

Table 9.9 DRASTIC Ranges and Ratings for Aquifer Media

Aquifer media		
Range	**Rating**	**Typical rating**
Massive shale	1–3	2
Metamorphic/igneous	2–5	3
Weathered metamorphic/igneous	3–5	4
Thin bedded sandstone, limestone, shale sequences	5–9	6
Massive sandstone	4–9	6
Massive limestone	4–9	6
Sand and gravel	6–9	8
Basalt	2–10	9
Karst limestone	9–10	10
Weight: 3	Agricultural weight: 3	

Source: Aller et al. (1985).

Table 9.10 DRASTIC Ranges and Ratings for Soil Media

Soil media	
Range	Rating
Thin or absent	10
Gravel	10
Sand	9
Shrinking and/or aggregated clay	7
Sandy loam	6
Loam	5
Silty loam	4
Clay loam	3
Nonshrinking and nonaggregated clay	1
Weight: 2	Agricultural weight: 5

Source: Aller et al. (1985).

9.5 CONTAINMENT SITE IMPROVEMENT

The first option is usually to avoid sites that are not suitable for location of waste containment systems. From the standpoint of engineering and environmental protection, suitable sites are those in which the present and future costs of implementing an effective control scheme for contaminants are acceptable. It is worth noting that some of the costs may not be readily quantifiable in financial terms. In some cases, the contaminant plume exists already. The containment objective may be

Table 9.11 DRASTIC Ranges and Ratings for Topography

Topography (percent slope)	
Range	Rating
0–2	10
2–6	9
6–12	5
12–18	3
18+	1
Weight: 1	Agricultural weight: 3

Source: Aller et al. (1985).

Table 9.12 DRASTIC Ranges and Ratings for Impact of the Vadose Zone Media

Impact of vadose zone media		
Range	Rating	Typical rating
Silt/clay	1–2	1
Shale	2–5	3
Limestone	2–7	6
Sandstone	4–8	6
Bedded limestone, sandstone, shale	4–8	6
Sand and gravel with significant silt and clay	4–8	6
Metamorphic/igneous	2–8	4
Sand and gravel	6–9	8
Basalt	2–10	9
Karst limestone	8–10	10
Weight: 5	Agricultural weight: 4	

Source: Aller et al. (1985).

focused on minimizing the rate of plume movement in the subsurface. In other cases, such as landfill siting, it may be necessary to implement a redundant containment scheme for control of leachate that may be generated by the landfill.

The analysis presented in detail in Chapter 11 indicates that the contaminant transport rate through porous media (such as those present in the subsurface) is directly proportional to the permeability of the media. The Kozeny relationship,

Table 9.13 DRASTIC Ranges and Ratings for Hydraulic Conductivity

Hydraulic conductivity (gpd/ft^2)	
Range	Rating
1–100	1
100–300	2
300–700	4
700–1000	6
1000–2000	8
2000+	10
Weight: 3	Agricultural weight: 2

Source: Aller et al. (1985).

Table 9.14 Illustration of the Application of the DRASTIC Method Using a Geological Setting Comprising Glacial Till over Bedded Sedimentary Rocks

Setting (7 Aa glacial till over bedded sedimentary rock)			General	
Feature	Range	Weight	Rating	Number
Depth to water table	30–50	5	5	25
Net recharge	4–7	4	6	24
Aquifer media	Thin bedded SS, LS, SH Sequences	3	6	18
Soil media	Clay loam	2	3	6
Topography	2–6%	1	9	9
Impact vadose zone	Silt/clay	5	1	5
Hydraulic conductivity	100–300	3	2	6
		DRASTIC index		93

Source: Aller et al. (1985).

shown as Eq. (9.4), is one of several correlations between permeability and porous media characteristics.

$$K = \frac{n^3}{rs^2} \tag{9.4}$$

where

K = soil permeability (L^2/T)
n = soil porosity (dimension less fraction)
s = specific surface area of the soil (L^{-1})
r = Kozeny constant (approximately 5.0)

It is then apparent that a reduction in the porosity n of a porous medium is an effective measure for reducing contaminant transport rate. Porosity can be reduced by stabilization of a loose medium or by compaction. The first approach forms the basis for the application of grout barriers, which are discussed in Chapter 10. The second approach can be implemented most effectively through dynamic compaction (otherwise referred to as heavy tamping). As described by Inyang (1992) with the illustrations presented as Figures 9.8 and 9.9, heavy tamping can be used to redirect contaminant flow from sensitive to less sensitive areas.

The heavy tamping process involves the elevation of a heavy hammer which is then allowed to drop onto the ground surface. This is repeated in a

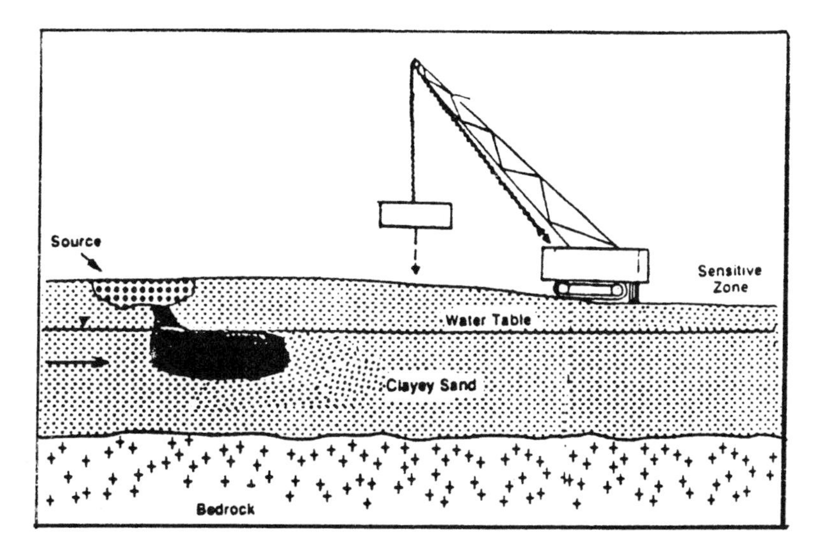

Figure 9.8 Illustration of the application of heavy tamping to control contaminant migration in the subsurface (Inyang, 1992).

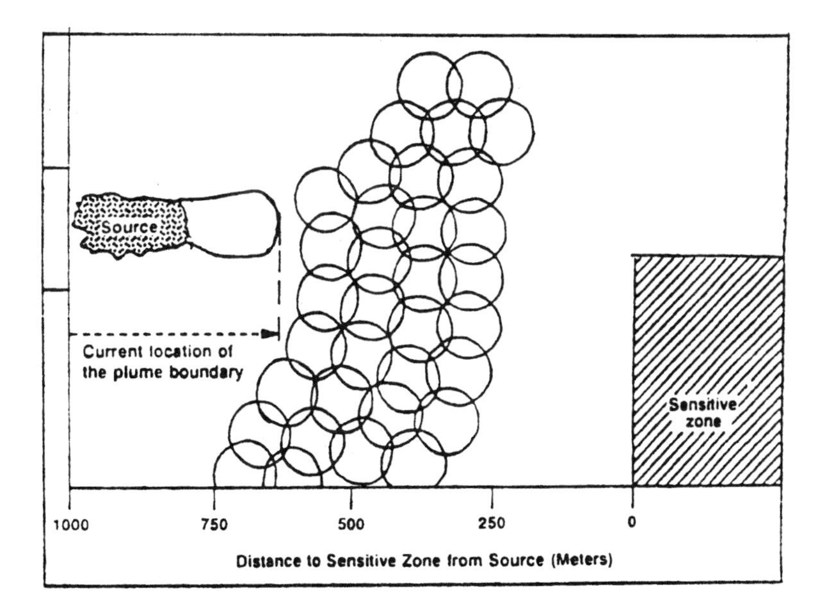

Figure 9.9 Possible arrangement of impact points to control contaminant migration rate toward a sensitive zone (Inyang, 1992).

suitable pattern until the soil is adequately densified. The energy imposed by the hammer on the soil medium is dissipated through the following processes, the relative intensities of which vary by soil type, energy level, and degree of saturation:

Elastic wave propagation
Plastic compression of soil
Crushing of soil particles
Pore pressure buildup in soil

In essence, the soil particles are shaken to a denser structure even at locations that may be several meters away from the zone of hammer contact. Typical ranges of magnitudes for dynamic compaction parameters are presented in Table 9.15. Some relationships among these parameters have been proposed by Menard and Broise (1975), and by Charles et al. (1981), as shown in Eqs. (9.5) and (9.6), respectively. Both relationships can be used to estimate the depth of influence of the hammer force for a given set of soil strength, hammer weight, and drop height values.

Table 9.15 Suggested Numerical Ranges of Heavy Tamping Design Factors for Contaminant Migration Rate Control

Factor	Numerical range	Comments
Required depth of compaction (m)	<12	Depends on the thickness of the loose layer
Applied impact energy energy (measure-m/m^2)	200–300	Applied impact energy depends on specified improvement in relative density of soil
Tamper diameter (m)	1.0–3.0	
Drop height (m)	10–30	
Drop weight (tonnes)	5–40	
Repetitions (impacts per point)	2–8	Depends on specified improvement in relative density
Spacing of impact points (m)	1.0–4.0	Depends on desired overlap of the tributary areas
Minimum depth to water table (m)	2.0–3.0	Temporary dewatering may be required to attain this condition
Required improvement in relative density (%)	>40	Depends on desired decrease in soil permeability
Peak particle velocity at nearby structures (cm/s)	<4.0	Should be evaluated prior to full-scale project implementation

Source: Inyang (1992).

$$D = R\,(WH)^{0.5} \tag{9.5}$$

$$D = 0.4d\left[\left(\frac{E}{A}\right)\left(\frac{1}{d}\right)\left(\frac{1}{c}\right)\right]^{0.5} \tag{9.6}$$

where

D = depth of soil improvement (L)
R = empirical factor that depends on soil type, ranging from about 0.35 for fine-grained soils to 0.6 for granular free-draining soils; 0.5 is frequently used as an approximate value
W = weight of hammer (g)
h = hammer drop height (L)
d = diameter of hammer (L)
E = applied impact energy (g-L)
A = area of impact (L^2)
C = undrained shear strength of soil (g/L^2)

It should be noted that dynamic compaction may not always produce a permanent barrier to contaminant migration.

REFERENCES

Aller, L., Bennett, T., Lehr, J. H., and Petty, R. J. (1985). DRASTIC: A standardized system for evaluating groundwater pollution potential using hydrogeologic settings. Technical Report under Cooperative Agreement CX-819715-01, Office of Research and Development, U.S. Environmental Protection Agency, Ada, Oklahoma, UK.

Anandalingam, G., and Westfall, M. (1989). Selection of hazardous waste disposal alternative using multi-attribute utility theory and fuzzy set analysis. *J. Environ. Syst.*, 18(1):69–85.

Armour, A. (1986). Social impact assessment of new hazardous waste facilities. Technical document developed for the Hazardous Waste Facilities Siting Board, Annapolis, MD.

Bain, G. L., and Hansen, P. (1981). Geologic and hydrologic implications of hazardous wastes management rules on siting in North Carolina. *Bull. Assoc. Eng. Geol.*, XVIII(3):267–275.

Charles, J. A., Burford, D., and Watts, K. S. (1981). Field studies of the effectiveness of dynamic consolidation. *Proc. 10th Int. Conf. on Soil Mechanics and Foundation Engineering, I.* Stockholm, pp. 617–622.

Charnpratheep, K., Zhou, Q., and Garner, B. (1997). Preliminary landfill site screening using fuzzy geographical information systems. *Waste Manage. Res.*, 15:197–215.

English, M., Barkenbus, J., and Wilt, C. (1993). Solid waste facility siting: Issues and trends. *J. Air Waste Manage. Assoc.*, 43:1345–1350.

Freeze, R. A., and Cherry, J. A. (1979). *Groundwater*. Prentice-Hall, Englwood Cliffs, NJ.

Inyang, H. I. (1992). Application of dynamic compaction to contaminant migration control. *Proc. 13th Annual Natl. Conf. and Exhibition on Hazardous Materials Control*, Washington, DC, pp. 625–630.

Jackson, R. E. (1980). Aquifer contamination and protection. Project 8.3 of the International Hydrological Programme, United Nations Educational, Scientific and Cultural Organization, Paris, France.

Lemme, G., Carlson, C. G., Dean, R., and Khakurai, B. (1990). Contamination vulnerability indexes: A water quality planning tool. *J. Soil Water Conserv.*, March–April: 349–350.

Mayernik, J. A., and Fehrenkamp, K. (1992). A new model for conducting quantitative ecological risk assessments at hazardous waste sites. *Proc. 13th Annual Natl. Conf. on Hazardous Materials Control*, Washington, DC, pp. 813–819.

Menard, L., and Broise, Y. (1975). Theoretical and practical aspects of dynamic consolidation. *Geotechnique*, 25(1):3–18.

Mustalish, R. W., and Costanzo, F. (1991). An application of the DRASTIC model to the siting of waste facilities. *Proc. 12th Natl. Conf. on Hazardous Materials Control*, Washington, DC, pp. 51–53.

U.S. EPA. (1985). Leachate plume management, EPA/540/2-85/004. Office of Emergency and Remedial Response, U.S. Environmental Protection Agency, Washington, DC.

Vita, C. L. (1987). Generalized risk-based decision making for hazardous waste sites. *Proc. Natl. Conf. on Hazardous Wastes and Hazardous Materials*, Washington, DC, pp. 32–37.

Wright, F. G., Inyang, H. I., and Myers, V. B. (1993). Risk reduction through regulatory control of waste disposal facility siting. *J. Environ. Syst.* 22(1):27–35.

10 Configurations of Containment Systems

The configuration of a containment system for wastes plays a significant role with respect to the effectiveness of the system in meeting its design function during the design life. Essentially, configuration is the arrangement of the components of the system. It influences the extent to which the components are exposed to environmental stresses, contaminant and moisture contact, and anthropogenic activities. Also, the configuration of components within a multicomponent system affects the distribution and propagation of stresses and deformation from one component to another. The objective of this chapter is to introduce the reader to the most common configurations of waste containment systems. Design approaches and quantitative expressions for various components of containment systems are provided in Chapters 9, 11, and 12.

10.1 LANDFILLS

In general, the overall design goal of landfills is minimization of the release of stored hazardous materials into the environment. Two design approaches tie in with this goal:

Minimization of the quantity of leachates generated within the landfill through adequate design of the cover system and surface drainage components

Minimization of the quantity of hazardous materials that migrate from the landfill into the environment through the incorporation of well-designed liners and leachate collection systems below the stored waste

Performance requirements and associated design standards for landfills are specified by the U.S. EPA (1985a, 1985b) for U.S. practice. Figure 10.1 shows the overall configuration of a landfill. Figures 10.2 and 10.3 show the components in greater detail. The reader should note that the configurations shown may not be the most optimal for all countries and climatic regions. It may be necessary to formulate configurations for specific regions on the basis of available construction materials, waste chemistry and generation characteristics, monitoring requirements, climatic factors, and design service life. Use could be made of design equations summarized by Inyang and Tumay (1995), among others.

The major features of the most common set of landfill component layers are as follows.

Synthetic/composite double liner system. The primary leachate collection and recovery (LCR) system keeps the leachate head on the primary membrane liner at a minimum. Granular drainage layers must have the following *minimum* characteristics: 30 cm (12 in.) thickness, minimum hydraulic conductivity of 1.00 cm/s, and a slope greater than or equal to 2%. It also retains the leachate for removal during the postclosure care period.

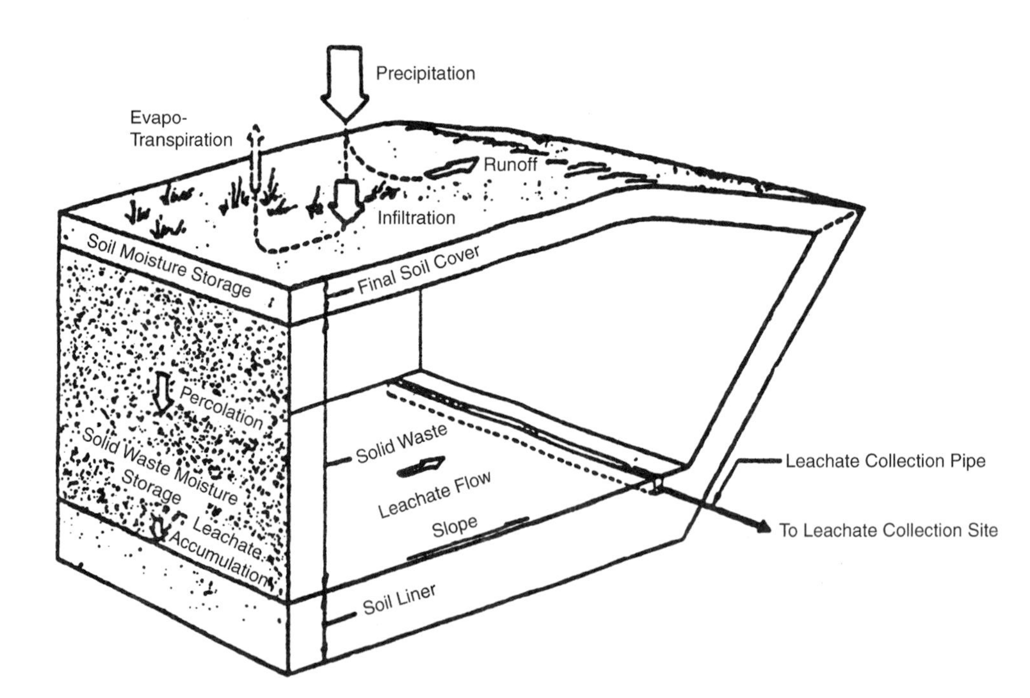

Figure 10.1 Illustration of a landfill as a waste containment chamber.

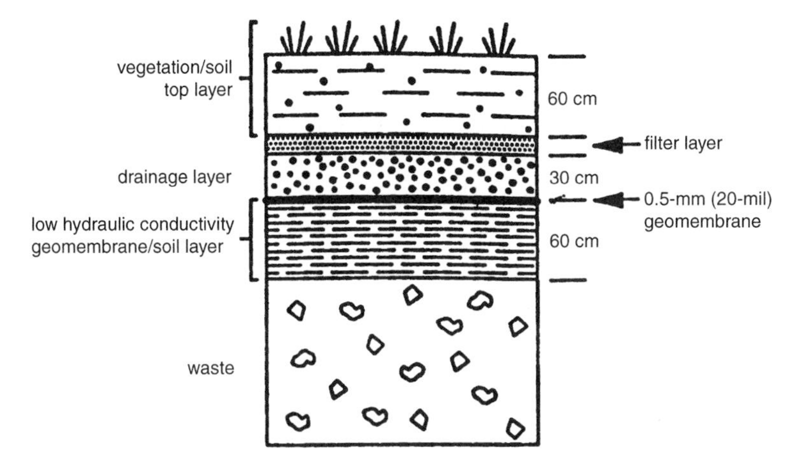

Figure 10.2 Recommended cover system for landfills (U.S. EPA, 1991).

The primary flexible membrane liner (FML) is designed and constructed with adequate quality assurance to minimize the release of leachates during the active life of the landfill and the postclosure care period. The secondary LCR systems allow the detection and removal of leachates which leak through the primary FML. The secondary FML acts in tan-

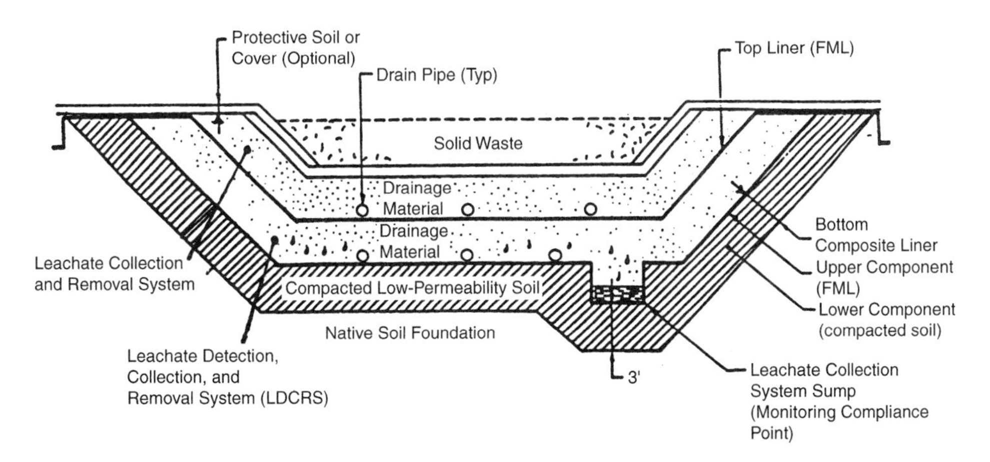

Figure 10.3 A double liner and leachate collection system for a landfill (U.S. EPA, 1989).

dem with the compacted clay liner to prevent the release of leachates in quantities that exceed specified limits.

Synthetic/clay double liner system. The leachate collection systems serve the same functions as described above. However, only one FML layer is incorporated. A clay liner is included, the thickness of which depends on the calculated breakthrough time for a single leachate constituent to pass from the top of the clay liner.

As illustrated in Figure 10.2, the cover system is multilayered. It is usually designed to promote surface drainage, withstand waste settlement, and accommodate soil profile subsidence. The top vegetative cover is intended to minimize erosion and promote evapotranspiration. An optional gas collection layer may be incorporated. A low-permeability clay layer (at least 2 ft thick) and a 20-mil flexible membrane cap (FMC) are also included. On top of the FMC, a surface-water drainage layer is usually placed. It may consist of three components: a bedding layer which protects the FMC; the actual drainage layer; and a filter layer that prevents the intrusion of vegetative cover soil into the drainage component.

Additional literature that addresses various design and service aspects of landfill components are Haikola et al. (1995), Bonaparte (1995), Benson and Othman (1993), Inyang (1994), Inyang et al. (1997a, 1997b), Koerner (1990), Richardson and Koerner (1989), Ware and Jackson (1978), Duvel (1979), Reid et al. (1971), Thornton and Blackwell (1976), Weston (1989), Emcon Associates (1982), Haxo (1980), Hughes (1975), Cope (1987), Koerner and Whitty (1984), Duplancic (1987), Goodall and Quigley (1987), Giroud and Ah-Line (1984), Cosler and Snow (1984), Glebs (1980), Gordon and Huebner (1983), Quinn (1983), and Bass et al. (1984).

10.2 SLURRY WALLS

Slurry walls are subsurface trenches that are filled with low-permeability soil materials and possibly other additives to retard the migration of contaminants into the surrounding soil/groundwater environment. Slurry walls are particularly useful in loose subsurface materials, where they can divert uncontaminated groundwater away from waste or contaminant plume, reduce the flow rate of leachate into the subsurface, increase the retardation or decay of contaminants through lengthening of their flow path from source to sink, and accumulate groundwater or leachate for removal through extraction systems.

The configuration of a slurry wall can be described in terms of the horizontal and vertical arrangements of its components. In plan view, the slurry wall can be linear, arcuate, or rectilinear, as shown in Figures 10.4 and 10.5.

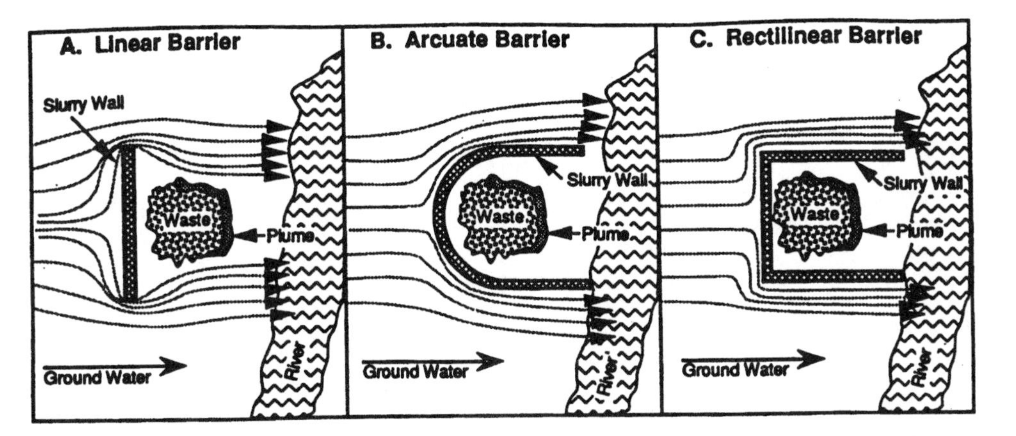

Figure 10.4 Plan views of upgradient slurry walls for partial containment of buried waste and/or plume.

Depending on the design objective, the wall can be placed before the contaminant source to minimize the potential for source submergence by groundwater, or after the waste to reduce the migration rate and promote the decay of contaminant plume. More commonly, the waste source is entirely enclosed by the slurry wall. Figures 10.6 and 10.7 show vertical configurations of slurry walls.

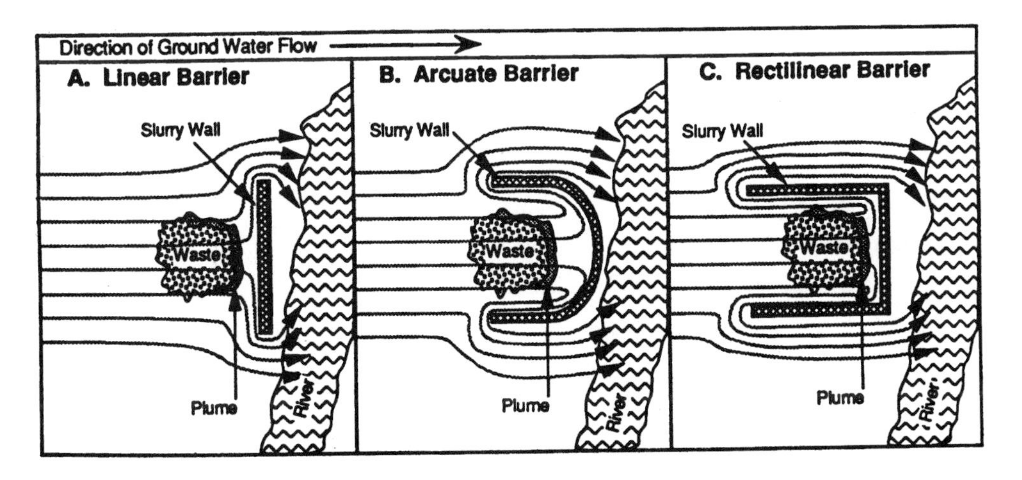

Figure 10.5 Plan view of downgradient slurry walls for partial containment of buried waste and/or plume.

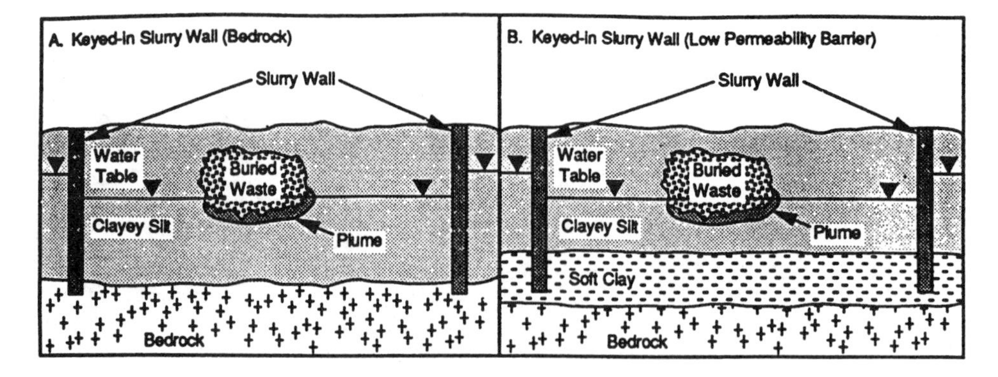

Figure 10.6 Vertical sections through keyed-in slurry wall containment systems (Inyang, 1992).

In Figure 10.6, the walls are embedded (keyed in) in bedrock or a low-permeability soil layer to effect a complete isolation of the waste by design. In Figure 10.7, a hanging slurry wall is illustrated. Usually, this configuration is used when the low-permeability soil layer or bedrock is too deep for keying in of the wall and the contaminant floats on a relatively high water table. The advantages and disadvantages of the three principal arrangements of keyed-in slurry walls are summarized in Table 10.1. In addition to the advective-dispersive equations presented in Chapters 4 and 11 of this book, design and performance evalua-

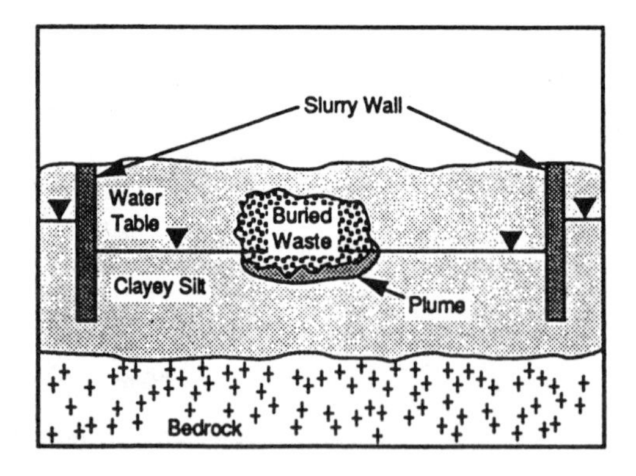

Figure 10.7 A vertical section through a hanging slurry wall containment system for buried wastes (Inyang, 1992).

Table 10.1 Advantages and Disadvantages of Various Horizontal Configurations of Keyed-in Slurry Walls

Configuration	Advantages and disadvantages
Total enclosure	Completely diverts groundwater away from waste in all directions Minimizes escape of pollutants from the containment into the surrounding media For extensive wastes, it may be expensive
Partial barrier: upgradient	Diverts groundwater partially around waste, especially where the gradient may be high May require relatively high precision in the prediction of the direction of groundwater flow Cheaper than total enclosure Minimizes leachate generation but is ineffective for control of its transport from site
Partial barrier: downgradient	May retain leachate for retrieval by extraction systems Increases the length of the flow path of leachate from source to sensitive locations Cheaper than total enclosure Minimizes leachate migration but is ineffective for control of leachate production May require relatively high precision in the prediction of the direction of groundwater flow

tion methods for slurry walls have been described by Manassero and Pasqualini (1995), Tamaro et al. (1993), Evans et al. (1995) and Manassero et al. (1995).

10.3 DRAINAGE TRENCHES AND WELLS

As an alternative to the use of physical barriers to control the movement of groundwater and contaminants in the subsurface, drainage trenches and wells can also be used. For details on well and trench hydraulics, the reader is referred to standard groundwater hydrology textbooks that focus on those issues. In this section, useful configurations of drainage trenches and wells are presented to enable the reader to appreciate the potential effectiveness of this contaminant migration control approach.

In this approach, the source of the contaminant is not removed directly, but the leachate is forced to drain either by gravity or imposed hydraulic force

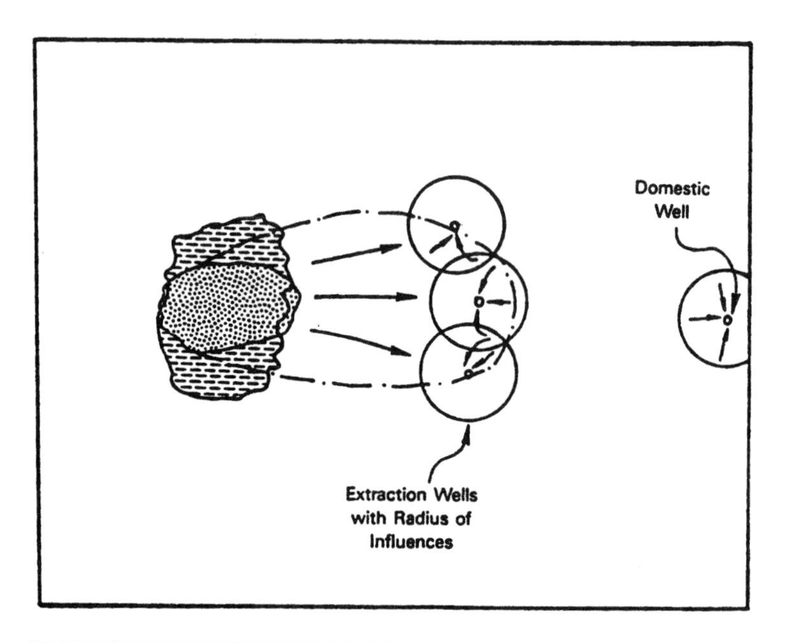

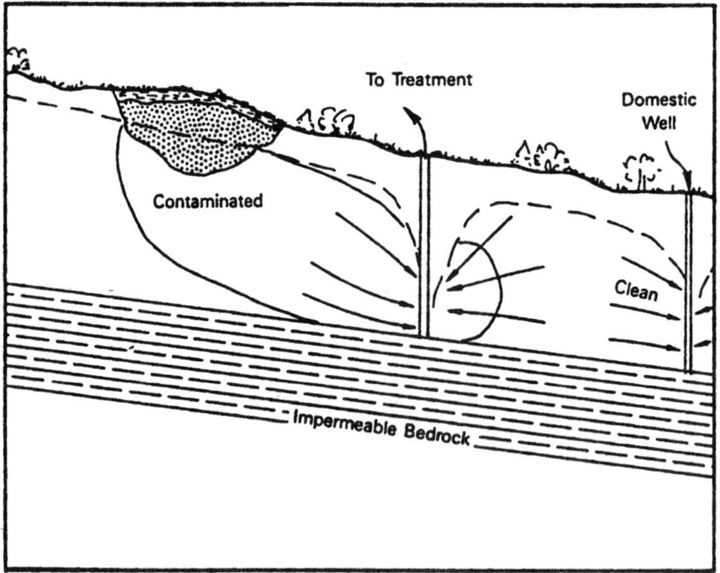

Figure 10.8 Plane and cross-sectional views of an extraction well system for containment of contaminants in the subsurface (U.S. EPA, 1985c).

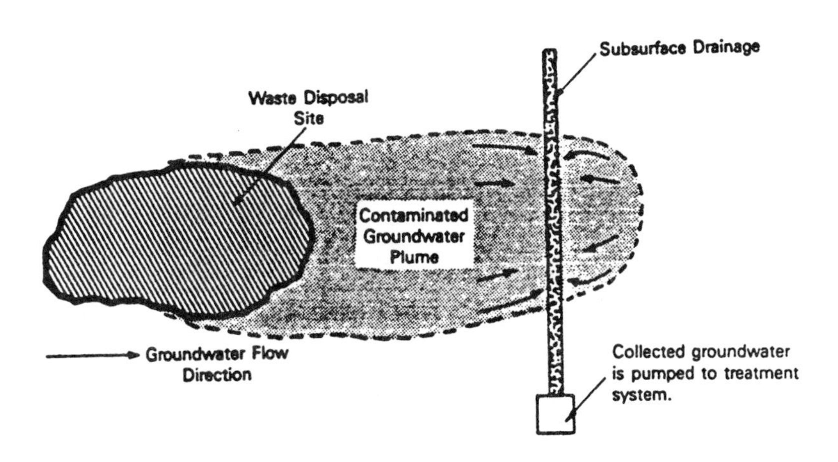

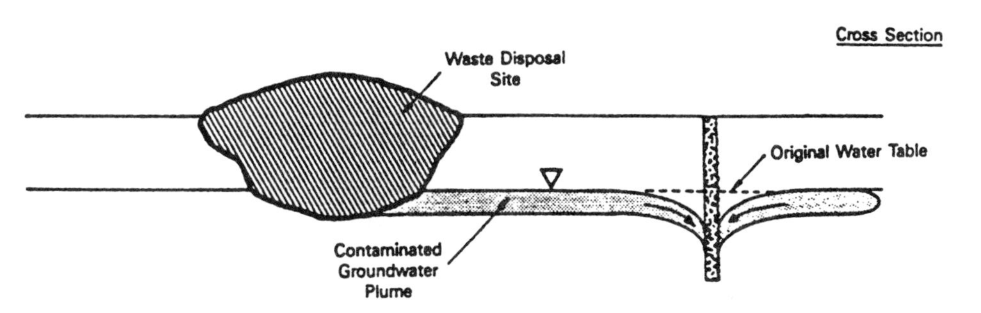

Figure 10.9 Use of a subsurface trench for containment of a contaminant plume (U.S. EPA, 1985c).

to a trench or wells from which it is removed by pumping. Thus, this method of containment requires the use of above-ground treatment systems. Injection wells may be used to flush the contaminant source and increase the hydraulic potential for migration of contaminated water to extraction wells or trenches. As illustrated in Figures 10.8 and 10.9, each well has its radius of influence. Their spacing should be such that their zones of influence overlap to promote effective containment. A system of collector pipes can also be used with a main collector pipe as illustrated in Figure 10.10 to enhance areal coverage and the effectiveness of the plume containment.

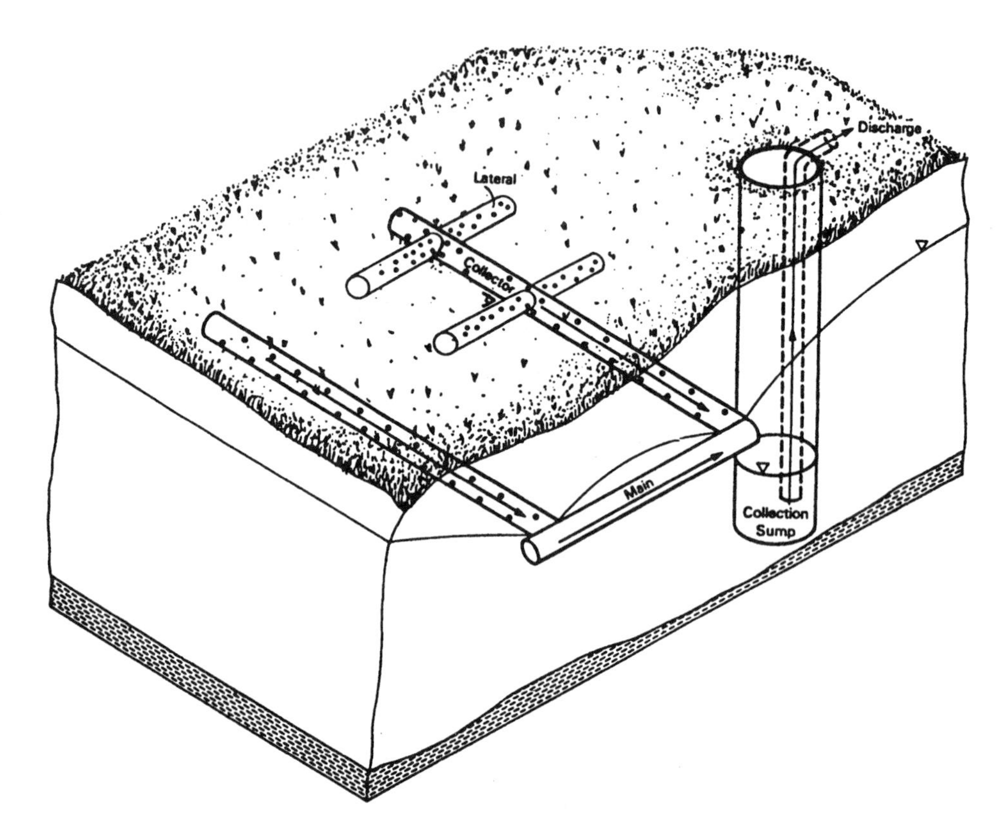

Figure 10.10 Configuration of collector pipes and the main discharge pipe for extraction of contaminated groundwater (adapted from U.S. EPA, 1985c).

10.4 SURFACE IMPOUNDMENTS

Surface impoundments are natural or man-made topographic depressions or diked areas, bounded primarily by earthen materials and other engineered materials, and designed to hold liquid wastes or other wastes that contain free liquids. Impoundments are often lined with clay, polymeric materials, or both. Examples of surface impoundments are lagoons, settling and storage pits, and ponds. Concrete-lined basins are not ordinarily classified as impoundments. Impoundments may range in size from a few square meters to a few hundred acres. They can be made to extend below the ground surface as in Figure 10.11A, or they can be created on the ground surface through a system of well-configured dikes as shown in Figure 10.11B.

Surface impoundments may be temporary or permanent. Typically, they

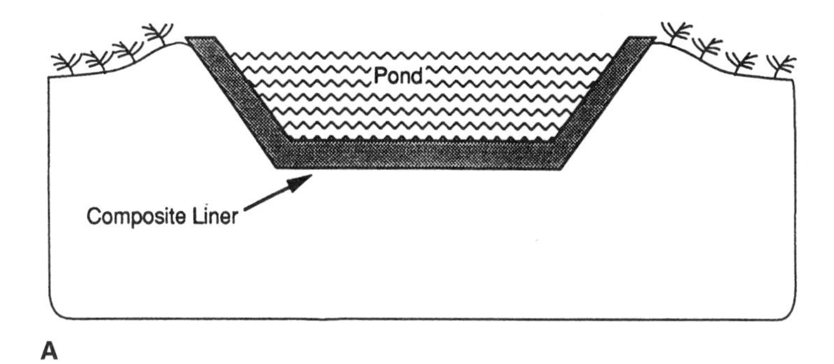

A

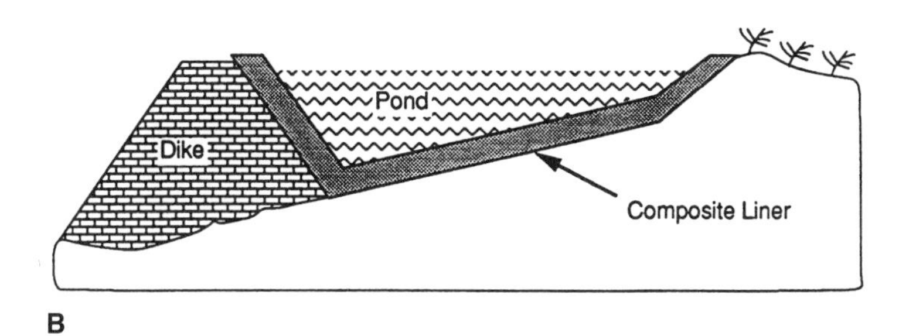

B

Figure 10.11 **A.** An entrenched surface impoundment. **B.** A surface impoundment that incorporates a dike.

serve as open containments for waste waters and sludges exuded from treatment processes.

Temporary impoundment. Wastes are usually removed and the sites treated after closure of a temporary impoundment. A single liner, made of natural or synthetic materials, is specified. The unit is designed to prevent breakthrough during the life of the impoundment.

Permanent surface impoundments. Permanent impoundments are usually closed after their active lives. Before closure, the liquid fraction of the waste must be removed and the solids stabilized.

10.5 GROUT CURTAINS

Grout curtains are barriers that are formed by pumping pore-filling materials into the ground, thereby reducing the rate at which fluids can subsequently travel through the ground. Typically, the grout mixture is injected under pressure,

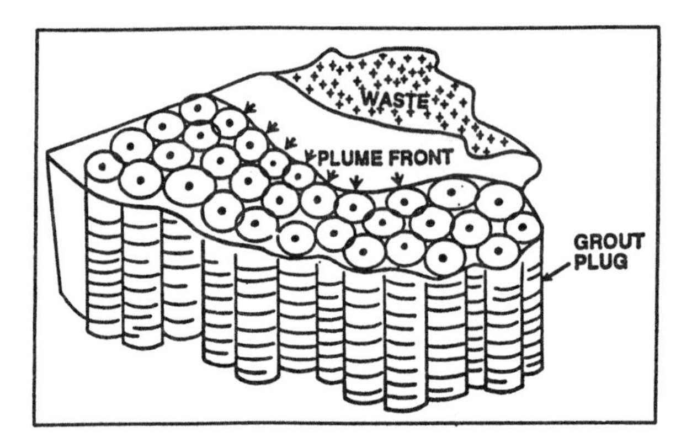

Figure 10.12 Configuration of a grout curtain designed to contain a contaminant plume.

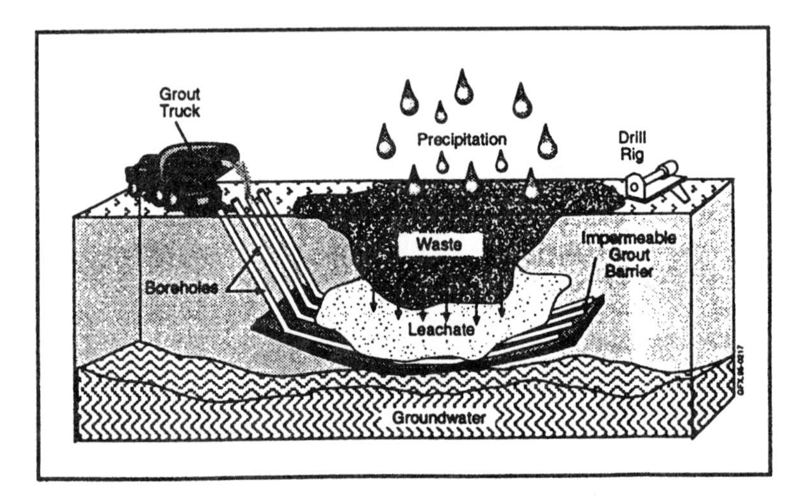

Figure 10.13 An innovative emplacement method for grout bed installation underneath a waste mass (U.S. DOE, 1995).

through a pipe that penetrates the soil or rock strata that are to be improved. Each injection point has its own radius of influence, the dimension of which depends on the injection pressure and duration, characteristics of the grout, and the hydraulic properties of the subsurface geomedia. The spacing of injection points should not allow windows to exist between the influence areas of adjacent injection holes. The configuration of a grout curtain is illustrated in Figure 10.12. Within the past few years, it has been possible to emplace grout beds horizontally beneath waste masses without drilling through wastes. This is illustrated in Figure 10.13.

10.6 COMPOSITE SYSTEMS

Site contamination scenarios may warrant the application of two or more types of waste containment structures to increase the effectiveness of the overall waste containment system. There are potentially numerous configurations of such composite containment systems. A few of these systems are illustrated in Figure 10.14. In Figure 10.14A, a cover is combined with a slurry wall. The cover system minimizes infiltration of moisture (from precipitation) into the waste. The wall controls the flow of groundwater into the enclosure, thus ensuring that the waste is not submerged. In Figure 10.14B, the water table can be maintained at a level below the waste through the use of extraction wells. The slurry walls separate external groundwater, thus enabling the internal water table to be lowered effectively. In Figure 10.14C, a grout is used to seal fractures in rock to block the

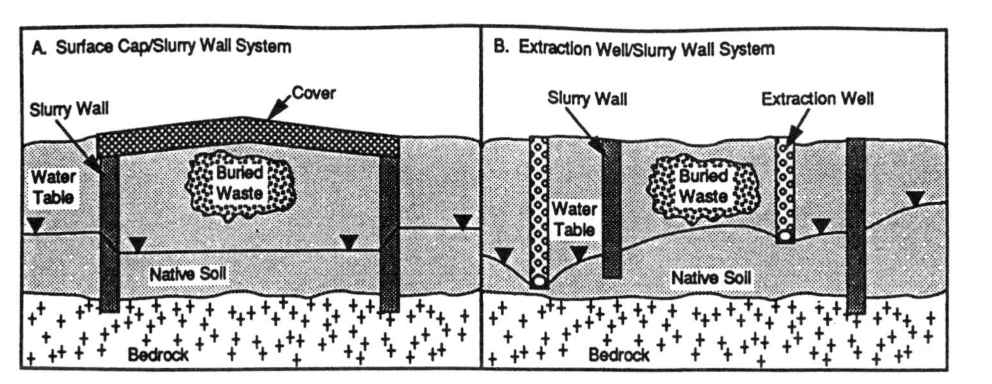

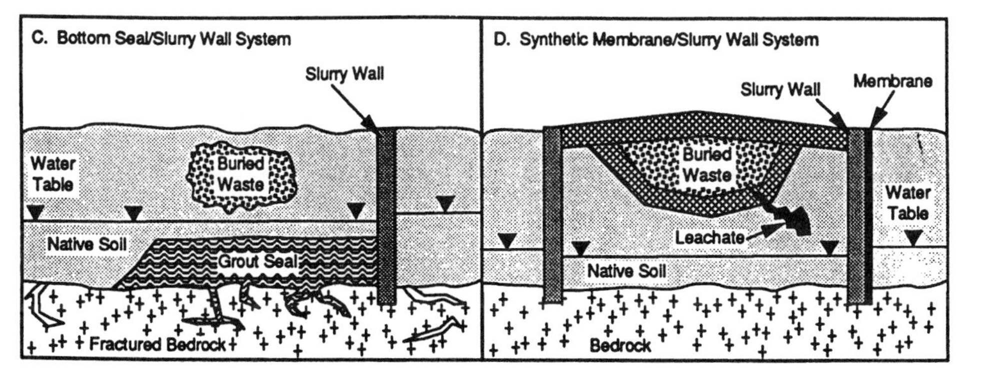

Figure 10.14 Examples of configurations in which slurry walls are combined with ancillary measures.

contamination of the fractures and subsequent feedback of contaminants to groundwater. The slurry wall isolates external groundwater from the potentially contaminated zone. In Figure 10.14D, the use of a slurry wall to control possible groundwater contamination by releases from a landfill is illustrated. The external surface of a section of the slurry wall is covered by a geomembrane which provides additional effectiveness against gas migration.

REFERENCES

Bass, J. M., Ehrenfeld, J. R., and Valentine, J. R. (1984). Potential Clogging of Landfill Drainage Systems. EPA-600/S2-83-1090. U.S. Environmental Protection Agency, Municipal Environmental Research Lab., Cincinnati, OH.

Benson, C. H., and Othman, M. A. (1993). Hydraulic conductivity of compacted clay frozen and thawed in-situ. *J. Geotech. Eng*. 19(2):276–294.

Bonaparte, R. 1995. Long-term performance of landfills. *Proc. ASCE Geoenvironment 2000 Conf*., New Orleans, LA, pp. 514–553.

Cope, F. W. (1987). Design of waste containment structures. *Proc. Specialty Conf. Geotechnical Practice for Waste Disposal*, Ann Arbor, MI, pp. 1–20.

Cosler, D. J., and Snow, R. E. (1984). Leachate collection system performance analyses. *J. Geotech. Eng*. vol. 110(8):1024–1041.

Duplancic, N. 1987. Hazardous waste landfill cap system stability. *Proc. Specialty Conf. on Geotechnical Practice for Waste Disposal*, Ann Arbor, MI, pp. 432–446.

Duvel, W. A. (1979). Solid waste disposal: Landfilling. *Chem. Eng*. 86(14):77–86.

Emcon Associates. (1982). Field Assessment of Site Closure, Boone County, Kentucky. U.S. Environmental Protection Agency, Cincinnati, OH.

Evans, J. C., Costa, M. J., and Cooley, B. (1995). The state of stress in soil-bentonite slurry trench cut off walls. *Proc. ASCE Geoenvironment 2000 Conf*., New Orleans, LA, pp. 1173–1191.

Giroud, J. P., and Ah-Line, C. (1984). Design of earth and concrete covers for geomembranes. *Proc. Int. Conf. on Geomembranes*, Denver, CO, pp. 487–492.

Glebs, R. T. (1980). Under right conditions, landfills can extend below ground water table. *Solid Waste Management*, 23(2):50–59.

Goodall, D. C., and Quigley, R. M. (1977). Pollutant migration from two sanitary landfill sites near Sarnia, Ontario. *Can. Geotech. J*. 14(2):223–236.

Gordon, M. E., and Huebner, P. M. (1983). An evaluation of the performance of zone of saturation landfills in Wisconsin. *Proc. 6th Annual Madison Conf. of Applied Research and Practice on Municipal and Industrial Waste, University of Wisconsin, Madison*.

Haikola, B. M., Loehr, R. C., and Daniel, D. E. (1995). Hazardous waste landfill performance as measured by primary leachate quantity. *Proc. ASCE Geoenvironment 2000 Conf*., New Orleans, LA, pp. 554–567.

Haxo, H. E. (1980). Interaction of selected liner materials with various hazardous wastes. In *Disposal of Hazardous Wastes, Proc. 6th Annual Research Symp*., EPA-600/9-80-010. U.S. Environmental Protection Agency, Cincinnati, OH, pp. 160–180.

Hughes, J. (1975). Use of bentonite as a soil sealant for leachate, control in sanitary landfills. Soil Lab. Report 280-E, American Colloid Co., Skokie, IL.

Inyang, H. I. (1992). Selection and design of slurry walls as barriers to control pollutant migration. Draft Technical Guidance Document, Office of Solid Waste and Emergency Response, U.S. Environmental Protection Agency, Washington, DC.

Inyang, H. I. (1994). A Weibull-based reliability analysis of waste containment systems. *Proc. First Int. Congress on Environmental Geotechnics*, Alberta, Canada, pp. 273–278.

Inyang, H. I., Fang, H. Y., Iskandar, A., and Choquette, M. R. (1997a). Chemical and mineralogical analyses of clayey barrier materials. Encyclopedia of Environmental Analysis and Remediation, Vol. 8, pp. 5131–5141. Wiley, New York.

Inyang, H. I., Iskandar, A., and Parikh, J. M. (1997). Physico-chemical interactions in waste containment barriers. In. *Encyclopedia of Environmental Analysis and Remediation*, Vol. 2, pp. 1158–1165. Wiley, New York.

Inyang, H. I., and Tumay, M. T. (1995). Containment systems for contaminants in the subsurface. In *Encyclopedia of Environmental Control Technology*, pp. 175–215. Gulf Publishing Company.

Koerner, R. M. (1990). *Designing with Geosynthetics*, 2nd ed. Prentice-Hall.

Koerner, R. M., and Whitty, J. E. (1984). Experimental friction evaluation of slippage between geomembranes, geotextiles, and soils. *Proc. Int. Conf. on Geomembranes*, Denver, CO, pp. 191–196.

Manassero, M., Fratalocchi, E., Pasqualine, E., Spanna, C., and Verga, F (1995). Containment with vertical cut off walls. *Proc. ASCE Geoenvironment 2000 Conf.*, New Orleans, LA, pp. 1142–1172.

Manassero, M., and Pasqualini, E. (1992). Ground pollutant containment barriers. Proc. Mediterranean Conf. on Environmental Geotechnology, Cesme, Turkey, pp. 195–204.

Quinn, K. J. (1983). Numerical simulation of zone of saturation landfill designs. *Proc. 6th Annual Madison conf. on Applied Research and Practice on Municipal and Industrial Waste*, University of Wisconsin, Madison.

Reid, G., Streebin, L. E., Canter, L. W., Robertson, J. M., and Klehro, E. (1971). Development of specification for liner materials for use in oil-brine pits, lagoons, and other retention systems. Draft, Oklahoma Economic Development Foundation, Norman, OK.

Richardson, G. N., and Koerner, R. M. (1989). Geosynthetic Design Guidance for Hazardous Waste Landfill Cells and Surface Impoundments. Geosynthetic Research Institute, Drexel University, Philadelphia.

Tamaro, M., Pamukai, S., and Lopez, P. (1993). Prediction of structural slurry wall behavior. *Proc. Third Int. Conf. on Case Histories in Geotechnical Engineering*, St. Louis, MO, pp. 695–701.

Thornton, D. E., and Blackwell, P. (1976). Field Evaluation of Plastic Film Liners for Petroleum Storage Area in the McKenzie Delta. EPS-3-EC-76-13. Canadian Environmental Protection Service, Edmonton, Alberta.

U.S. DOE. (1995). Landfill Stabilization Focus Area. Technology Summary DOE/EM-0251. Office of Environmental Management and Technology Development, U.S. Department of Energy, Washington, DC.

U.S. EPA. (1991). Design and Construction of RCRA/CERCLA Final Covers. EPA/625/4-91/025. Office of Research and Development, U.S. Environmental Protection Agency, Washington, DC.

U.S. EPA. (1989). Requirement for Hazardous Waste Landfill Design, Construction and Closure. Seminar Publication. EPA/625/4-89/022. Office of Research and Development, U.S. Environmental Protection Agency, Washington, DC.

U.S. EPA. (1985a). Draft. Minimum Technology Guidance Document on Double Liner Systems for Landfills and Surface Impounds: Design, Construction, and Operation. U.S. Environmental Protection Agency, Washington, DC.

U.S. EPA. (1985b). Minimum Technology Guidance on Single Liner Systems for Landfills, Surface Impoundments, and Waste Piles—Design, Construction, and Operation. Draft. EPA/530-SW-85-013. U.S. Environmental Protection Agency, Washington, D.C.

U.S. EPA. (1985c). Leachate Plume Management. Publication EPA/540/2-85/004. Office of Emergency and Remedial Response, U.S. Environmental Protection Agency, Washington, DC.

Ware, S., and Jackson, G. (1978). Liners for Sanitary Landfills and Chemical Hazardous Waste Disposal Sites. EPA-600/9-78-005. U.S. Environmental Protection Agency, Cincinnati, OH.

Weston, R. F. Inc. (1989). Pollution Prediction Techniques for Waste Disposal Siting: A State-of-the-Art Assessment. EPA-SW-162C. U.S. Environmental Protection Agency, Cincinnati, OH.

11
Elements of Containment System Design

11.1 INTRODUCTION

A waste containment system is expected to perform two important functions: (1) it should prevent or minimize transport of contaminants from within itself to the surrounding soil and groundwater, and (2) it should retain its structural integrity during the design life period. The first function requires the application of mass transfer and transport principles (discussed in Chapter 4), whereas the second function requires the application of geotechnical aspects, primarily slope stability and settlement principles. These two functions are central not only to the overall siting and design of containment systems but also to the design of individual components such as top covers, bottom liners, drainage systems, slurry walls, etc. To fulfill the first function, we need to address the following questions:

1. What is the rate of infiltration into the waste material?
2. What proportion of the infiltrated water may be expected to cross the barrier?
3. What is the quality of the liquid that exits the barriers?

Fulfillment of the second function involves estimation of factor of safety against slope failure of the containment system, and settlement calculations. With the increased use of polymeric materials in barriers, this also involves checking for the stability of interfaces between geomembranes and soils.

We will take up these two functions in this chapter from system scale to component scale. Thus, we will look first at the water balance of the system, before discussing flow and transport through the barrier layers. In the discussion of stability issues, we will summarize the essential principles of slope stability and settlement analyses that are pertinent to waste containment systems. For a detailed discussion of these issues, the reader is advised to consult a standard textbook on geotechnical engineering.

11.2 LEACHATE GENERATION

As water from rainfall or runoff infiltrates into a waste containment system, it reacts with the solid and liquid constituents of the waste. During the percolation process, the infiltrating water is usually contaminated by the waste. Depending on the type of waste and the reactions between the percolating water and the waste, this percolated water, commonly termed *leachate*, may become highly contaminated. The most common reaction processes occurring within the waste are (1) dissolution of certain solid forms of the waste and subsequent precipitation; (2) decomposition and disintegration of the solids; and (3) reactions between the original liquids contained in the waste and the percolating water. The concentrations of various chemicals in the leachate are governed by a number of mass transfer processes operating simultaneously. For instance, an acidic pH condition of the liquid may trigger a number of processes, such as dissolution and precipitation, ion exchange, and sorption, simultaneously, as discussed in Chapter 4.

Apart from the reaction processes occurring within the waste, the concentrations of the leachate produced also depend on the waste disposal technique

Table 11.1 Ranges of Constituent Concentrations (in mg/L Unless Noted) in Leachate from Municipal Waste Landfills (EPA/530-SW-86-054, 1986)

Constituent	Concentration range	Constituent	Concentration range
COD	50–90,000	Hardness (as $CaCO_3$)	0.1–36,000
BOD	5–75,000	Total P	0.1–150
Total organic carbon (TOC)	50–45,000	Organic P	0.4–100
Total solids (TS)	1–75,000	Nitrate nitrogen	0.1–45
TDS	725–55,000	Phosphate (inorganic)	0.4–150
Total suspended solids (TSS)	10–45,000	Ammonia nitrogen (NH_3-N)	0.1–2,000
Volatile suspended solids (VSS)	20–750	Organic N	0.1–1,000
Total volatile solids (TVS)	90–50,000	Total Kjeldahl nitrogen (TKN)	7–1,970
Fixed solids (FS)	800–50,000	Acidity	2,700–6,000
AlKalinity (as $CaCO_3$)	0.1–20,350	Turbidity (Jackson units)	30–450
Total coliform (CFU/100 mL)	$0–10^5$	Cl	30–5,000
Fe	200–5,500	PH (dimensionless)	3.5–8.5
Zn	0.6–220	Na	20–7,600
Sulfate	25–500	Cu	0.1–9
Ni	0.2–79	Pb	0.001–1.44
Total volatile acids (TVA)	70–27,700	Mg	3–15,600
Mn	0.6–41	K	35–2,300
Fecal coliform (CFU/1,000 ml)	$0–10^5$	Cd	0–0.375
Specific conductance (mhg/cm)	960–16,300	Hg	0–0.16
Ammonium nitrogen (NH_4-N)	0–1,106	Se	0–2.7
		Cr	0.02–18

Table 11.2 Representative Hazardous Substances within Industrial Waste Streams

Industry	Hazardous substances: Arsenic	Cadmium	Chlorinated hydrocarbons[a]	Chromium	Copper	Cyanides	Lead	Mercury	Misc. organics[b]	Selenium	Zinc
Battery		×		×	×						×
Chemical manufacturing			×	×	×			×	×		
Electrical and electronic			×		×	×	×	×		×	
Electroplating and metal finishing		×		×	×	×					×
Explosives	×				×		×	×	×		
Leather				×					×		
Mining and metallurgy	×	×		×	×	×	×	×		×	×
Paint and dye		×		×	×	×	×	×	×	×	
Pesticide	×		×			×	×	×	×		×
Petroleum and coal	×		×				×				
Pharmaceutical	×							×	×		
Printing and duplicating	×			×	×		×		×	×	
Pulp and paper								×	×		
Textile				×	×				×		

[a] Including polychlorinated biphenyls.
[b] For example, acrolein, chloropicrin, dimethyl sulfate, dinitrobenzene, dinitrophenol, nitroaniline, and pentachlorophenol.
Source: Matrecon, Inc. (1980).

and time elapsed after the waste disposal. In general, the concentrations of chemicals in leachate are believed to reach a peak value some time after the placement of waste, and then decrease with time. This decrease in concentrations with time is due to dilution, biochemical processes leading to breakdown of chemicals, and continuous removal of leachate from a collection point (Rowe, 1991; Ehrig and Scheelhaase, 1993; Lu et al., 1985).

Because of these various factors, leachate quality in a containment system usually exhibits tremendous spatial and temporal variability. A knowledge of the predominant constituents in the leachate is essential in order to design the waste containment barriers. Table 11.1 shows the typical ranges of constituent concentrations found in leachate from municipal waste landfills. In contrast to municipal waste landfills, hazardous waste landfills typically contain elevated amounts of heavy metals, organic compounds, and other toxic substances. Table 11.2 shows representative hazardous constituents that may be expected in industrial leachate.

To assess the impact of leachate generation on the quality of soil and groundwater in the vicinity, one needs to estimate both the quantity and the quality of leachate exiting the containment system. In the next three sections, we will discuss methods commonly adopted to estimate leachate quantity and quality.

11.3 WATER BALANCE IN WASTE CONTAINMENT SYSTEMS

The total quantity of leachate generated at a given waste containment system is primarily a function of the quantity of water infiltrated into the system and quantity of fluids generated within the waste. The former is in turn dependent on a number and intensity of climatologic and hydrologic processes, primarily rainfall, runoff, and evaporation. Thus, to estimate the quantity of leachate, one needs to conduct a water balance for the entire system. This essentially involves a bookkeeping procedure to account for the final disposition of total precipitation at a given site, through a number of pathways such as evaporation, infiltration, and runoff.

Figure 11.1 shows the various pathways through which precipitation at a waste containment system is disposed of. Precipitation in the form of snow or rainfall is partitioned into interception by vegetation (for subsequent evapotranspiration), temporary storage followed by subsequent runoff from the surface of the system, and infiltration into the cover. The proportion of the precipitation that infiltrates into the system will alter the water storage in the topsoil, which will undergo possible evapotranspiration from the vegetation and the soils. In addition, a proportion of the infiltrated water may be carried through a lateral drainage system if one is provided in the surface cover system. Eventually, a portion of the infiltrated water may travel down into the waste, past the topsoil cover and barrier, and contribute to leachate percolation.

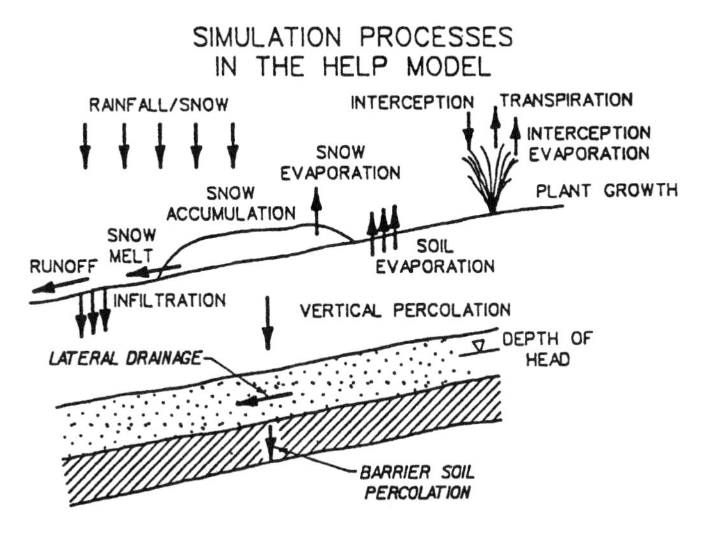

Figure 11.1 Water pathways for a waste containment system (U.S. EPA, 1991).

It is possible to estimate the water consumed in each of the above pathways and obtain the quantity of percolation through the waste using water routing methods of hydrology. Fenn et al. (1975) proposed one of the earlier comprehensive water balance methods for predicting leachate generation for solid-waste disposal sites. This method employs simple expressions to estimate runoff and evapotranspiration, and is suitable for hand calculations. Subsequently, a computer-based water budget model, known as Hydrologic Evaluation of Landfill Performance (HELP), was developed. The former method is helpful in understanding the individual processes accounted for in the latter; therefore, we start our discussion with it.

11.3.1 Monthly Water Balance Method

The monthly water balance method involves using monthly hydrologic and climatologic data to track changes in infiltration, evapotranspiration, and soil water storage. Fenn et al. (1975), Lutton et al. (1979), Lu et al. (1985), and Oweis and Khera (1990) demonstrated its use in the context of waste containment systems. Here, we will closely follow the approach used by Oweis and Khera (1990). The water balance for the entire system is analyzed using the known retention and transmission characteristics of the soil cover and refuse in a method proposed by Thornthwaite and Mather (1957). Based on water conservation at the site, the quantity of the infiltrated water I, is expressed as

$$I = P + SR - R \tag{11.1}$$

where P = precipitation, SR = water carried by surface runoff into the system, and R = surface runoff out of the system. The pathway of interception by vegetation is ignored in Eq. (11.1). R may be estimated by any of the empirical methods available in the hydrology literature. Rational formula offers a simple way of estimating R as a proportion of precipitation. That is,

$$R = CP \tag{11.2}$$

where C = runoff coefficient, which is a function of soil type, vegetation, and surface topography. Table 11.3 shows typical values of C. Once the quantity of infiltrated water I is known via Eq. (11.1), the quantity percolated out of the soil cover into the waste can be estimated applying water conservation principles to the soil cover. Thus,

$$PER = I - AET - \Delta S_c \tag{11.3}$$

where PER = quantity of water percolated out of the soil cover, AET = actual evapotranspiration from the cover, and ΔS_c = change in storage of the cover as a result of infiltration. We make a distinction here between the potential evapotranspiration (PET) and actual evapotranspiration (AET). PET occurs when more than adequate moisture is available to meet the evaporative demand of the atmosphere. AET, on the other hand, is the actual amount of evapotranspiration that takes place when the soil is dry ($I < PET$), therefore the evaporation demand cannot be met. Thus, AET is always less than or equal to PET.

The monthly potential evapotranspiration may be estimated using the Thornthwaite equation,

$$PET\ (\text{mm}) = 16\left(\frac{10t}{TE}\right)^a \tag{11.4}$$

Table 11.3 Runoff Coefficients as Affected by Cover Material and Slope

	Runoff coefficient, C		
Type of area	Flat: slope <2%	Rolling: slope 2–10%	Hilly: slope >10%
Grassed areas	0.25	0.3	0.3
Earth areas	0.6	0.65	0.7
Meadows and pasture lands	0.25	0.3	0.35
Cultivated land:			
Impermeable (clay)	0.5	0.55	0.6
Permeable (loam)	0.25	0.3	0.35

Source: Perry (1976).

where t = temperature of the month under consideration, and TE = temperature efficiency index, given as the summation of the heat indices of the 12 months in the year. Considering that heat index for a given month is expressed as $(t/5)^{1.514}$, TE may be written as

$$TE = \sum_{i=1}^{12} \left(\frac{t_i}{5}\right)^{1.514} \tag{11.5}$$

In Eq. (11.4), a is an empirical coefficient given by

$$a = 6.75 \times 10^{-7}(TE)^3 - 7.7 \times 10^{-5}(TE)^2 + 1.79 \times 10^{-2}(TE) + 0.492\,39 \tag{11.6}$$

PET, as given by Eq. (11.4), also depends on the hours of sunlight, in addition to the temperature and the heat index. To account for the unequal durations of sunlight (daylight times) during the year, PET is usually multiplied by an adjustment factor. Values of this factor, listed in Table 11.4, depend on the month of the year and latitude of the location under consideration.

Equation (11.3) is implemented as follows to determine the percolation quantities. The quantity $(I - PET)$ is first calculated. If this quantity is positive, the evaporative demand by the atmosphere is met, and the surplus goes down into the soil cover and makes it wetter. In this case, $AET = PET$. If the water content of the soil cover is already high—say, at its field capacity—the soil can store no additional water and the surplus will percolate into the waste (note that the field capacity is the maximum water content that a soil can retain under gravitational draining). On the other hand, if $(I - PET)$ is negative, the evaporation demand by the atmosphere is not met, and $AET < PET$. The soil gives up its water content to the atmosphere, if it is wet, and gets dryer. The amount of drying depends not only on the magnitude of $(I - PET)$, but also on the water content of the soil cover. Under these conditions, no percolation occurs until there arises a situation when $(I - PET)$ is again positive and the topsoil cover is brought to its field capacity.

The amount of drying from the soil cover that takes place during the dry spell (when there is no percolation) depends on the type of soil and the cumulative water deficit. This is essentially a problem of flow through unsaturated soil. Thornthwaite and Mather (1957) provide tables with which one can estimate the soil moisture retention after evapotranspiration occurred from a dry soil. Table 11.5 provides an abbreviated form as documented in Oweis and Khera (1990). The soil moisture storage required in this table may be obtained from Fig. 11.2, which summarizes the water-holding characteristics of various soils.

Table 11.4 Adjustment Factors for Potential Evapotranspiration Computed by the Thornthwaite Equation

Latitude	Jan.	Feb.	Mar.	Apr.	May	June	July	Aug.	Sep.	Oct.	Nov.	Dec.
0	1.04	0.94	1.04	1.01	1.04	1.01	1.04	1.04	1.01	1.04	1.01	1.04
10	1.00	0.91	1.03	1.03	1.08	1.06	1.08	1.07	1.02	1.02	0.98	0.99
20	0.95	0.90	1.03	1.05	1.13	1.11	1.14	1.11	1.02	1.00	0.93	0.94
30	0.90	0.87	1.03	1.08	1.18	1.17	1.20	1.14	1.03	0.98	0.89	0.88
35	0.87	0.85	1.03	1.09	1.21	1.21	1.23	1.16	1.03	0.97	0.86	0.85
40	0.84	0.83	1.03	1.11	1.24	1.25	1.27	1.18	1.04	0.96	0.83	0.81
45	0.80	0.81	1.02	1.13	1.28	1.29	1.31	1.21	1.04	0.94	0.79	0.75
50	0.74	0.78	1.02	1.15	1.33	1.36	1.37	1.25	1.06	0.92	0.76	0.70

Source: Chow (1964).

Table 11.5 Soil Moisture Retention After Potential Evapotranspiration Has Occurred

ΣNEG $(I\text{-}PET)$[a]	S_r (mm)[b]								
	25	50	75	100	125	150	200	250	300
0	25	50	75	100	125	150	200	250	300
10	16	41	65	90	115	140	190	240	290
20	10	33	57	81	106	131	181	231	280
30	7	27	50	74	98	122	172	222	271
40	4	21	43	66	90	114	163	213	262
50	3	17	38	60	83	107	155	204	254
60	2	14	33	54	76	100	148	196	245
70	1	11	28	49	70	93	140	188	237
80	1	9	25	44	65	87	133	181	229
90	1	7	22	40	60	82	127	174	222
100		6	19	36	55	76	120	167	214
150		2	10	22	37	54	94	136	181
200		1	5	13	24	39	73	111	153
250			2	8	16	28	56	91	130
300			1	5	11	20	44	74	109
350			1	3	7	14	34	61	92
400				2	5	10	26	50	78
450				1	3	7	20	41	66
500				1	2	5	16	33	56
600					1	3	10	22	40
700						1	6	15	28
800						1	4	10	20
1000							1	4	10

[a] $NEG(I\text{-}PET)$ is lack of infiltration water needed for vegetation.
[b] S_r, soil moisture storage at field capacity.
Source: Oweis and Khera (1990).

The final estimate of leachate, q, coming out of the refuse, may now be obtained by applying water conservation principle to the waste body. Thus,

$$q = PER + W_d - \Delta S_w \tag{11.7}$$

where W_d = water generated from waste decomposition, and ΔS_w = change in moisture storage of the waste. Until the waste attains its field capacity, no leachate may be expected at its bottom. The percolation will first satisfy the moisture-holding capacity of the waste and the surplus will then leach out of the waste. The water-holding capacity of waste is highly variable, as shown in Table 11.6 for some common types of refuse. Depending on the permeability of the barrier

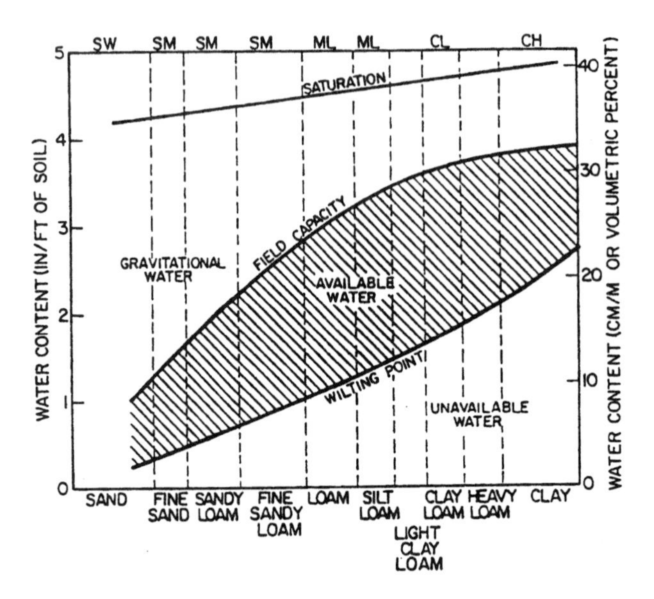

Figure 11.2 Water storage capacities of USDA soils (Lutton et al. 1979).

at the bottom of the refuse, a leachate mound may form, which provides a hydraulic head for leakage through the barrier.

As discussed in the preceding paragraphs, the monthly water balance method requires accounting of how the precipitation is apportioned, one month at a time. The simple expressions involved in the calculations make it suitable for spreadsheet programming. Detailed case studies using this method are given by Fenn et al. (1975), Lutton et al. (1979), Lu et al. (1985), and Oweis and Khera (1990).

11.3.2 Hydrologic Evaluation of Landfill Performance (HELP)

Hydrologic Evaluation of Landfill Performance (HELP) is a computer program developed by the U.S. Army Waterways Experiment Station, which was designed for water budget modeling of waste containment systems. Details of the program implementation, applications, and its limitations are provided by Schroeder et al. (1984a, 1984b) and the U.S. EPA (1991). Only an outline of this method is provided here, in order to highlight the important factors affecting leachate quantity estimates. The program has become popular during recent years primarily be-

Table 11.6 Water Absorption Ranges for Solid Waste Components

	Moisture content, percent dry weight					
	Water absorption capability			Total moisture-holding capability[a]		
Component	Maximum	Average	Minimum	Maximum	Average	Minimum
Newsprint[b]		290			290[c]	
Cardboard (solid and corrugated)[b]		170			170[c]	
Other miscellaneous paper	400		100	400[c]		100[c]
Lawn clippings (grass and leaves)	200		60	370		140
Shrubbery, tree prunings	100		10	250		10
Food waste (kitchen garbage)	100		0	300		0
Textiles (cloth of all types, rope)	300		100	300[c]		100[c]
Wood, plastic, glass, metal (all inorganic)		0			0	

[a] Calculated from water absorption plus initial moisture content in as-received samples.
[b] Sample variation was negligible.
[c] Initial moisture contents as-received were less than 6% in the laboratory tests; therefore, they were considered negligible compared to the variation in moisture absorbed.
Source: Stone (1974).

cause of its ability to incorporate a number of layers in a waste containment system and also to account for the lateral drainage in individual layers. Besides providing accurate estimates of water budget components, HELP is useful in evaluating and comparing alternative containment systems.

A typical hazardous-waste containment profile that HELP is capable of simulating is shown in Figure 11.3. The model uses methods similar to those described in the previous section. It apportions the precipitation into several water budget components as the water percolates through each of the layers. These components include not only surface runoff, evapotranspiration, and water content changes in the soils, but also the lateral drainage collected in each drain system, and percolation through each layer of the system. It can generate daily, monthly, annual, and long-term average water budgets. A number of physical and hydrologic methods are used in HELP to simulate individual water budget components. A short summary of these methods is given below. Prior understanding of these methods is essential to interpret the simulation results correctly.

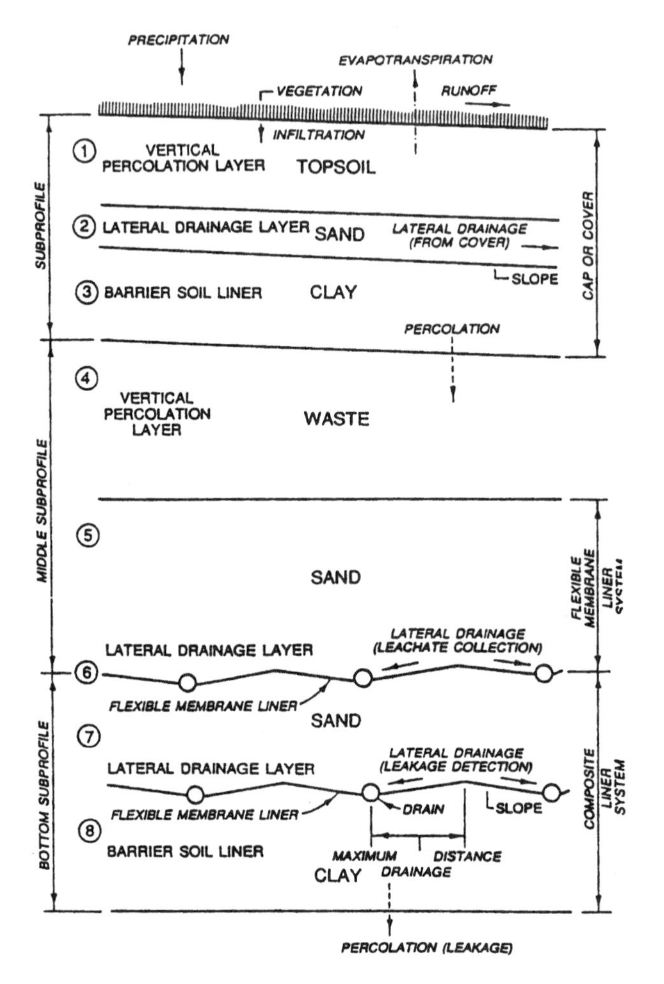

Figure 11.3 Waste containment profile that can be simulated in HELP (U.S. EPA, 1991).

Infiltration

Infiltration is estimated indirectly by subtracting the runoff and surface evapotranspiration from rainfall and snowmelt. Runoff and surface evapotranspiration are treated as functions of interception. Snowmelt is computed using a modified version of the simple degree-day method with 0°C as the base temperature (USDA, 1972). Interception is modeled after the work of Horton (1919). It approaches a maximum value exponentially as the rainfall increases to about 0.5

cm. The maximum interception is a function of the quantity of aboveground biomass or leaf area index and is limited to 0.13 cm. The intercepted water evaporates from the surface and decreases the evaporative demand of the atmosphere. The Soil Conservation Service (SCS) curve number method is used to estimate surface runoff, as presented by the USDA (1972). The SCS curve number method is an empirical method developed for small watersheds (12–200 hectares) with mild slopes (3–7%). It correlates daily runoff with daily rainfall for watersheds with a variety of soils, types of vegetation, land management practices, and antecedent moisture conditions.

Evapotranspiration

Evapotranspiration is treated as a summation of three components: evaporation of water from the surface, from the soil, and from the plants. Each of the components is computed separately. Evaporation of water from the surface is limited to the smaller of the potential evapotranspiration and the sum of the snow storage and interception. A modified version of the Penman method is used to compute potential evapotranspiration (Ritchie, 1972). To compute evaporation from soil, a two-stage square-root-of-time routine is used (Ritchie, 1972). In stage one, the soil evaporation equals the evaporative demand placed on the soil. Demand is based on energy and is equal to the potential evapotranspiration discounted for surface evaporation and shading from ground cover. A vegetative growth model is used to compute the total quantity of vegetation, both active and dormant. In stage two, low soil moisture and low rates of water vapor transport to the surface limit evaporation from the soil. Soil evaporation during this state is treated as a function of the square root of the time period during which the soil has been in dry condition. Transpiration from the plants is estimated following the methods of the CREAMS and SWRRB models (Knisel, 1980; Williams, et al., 1985). According to these models, the potential plant transpiration is a linear function of the potential evapotranspiration and the active leaf area index.

Subsurface Water Routing

Subsurface water routing processes modeled by HELP include vertical unsaturated drainage, percolation through saturated soil liners, leakage through geomembranes, and lateral saturated drainage. The unsaturated vertical drainage is computed assuming unit hydraulic gradient. The unsaturated hydraulic conductivity is calculated using the Campbell equation, which is based on the Brooks and Corey model (see Chapter 3). Percolation through saturated layers is calculated using Darcy's law and is assumed to occur only when there is a zone of saturation directly above the liner. Leakage through geomembranes is modeled as a reduction of the cross-sectional area of flow through the subsoil below the geomembrane. The rate of flow through the leaking subsoil is computed as the percolation rate through a saturated barrier soil liner. This method seems to provide good results for composite liners but not for a single geomembrane. The lateral drainage is simulated using a steady-state analytical approximation of the Boussinesq equation (McEnroe and Schroeder, 1988). The lateral drainage, percolation, and leakage through the geomembrane are solved simultaneously with an average depth of saturation using an implicit solution technique.

In order to simulate the above processes, the following input data are required by HELP:

1. Climatological data—daily precipitation, daily mean temperature, daily solar evaporation, maximum leaf area index, growing season, and evaporative zone depth
2. Soil and design data—porosity, field capacity, wilting point, and saturated hydraulic conductivity of each layer, SCS runoff curve number, surface area, number of layers in the profile, and thicknesses of the layers

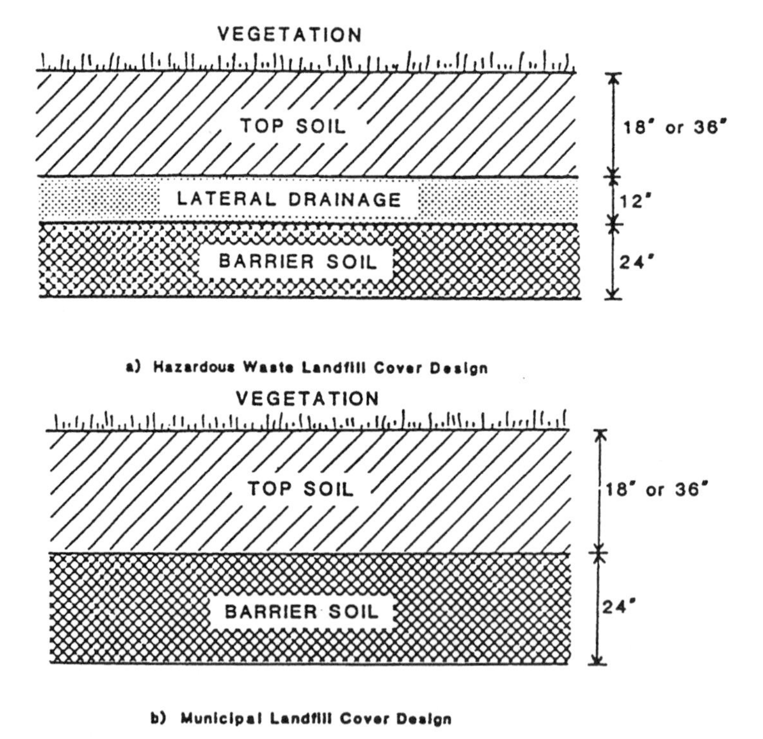

Figure 11.4 Two types of landfill covers tested for water balance using HELP (U.S. EPA, 1991).

The program includes default daily weather data for 102 U.S. cities and is capable of synthetically generating weather data for 183 U.S. cities. In addition, default data are included for 15 soil types as well as solid waste. Default SCS curve numbers are also provided. The output from the HELP consists primarily of percolation or leakage through each layer and depth of saturation on the surface of liners. Incremental and cumulative qualities of water budget for the various components are also provided.

The HELP model allows one to look at the sensitivity of the water balance to various design variables. Case studies were reported by the U.S. EPA (1991) demonstrating the sensitivity analyses. The climatologic regimes at three different locations (Santa Maria, CA; Schenectady, NY; and Shreveport, LA) were used to study the water balance for two typical landfill covers (Fig. 11.4). Two types of topsoils with different thicknesses were also studied. The results from the simulations are shown in Tables 11.7–11.9. Table 11.7 shows the effects of climate and vegetation with and without lateral drainage. It is seen that vegetation decreases runoff and increases evapotranspiration but tends to have little effect on the rest of the water balance. A three-layer cover design (which allows for lateral drainage) is definitely superior to a two-layer design, indicating that the design of the cover is far more important than the climatologic and vegetation factors. It should be noted that although vegetation is shown to have little effect

Table 11.7 Effects of Climate and Vegetation on Water Balance

	Two-layer cover design[a] location			Three-layer cover design[b] location		
	CA	LA (% precipitation)	NY	CA	LA (% precipitation)	NY
Poor grass						
Runoff	5.6	4.6	5.5	3.0	4.4	2.2
Evapotranspiration	51.8	53.0	52.1	51.6	51.9	50.3
Lateral drainage	—	—	—	41.2	40.6	44.0
Percolation	42.6	42.4	42.4	4.2	3.1	2.5
Good grass						
Runoff	3.1	0.2	3.5	0.0	0.2	0.0
Evapotranspiration	55.0	57.2	55.3	52.6	53.0	51.0
Lateral drainage	—	—	—	43.2	43.7	45.5
Percolation	42.9	42.6	41.2	4.2	3.1	2.5

[a] 900 mm of sandy loam topsoil and 0.6 m of 10^{-6} cm/s clay liner.
[b] 450 mm of sandy loam topsoil, 300 mm of 0.03 cm/s sand with 60 m drain length at 3% slope, and 0.6 m of 10^{-7} cm/s clay liner.
Source: U.S. EPA (1991).

Table 11.8 Effects of Topsoil Thickness on Water Balance

	45.7 cm of topsoil location			91.4 cm of topsoil location		
	CA	LA (% precipitation)	NY	CA	LA (% precipitation)	NY
Runoff	11.2	7.5	13.4	5.6	4.6	5.5
Evapotranspiration	51.9	56.9	54.5	51.8	53.0	52.1
Percolation	36.9	35.6	32.1	42.6	42.4	42.4

Sandy loam topsoil with a poor stand of grass underlain by 0.6 m of 10^{-6} cm/s clay liner.
Source: U.S. EPA (1991).

on percolation in these simulations, its importance lies elsewhere, in the context of erosion prevention.

Table 11.8 shows the effects of topsoil thickness on the water balance for the cover design where lateral drainage is absent. The effects of topsoil thickness are similar for all three locations. Runoff and evapotranspiration were greater for the thinner topsoil, indicating that the head above the barrier maintained higher moisture contents in the evaporative zone. The percolation was consequently less than for cases with greater topsoil thickness. Although thin topsoil is in general favorable for reduced percolation, it is important to provide adequate thickness for the top cover to support vegetation, maintain soil stability, and control erosion.

Table 11.9 shows the effects of topsoil type on water budget components. Clayey topsoil increased both runoff and evapotranspiration, which in turn greatly decreased lateral drainage and percolation. Considering the California site, runoff

Table 11.9 Effects of Topsoil Type on Water Budget Components

	Three-layer cover design[a]					
	Sandy loam location			Silty clayey loam location		
	CA	LA (% precipitation)	NY	CA	LA (% precipitation)	NY
Runoff	3.0	4.4	2.2	21.6	22.3	19.2
Evapotranspiration	51.6	51.9	50.3	61.2	64.4	58.6
Lateral drainage	41.2	40.6	44.0	15.0	11.3	20.3
Percolation	4.2	3.1	2.5	2.2	2.0	1.9

[a] 450 mm of topsoil with poor stand of grass, 300 mm of 0.03 cm/s sand with 60 m drain length at 3% slope, and 0.6 m of 10^{-7} cm/s clay liner.
Source: U.S. EPA (1991).

increased from 3% to 22% of the precipitation whereas evapotranspiration increased from 52% to 61% of precipitation, when the topsoil is changed from sandy loam to a silty clayey loam. The lower hydraulic conductivity of clayey topsoil slowed down the lateral drainage, which dropped from about 41% to 15% of precipitation. Because of these effects, percolation differed only slightly, from 4% to 2%, between the two soil types.

11.4 LEACHATE COLLECTION AND REMOVAL SYSTEMS (LCRS)

Flow of leachate through barriers can be minimized in two ways—by minimizing the leachate accumulated at the top of the barrier via a collection system, and by designing the barrier such that its permeability is very low. Design of the leachate collection system is an important element in the overall design of the waste containment system. It includes sizing and spacing of the pipes used to collect and remove the leachate. The total leachate percolated out of the waste is apportioned between the leachate collection system and the liner leakage. Thus the problem of leachate collection above the liner and that of flow through the barrier is an integrated one and cannot be decoupled. In practice, however, simpler methods are used to predict the height of the leachate mound, assuming either a constant leakage rate through the liner or no leakage at all. Leakage through the liner is often ignored for a conservative design of leachate collection and treatment systems.

The maximum height of leachate mound above the barrier depends on how the collection system is laid out. Figure 11.5 shows typical configurations for the collection system. The configuration shown in Figure 11.5a offers a simple expression for maximum height, h_{max}, of the leachate mound. Applying Darcy's law to estimate flow in the drainage layer, the height of the mound, $h(x)$, can be estimated as (see Table 3.5)

$$h(x) = \left[\left(\frac{q}{k_d} \right) (L - x) x \right]^{0.5} \tag{11.8}$$

where q = percolation rate of leachate coming onto the drainage layer, k_d = saturated hydraulic conductivity of the drainage layer, and L = drain spacing. The maximum height of the mound occurs at the midpoint between the two drains; therefore,

$$h_{max} = \frac{L}{2} \left(\frac{q}{k_d} \right)^{0.5} \tag{11.9}$$

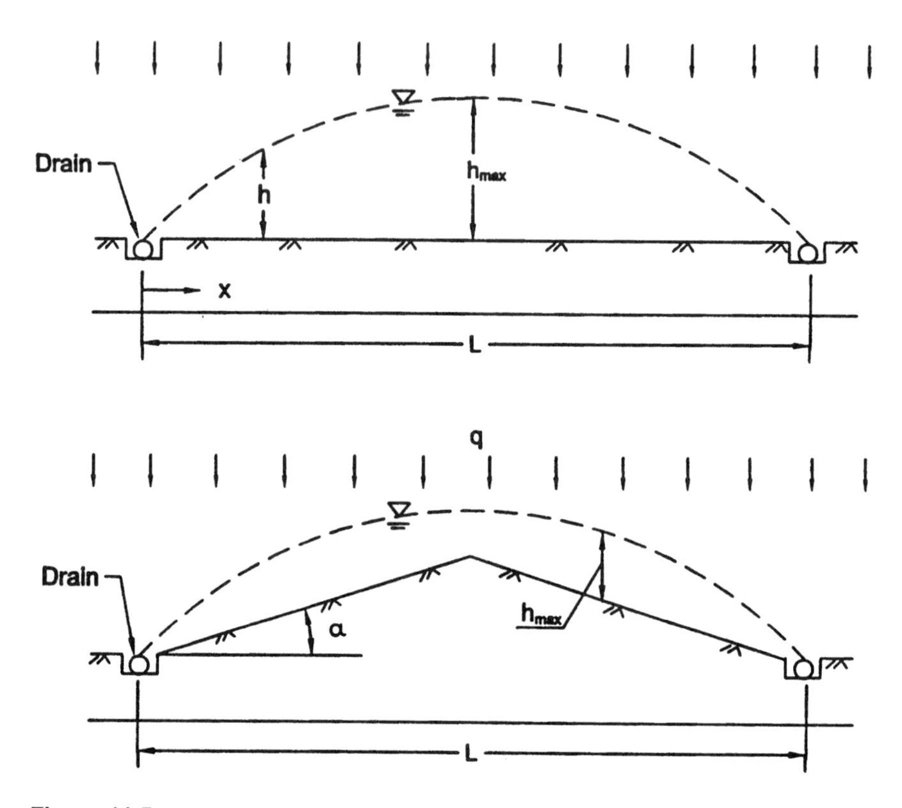

Figure 11.5 Leachate collection system configurations.

From a design standpoint, the drains must be capable of discharging (qL) per unit length, since the percolation over one-half of the drain spacing flows from either side of each drain. When one puts constraints on h_{max}, in order to ensure that the mound lies totally within the drainage layer, Eq. (11.9) allows estimation of the required spacing L or hydraulic conductivity k_d.

Figure 11.5b illustrates a more common configuration of a collection system, where the barriers are typically graded to enhance flow toward leachate collection drains. Moore (1980) developed a solution for this case and expressed $h_{\max}$ as (see Table 3.5)

$$h_{\max} = \left(\frac{L\sqrt{c}}{2}\right)\left(\frac{\tan^2 \alpha}{c} + 1 - \frac{\tan \alpha}{c}\sqrt{\tan^2 \alpha + c}\right) \tag{11.10}$$

where $c = q/k_d$, and α = slope angle of the liner. As stated earlier, Eq. (11.10) may also be used to obtain the required design parameters given h_{max}.

Other configurations for collections systems were also studied; for instance, McBean et al. (1982) considered a sloping collection system as in Figure 11.5b but with drains at the apex of the barrier as well. However, the closed-form expressions for these configurations (shown in Table 3.5) become more complicated than those presented above.

It should be noted that Eqs. (11.9) and (11.10) represent only approximate solutions, since the leakage through the barrier was not considered in their development. When the barrier leaks a portion of q, say q_i, the actual amount of leachate collected in the drains is equal to $(q - q_i)L$. Methods that couple leakage through the barrier and flow in the drainage layer are available in the literature (Wong, 1977; Dematracopoulos et al., 1984; Korfiatis and Demetracopoulos, 1986; McEnroe and Schroeder, 1988). Wong (1977) developed some of the earlier equations enabling apportionment of leachate between drainage collection and leakage through the barrier. For the landfill system shown schematically in Figure 11.6, Wong made the following assumptions to simplify the mathematical treatment.

1. The drainage layer and the refuse-cover mixture above the liner have the same conductivity.
2. All layers are at field capacity, so that any infiltration of precipitation into the refuse results in gravity drainage to the bottom of the landfill.
3. The percolation of leachate results in an instantaneous input onto the clay liner, so a saturated volume of rectilinear shape is formed.
4. The saturated volume above the liner retains its rectilinear shape as drainage toward the drains and leakage through the barrier take place.

Under these assumptions, Wong arrived at the following equations to describe the movement of leachate:

$$\frac{s}{s_0} = 1 - \left(\frac{t}{t_1}\right) \tag{11.11}$$

$$\frac{h}{h_0} = \left(1 + \frac{d}{h_0 \cos \alpha}\right) e^{-\kappa t/t_1} - \frac{d}{h_0 \cos \alpha} \qquad 0 \le t \le t_1 \tag{11.12}$$

where

$$t_1 = \frac{s_0}{k_1 \sin \alpha} \tag{11.13}$$

$$\kappa = \left(\frac{s_0}{d}\right)\left(\frac{k_2}{k_1}\right)\cot\alpha \tag{11.14}$$

s = length of saturated volume of soil at time t (see Fig. 11.6); h = thickness of saturated volume at time t; s_0, h_0 = initial dimensions of saturated volume; k_1 = hydraulic conductivity of material above the liner; k_2 = hydraulic conductivity of the clay liner; α =slope angle of the liner; and d = thickness of the liner. Note that the equations are applicable as long as s/s_0 and h/h_0 are positive values and the time is measured from the instant when the saturated volume appears on the liner.

The solution expressed in Eqs. (11.11) and (11.12) is shown in Figure 11.7. The information shown in Figure 11.7 may be used to apportion the leachate volume as follows. At any time t, the volume that remains saturated is $V = sh\cos\alpha \approx sh$, since $\cos\alpha \approx 1.0$. Therefore, the rate of volume change is

$$\frac{dV}{dt} = h\frac{ds}{dt} + s\frac{dh}{dt} \tag{11.15}$$

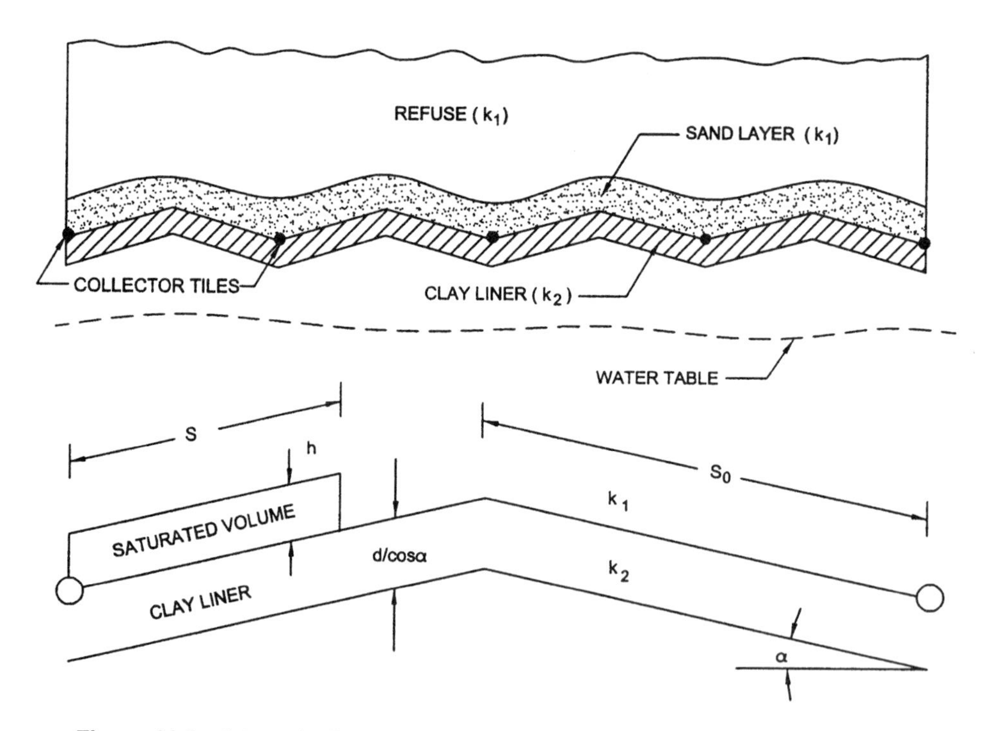

Figure 11.6 Schematic diagram of landfill model showing parameters relevant to leachate collection system (Wong, 1977).

The first term on the right-hand side of Eq. (11.15) denotes the rate at which leachate is drained to the collection pipe and the second term denotes the rate at which saturated volume decreases due to leakage through the liner. Integrating these two terms separately and normalizing with respect to the initial saturated volume $V_0(= s_0 h_0)$,

$$\frac{V_1}{V_0} = \int \frac{h}{h_0} d\left(\frac{s}{s_0}\right) \tag{11.16}$$

and

$$\frac{V_2}{V_0} = \int \frac{s}{s_0} d\left(\frac{h}{h_0}\right) \tag{11.17}$$

where V_1 = volume drained away by the collection system and V_2 = volume leaked through the liner. Equations (11.16) and (11.17) imply that the area *SDBC* in Figure 11.7 represents V_1/V_0, the fraction of the initial saturated volume drained

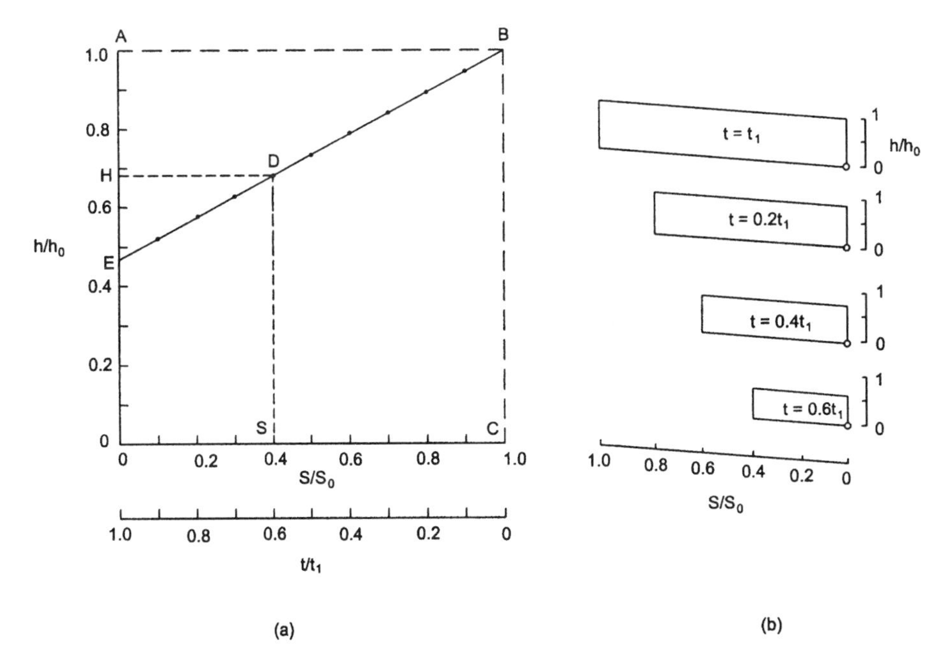

Figure 11.7 Solution for the coupled barrier leakage and leachate collection problem: (a) h/h_0 as a function of t/t_1 and S/S_0; (b) schematic diagram showing the physical dimensions of saturated volume on the liner at various times. (From Wong, 1977.)

away by the leachate collection system, while the area *ABDH* represents V_2/V_0, the fraction of the initial saturated volume that has leaked through the liner. Thus, Figure 11.7 serves as a graphical solution for the apportionment of initial saturated volume at time $t = 0.6t_1$ (point *S*). The area *OHDS* represents the fraction of the initial saturated volume that still remains on the liner. V_1/V_0 is often referred as the collection efficiency of the drains. Thus, the closer the point *E* is to point *A*, the greater is the collection efficiency. In reality, *E* will never coincide with *A* because of the leakage through the liner. From a design standpoint, however, the parameters involved in Eq. (11.12) must be chosen such that *E* is as close to *A* as possible. Despite the numerous assumptions made in Wong's model, it is one of the useful models available at present to apportion leachate. A detailed evaluation of the model and the effects of various design parameters on the apportionment are provided by Wong (1977) and Kmet et al. (1981).

11.5 FLOW AND TRANSPORT THROUGH BARRIERS

As stated earlier, the apportionment of leachate between the barrier leakage and the collection system is a single problem. However, it is customary to separate the leachate collection system design from the barrier design. While designing the barrier, the head of leachate mound is assumed to be constant, to simplify the flow problem. For a conservative estimate of leakage through the barrier, the leachate mound may be assumed to be at its maximum elevation, since the leakage estimate will be higher under this assumption.

When considering the leakage through the barrier, it is important to keep in mind that both quality and quantity of leakage govern the impact of the waste containment system on the surrounding soil and groundwater resources. Thus, a significant leakage of ''relatively clean'' water might be more acceptable than an insignificant leakage of highly contaminated water. Earlier regulations of waste containment systems were based on fixing the quantity of leakage through the barrier, viz., the leakage rate should not exceed 1.65 in./year, the hydraulic conductivity of the barrier should not exceed 1×10^{-7} cm/s, etc. As such, these quantity-based criteria value the advective transport of contaminants through the barrier with little regard to the effects of dispersion mechanism. More recent criteria for the performance of waste containment systems consider all the mass transport mechanisms. The performance criteria for waste containment systems may thus be studied under two broad categories:

1. *Quantity-based criteria*, which limit the overall leakage rate through the barrier, perhaps by fixing the hydraulic conductivity and the thickness of the barrier

2. *Quality-based criteria*, which limit the exit leachate concentrations or fluxes during the design life-period of the system

We will consider these two by considering the flow and transport phenomena through barriers separately.

11.5.1 Flow Through Barriers

For a simple configuration of a soil barrier underlain by subsoil (Fig. 11.8), Darcy's law can be used to estimate the flow rate. The pore velocity through the barrier, V_a, will be maximum under fully saturated conditions; therefore,

$$V_a = \frac{k_s i}{n} \tag{11.18}$$

where k_s = saturated hydraulic conductivity of the barrier [L/T], i = hydraulic gradient (dimensionless), and n = porosity of the barrier (dimensionless). The transit time through the barrier, t, may therefore be expressed as

$$t = \frac{d}{V_a} = \frac{dn}{k_s i} \tag{11.19}$$

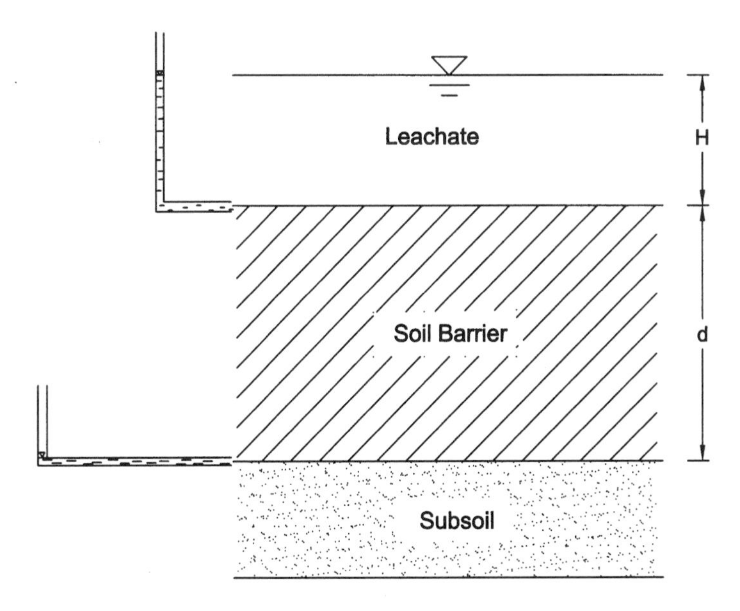

Figure 11.8 Schematic of a soil barrier layer underlain by a subsoil.

where d = thickness of the barrier [L]. From a design standpoint, if transit time t is regulated, then the required thickness of the barrier is

$$d = \frac{k_s i t}{n} \tag{11.20}$$

Caution should be exercised in identifying the gradient i. While the pressure head at the top of the liner may be well defined, the condition at the bottom of the liner is uncertain. A zero-pressure head is often assumed at the bottom of the liner. Under these conditions, the hydraulic gradient is

$$i = \frac{H + d}{d} \tag{11.21}$$

where H = height of leachate mound on the barrier [L]. The expanded expressions for t and d then become

$$t = \frac{dn}{k_s}\left(\frac{d}{H + d}\right) \tag{11.22}$$

and

$$d = 0.5\left\{\left(\frac{k_s t}{n}\right) + \left[\left(\frac{k_s t}{n}\right)^2 + \left(\frac{4k_s tH}{n}\right)\right]^{0.5}\right\} \tag{11.23}$$

The above equations are valid for barriers under fully saturated conditions. Although it may be conservative from a design perspective to assume that the barrier is fully saturated, it may sometimes be necessary to estimate the transit time through an initially unsaturated barrier. The Green-Ampt wetting-front model is suitable for this purpose. This is a phenomenological model based on the concept of a sharp wetting front, shaped like a square wave, moving down through the unsaturated soil (Fig. 11.9). Above the wetting front, the soil is assumed to be fully saturated; and below it, the moisture content is assumed to be at its initial level. The energy required for the movement of the wetting front is provided by the water suction below the front. The flow rate per unit area of the barrier, q, may be obtained using Darcy's law and the mass conservation principle. Thus,

$$q = k_u\left[\frac{(H + L + H_d)}{L}\right] \tag{11.24}$$

$$= (w_s - w_i)\frac{dL}{dt} \tag{11.25}$$

where k_u = unsaturated hydraulic conductivity at the wetting front [L/T], L = location of the wetting front [L], H_d = capillary suction head below the wetting

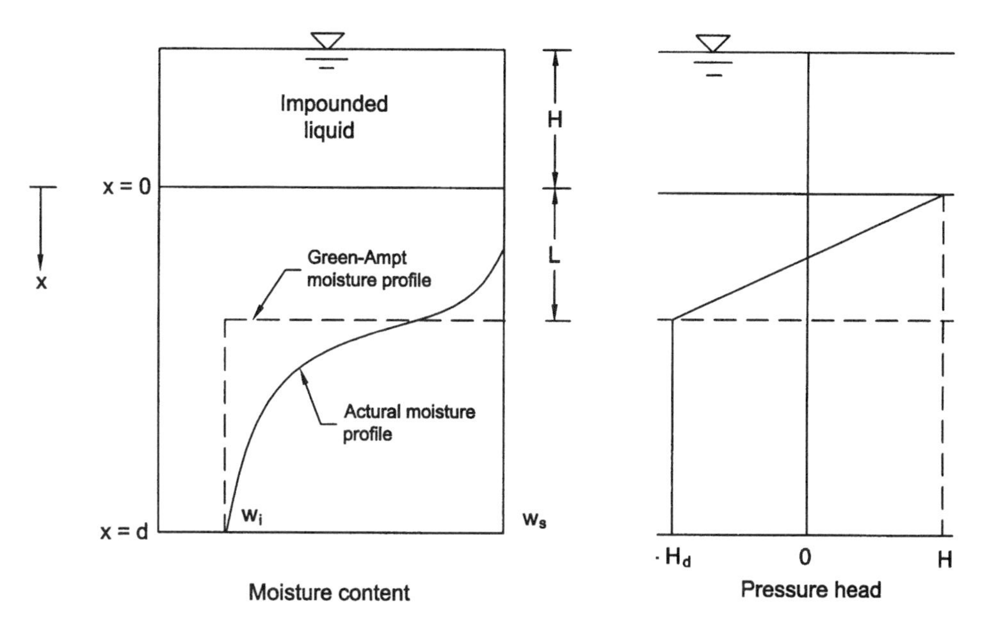

Figure 11.9 Representation of moisture profile in the barrier using Green-Ampt wetting front model.

front [L], w_s = saturated moisture content of the barrier soil (dimensionless), w_i = initial moisture content of the barrier soil, and t = time [T]. Combining Eqs. (11.24) and (11.25), and integrating by parts over $L = 0$ to d, the transit time t may be expressed as

$$t = \left(\frac{w_s - w_i}{k_u}\right)\left[d - (H + H_d)\ln\left(1 + \frac{d}{H + H_d}\right)\right] \quad (11.26)$$

It should be noted that the above approach is a simple idealization of unsaturated flow and as such is only approximate. More accurate estimates of transit time through the barrier are possible only through numerical solution of the unsaturated flow equations described in Chapter 3.

11.5.2 Mass Transport Through Barriers

As discussed in Chapter 4, mass transport through soils is governed mainly by two processes, advection and dispersion. The advection component is governed by the flow rate, discussed in the preceding section. If only advection controls the transport process, the leachate flux f may be obtained as

$$f = VC_0 \tag{11.27}$$

where f = leachate flux [M/L^{2}T], V seepage velocity [L/T], and C_0 = source concentration [M/L^3] at the top of the barrier, assumed to be constant. Taking the other extreme position that dispersion is the only process governing the leachate flux, f may be obtained using Fick's first law as

$$f = -nD\frac{\partial C}{\partial L} = nD\frac{C_0}{d} \tag{11.28}$$

where D = dispersion coefficient [L^2/T], which accounts for both molecular diffusion and hydrodynamic dispersion, and $\partial C/\partial L$ is the concentration gradient, which is assumed to be steady under a constant concentration of C_0 at the top and zero concentration at the bottom of the barrier. It should be noted that the concentrations at the boundaries may be such that steady-state concentration gradients may never exist. For instance, in the cases where the bottom of the barrier is not flushed with clean water, the outlet concentrations will keep on increasing until they approach the inlet concentration C_0. Beyond this stage, the concentration gradients (and consequently the diffusion process) cease to exist in the liner. Also, it may not be appropriate to assume that the inlet concentrations C_0 are constant, since the contaminant mass is finite and the source is exhausted as the leachate travels out of the barrier. Even if the boundary conditions at the inlet and outlet of the barrier are favorable for steady concentration gradients, a significant time period elapses before such steady-state conditions occur.

To account for time-dependent transport of contaminants through the barrier, Fick's second law may be used. According to this law,

$$\frac{\partial C}{\partial t} = D\frac{\partial^2 C}{\partial x^2} \tag{11.29}$$

where D = dispersion coefficient, which accounts for both molecular diffusion and hydrodynamic dispersion. The solution for Eq. (11.29), discussed in Chapter 4, is expressed as

$$\frac{C}{C_0} = \operatorname{erfc}\left(\frac{x}{2\sqrt{Dt}}\right) \tag{11.30}$$

This solution is shown schematically in Figure 11.10. Under pure dispersion conditions, Eq. (11.30) may be used to obtain the time it takes for the concentrations at the outlet of the barrier to reach a specified value.

When both advection and dispersion are to be considered in the mass transport modeling, the appropriate solutions discussed in Chapter 4 for the advection-dispersion equation may be used. The solution commonly employed is that of Ogata and Banks (1961), which gives concentrations C as a function of time t

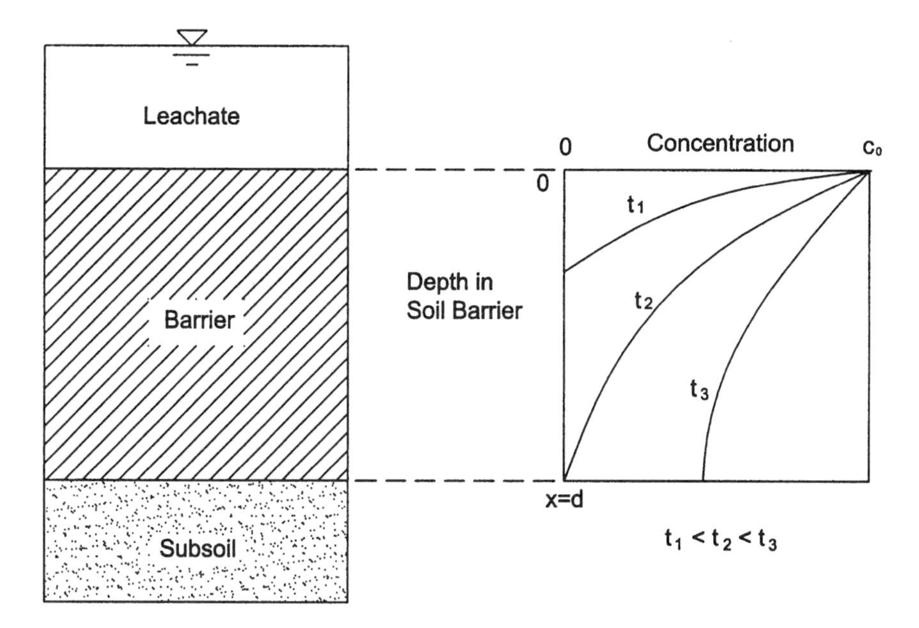

Figure 11.10 Schematic profiles of leachate concentrations at various times in the soil barrier.

and depth x beneath the surface of a barrier, when it is assumed to be infinitely deep and subject to the following initial and boundary conditions:

$$C(x, 0) = 0 \qquad x > 0 \tag{11.31a}$$

$$C(0, t) = C_0 \qquad t \geq 0 \tag{11.31b}$$

$$C(\infty, t) = 0 \qquad t \geq 0 \tag{11.31c}$$

In the absence of decay, the solution to the advection-dispersion equation is the same as Eq. (4.70), which is rewritten here.

$$C(x, t) = \frac{C_0}{2}\left[\operatorname{erfc}\left(\frac{Rx - V_x t}{2\sqrt{RD_x t}}\right) + \exp\left(\frac{V_x x}{D_x}\right)\left\{\frac{Rx + V_x t}{2\sqrt{RD_x t}}\right\}\right] \tag{4.70}$$

This solution is often expressed in terms of dimensionless variables as follows:

$$C(x, t) = \frac{C_0}{2}\left[\operatorname{erfc}\left(\frac{1 - T}{2\sqrt{T/P}}\right) + \exp(P)\left(\frac{1 + T}{2\sqrt{T/P}}\right)\right] \tag{11.32}$$

where

$$T = \frac{V_x t}{Rx} = \left.\frac{V_x t}{Rd}\right|_{\text{at } x = d} \tag{11.33}$$

and

$$P = \frac{V_x x}{D_x} = \left.\frac{V_x d}{D_x}\right|_{\text{at } x = d} \tag{11.34}$$

T indicates effective pore volumes of flow, and P is the Peclet number defined with respect to the barrier thickness d.

Once the concentrations $C(x, t)$ are known at the outlet of the barrier, the total flux f may be obtained by summing the advection and dispersion components. Shackelford (1990) expressed the total flux in a dimensionless form using a dimensionless flux number, F_N, as follows:

$$F_N = \frac{fd}{nC_0 D_x} = \frac{1}{2}(PQ_1 + Q_2) \tag{11.35}$$

where

$$Q_1 = \text{erfc}\left(\frac{1 - T}{2\sqrt{T/P}}\right) \tag{11.36a}$$

and

$$Q_2 = \frac{2 \exp\left\{-\left[(1 - T)/2\sqrt{T/P}\right]^2\right\}}{\sqrt{\pi T/P}} \tag{11.36b}$$

To aid in the design process, Eqs. (11.32) and (11.35) are reduced to a graphical form (Shackelford, 1990). Using these graphs, $C(x, t)$ and F_N may be obtained for specified values of T and P. Shackelford (1990) suggests an iterative procedure using these solutions to obtain a design thickness of the barrier. The procedure involves the following steps:

1. Assume a liner thickness, d.
2. Calculate P using Eq. (11.34).
3. Determine T for a specified value of either C/C_0 or f using Eq. (11.32) or (11.35).
4. Use Eq. (11.33) to determine the transit time t.
5. Repeat steps 1–4 until t is greater than or equal to the design life of the containment system.

The above solutions are valid for simplified initial and boundary conditions described in Eq. (11.31). The boundary condition $C = C_0$ at the inlet is not an accurate one, since it represents discontinuity at the liquid–soil interface and violates the mass conservation principle. Solutions for more realistic boundary conditions are given by Van Genuchten and Alves (1982). However, the solutions given above are fairly accurate with differences up to only 5% when compared with the more accurate solutions (Gershon and Nir, 1969).

Booker and Rowe (1987) considered the effect of finite nature of contaminant source on the mass transport through barrier. In contrast to the boundary condition represented in Eq. (11.31b), they assumed that the finite mass of contaminant within the landfill could be represented in terms of an equivalent height of leachate, H_f. The analytical solution obtained by Booker and Rowe (1987) for contaminant transport in an infinitely deep deposit as the source concentration varied with time is as follows:

$$C(x, t) = \frac{C_0 \exp(ab - b^2 t)}{(b - c)}[bf(b, t) - cf(c, t)] \tag{11.37}$$

where

$$f(b, t) = \exp(ab + b^2 t) \operatorname{erfc}\left(\frac{a}{2\sqrt{t}} + b\sqrt{t}\right) \tag{11.38}$$

$$f(c, t) = \exp(ac + c^2 t) \operatorname{erfc}\left(\frac{a}{2\sqrt{t}} + c\sqrt{t}\right) \tag{11.39}$$

$$a = x\left(\frac{n + \rho_b K_d}{nD_x}\right)^{1/2} \tag{11.40}$$

$$b = V_x\left[\frac{n}{4D_x(n + \rho_b K_d)}\right]^{1/2} \tag{11.41}$$

$$c = \frac{nD_x}{H_f}\left[\frac{n + \rho_b K_d}{nD_x}\right]^{1/2} - b \tag{11.42}$$

where ρ_b and K_d are dry mass density and distribution coefficient, respectively. The function $f(p, t)$ (for $p = b$ or $p = c$) may be simplified as

$$f(p, t) = \exp\left(\frac{-a^2}{4t}\right)\phi(\alpha) \tag{11.43}$$

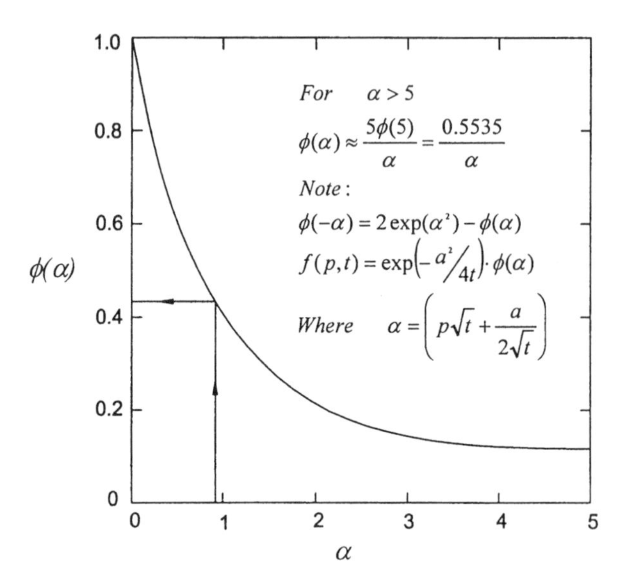

Figure 11.11 $\phi(\alpha)$ as a function of α in Booker and Rowe's solution.

where $\alpha = (p\sqrt{t} + a/2\sqrt{t})$ and the function $\phi(\alpha)$ is given in Fig. 11.11. Note that with some algebraic manipulation, it could be shown that Eq. (11.37) reduces to Eq. (4.70) for $H_f \rightarrow \alpha$ (i.e., $c = -b$).

11.5.3 Relative Importance of Advection and Dispersion

Considering the impermeable nature of barrier materials, it is tempting to ignore advection process and assume that only dispersion governs the leachate transport. However, several studies illustrated that neglecting advection may lead to erroneous and unconservative estimates of leachate flux for typical landfill scenarios (Rowe, 1987; Shackelford, 1988). Figure 11.12, for instance, shows the concentration profiles in and chemical flux through a 1.2-m clay liner with and without advective component of mass transport. The seepage velocity (0.006 m/year) considered in this study is representative of clay liners with hydraulic conductivities less than 1×10^{-7} cm/s and subjected to hydraulic gradients less than 0.2. It is seen in the figure that not only are the concentrations in the liner greater when advection is considered, the concentration gradients (which govern the leachate fluxes) at the bottom of the liner are also greater. In general, as the seepage velocity is increased with all other factors kept constant, the concentrations in the liner as well as the fluxes exiting the liner are increased. For the same conditions, Figure 11.12b illustrates the time-dependent fluxes exiting the liner under

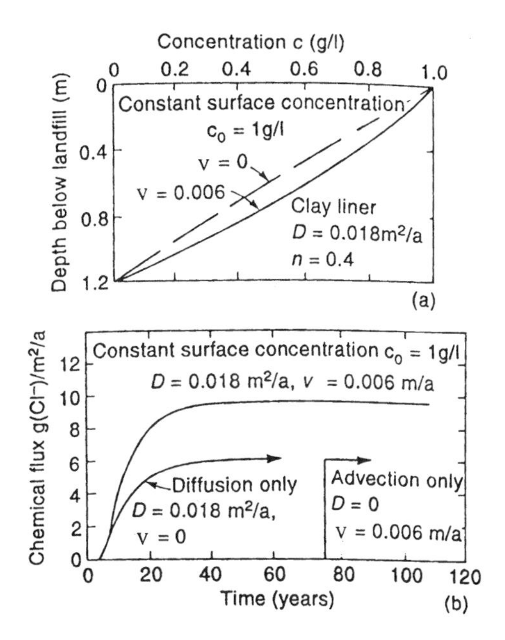

Figure 11.12 Mass transport through clay liners: (a) steady-state concentration profiles in a 1.2-m thick clay liner; (b) time-dependent leachate fluxes exiting the liner. (From Rowe, 1987).

pure advection, diffusion, and combined advection and diffusion. It is seen that consideration of pure advection alone resulted in an estimate of 75 years for the flux to exit the liner (the same time as it takes for the seepage front to arrive at the bottom under plug-flow conditions). This estimate is considerably greater than what pure diffusion conditions suggest. Consideration of both advection and diffusion gave a substantially higher flux, with the peak flux being 55% greater than that obtained under pure diffusion or advection.

Calculations such as those shown in Figure 11.12 are valuable in assessing the relative significance of advection and diffusion in the evaluations of transit time and exit flux. Using such calculations, Rowe (1987) arrived at Figure 11.13, which shows the range of velocities over which dispersion or advection may control the magnitude of exit flux for a 1.2-m-thick clay liner. The results shown in Figure 4.2 were used to assess the relative dominance of molecular diffusion and mechanical dispersion over a range of seepage velocities. Shackelford (1988) used similar calculations to show the relative significance of the transport mechanisms in the assessment of transit times (Fig. 11.14). It is clearly demonstrated in this study that diffusion shortens the transit time even for hydraulic conductivi-

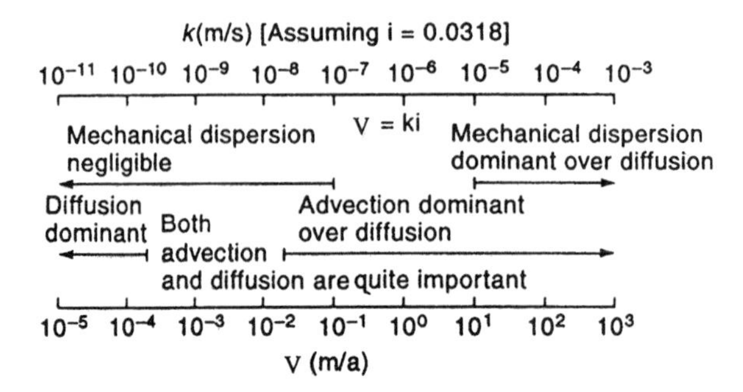

Figure 11.13 Range of Darcy velocities over which diffusion, mechanical dispersion, and advection control the magnitude of the exit flux (Rowe, 1987).

ties of the order of 10^{-7} cm/s. It becomes a dominant transport process at hydraulic conductivities less than 2×10^{-8} cm/s.

It should be noted that the directions in which mass transport occurs due to advection and dispersion may not always be the same. Figure 11.15 shows three transport scenarios, which warrant consideration when assessing the transit

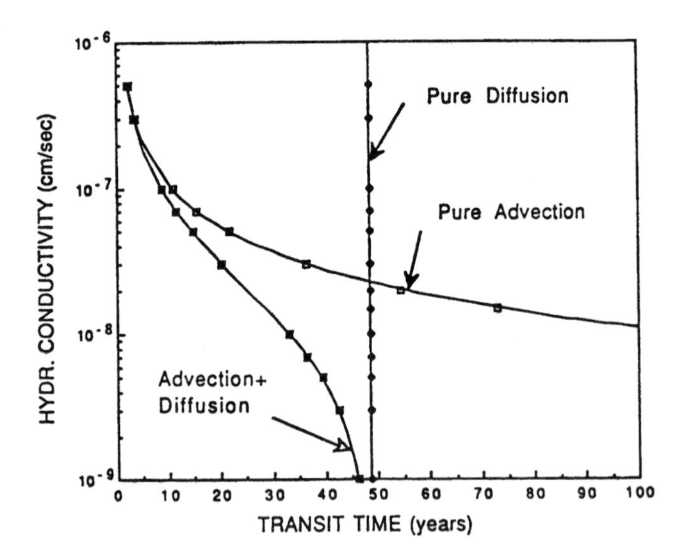

Figure 11.14 Transit times corresponding to $C/C_0 = 0.5$ as a function of hydraulic conductivity for a hypothetical 3-ft clay liner with an average porosity of 0.50 and subjected to a hydraulic gradient of 1.33. A diffusion coefficient of 6×10^{-6} cm^2/s was assumed. (From Shackelford, 1988.)

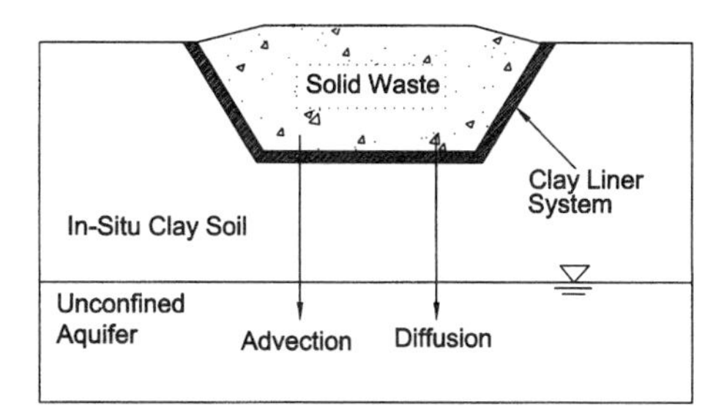

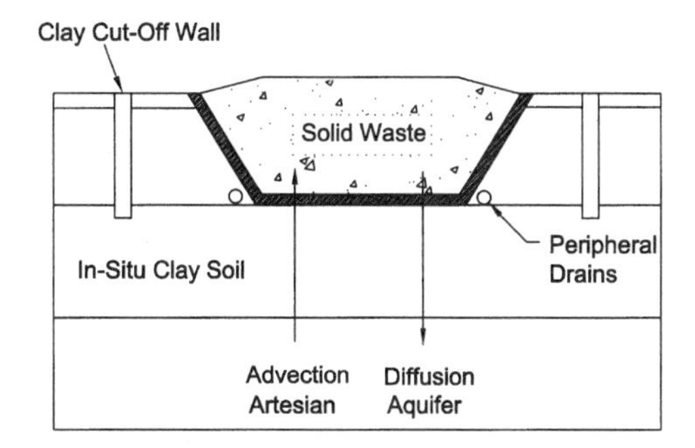

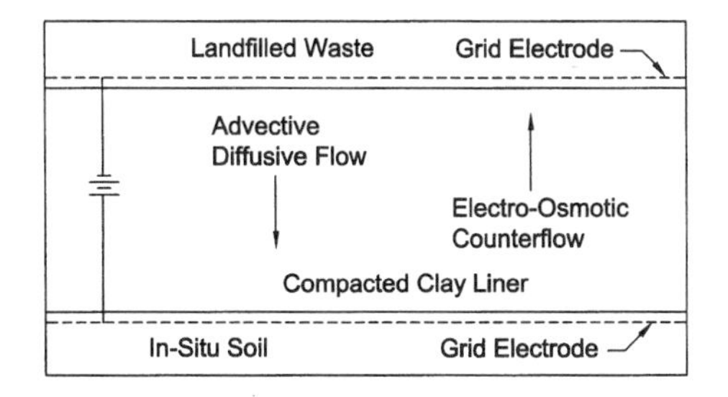

Figure 11.15 Transport scenarios involving different directions of advection and diffusion.

times and exit fluxes. Figure 11.15a shows the situation where both advection and diffusion follow the same direction and the above discussion is valid. Figure 11.15b illustrates a more favorable scenario, where the hydrogeology at the containment system is such that flow is directed into the waste. An example is the placement of the containment system below the groundwater table and controlling the flow into the waste using an internal leachate collection system. The advection process in this case opposes the outward diffusion flux of contaminants. Gradient reversal may also be achieved using a continuous or periodic application of electric gradients across the barrier (Fig. 11.15c). An electroosmotic counterflow is generated to oppose the flow due to advection and dispersion. Gradient reversals are often viewed as attractive methods in dealing with cracks and macropores in the barriers, which may participate in the counterflow and prevent outward leakage. Evaluation of transit time and exit flux magnitudes in these cases requires a good handle on the mechanisms that cause gradient reversals.

11.6 STABILITY OF WASTE CONTAINMENT SYSTEMS

In addition to minimizing the flow of leachate into the surrounding soil and groundwater, a waste containment system should retain its structural integrity during its design life. The structural stability of the containment system as a whole or of its individual components controls its key function of containing the waste. For instance, excessive settlements of the landfill material might disrupt the top cover of the system. The settlements of foundation soils might damage the leachate collection system and bottom liner, and allow leachate to exit the system freely. Similarly, slope failures of the structure, either local or systemic, would lead to catastrophic consequences.

It should be noted that the conditions that are favorable for low hydraulic conductivity may not necessarily be favorable from a stability point of view. For instance, the compaction parameters leading to low hydraulic conductivity of compacted clays may not be favorable for high interface strengths between the clays and geomembranes (Seed and Boulanger, 1991; Stark and Poeppel, 1994). Thus, in the design phase of a waste containment system, both flow and stability should be given equal importance, and construction parameters acceptable from both viewpoints must be chosen.

Stability analyses for waste containment systems are complicated primarily because of the following reasons.

1. The properties of the waste are both spatially and temporally variable. It is difficult to obtain representative samples and conduct routine geotechnical tests to determine their properties. Some studies even suggest that a Mohr-Coulomb representation of shear strength, commonly used in geotechnical engineering, may not be appropriate for waste materials (Mitchell et al., 1995).
2. A given containment system may contain components made up of a number of different materials, which are diverse in their stress–strain properties as shown in Figure 11.16. The differences in these properties create compatibility problems at the interfaces between various components. The deformations required to mobilize the peak strengths of the individual materials could be significantly different.

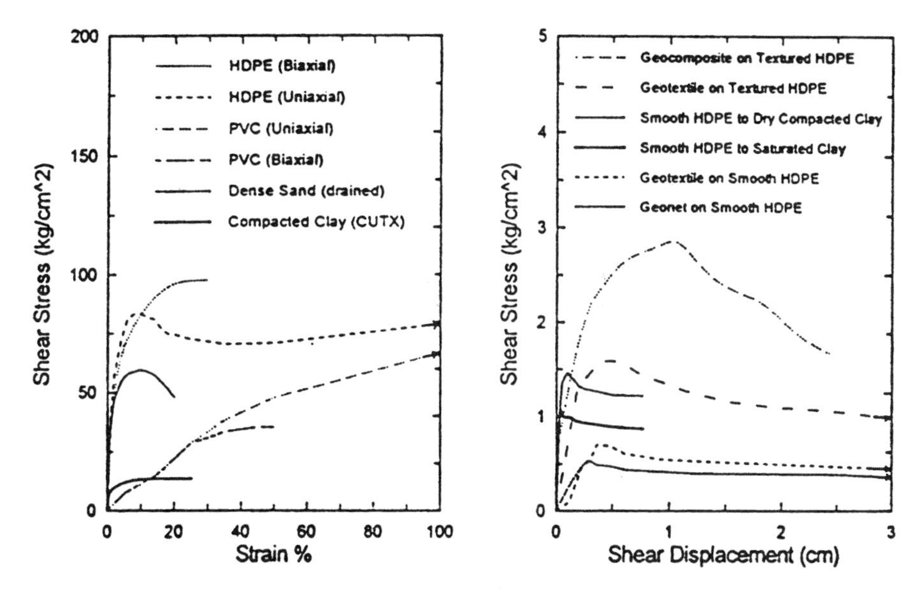

Figure 11.16 Diverse stress–strain properties of material components in a containment system (Mitchell et al., 1995).

3. The modes of failure are not well established. For instance, slope failure for a containment system may not follow the rotational mode commonly used in geotechnical engineering.

Because of these complexities, stability analyses for waste containment structures have not been well established. Initial efforts involve extrapolation of existing analyses available for pure soils in geotechnical engineering. Three issues of stability are of broad importance in the case of waste containment systems:

1. Local and systemic failures of slopes
2. Stability of the interfaces between various components, primarily between geotextile/geomembrane and soil
3. Total and differential settlements of the system

A detailed discussion of these issues is provided in the following subsections.

11.6.1 Slope Stability

Slope failures in containment systems fall under three broad categories, as shown in Figure 11.17. The failures may occur either within the waste material or across

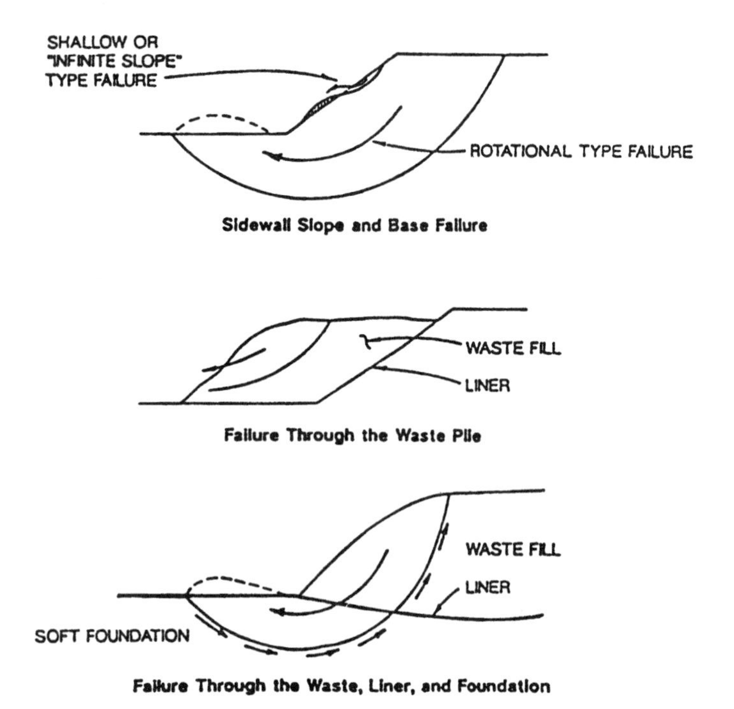

Figure 11.17 Potential slope failure modes in landfills (Mitchell et al., 1995).

the foundation soil and soil–liner interfaces. The mode of failure is uncertain in these types of failures, but generally, a rotational type of failure shown in the figure is considered to be the most probable mode. With this idealization, the stability analysis will reduce to the common geotechnical problem of assessing the stability of a natural slope or a man-made embankment.

Traditional methods of slope stability analysis evaluate shear stresses causing failure along an assumed surface and compare them with the shear strength mobilized. The ratio of shear strength to shear stress is taken as the factor of safety against failure. In the method of slices, commonly used to assess the stability of finite slopes, the mass of soil is divided into a number of segments and the forces on each segment are evaluated (Fig. 11.18). Considering the slice *abcd*, for instance, the forces consist of the weight of the slice, the normal and shear forces acting on the failure surface *cd*, and the normal and shear forces acting on the vertical faces *ad* and *bc*. An approximate solution is possible when certain assumptions are made concerning the forces on the vertical faces. Assuming that the resultants of P_i and T_i are equal to P_{i+1} and T_{i+1}, the normal and shear forces on surface *cd* may be obtained as

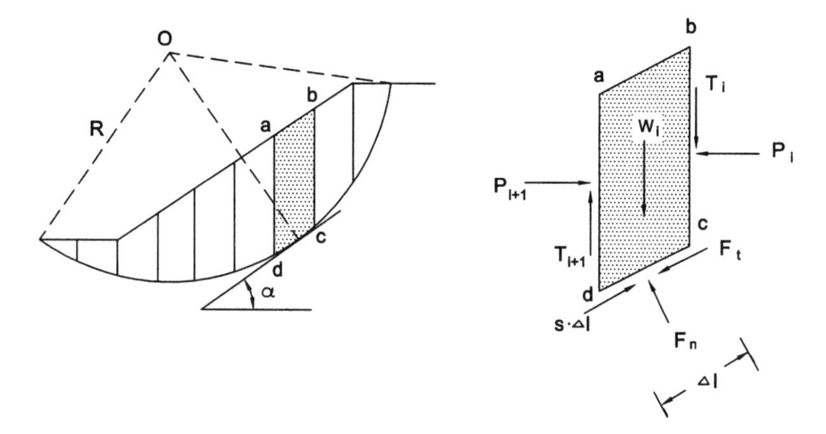

Figure 11.18 Slope stability analysis by the method of slices.

$$F_n = W_i \cos \alpha \tag{11.44}$$

and

$$F_t = W_i \sin \alpha \tag{11.45}$$

where W_i = weight of the *i*th slice. The shearing resistance s per unit area on the segment *cd* may be expressed using Mohr-Coulomb theory as

$$s = c + \sigma \tan \phi \tag{11.46}$$

where c = cohesion, σ = normal stress on *cd*, and ϕ = friction angle at the interface. The total shear strength available on the surface segment *cd* may therefore be written as

$$s\,\Delta l = (c + \sigma \tan \phi)\,\Delta l = c\,\Delta l + W_i \cos \alpha \tan \phi \tag{11.47}$$

where Δl = length of the segment *cd*, and F_n from Eq. (11.44) is substituted for $(\sigma \Delta l)$. The factor of safety, *FS*, against slope failure may now be expressed as a ratio of the moments about point *O* of the shear stresses F_t and those of shear strengths $(s\,\Delta l)$ for all slices. Thus,

$$FS = \frac{\sum Rs\,\Delta l}{\sum RF_t} = \frac{\sum (C\,\Delta l + W_i \cos \alpha \tan \phi)}{\sum W_i \sin \alpha} \tag{11.48}$$

The preceding analysis offers a simple and approximate way to assess the stability of a slope; for more accurate analyses, the reader is referred to standard textbooks on geotechnical engineering (Terzaghi et al., 1996; Wu, 1976).

11.6.2 Stability of Interfaces

Stability of the interfaces is a major consideration for systems that contain layers of soil and polymeric materials such as geotextiles and geomembranes. In general, low interface strengths are considered to be the weak link in the stability of containment systems (Mitchell et al., 1995). As shown in Figure 11.19, failure at the interface may occur either as a pullout of the fabric component from the anchor trench or as a slide-out of the waste and cover materials from the fabric.

As a simple case, consider a soil cover that rests on a polymer fabric (Fig. 11.20). The factor of safety FS against sliding of the soil cover on the fabric may be computed as the ratio of the resisting force to the driving force. Thus,

$$FS = \frac{N \tan \delta}{W \sin \theta} = \frac{W \cos \theta \tan \delta}{W \sin \theta} = \frac{\tan \delta}{\tan \theta} \tag{11.49}$$

where W = weight of the soil, N = reaction force normal to the sliding plane, θ = angle of inclination of the sliding plane, and δ = friction angle at the interface. An inherent assumption of this analysis is that the soil layer is of infinite extent. A more realistic analysis should include a finite extent of failure surface. This is possible using the translational wedge method. In this method, the failure surface is approximated by a number of planar segments. The soil mass above each segment forms a potential sliding wedge. Figure 11.21 shows a generalized case where seepage forces exist in the soil layer above the interface and a reinforcement is provided in the cover with a tensile resistance of magnitude, T. The

Pullout of Liner System Components from Anchor Trenches

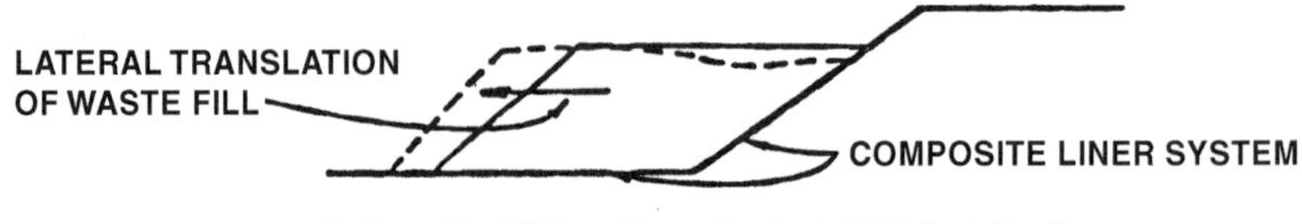

Failure By Sliding Along the Landfill Liner System

Figure 11.19 Potential interfacial failure modes (Mitchell et al., 1995).

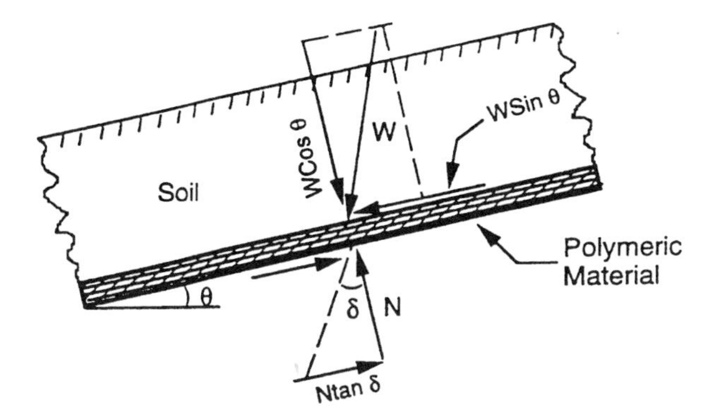

Figure 11.20 Forces at the interface between a polymer fabric and soil cover.

failure zone may consist of three wedges. The active wedge at the top of the slope tends to push the central wedge downward, whereas the passive wedge at the bottom of the slope tends to resist this movement. The angles α and β may be chosen such that the driving force of the active wedge is maximized and the resisting force of the passive wedge is minimized. Considering the equilibrium of the active and passive wedges, angles α and β may be expressed as (Oweis, 1993)

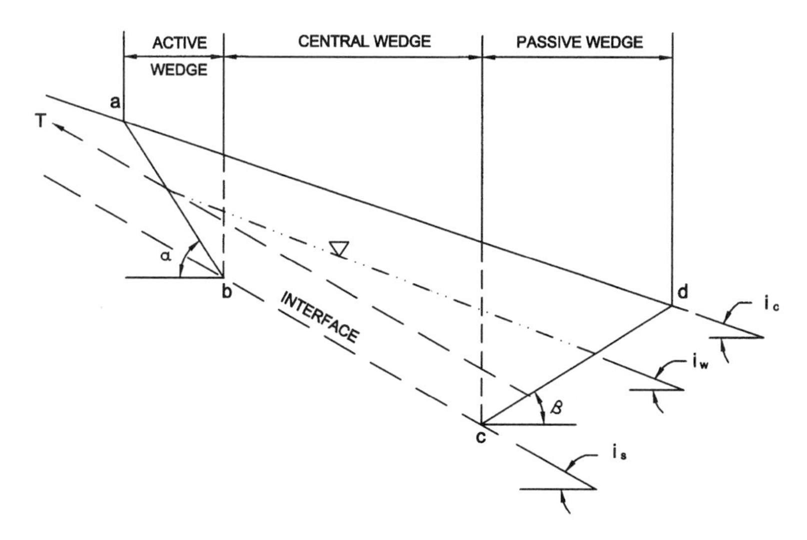

Figure 11.21 Stability of the soil cover using translational wedge method (Oweis, 1993).

$$\tan \alpha = \tan \phi + \sqrt{1 + \tan^2 \phi - \tan i_c / \sin \phi \cos \phi} \tag{11.50}$$

and

$$\tan \beta = -\tan \phi + \sqrt{1 + \tan^2 \phi - \tan i_c / \sin \phi \cos \phi} \tag{11.51}$$

Note that for a horizontal surface ($i_c = 0$), the angles reduce to $\alpha = 45 + \phi/2$, and $\beta = 45 - \phi/2$, which must be familiar to students of geotechnical engineering. The factor of safety against sliding at the interface may be obtained by considering the force balance of the central wedge. The forces acting on the central wedge are essentially the active and passive forces on the sides of the wedge, the weight of the wedge, and the resistance available at the failure surface *bc*. For the simplest, and also the most practical, case of a soil cover of uniform thickness with a groundwater level parallel to the interface, the stability equation for the central wedge will yield the factor of safety as (Oweis, 1993)

$$FS = \frac{1}{(1 - t)} \left[\frac{c}{\gamma h \sin i_s \cos i_s} + \frac{\tan \delta_i}{\tan i_s} \left(1 - \frac{\gamma_w h_w}{\gamma h} \right) \right] \tag{11.52}$$

where

$$t = \frac{T}{W_c \sin i_s} \tag{11.53}$$

W_c = weight of the central wedge
c = cohesion of the soil
γ = unit weight of the soil
h = thickness of the cover
δ_i = angle of friction at the interface
γ_w = unit weight of water
h_w = uniform thickness of the saturated layer of the cover.

Stability analyses for the soil cover are also available for other simplified conditions of the failure surface. For instance, Giroud and Beech (1989) and the U.S. EPA (1991) document analyses for the condition of a horizontal failure surface in the passive wedge of the soil. These analyses are useful from a design standpoint to determine the thickness of the soil cover, and whether and how much tensile reinforcement is required to achieve a desired factor of safety against slope failure.

Another important consideration regarding the interfacial stability is the compatibility of frictional stresses developed at the top and bottom faces of a geomembrane. Typically, geomembranes are located at the interface of two different materials, and the differences in frictional stresses may cause excessive shearing stresses in the geomembrane. The tensile strength of the geomembrane should be such that it can withstand these differences with an adequate factor of safety.

11.6.3 Settlements

There are significant differences between soil and waste materials in terms of the settlement mechanisms. Settlement analyses for waste materials are complicated because of the inherent heterogeneity, their decomposition, and short-term and long-term environmental conditions affecting the composition of the materials. The mechanisms involved in waste settlements are several (Edil et al., 1990; Sowers, 1973): (1) particle reorientation; (2) raveling, or movement of fine particles into large voids; (3) physicochemical changes including corrosion, combustion, and oxidation of the waste materials; (4) biochemical decomposition; (5) dissolution of soluble materials; and (6) plastic flow and creep. Edil et al. (1990) noted that refuse fills typically settle from 5% to 30% of their original thickness under self-weight, and the magnitude of the settlement depends on factors such as degree of initial compaction, waste composition, and environmental conditions. The primary settlement is generally known to occur relatively rapidly, between 1 month and 5 years after the waste placement. Unlike the case of pure soils, the secondary settlements are known to be relatively significant.

Although initial efforts (Sowers, 1973; Oweis and Khera, 1990) relied on one-dimensional consolidation theory to approximate settlements of waste fills [Eqs. (1.16)–(1.18)], application of the theory to waste materials is generally established to be of limited validity. This is because the e versus log σ curve may exhibit nonlinearity, and several of the parameters required, such as compression and rebound indices, change with time and cannot be determined adequately (Fassett, et al., 1993). Edil et al. (1990) evaluated two models, the Gibson-Lo model and the power creep law, using compressibility data from five sites. The former is a rheological representation of the soil using a spring to represent primary compression and a parallel dashpot to represent secondary compression. The time-dependent settlement based on the Gibson and Lo model may be expressed as

$$S(t) = H\,\Delta\sigma\left\{a + b\left[1 - \exp\left(\frac{-\lambda}{bt}\right)\right]\right\} \tag{11.54}$$

where

$S(t)$ = settlement at time t
H = initial height of waste material
$\Delta\sigma$ = change in overburden pressure

a = primary compressibility parameter
b = secondary compressibility parameter
$\frac{\lambda}{b}$ = rate of secondary compression

The power creep law is useful to represent the creep behavior of materials and is generally expressed as

$$S(t) = H\,\Delta\sigma\, m\left(\frac{t}{t_r}\right)^n \tag{11.55}$$

where

m = reference compressibility
n = compression rate
t_r = reference time
t = time since load application

Based on the limited data, Edil et al. (1990) reported that the Gibson and Lo model estimated settlements within 2% and 20% of actual settlements, and the power creep law estimated settlements within 0–14%.

An important effect of the waste settlement is to deform the cover system and strain the polymeric barriers. Estimation of these deformations is difficult and necessarily involves assumptions related to settlement trough geometry and waste-volume reduction rate. The soil arching theory and the tensioned membrane theory may be used to estimate stresses that develop in geomembranes placed within soil covers overlying void spaces (Giroud et al., 1988; Knipshield, 1985). Knipshield (1985) developed a semiempirical procedure for estimating the design ratio (analogous to the factor of safety) for geomembranes that settle. The design ratio (DR) is estimated as

$$DR = \frac{\gamma_r}{\gamma_u} \tag{11.56}$$

where

γ_r = geomembrane rupture strain (%)
γ_u = uniform strain on the geomembrane (%)

The rupture strain, γ_r, of the geomembrane is determined from a load versus strain curve as shown in Figure 11.22. The uniform strain, γ_u, may be obtained from Figure 11.23. The settlement, S, may be estimated as a fraction of the thickness of the waste, and the width of the waste storage cell may be used as the width of the settled area ($2L$ in Fig. 11.23). As seen in the figure, the uniform

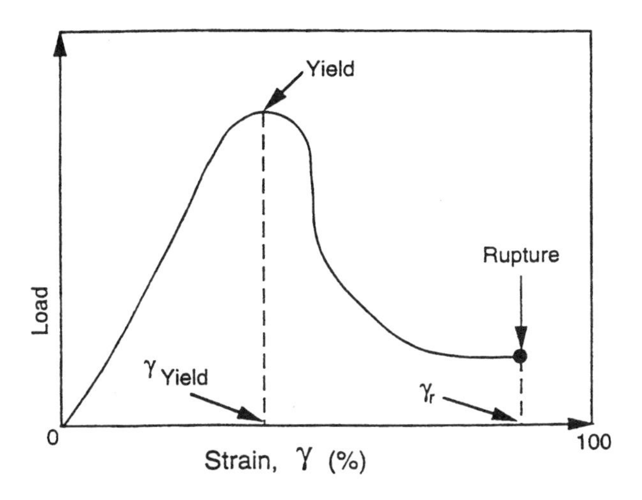

Figure 11.22 A typical load–strain curve for geomembranes.

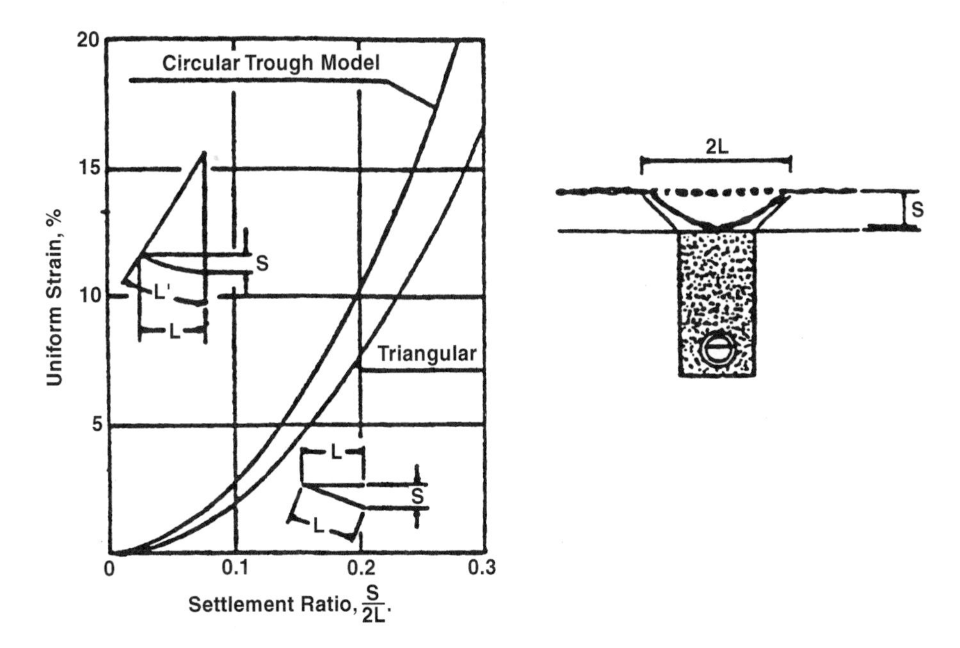

Figure 11.23 A graphical method to estimate uniform strain under triangular and circular trough assumptions (Knipshield, 1985).

strain estimate is dependent on the approximation of the trough geometry, triangular or circular.

11.6.4 Geotechnical Properties of Waste

The accuracy of the stability analyses outlined above is limited by the uncertainty and variability involved in the properties of waste. Several attempts have been reported in the literature to determine the engineering properties of waste. The properties of materials required in stability analyses of waste containment systems may be discussed under four categories:

1. Unit weight of waste
2. Shear strength of waste
3. Shear strength properties at the interfaces between various components
4. Compressibility of the waste

The unit weight of waste materials varies within a broad range, because of several factors: (1) type and origin of the waste; (2) mode of operation of the containment system and placement techniques; (3) degree of compaction used; (4) age of the waste after placement, which governs the environmental effects such as biochemical decomposition; and (5) depth at which waste is located in the system. In general, the unit weight of the waste is believed to increase with depth in the containment system (Fassett et al., 1993). The ranges of unit weight values reported in the literature are summarized in Table 11.10. It is important to note that these values were obtained based on a limited number of site and waste-specific observations, and as such, are useful only in preliminary evaluations.

As mentioned earlier, it is difficult to assess the shear strength of waste materials because of the questionable validity of the Mohr-Coulomb theory itself. However, due to lack of stress–strain characterizations of the materials, the industry continues to adopt this theory as an approximate guide in the design of containment systems. A number of studies have been reported in the literature that involve determination of cohesion and friction angle of waste materials. The parameters were estimated using one of three approaches: (1) direct laboratory testing on waste materials using direct shear and triaxial tests (Fang et al., 1977; Earth Technology Corporation, 1988); (2) back-calculation from field tests (Converse, Davis, Dixon Associates, 1975); and (3) in-situ testing of the materials using vane shear test and by correlating SPT values to strength properties (Earth Technology Corporation, 1988). Based on a summary of the various studies, Singh and Murphy (1990) documented the range of recommended strengths of municipal solid waste in terms of a relationship between cohesion and friction angle (Fig. 11.24). Due to the large scatter in the data, it is advisable to use the lower bound of the shaded area in preliminary investigations. Similar to the municipal waste, mineral

Table 11.10 Summary of Refuse Unit Weights Reported in the Literature

Refuse placement condition	Unit weight (kN/m^3)	Reference
Sanitary refuse: range for various compaction efforts	4.72–9.42	Sowers (1968)
Refuse landfill (refuse-to-soil cover ratio varied from 2:1 to 10:1)	7–14	Landva and Clark (1990)
Fills for poor compaction	3–9	Fassett et al. (1993)
Fills for moderate compaction	5–8	
Fills for good compaction	9–10.5	
Municipal waste shredded or baled	2.83–10.5	Oweis and Khera (1990)
Municipal refuse:		
In landfill	6.91–7.54	NSWMA (1985)
After degradation and settlement	9.9–11	
Sanitary landfill		
Not shredded		
Poor compaction	3.1	U.S. Department of the Navy (1983)
Good compaction	6.29	
Best compaction	9.43	
Shredded	8.64	

waste has also been shown to exhibit significant variability in the properties of cohesion and friction angle.

The shear strength properties at the interface usually reflect the weakest link in the stability of the waste containment systems. However, studies addressing these properties are relatively few. The friction angles between sand and geotextile/geomembrane are typically lower than those of pure sands. Studies by Martin et al. (1984) showed that the friction angle for sand–geotextile/geomembrane interfaces ranged from 17° to 30° based on tests conducted on three types of sands and four types of geomembranes and geotextiles. In the same studies, it was shown that the friction angle between geomembranes and geotextiles varied within a broad range of 6° to 28°. Mitchell et al. (1990) presented extensive set of results from their failure investigations on Kettleman Hills landfill unit in California. Their results indicated that the geosynthetic–geosynthetic interfaces possess friction angle varying from 6° to values greater than 20°. The interface friction angle seemed to depend on the type of geosynthetic combination, dry or submerged conditions, polishing of interfaces as a result of prior slip, and the orientation of the geonets relative to the direction of sliding along a geomembrane.

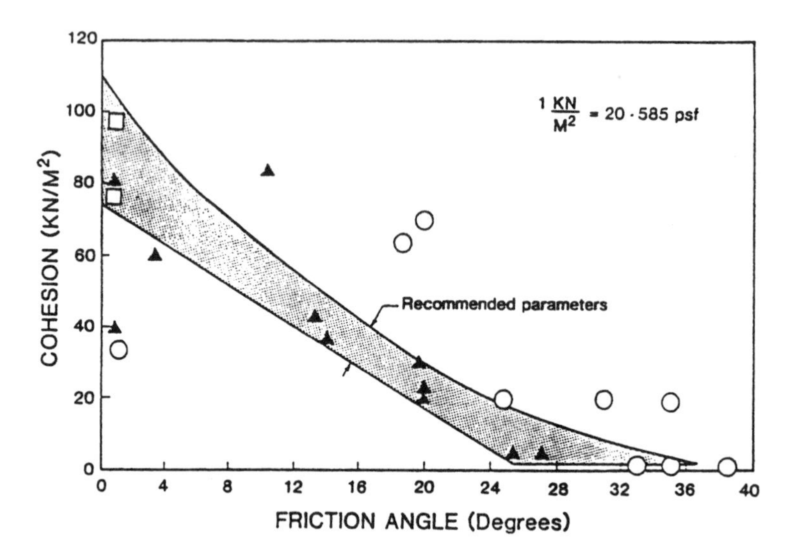

Figure 11.24 Recommended strength parameters for municipal refuse (Singh and Murphy, 1990).

The strength of interfaces between compacted clay and polymeric materials depends on the compaction conditions. Tests conducted on HDPE-compacted clay interfaces showed that the strength, represented using equivalent friction angle, varied greatly over the range of compaction conditions (Seed and Boulanger, 1991). Three zones may be identified over the range of compaction conditions as shown in Figure 11.25. It is important to note that the compaction conditions of Zone I, which are favorable to yield low hydraulic conductivities, yield the lowest strengths. Conversely, the compaction conditions of Zone III, which are not favorable to yield low hydraulic conductivities, yield highest strengths. These results reveal that fulfillment of flow and stability criteria may often conflict with each other during selection of compaction conditions for the liners.

Similar to the properties discussed above, the compressibility of waste materials varies within a wide range. Most of the studies reported in the literature used one-dimensional consolidation theory and obtained values for compression and rebound indices to characterize the compressibility of waste. In general, the indices are believed to depend on organic matter content and the environmental conditions, which may or may not promote decomposition of the waste. For instance, Sowers (1968, 1973) documented that the compression index of refuse (equivalent to C_c of clays) varied from 0.15 to 0.55 times the initial void ratio, depending on the organic content of the waste. Because of the tremendous vari-

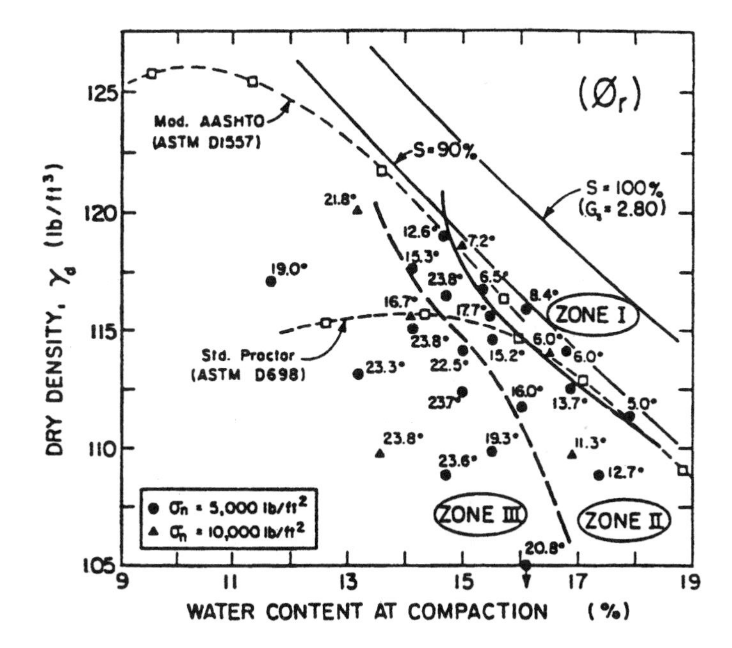

Figure 11.25 Measured shear strengths for smooth HDPE—compacted clayey till (with 5% bentonite) interfaces as function of compaction conditions (Seed and Boulanger, 1991).

ability of the refuse materials, such observations should not be generalized and should be used only in preliminary evaluations.

REFERENCES

Booker, J. R., and Rowe, R. K. (1987). One dimensional advective-dispersive transport into a deep layer having a variable surface concentration. *International Journal for Numerical and Analytical Methods in Geomechanics*, 11(2):131–142.

Chow, V. T. (1964). *Handbook of Applied Hydrology*. McGraw-Hill, New York.

Converse, Davis, Dixon Associates. (1975). Slope stability investigation, proposed final slope adjacent to Pomona Freeway. Operating Industries Disposal Site.

Demetracopoulos, A. C., Korfiatis, G. P., Bourodimos, E. L., and Nawy, E. G. (1984). Modeling for design of landfill bottom liners. *ASCE J. Environ. Eng.*, 110(6): 1084–1098.

Earth Technology Corporation. (1988). Instability of landfill slope, Puente Hills Landfill, Los Angeles County, California. Report submitted to Los Angeles County Sanitation District.

Edil, T. B., Ranguette, V. J., and Wuellner, W. W. (1990). Settlement of municipal refuse. In *Geotechnics of Waste Fills: Theory and Practice*, ASTM STP 1070, A. Landva and G. D. Knowles, eds., American Society for Testing and Materials, Philadelphia.

Ehrig, H. J., and Scheelhaase, T. (1993). Pollution potential and long term behaviour of sanitary landfills. *Proc. Fourth Int. Landfill Symp.*, Cagliari, Italy, pp. 1204–1225.

Fang, H. Y., et al. (1977). Strength testing of bales of sanitary landfill. Fritz Engineering Laboratory, Department of Civil Engineering, Lehigh University.

Fassett, J. B., Leonards, G. A., and Repetto, P. C. (1993). Geotechnical properties of municipal solid wastes and their uses in landfill design. *Proc. Waste Tech 93 Conf.*

Fenn, D. G., Hanley, K. J., and DeGeare, T. V. (1975). Use of the water balance method for predicting leachate generation from solid waste disposal sites, EPA/530/SW-168. U.S. Environmental Protection Agency, Cincinnati, OH.

Gershon, N. D., and Nir, A. (1969). Effects of boundary conditions of models on tracer distribution in flow through porous mediums. *Water Resources Research*, 5(4):830–839.

Giroud, J. P., and Beech, J. F. (1989). Stability of soil layers on geosynthetic lining systems. *Geosynthetics '89*, IFAI, San Diego, CA.

Giroud, J. P., Bonaparte, R., Beech, J. F., and Gross, B. A. (1988). Load-carrying capacity of a soil layer supported by a geosynthetic overlying a void. *Proc. Int. Symp. on Theory and Practice of Earth Reinforcement*, Fukuoka, Japan, pp. 185–190.

Horton, R. E. (1919). Rainfall interception. *Monthly Weather Review, U.S. Weather Bureau*, 47(9):603–623.

Kmet, P. (1982). EPA's water balance method—Its uses and limitations. Wisconsin Department of National Resources, Bureau of Solid Waste Management, Madison, WI.

Kmet, P., et al. (1981). Analysis of design parameters affecting the collection efficiency of clay lined landfills. *Proc. 4th Annual Madison Conf. of Applied Research and Practice on Municipal and Industrial Waste*.

Knipshield, F. W. (1985). Material selection and dimensioning of geomembranes for groundwater protection. *Waste and Refuse*, 22.

Knisel, W. G., ed. (1980). CREAMS, a field-scale model for chemical runoff and erosion from agricultural management systems, Vols. I, II, and III. USDA-SEA-AR Conservation Research Report 26.

Korfiatis, G. P., and Demetracopoulos, A. C. (1986). Flow characteristics of landfill leachate collection systems and liners. *ASCE J. of Environ. Eng.*, 112(3):538–550.

Landva, A. O., and Clark, J. I. (1990). Geotechnics of waste fills. In *Geotechnics of Waste Fills—Theory and Practice*, ASTM Special Technical Publication 1070, A. Landva and G. Knowles, eds. American Society for Testing and Materials, Philadelphia.

Lu, J. C. S., Eichenberger, B., and Stearns, R. J. (1985). Leachate from municipal landfills, production and management, *Pollution Technology Review No. 119*, Noyes Publications, Park Ridge, NJ.

Lutton, R. J., Regan, G. L., and Jones, L. W. (1979). Design and construction of covers for solid waste landfills, EPA-600/2-79-165, Office of Research and Development, U.S. Environmental Protection Agency, Cincinnati, OH.

Martin, R. B., Koerner, R. M., and Whitty, J. E. (1984). Experimental friction evaluation of slippage between geomembranes, geotextiles, and soils. *Proc. Int. Conf. on Geomembranes*, Denver, CO, June 20–23, pp. 191–196.

Matrecon, Inc. (1980). Lining of waste impoundment and disposal facilities, EPA 530/ SW-870. U.S. Environmental Protection Agency, Cincinnati, OH.
McBean, E. A., Poland, R., Rovers, F. A., and Crutcher, A. J. (1982). Leachate collection design for contaminant landfills. *ASCE J. Environ. Eng.*, 108:204.
McEnroe, B. M., and Schroeder, P. R., (1988). Leachate collection in landfills: Steady case. *ASCE J. Environ. Eng.*, 114(5):1052–1062.
Mitchell, J. K., Bray, J. D., and Mitchell, R. A. (1995). Material interactions in solid waste landfills. *Proc. Geoenvironment—2000*, ASCE Geotechnical Special Publication No. 46, vol. 1, pp. 568–590.
Mitchell, J. K., Seed, R. B., and Seed, H. B. (1990). Kettleman hills waste landfill slope failure. I: Liner system properties. *ASCE J. Geotech. Eng.*, 116(4):647–668.
Moore, C. A. (1980). Landfill and Surface Impoundment Performance Evaluation. US EPA SW-869.
National Solid Waste Management Association (NSWMA). (1985). Basic data: Solid waste amounts, composition and management systems, Technical Bulletin 85-6. National Solid Waste Management Association, October 1985, p. 8.
Ogata, A., and Banks, R. B. (1961). A solution of the differential equation of longitudinal dispersion in porous media. U.S. Geol. Surv. Prof. Paper 411-A.
Oweis, I. S. (1993). Stability of landfills. In *Geotechnical Practice for Waste Disposal*, D. E. Daniel, ed. Chapman & Hall, London, Chap. 11.
Oweis, I. S., and Khera, R. P. (1990). *Geotechnology of Waste Management*. Butterworth, London.
Perry, R. H. (1976). *Engineering Manual*, 3rd ed. McGraw-Hill, New York.
Ritchie, J. T. (1972). A model for predicting evaporation from a row crop with incomplete cover. *Water Resources Res.* 8(5):1204–1213.
Rowe, R. K. (1987). Pollutant transport through barriers. *Geotechnical Practice for Waste Disposal '87*, Proc. Specialty Conf. sponsored by the Geotechnical Engineering Division of the ASCE, University of Michigan, Ann Arbor, June 15–17, 1987, pp. 159–181.
Rowe, R. K. (1991). Contaminant impact assessment and the contaminating lifespan of landfills. *Can. J. Civil Eng.*, 18(2):244–253.
Schroeder, P. R., Morgan, J. M., Walski, T. M., and Gibson, A. C. (1984a). The hydrologic evaluation of landfill performance (HELP) model; vol. 1, User's guide for Version 1, Technical Resource Document, EPA/530-SW-84-009. U.S. Environmental Protection Agency, Cincinnati, OH.
Schroeder, P. R., Gibson, A. C., and Smolen, M. D. (1984b). The hydrologic evaluation of landfill performance (HELP) model, vol. II, Documentation for Version 1, Technical Resource Document, EPA/530-SW-84-010. U.S. Environmental Protection Agency, Cincinnati, OH.
Seed, R. B., and Boulanger, R. W. (1991). Smooth HDPE-clay liner interface shear strengths: Compaction effects. *ASCE J. Geotech. Eng.*, 117(4):686–693.
Shackelford, C. D. (1990). Transit-time design of earthen barriers. *Eng. Geol.*, Elsevier Publ., Amsterdam, 29, pp. 79–94.
Shackelford, C. D. (1988). Diffusion as a transport process in fine-grained barrier materials. *Geotech*, News, 6(2):24–27.
Singh, S., and Murphy, B. J. (1990). Evaluation of the stability of sanitary landfills. In

Geotechnics of Waste Fills: Theory and Practice, ASTM STP 1070, A. Landva and G. D. Knowles, eds. American Society for Testing and Materials, Philadelphia, pp. 240–258.

Sowers, G. F. (1968). Foundation problems in sanitary landfills. *ASCE J. Sanitary Eng.*, 94(SA1).

Sowers, G. F. (1973). Settlement of waste disposal fills. *Proc. Eighth Int. Conf. on Soil Mechanics and Foundation Engineering*, Moscow, vol. 2, pp. 207–210.

Stark, T. D., and Poeppel, A. R. (1994). Landfill liner interface strengths from torsional-ring shear tests. *ASCE J. Geotech. Eng.*, 120(3):597–615.

Stone, R. (1974). Disposal of sewage sludge into a sanitary landfill, EPA-SW-71d. U.S. Environmental Protection Agency, Cincinnati, OH.

Terzaghi, K., Peck, R. B., and Mesri, G. (1996). *Soil Mechanics in Engineering Practice*, 3rd edition. Wiley, New York.

Thornthwaite, C. W., and Mather, J. R. (1957). Instruction and tables for computing potential evapotranspiration and the water balance. *Publications in Climatology*, 10(3): 185–311, (Drexel Institute of Technology, Centerton, NJ).

USDA, Soil Conservation Service. (1972). Section 4, Hydrology. In *National Engineering Handbook*. U.S. Government Printing Office, Washington, DC.

U.S. Department of the Navy. 1982. *Soil Mechanics Design Manual 7*. Naval Facilities Engineering Command, Alexandria, VA.

U.S. Department of the Navy. 1983. *Soil Dynamics, Deep Stabilization, and Special Geotechnical Construction*, Design Manual 7.3, NAVFAC DM 7.3M. Naval Facilities Engineering Command, Alexandria, VA, pp. 7.3–7.9.

U.S. EPA. (1986). Subtitle D Study: Phase I Report, EPA/530-SW-86-054.

U.S. EPA. (1991). Design and Construction of RCRA/CERCLA Final Covers, EPA/625/4-91/025.

Van Genuchten, M. Th., and Alves, W. J. (1982). Analytical solutions of the one-dimensional convective-dispersive solute transport equation. U.S. Department of Agriculture, Technical Bulletin No. 1661.

Williams, J. R., Nicks, A. D., and Arnold, J. G. (1985). SWRRB, a simulator for water resources in rural basins. *J. Hydraulic Eng., ASCE*, 111(6):970–986.

Wong, J. (1977). The design of a system for collecting leachate from a lined landfill site. *Water Resources Res.*, 13(2):404–410.

Wu, T. H. (1976). *Soil Mechanics*. Allyn and Bacon, Boston.

12
Barrier Composition and Performance

12.1 CONTAINMENT SYSTEM PERFORMANCE ELEMENTS

Containment systems are usually designed to perform at an acceptable level of service for time periods ranging from about 10 years for a slurry wall to more than 1000 years for nuclear waste repositories. The effectiveness of a containment system with respect to meeting the design objectives can be expressed generically as a function of several factors, represented as follows:

$$E_t = f[(\mathrm{D} + P_m + C) + (Q) + (M_o + M_e) + (S_g + S_h) + (L_t)] \quad (12.1)$$

where

E_t = effectiveness of the containment system at a particular time after its construction
D = dimensions of components of the containment structure
P_m = material properties of the structural components
C = configuration (arrangement) of components
Q = quality assurance practices during construction
M_o = design and operational efficiency of the monitoring system
M_e = maintenance effectiveness in terms of the type and frequency of maintenance activities
S_g = geotechnical/geological characteristics of the site
S_h = Hydrologic/climatic characteristics of the site
L_t = Time-variable loads and phenomena that induce stresses within and/or among the components of the structure

A change in design affects primarily D, P_m, and C; change in location changes S_g and S_h; and a change in maintenance and monitoring plans affects M_e and M_o.

The performance of a containment system can be assessed in a variety of ways, ranging from general assessments to estimates of system failure probabilities for specific time frames. Detailed analysis requires the adoption of a precise definition for "performance" and the establishment of levels of system or component performance below which the system or component can be considered to fail. The performance of a containment system is the extent to which it can meet the functional design objectives. For most containment systems, the functional design objectives fall into one or both of the following categories:

Reduction in the rate of flow and/or total volume of contaminants released into the protected environment or moisture that infiltrates into contained waste

Reduction in the concentrations of specific contaminants outside the containment system

The performance can be deemed a failure when maximum levels in flow volumes or contaminant concentrations are exceeded at specified points within and outside the containment system. Structural failure of a containment system or one or more of its components is different from functional failure. The failure of a component (e.g., a secondary liner of a landfill) may not necessarily lead to the functional failure of the containment system. However, the structural failure of a critical component may lead to its functional failure and, possibly, to system functional failure. These distinctions are important with respect to the formulation of multicomponent, time-dependent reliability models for assessing the long-term performance of waste containment systems.

12.2 SYSTEM PERFORMANCE PATTERN

The term *effectiveness*, E_t, represented nominally by Eq. (12.1), has many surrogates, the most notable of which are reliability and long-term performance. Although these two terms are commonly used in attempts to analyze the long-term effectiveness of waste containment systems, often, very few of the factors that determine effectiveness are covered by existing models. For a long-term performance model to be complete, it must address the following factors at a minimum.

The fact that most containment systems are composite systems with many components that are susceptible in varying degrees, to externally and internally generated physical, chemical, and biological stresses

Changes in the ratios of stress/resistance with time for each of the critical components of the system and how such changes affect the magnitudes of system performance measures, relative to specified requirements

A single parameter such as contaminant concentration outside a barrier or containment system may be used for indexing the performance of a barrier with respect to time. A composite reliability index that is not focused on a single performance parameter may also be used. In the first case, the time variable is the concentration value itself; in the second case, the time variable is an index that has a possible range of 0–1.0 or 0–100%.

For the first case, the patterns presented in Figure 12.1 represent the diversity of scenarios. The source term concentration, C_o, that was previously described in Eq. (9.1), is a time-dependent parameter. Hence,

$$C_o = f(t) \tag{12.2}$$

Also, the role of time is such that the reliability of a component (the system may comprise only one component) or multi-component system usually changes. Thus, t in Eq. (12.2) can be replaced by R_t, the time-varying reliability:

$$C_o = f(R_t) \tag{12.3}$$

where

C_o = contaminant source term concentration at the measurement point outside the containment system (M/L^3)

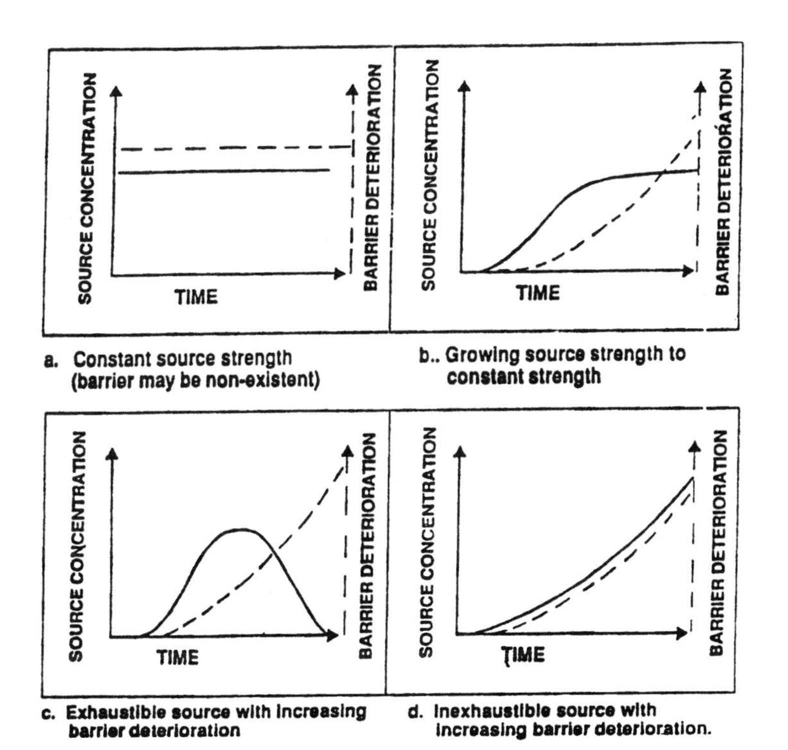

Figure 12.1 Contaminant source strength patterns for fate-and-transport modeling.

t = time since the beginning of contaminant containment (T)
R_t = reliability of the containment system or component with respect to meeting performance standards (0–1.0 or 0–100%)

The catalog of patterns shown as Figure 12.1 represents Eq. (12.2). Also, Eq. (12.3) is indirectly represented, noting that R_t has an inverse relationship with the barrier deterioration pattern that is illustrated schematically in Figure 12.1. In Figure 12.1a, the source term concentration C_o has a finite value but is time-invariant. This represents a case in which there is no barrier and the contaminant is supplied (presumably through leaching) infinitely at the same rate. This case is rare and could occur over very short time intervals. In Figure 12.1b, the barrier deteriorates exponentially, and the source term concentration grows to a near-constant value. As in case (a), this also implies an infinite source of the contaminant. It should be noted that both the zero deterioration and zero concentration points are displaced forward along the time axis. These displacements account conceptually for the time it takes a newly constructed containment system to begin significant deterioration and the travel time of contaminants to exit points, respectively. For more practical purposes, a more complete illustration of the relationship among facility operating period, postclosure period, and failure time frame (herein called the significant deterioration period) is presented in Figure 12.2 (after Mulkey et al., 1989). The concentration at a well down-gradient from the hypothetical facility (C_{exp}) is shown schematically to be less than the concentration that is typically obtained from laboratory-based leaching tests. Contaminant attenuation factors along the travel path account for most of the differences (in addition to the inability of leaching test protocols to simulate in-situ leaching processes).

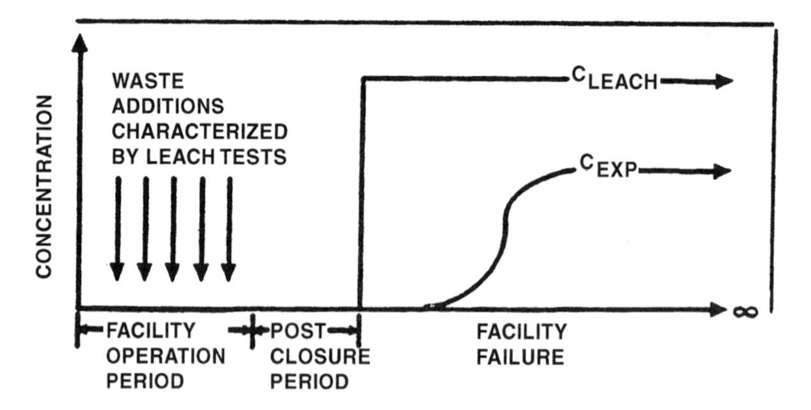

Figure 12.2 A generalized time history of contaminant concentration at an exposure point downgradient from a facility (Mulkey et al., 1989).

The rise and decay of contaminant concentration C_o with distance and time is illustrated (Rowe and Booker, 1991) in Figure 12.3 for monitoring points below a landfill. Figure 12.4 illustrates estimates of the time-varying production of gas from a landfill under two moisture conditions (after Tchobanoglous et al., 1993; and Bonaparte, 1995). The decay of contaminant quantities with time at each measurement point tends to follow the pattern shown in Figure 12.1c. The pattern in Figure 12.1d implies an infinite source which is uncommon especially for long time frames.

Inyang (1994) developed a conceptual model for describing the time variation of containment system performance. As shown in Figure 12.5, from an initial level of E_{to}, the effectiveness decays along curve 1 if the pattern is not interrupted by transient catastrophic events (e.g., a damaging earthquake) or maintenance activity. The effective service life of the containment system may be defined by t_r, which corresponds to a specified minimum level of effectiveness, E_{tr}. At an intermediate time, t_g, a damaging event may instantaneously reduce the system effectiveness from E_1 to E_{tg}, thus generating the new decay pattern represented as curve 2. A maintenance activity at time t_m has the net effect of raising the system effectiveness from E_2 to E_{tm} and generating the new decay pattern shown as curve 3.

The Inyang conceptual model provides a framework for the application

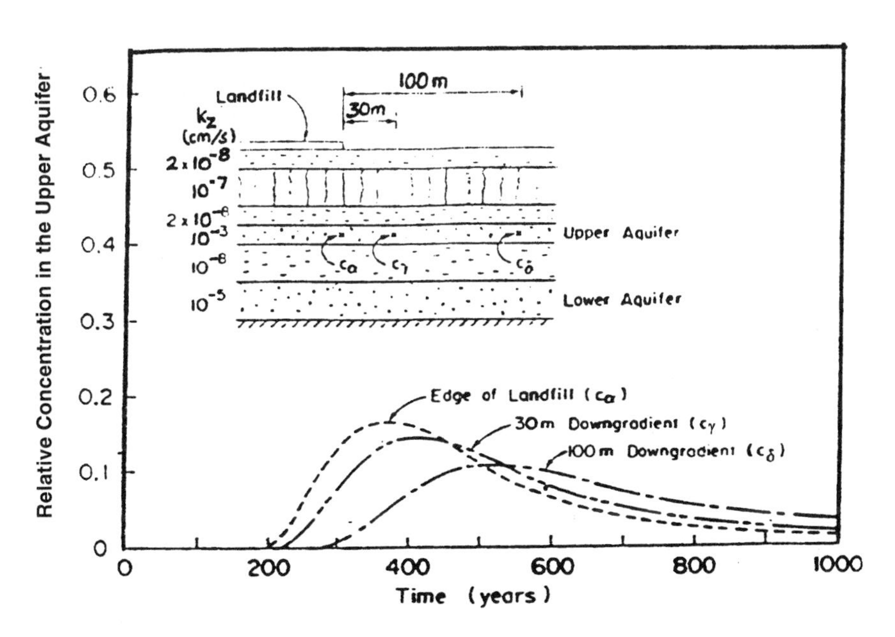

Figure 12.3 Contaminant concentration-versus-time relationships predicted for specific locations below a landfill (Rowe and Booker, 1991).

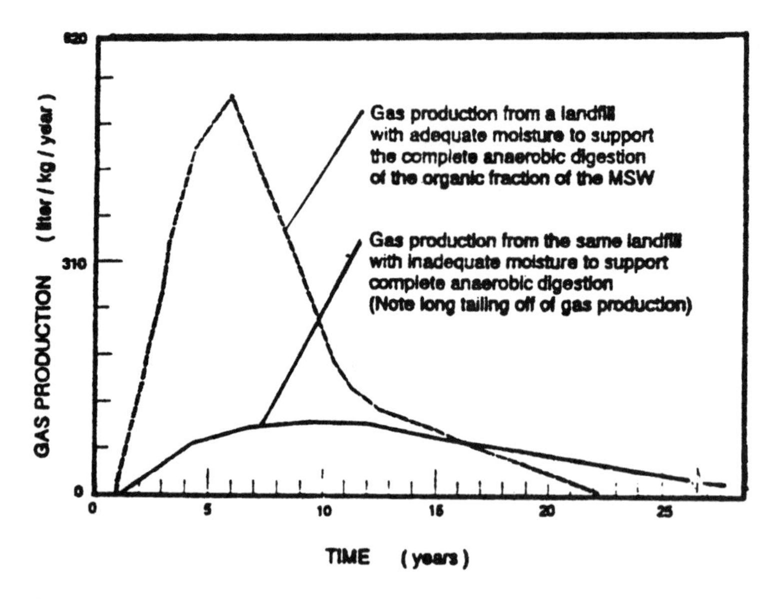

Figure 12.4 Conceptual models for gas production at municipal waste landfills (adapted from Tchobanoglous et al., 1993; and Bonaparte, 1995).

of system deterioration models and development of single-component parameter tracking models akin to those represented in Figure 12.1. For the composite systems, the Weibull reliability model for barrier systems (Inyang, 1994) is an example of the time-scaling framework for waste containment system performance. This presents reliability as

$$R_t = \exp\left[-\int_0^t \lambda(t)\,\partial t\right] \tag{12.4}$$

where

R_t = reliability of a containment system with respect to the performance of a specified design function at a given time
λ = containment system failure rate
t = time at which reliability is sought

Equation (12.4) implies that R_t can be expressed in terms of the rate for the system to perform a given function (e.g., minimization of contaminant release to the level of 10^{-3} cm^3/m^2/s). With the assumption that the rate of failure increases with time (as is the case with almost all constructed systems), R_t can be expressed as Eq. (12.5):

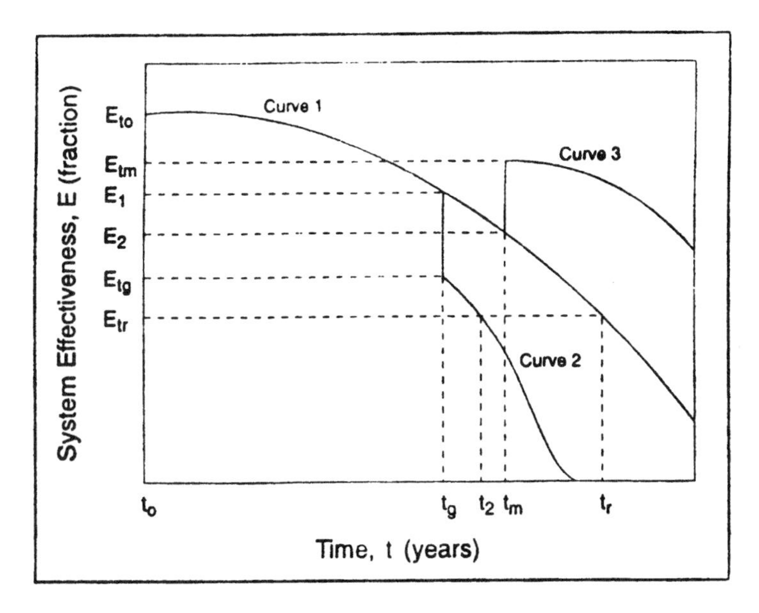

Figure 12.5 A Conceptual long-term deterioration pattern and maintenance scheme for waste containment systems (Inyang, 1994).

$$R_t = \exp\left\{-\left[\left(\frac{t - t_0}{n}\right)\right]^b\right\} \tag{12.5}$$

where

t_0 = location parameter = time at the beginning of degradation
b = shape parameter (dimensionless)
n = scale parameter (dimensionless)

In order to match the reality that systems deteriorate with time, b must have a positive value. Mathematically, b defines the geometry of the reliability versus-time curve and should range from 2.0 to 5.0 for containment systems. The scale parameter n is the normalization factor for the time interval considered. Theoretically, it is a time duration that corresponds to

$$F_t = 1 - R_t = 1 - e^{-1} = 0.632 \tag{12.6}$$

Thus, to obtain n, the failure function of the containment system must be known or estimated. From a plot of failure probability (F_t) versus time, the time value

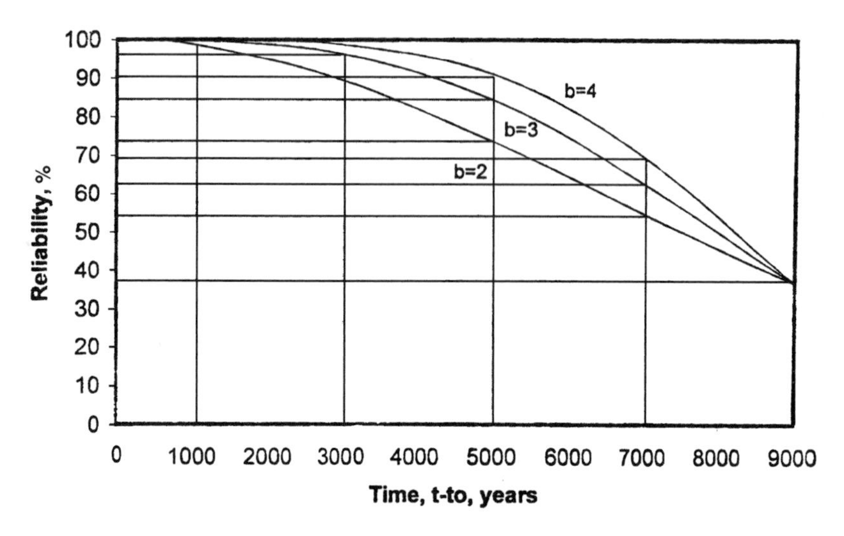

Figure 12.6 A plot of composite reliability decay for a hypothetical nuclear waste repository using the Inyang Weibull model (Inyang, 1994).

that corresponds to $F_t = 0.632$ is taken as n. Figure 12.6 shows plots of R_t for a hypothetical nuclear waste repository using b values of 2, 3, and 4, and an n value of 9000 years.

12.3 TYPES OF BARRIER MATERIALS

Several materials are used as barriers in waste containment systems. The criterion for the selection of materials for each type of containment system can be one or more of the following requirements:

- Materials that have adequately high sorption capacity
- Materials that can be compacted or cemented to high densities and effective porosities that are low enough to minimize the flow of contaminants through them
- Materials that can react chemically with the contaminants that permeate them, such that the concentrations of the contaminants are effectively reduced
- Materials that can plug the pore spaces in natural soil such that large channels for contaminant transport are effectively eliminated

In terms of chemical and mineralogical characteristics, barrier materials that perform one or more of these functions fall into the categories described below.

12.3.1 Clayey Barrier Materials

Clayey soils are generally available in large quantities for use as barrier materials. The relatively high physicochemical interaction between clayey materials and contaminants results in high values of distribution coefficient for most contaminants. Also, soils that have a high content of clay particles ($\leq 2\ \mu m$ in diameter) tend to have low permeability. Among the most commonly utilized clay minerals in small weight fractions (2–6%) are those described below.

Montmorrillonite: Montmorrillonite (particularly when it has Na^+ as the exchangeable cation) has the capacity to expand when wet. The Ca^+ variety expands less. Expansion generally results in reduced permeability. Montmorrillonite is known to readily sorb polar organic compounds, positively charged organic groups, and inorganic ions.

Kaolinite: This mineral has a relatively small negative surface charge. Hydrogen bonding across its interlayers inhibits expansion. Its capacity to sorb contaminants is generally less than that of Na-montmorrillonite.

Illite: Potassium is the primary exchange cation in illite. Potassium is of a size and shape that fits into the hexagonal hole of interlayers in illite. It does not expand significantly. However, its cation-exchange capacity is higher than that of kaolinite.

Some information on these three principal clay minerals is provided in Table 12.1 (Brown and Anderson, 1980). Other clays that have been investigated as barrier

Table 12.1 Physicochemical Characteristics of the Three Most Common Barrier Soil Minerals

Clay mineral	Particle thickness: Contracted (nm)	Particle thickness: Hydrated[a] (nm)	Volume change (%)	Charge per formula weight[b]	Surface area (m^2/g)	Exchange capacity (Meq/100 g): Cation	Exchange capacity (Meq/100 g): Anion
Kaolinite (nonexpansive 1:1 lattice)	200.0	202.0	1	0	8	10	pH dependent
Illite (nonexpansive 2:1 lattice)	20.0	22.0	10	1.0	80	15	pH dependent
Montmorillonite (expansive 2:1 lattice)	2.0	6.0	200	0.5	800	100	pH dependent <5

[a] Four water layers absorbed for each available basal surface.
[b] Units are multiples of electro-static units (esu). One charge = 4.8029×10^{-7} esu.
Source: Brown and Anderson (1980), p. 124.

materials are attapulgite and vermiculite. Clays are commonly used mix components of covers, liners, and slurry walls. Often, they are mixed with silts and sands.

12.3.2 Mineral Oxides, Organics, and Polymer Liquids

Several oxides exhibit high sorption and reactive capacities for both organic and inorganic contaminants. Mineral oxides such as zero-valent iron (Fe^0), hematite, fly ash, and geothite are usually utilized as additives to clayey materials. Fe^0 is the most common material used in permeable reactive barriers for chlorinated organic compounds. Organics and cementitious materials are used as grouts, examples of which are summarized in Table 12.2.

12.3.3 Polymeric Membranes

Membranes are used as barrier materials in impoundment/landfill liners, covers, and slurry walls. They lack the contaminant attenuation capacity of clays and minerals, but they have higher containment capacity for vapors. Their flexibility makes it possible to maintain the waste containment function under moderate deformation conditions. Additives such as carbon black, pigments, fillers, plasticizers, processing aids, cross-linking chemicals, antidegradants, and biocides are usually combined with raw polymers to produce a variety of polymeric sheets. The majority of polymeric materials used in waste containment systems are pre-

Table 12.2 Some Key Properties of Barrier Additives and Chemical Grouts

Grout	Hydraulic conductivity (cm/s)	Resistance to acids	Resistance to bases	Resistance to organics	Expected lifetime (years)
Sodium silicate	10^{-5}	Fair	Poor	Fair	10–20
Acrylate gels	10^{-7}–10^{-9}	Poor	Good	Fair	10–20
Colloidal silica	10^{-8}	Good	Poor	Good	>25
Iron hydroxide	10^{-7}	Poor	Good	Good	>25
Montan wax	10^{-4}–10^{-7}	Fair	Fair	Fair	25
Sulfur polymer cement	10^{-10}	Good	Poor	Fair	>25
Epoxy	10^{-10}	Good	Good	Good	>25
Polysiloxane	10^{-10}	Good	Good	Good	>25
Furan	10^{-8}–10^{-10}	Good	Good	Good	>25
Polyester styrene	10^{-10}	Good	Good	Fair	>25
Vinylester styrene	10^{-10}	Good	Good	Good	>25
Acrylics	10^{-9}–10^{-11}	Good	Good	Good	>25

Source: Whang and Committee, (1995).

sented in Table 12.3. Also, their compositions are summarized in Table 12.4 (after Haxo et al., 1985). The characteristics of each of the polymers listed, including their durability factors are discussed in detail by the U.S. EPA (1980).

12.3.4 Geosynthetic-Clay Materials

Geosynthetic-clay liner materials are composites of sodium bentonite in layers of geosynthetics. The geosynthetic layers are sewn such that bentonite is sandwiched among them. The composite material performs better than either component (Daniel and Koerner, 1993). Essentially, the geosynthetics provide increased crack resistance under differential settlement and cycles of wetting/drying and freezing and thawing. The clay material swells in contact with moisture such that permeability is low. It also provides contaminant attenuation capacity to the composite material.

Table 12.3 Polymeric Materials Used in Liners

Polymer	Abbreviation	Type of compound[a]	Thickness (mm)	Fabric reinforced
Butyl rubber	IIR	XL	0.9–2.3	Yes, no
Chlorinated polyethylene	CPE	TP	0.5–1.0	Yes, no
Chlorosulfonated polyethylene	CSPE	TP,XL	0.6–1.0	Yes, no
Elasticized polyolefin	ELPO	CX	0.6	No
Elasticized polyvinyl chloride	PVC-E	TP	0.8–0.9	Yes
Epichlorohydrin rubber	ECO	XL	1.67	Yes, no
Ethylene propylene rubber	EPDM	XL, TP	0.5–1.6	Yes, no
Neoprene	CR	XL	0.5–1.8	Yes, no
Nitrile rubber	NBR	XL	0.8	Yes
Polyester elastomer	PEEL	CX	0.18	No
Polyethylene				
Low-density	LDPE	CX	0.3–0.8	No
Linear low-density	LLDPE	CX	0.5–0.8	No
High-density	HDPE	CX	0.5–3.1	No
Polyvinyl chloride	PVC	TP	0.3–0.9	Yes, no
Polyvinyl chloride, oil resistant	PVC-OR	TP	0.8	Yes, no
Thermoplastic elastomer	TPE	CS	1.0	No

[a] TP = thermoplastic; CX = partially crystalline; XL = cross-linked.
Source: Haxo et al. (1985).

Table 12.4 Basic Compositions of Polymeric Membrane Liner Compounds

Component	Composition of compound type, parts by weight		
	Cross-linked	Thermoplastic	Crystalline
Polymer or alloy	100	100	100
Oil or plasticizer	5–40	5–55	0–10
Fillers:			
Carbon black	5–40	5–40	2–5
Inorganics	5–40	5–40	—
Antidegradants	1–2	1	1
Cross-linking system:			
Inorganic	5–9	a	—
Sulfur	5–9	—	—

[a] An inorganic curing system that cross-links over time is incorporated in CSPE liner compounds.
Source: Haxo et al. (1985).

12.3.5 Other Barrier Materials

Other materials used in various waste containment system configurations include Portland cement concrete, paper mill sludges (Moo-Young and Zimmie, 1995), coal combustion flyash (Sarsby and Finch, 1995; Fleming and Inyang, 1995); limestone (Artiola and Fuller, 1980); slags (Khera, 1995); and organobentonites (Smith and Li, 1995). Evans et al. (1990) have summarized experiences with zeolites, organically modified clays, flyash, bentonite, and natural clay in composite liners for waste containment.

12.4 MATERIAL DETERIORATION MECHANISMS

It is important to distinguish between material deterioration, which depends on the durability of the material, among other factors, and the deterioration of a containment system that comprises the material. Material deterioration analysis focuses on the physicochemical characteristics of the material and how they change with time in response to environmental conditions. The dimensions of the component that comprises the material, and the overall configuration of the containment system are usually not factored into expressions of material durability or deterioration. System deterioration or degradation covers both material durability and interactions among the components of the overall system. This requires consideration of component dimensions and system configuration as well.

Natural deterioration phenomena of barrier materials can be divided into two major categories. Some phenomena may be accelerated by the intensification of other phenomena within the same category or the other category.

12.4.1 Physicochemical Interactions

Physicochemical interactions are relatively slow processes which may cause progressive degradation of materials. The most significant physicochemical interactions are briefly discussed below.

Flocculation of Barrier Materials

The microstructure of densified clay barrier may change in response to contact with contaminants. As shown in Figure 12.7, clays usually exhibit dispersed structure when placed at moisture contents that are wet-of-optimum. Subsequent interaction with pure or dissolved contaminants of appropriate concentrations can flocculate the clay as shown in Figure 12.7. If flocculation occurs, vertical permeability usually increases. Flocculation is most commonly explained through the application of the Gouy-Chapman theory in which the thickness of the double layer that surrounds each clay particle is represented by Eq. (12.7).

$$t_d = \left[\frac{\epsilon k T}{8 \pi n_0 c^2 v^2} \right]^{1/2} \tag{12.7}$$

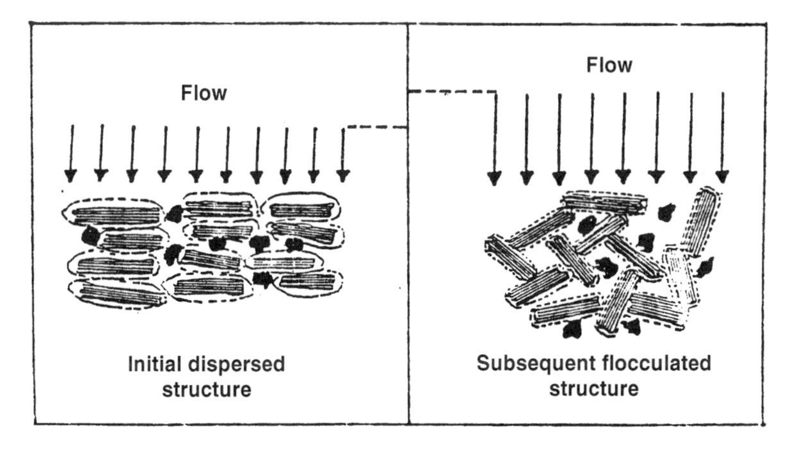

Figure 12.7 Schematic of structural changes in clay barrier due to change in permeant chemistry (Kargbo et al., 1993).

where

t_d = thickness of the double layer
ϵ = dielectric constant of the permeating fluid
k = Boltzmann's constant
T = absolute temperature
n_0 = ion concentration
c = unit electronic charge
v = valence of the cation on exchange sites on the clay

An inspection of Eq. (12.7) indicates that replacement of Na^+, which usually occupies the majority of exchange sites in the most common barrier material (bentonite), by a cation of higher valency would cause decrease in double-layer thickness. As illustrated in Figure 12.8, the net effect of this cation exchange is a reduction in the repulsive forces among the clay platelets such that the attractive forces (mostly van der Waals) exceed the repulsive forces. The resulting haphazard attraction produces a redistribution of pore sizes from smaller to larger pores. The vertical permeability is increased at the expense of horizontal permeability. Replacement of the initial pore fluid by concentrated organics or aqueous-phase organics also cause flocculation. In this case, the causative factor is a reduction in the dielectric constant ϵ, which results in a lower magnitude of t_d. Pure water has a dielectric constant of about 80 at 20°C. The dielectric constants of some

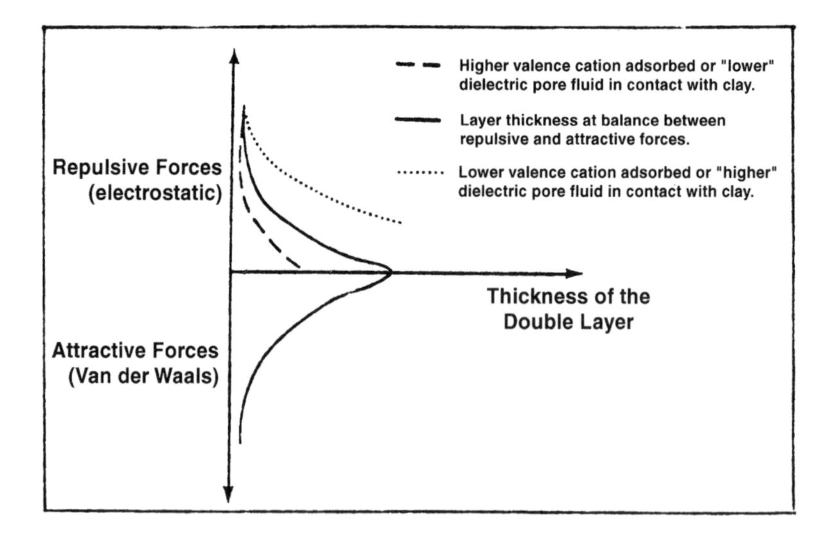

Figure 12.8 Effects of cation valence and pore fluid dielectricity on clay platelet interactions (Kargbo et al. 1993).

common organic substances compiled from literature are summarized in Table 12.5. For inorganics, the exchange order follows an electrochemical series in which higher valency is the first criterion and larger ionic radius (at the same valency) is the second criterion. Thus the replacement order for common ions is

$$\begin{array}{cccccccc} Al^{3+} > & Ca^{2+} > & Mg^{2+} > & NH_4^{+} > & K^{+} > & H^{+} > & Na^{+} > & Li^{+} \\ 0.57 & 1.06 & 0.78 & — & 1.33 & — & 0.98 & 0.7 \end{array} \qquad (12.8)$$

The numbers indicated below the cations indicate ionic size in angstroms.

Table 12.6 shows changes in clayey soil permeability after the replacement of pore fluid by permeants that contained dissolved inorganic substances. Mitchell and Madsen (1987), Bowders (1985), Evans et al. (1985), and Uppot and Stephenson (1988) have documented similar test results for barrier pore fluid replacement by organic permeants. However, it should be noted that effective stress of the order of 40 kPa or greater may suppress the flocculation of clay barriers by aggressive permeants. Fernandez and Quigley (1991) investigated the

Table 12.5 Dielectric Constants of Some Materials

Substance	Dielectric constant at 20°	Reference
Water	78–81	
Benzene	2.284	Abdul et al. (1990)
Nitrobenzene	34.82	Abdul et al. (1990)
Tuolene	2.438	Abdul et al. (1990)
Trichloroethylene	3.40	Abdul et al. (1990)
p-Xylene	2.270	Abdul et al. (1990)
Heptane	1.905	Budhu et al. (1991)
Dioxane	2.2	Budhu et al. (1991)
Acetone	20.7	Budhu et al. (1991)
Aniline	6.89	Budhu et al. (1991)
Methanol	32.63	Budhu et al. (1991)
Ethanol	24.30	Budhu et al. (1991)
Phenol	9.78	Abdul et al. (1991)
Ethylene glycol	37.0	Budhu et al. (1991)
Acetic acid	6.16	Budhu et al. (1991)
Chloroform	4.8	Murray and Quirk (1982)
t-Butanol	12.7	Murray and Quirk (1982)
Nitromethane	37.0	Murray and Quirk (1982)
n-Hexane	1.9	Murray and Quirk (1982)
Isopropanol	18.3	Fernandez and Quigley (1985)
Soil solids	2–4	Selig and Mansukhani (1975)

Table 12.6 Ratio of Final to Initial Permeabilities of Clays and Clayey Materials Following Pore Fluid Replacement by Inorganics

Pollutant	Concentration	Increase ratio	Material	Reference
Salts				
NaCl	5.85%	2.8	Bentonite	Pavilonsky (1985)
NaCl	High %	1.8–2.7	Bentonite	D'Appolonia (1982)
CaCl	11.1%	4.7	Bentonite	Pavilonsky (1985)
Ca^{2+} and Mg^{2+} (general)	10,000 ppm	2.9–3.2	Bentonite	D'Appolonia (1982)
$FeCl_3$	500 mg/liter	1.5–2.0	Clay: PI = 17–28	Peirce et al. (1987)
$Ni(NO_3)_2$	50 mg/liter	1.0–1.3	Clay: PI = 17–28	Peirce et al. (1987)
$Ni(NO_3)_2$	500 mg/liter	1.5–3.0	Clay: PI = 17–28	Peirce et al. (1987)
NH_4NO_3	10.000 ppm	1.8–2.7	Bentonite	D'Appolonia (1982)
Bases				
NaOH	0.4%	1.3	Loam: PI = 10	Pavilonsky (1985)
NaOH	0.4%	1.4	Bentonite	Pavilonsky (1985)
NaOH	1.0%	<5	S-B Backfill	D'Appolonia (1982)
NaOH	4.0%	302	Loam: PI = 10	Pavilonsky (1985)
NaOH	5.0%	2–5	S-B Backfill	D'Appolonia (1982)
$Ca(OH)_2$	1.0%	2–5	S-B Backfill	D'Appolonia (1982)
Strong bases	pH > 11	>5	S-B Backfill	D'Appolonia (1982)
Weak bases	pH < 11	<5	S-B Backfill	D'Appolonia (1982)

interaction of a mixture of municipal solid waste leachate and ethanol with unstressed and prestressed water-wet clay. As shown by the results of Figure 12.9, effective stress greater than 40 kPa suppressed an increase in hydraulic conductivity.

Dissolution of Barrier Soils

Barrier soils are primarily made up of aluminosilicates (clays) and silica (sand and silts). Silica is soluble in highly alkaline liquids. Strong acids may attack the structure of clay minerals. For example, experiments conducted by Pask and Davis (1945) in which clay minerals were boiled in acid produced dissolutions of 3%, 11%, and 33% for kaolinite, illite, and montmorrillonite, respectively. Tests conducted by Grubbs et al. (1972) indicate that injection of acid waste into clay minerals dissolve or completely alter them. The net effect of mineral dissolution is the removal of material from the walls of internal pores in compacted systems. Dissolution rates of earthen materials are very low. Thus, barrier

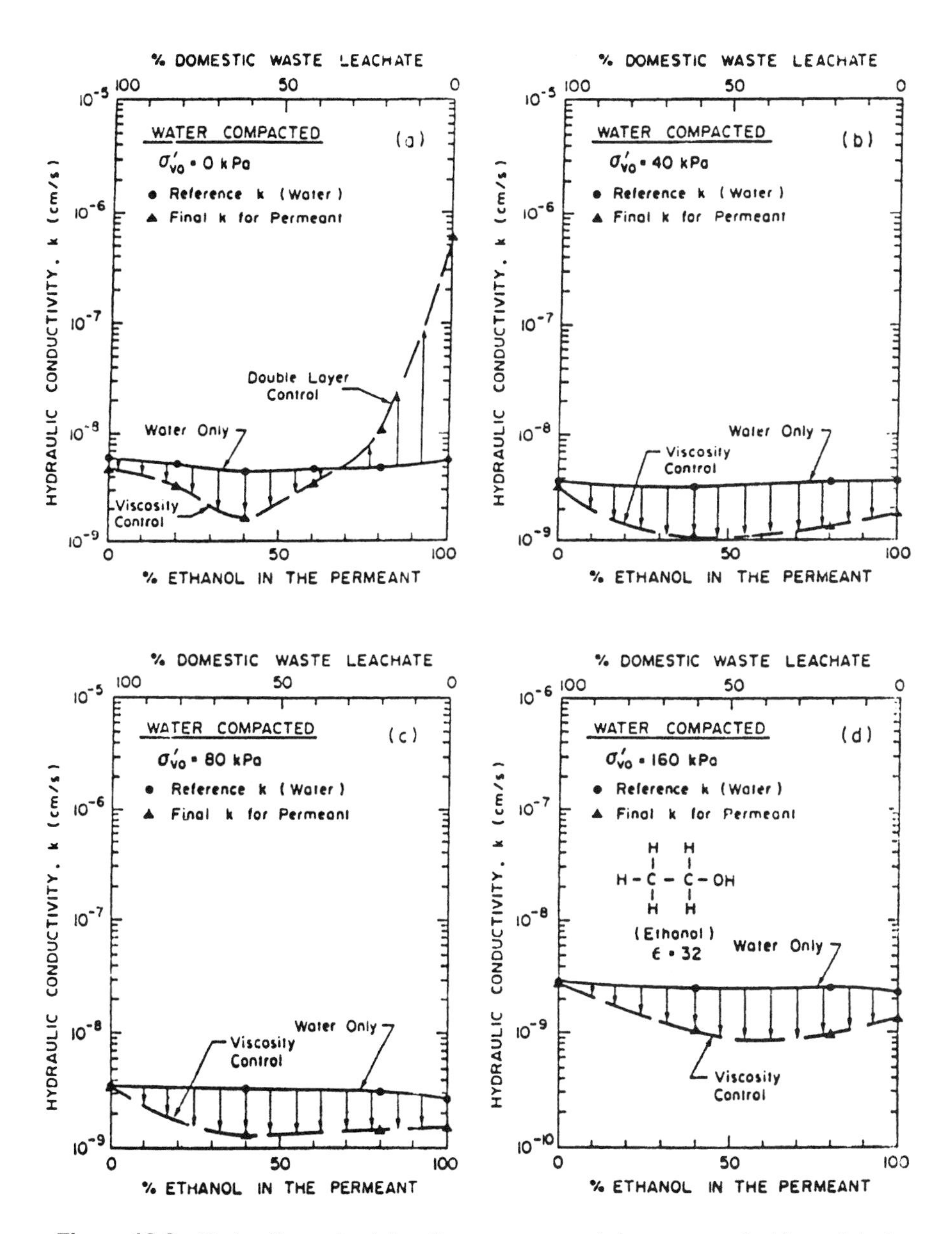

Figure 12.9 Hydraulic conductivity of water-compacted clay permeated with municipal solid waste leachate–ethanol mixtures subsequent to prestressing water-wet clay at various levels of vertical effective stress (Fernandez and Quigley, 1991).

materials require very long residence times of an aggressive liquid to dissolve significantly.

Cracking of Densified Barrier Materials

Cracking can be caused by a variety of stress-inducing phenomena. Among such phenomena are desiccation, cycles of heating and coaling, wetting and drying and freezing and thawing, geostatic and geodynamic loading, and even chemical attack. The propensity of barrier materials to crack is higher near the ground surface, where thermal, moisture, and other gradients are higher in magnitude and frequency. Fang (1994) developed a time-distribution scheme for various types of cracks in geomedia. As shown in Figure 12.10, the scheme identifies shrinkage, thermal loading, and loading as the initial set of stressors that produce cracks. During an intermediate stage called the degradation stage, the fractures grow dramatically. Crack growth is exponential, such that a limiting cracking stage which defines the useful life of the material or system is soon attained. Beyond this stage, the material is considered to be in a state of failure. In cold regions, cracks can be initiated and extended in barrier materials by freeze/thaw

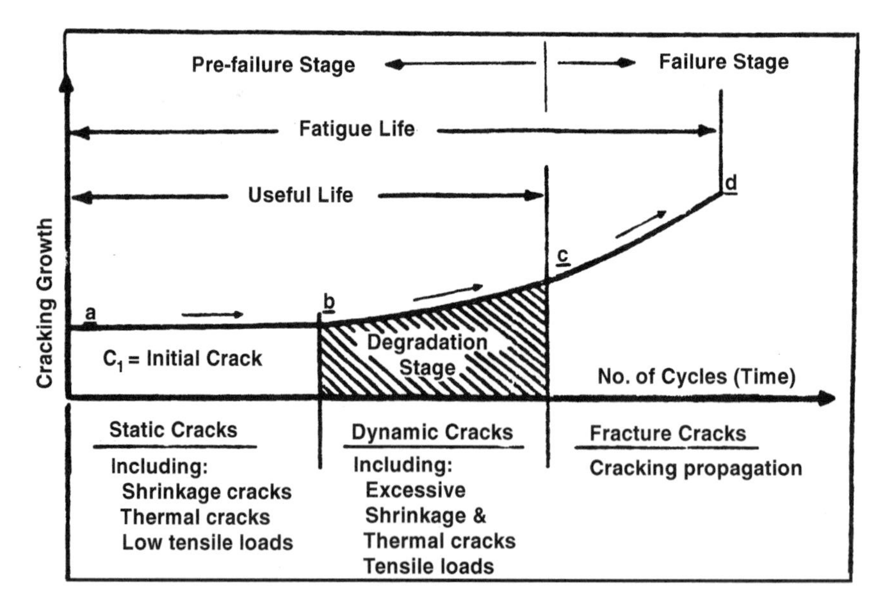

Figure 12.10 Schematic diagram illustrating the crack–time relationship at various cracking stages (Fang, 1994).

cycles. Figure 12.11 shows the effects of freeze/thaw cycles on the permeabilities of four compacted clays. The permeability tends to increase dramatically within the first 1–4 cycles. Thereafter, increase in permeability is less significant. Apparently, textural changes in soil are more difficult to achieve after the initial clogging of larger pores.

Exposed Slope Failures by Creep

Some containment systems, such as landfills and waste piles, have components that have exposed slopes. Much attention has been paid to the reliability of such slopes. The reliability of a landfill slope or any component of the landfill should be distinguished from the reliability of the entire landfill (a composite system). With respect to the deterioration of materials that make up a landfill slope, the soil may deform progressively under its weight or imposed stresses until the slope fails.

For cover systems, unconfined compressive strength is the most common strength parameter used in soil material performance assessment. As an early development in soil mechanics, Casagrande and Wilson (1951) conducted creep

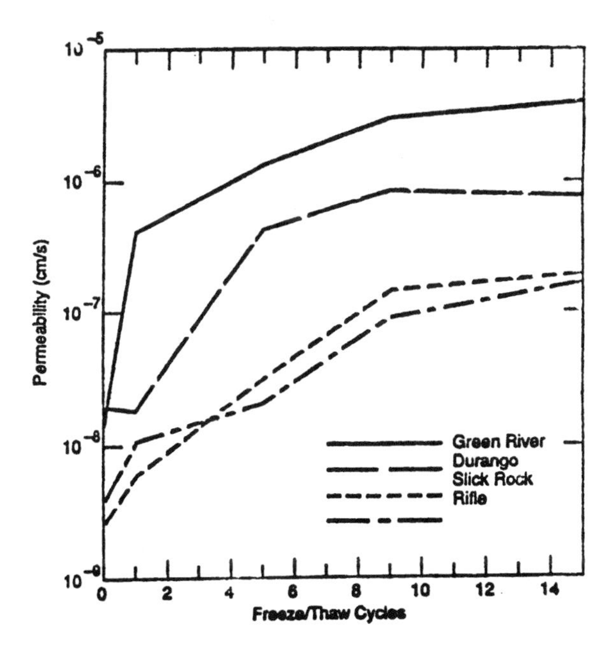

Figure 12.11 Summary of freeze/thaw permeability test results for four compacted clays (Chamberlain et al., 1990).

tests to assess the long-term unconfined compressive strength of clays and clay shales at constant moisture content. In general, the following equation can be used to express the rate of strength decrease with time in their tests:

$$\sigma_t = \sigma_a - \sigma_o \log \frac{t}{t_a} = \sigma_a \left(1 - \rho_a \log \frac{t}{t_a}\right) \tag{12.9}$$

where

σ_t = unconfined strength of saturated soils (F/L^2)
σ_o = decrease in strength for logarithmic time cycle (F/L^2)
ρ_a = coefficient of rheological decrease in strength = σ_o/σ_a (dimensionless fraction)
σ_a = reference (set initial) strength (F/L^2)
t_a = reference (set initial) time (T)
t = time in the future for which strength is sought (T)

If t_a is fixed at 1000 min, the envelop of magnitudes for σ_a is about 0.05–0.15. Closely related to the deterioration of cover slopes of containment systems is the increased potential of interface failures as moisture accumulates and drains downward at the earthen material/polymeric sheet interface. The lubricating effect could reduce interface friction such that portions of the cover fail.

Deterioration of Polymeric Materials

Sustained exposure to sunlight (ultraviolet radiation), chemicals, high temperatures, and cycles of these and other stresses can reduce the strength and generate potential flow channels in polymeric materials. An Arhenius-type model is commonly used to estimate the rate constant for the deterioration of polymeric materials using data from laboratory-based accelerated aging tests. The most significant molecular-level chemical attack processes on geomembranes (Koerner and Richardson, 1987) are breaking of carbon–carbon bonds (metathesis); breaking of carbon–non-carbon bonds in the amorphous phase (solvolysis); loss of electrons to oxygen by molecular constituents of the polymeric material (oxidation); and dissolution of the polymer, often preceded by swelling.

Polymer oxidation is enhanced by the release of a free radical. As explained by Koerner et al. (1990), the free radical can combine with oxygen to form a hydroperoxy radical which traverses the molecular structure of the polymer and reacts with other units. More radicals are released such that these oxidative reactions are intensified. This chain reaction is represented by Eqs. (12.10) and (12.11).

$$R^{\cdot} + O_2 \rightarrow ROO^{\cdot} \tag{12.10}$$

$$ROO^{\bullet} + RH \rightarrow ROOH + R^{\bullet} \tag{12.11}$$

where

$R^{\bullet}$ = free radical
$ROO^{\bullet}$ = hydroperoxy free radical
RH = polymer chain
$ROOH$ = oxidized polymer chain

The addition of anti-oxidation additives to geomembranes may reduce oxidative damage by reacting with the radicals and eliminating them as free-roaming substance.

Using the landfill simulator illustrated in Figure 12.12, the effects of landfill leachate, generated by a controlled infiltration of 25 in. year, on the swelling of a variety of polymeric liner materials were evaluated. The results obtained from these tests (U.S. EPA, 1980) are summarized in Table 12.7. Of the polymeric materials tested, chlorosulfonated polyethylene swelled the most. Perhaps the most important set of data on this test is strength retention after exposure. This

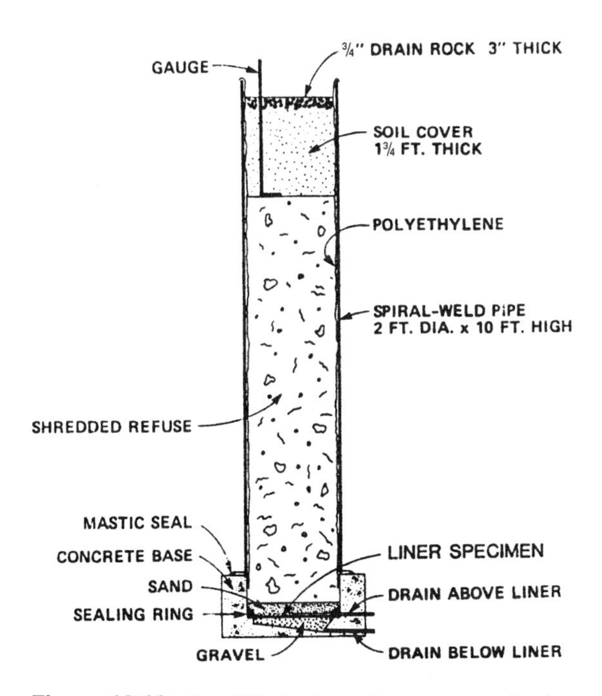

Figure 12.12 Landfill simulator for exposure of polymeric materials to leachate U.S. EPA, 1980).

Table 12.7 Swelling of Membrane Liners in Contact with Landfill Leachate (Weight % Increase)

	In landfill for 12 months		Immersed in leachate	
Membrane liner	Primary specimens	Buried specimens	8 mo.	19 mo.
Butyl rubber	2.0	1.8	1.4	2.6
Chlorinated polyethylene	6.8	9.0	7.9	14.4
Chlorosulfonated polyethylene	—	20.0	18.6	22.8
	12.8	13.6	12.1	14.9
Ethylene propylene rubber	5.5	6.0	2.9	3.8
Polybutylene	—	0.3	−0.2	0.7
Polyethylene	0.02	0.3	0	0.2
Polyvinyl chloride	—	5.0	2.4	4.4
	3.6	3.3	2.3	4.4
	—	0.8	0.9	1.9

Source: U.S. EPA (1980).

information is summarized in Table 12.8. Table 12.9 shows the chemistry of the leachate to which the materials were exposed.

12.4.2 Transient (Catastrophic) Events

Transient events such as earthquakes, floods, and (for nuclear waste repositories) global warming may have significant impacts on the deterioration of waste barrier materials. The longer the time frame of consideration, the higher is the probability of occurrence of transient events. Equation (12.12) shows the relationship between the probability of occurrence of a damaging event of specified magnitude in a given unit of time and the average time interval among such events:

$$P_s = \frac{1}{T_r} \tag{12.12}$$

where

P_s = probability of occurrence of a damaging event of specified magnitude in a unit time interval (for example, 1 year)
T_r = average return time interval (recurrence period) of the specified transient event of the same magnitude

Also,

$$P_n = 1 - (1 - P_s)^n \tag{12.13}$$

Table 12.8 Retention Modulus of Polymeric Membrane Liner Materials on Immersion in Landfill Leachate[a]

Polymer	Liner no.	S-200, psi			Retention % of original value	
		Unexposed	8 mo.	19 mo.	8 mo.	19 mo.
Butyl rubber	44	685	590	620	86	90
Chlorinated polyethylene	12	1330	1130	1190	89	89
	38	1205	1070	1090	89	90
	86	810	790	860	98	106
Chlorosulfonated polyethylene	3	735	395	335	54	46
	0[b]	1550	1775	2070	114	134
	85	1770	1360	1920	77	108
Elasticized polyolefin	36	970	1005	1050	104	109
Ethylene propylene rubber	8	690	870	860	126	124
	18	760	840	830	110	109
	83[b]	845	905	900	107	107
	91	855	775	790	91	92
	41	1040	1035	1030	100	99
Neoprene	9	1235	975	950	79	77
	37	1635	1630	1640	100	100
	42[b]	—	—	—	—	—
	90	1190	1245	1350	105	114
Polybutylene	98	3120	3160	3150	101	101
Polyester elastomer	75	2735	2790	2670	102	98
Polyethylene	21	1260	1340	1285	106	102
Polyvinyl chloride	11	2120	1845	1805	87	85
	17	1965	1570	1640	80	84
	19	1740	1545	1645	89	94
	40	1720	1560	1570	91	91
	67	1700	1570	1780	92	105
	88	2400	1900	2105	79	88
	89	2455	2370	2330	96	95
Polyvinyl chloride + pitch[c]	52	1020	870	880	85	86

[a] Average of stress at 200% elongation (S-200) measured in machine and transverse directions.
[b] Fabric reinforced.
[c] S-100 values given; original and subsequent exposed specimens failed at less than 200% elongation.
Source: U.S. EPA (1980).

where

P_n = probability of occurrence of the specified event during an extended time interval
n = extended time interval for which the analysis is conducted

As discussed in Sections 12.1 and 12.2, most containment systems comprise many components, each of which is susceptible to different degrees of

Table 12.9 Chemistry of Leachate Used to Test Polymeric Membrane Materials[a]

Test	Value
Total solids, %	3.31
Volatile solids, %	1.95
Nonvolatile solids, %	1.36
Chemical oxygen demand (COD), g/liter	45.9
pH	5.05
Total volatile acids (TVA), g/liter	24.33
Organic acids, g/liter	
Acetic	11.25
Propionic	2.87
Isobutyric	0.81
Butyric	6.93

[a] At the end of the first year of operation when the first set of liner specimens were recovered.
Source: U.S. EPA (1980).

stresses that can be imposed by transient events. Thus, the expression for the composite system must account for the composite nature of the system and the conditional probability of failure of each component. The relevant general expression for a case in which the failure of a component is tantamount to the failure of the system is given by Eq. (12.14).

$$P_c = P_n[P(f/o)_a + P(f/o)_b \text{---------} P(f/o)_k] \tag{12.14}$$

where

P_c = probability of failure of a containment system given that the design transient event has the probability P_n of occurrence during the considered time duration
$P(f/o)_i$ = conditional probability that a specific component, i, of the system will fail, given the occurrence of the design transient event ($i = a, b, \text{---}\ k$).

T_r can be obtained from historical data such as seismic intensity records and flood history. For example, Bollinger (1976) observed that for earthquake intensities ranging from V to VIII for the southeastern United States, the relationship expressed as Eq. (12.15) can be used to estimate T_r. Obviously, this equation, like all correlation equations, may require updating because, as more earthquakes occur, the correlative relationship may change.

$$\log(1/T_r) = 3.01 - 0.59I_o \tag{12.15}$$

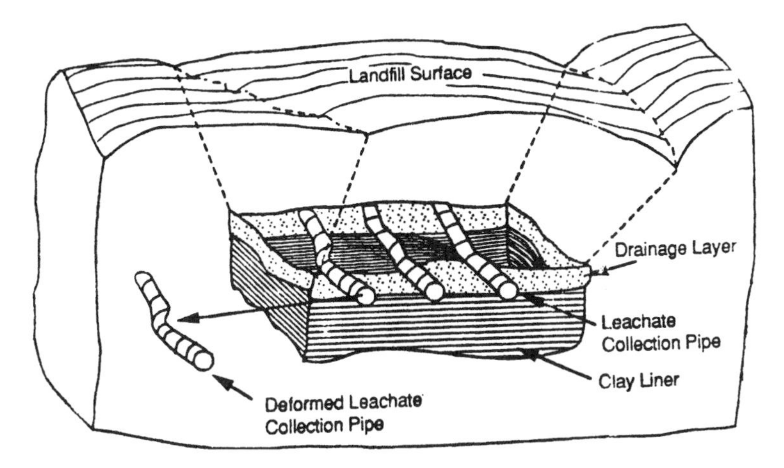

Figure 12.13 Illustration of the deformation of the components of a landfill by seismic activity (Inyang, 1992).

where I_o = earthquake intensity and T_r is, as previously defined, measured in years. Computation of $P(f/o)_i$ for real systems requires stability or fluid flow analysis of each of the components of the system. Methods for such analyses are described briefly in Chapter 11 of this text and in more detail in traditional geotechnical engineering texts. The definition of failure state for a component of the system [which is a requirement for the quantification of $P(f/o)_i$] is prone to confusion because functional failure and structural failure may not occur concurrently, at least initially. An example of the development of flaws in a containment system due to a transient event is illustrated in Figure 12.13 (Inyang, 1992) for a landfill impacted by seismic forces. The generation of flaws may result in acceleration of barrier material degradation.

12.5 SYSTEM PERFORMANCE MONITORING TECHNIQUES

Containment system monitoring techniques can be classified on the basis of technology or approach. The utilities of monitoring techniques with respect to each of the classifications are summarized in Table 12.10. There are three types of approaches to containment system performance monitoring: external barrier integrity monitoring, barrier permeation monitoring, and functional monitoring.

Table 12.10 General Application of Containment System Monitoring Techniques

	Monitoring method				
Barrier monitoring approach	Well network	Geophysical methods	Electrochemical methods	Mechanical and electrochemical methods	Electrical methods
1. Barrier integrity monitoring	U	G	R	C	G
2. Barrier permeation monitoring	U	R	G	R	C
3. External monitoring	C	G	G	R	G

Key: C = conventional; G = growing application; R = rare; U = unfeasible.
Note: These methods comprise several specific techniques, some of which may not necessarily fit into these three categories.

In barrier integrity monitoring, the objective is to detect the excessive deformation of barrier system components or flaws which may serve as flow channels for moisture and/or contaminants. In Table 12.10, examples of barrier integrity monitoring are listed as numbers 1, 2, and 3. Barrier permeation monitoring focuses on the detection of contaminants and moisture and measurement of their flow rates through critical components of the system. This type of monitoring is closely associated with performance assessment of the design of specific components of the barrier system.

In Table 12.11, examples of barrier permeation monitoring are numbered 4, 5, 7, and 8. External monitoring is conducted for assessment of the extent to which the entire containment system meets its overall design function of minimizing the release of contaminants from a containment system into the surrounding environment. As indicated in Table 12.10, well networks, geophysical methods, electrochemical sensing, and electrical sensing can be used in external monitoring. For details about specifics of these techniques, the reader is referred to Inyang et al. (1995) and other publications such as those of the U.S. EPA (1991, 1993a, 1993b) and Walther et al. (1986). Examples 6 and 9 of Table 12.11 fall within the external monitoring approach. In Figure 12.14, a common external monitoring scheme in which groundwater wells are utilized is shown. The extent of contaminant plume migration is directly proportional to the deterioration rate of the liner illustrated. For the reference time shown in Figure 12.14, if the barrier deterioration rate follows the most gradual of the three curves, the plume is not detectable at monitoring wells B and C. For the same time frame, if the barrier deterioration rate follows the highest curve, the target contaminants will be detected at all three wells, albeit at different concentrations.

Table 12.11 Examples of Traditional and Innovative Monitoring Techniques for Specific Waste Containment Problems

Waste containment problem	Monitoring techniques
1. Cover slope deformation and failure	Inclinometers, electrical resistance shear strips, and vibrating wire extensiometers
2. Lateral deformation of vertical walls	Electrical inclinometers, extensiometers, and tiltmeters
3. Excessive settlement and cracking of soil and concrete covers	Sealed Borros anchor, inclinometers, settlement platforms, electrical crack guages, convergence gauges, microwave sensors, dye-tracing
4. Excessive infiltration of cover systems	Tensiometers, gypsum blocks, psychrometers, neutron gauges and piezometers in covers, thermoelectric flow velocity gauges in drainage layers, and fiber-optic moisture sensors in covers
5. Voids and high permeability zones in emplaced and natural barriers	Crosshole seismic surveys, ground-penetrating radar surveys, seismic tomography, surface-based seismic refraction surveys, soil gas surveys, temperature logging, cone penetration testing, and dye tracing
6. Leachate plume migration from waste containment systems; and plume capture by permeable reactive walls	Electrochemical sensing cables and fiber-optic containment structure, DC electrical and electromagnetic resistance surveys above ground, ground-penetrating radar surveys, dye-tracing, and groundwater monitoring
7. Contaminant flow through bottom barriers and low-hydraulic conductivity walls	Electrochemical sensing cables and fiber-optic chemical sensors within, upgradient, and downgradient from barriers, and pumping tests
8. Leaks in geomembranes and sheet pile joints	Electrical leak detection systems and fiberoptic monitoring systems for joints
9. Flow of leachate through reactive, permeable walls	Thermal flow sensors upstream, within, and downstream from wall, electrochemical sensors, pH and Eh sensors

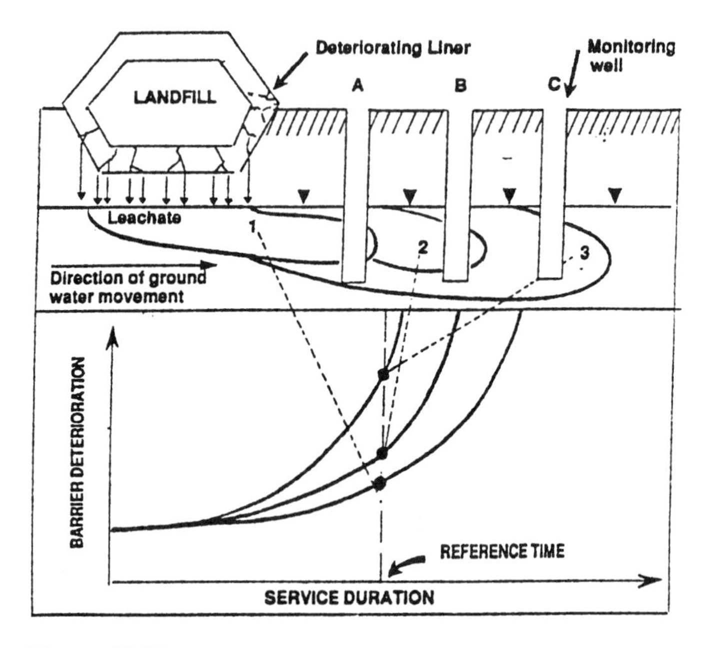

Figure 12.14 Relationship among barrier deterioration rate, extent of contaminant plume migration, and plume detection at monitoring wells.

REFERENCES

Abdul, A. S., Gibson, T. L., and Rai, D. N. (1990). Laboratory studies of flow of some organic solvents and their aqueous solutions through bentonite and kaolin clays. *Ground Water*, 28(4):524–533.

Artiola, J., and Fuller, W. H. (1980). Limestone liner for landfill leachates containing beryllium, cadmium, iron, nickel and zinc. *Soil Sci.*, 129(3):167–179.

Bollinger, G. A. (1976). The seismic regime in a minor earthquake zone. *Proc. ASCE Conf. on Numerical Methods in Geomechanics*, 11:917–937.

Bonaparte, R. 1995. Long-term performance of landfills. *Proc. Conf. (Geoenvironment 2000) on Characterization, Containment, Remediation and Performance in Environmental Geotechnics, American Society of Civil Engineers*, New Orleans, LA, pp. 514–553.

Bowders, J. J. (1985). The influence of various concentrations of organic liquids on the conductivity of compacted clay. Geotechnical Engineering Dissertation GT85-2, University of Texas, Austin.

Brown, K. W., and Anderson, D. (1980). Effect of organic chemicals on clay liner permeability. *Proc. 6th Annual Research Symp*. EPA-600/0-80-010. U.S. Environmental Protection Agency, Cincinnati, OH, pp. 123–134.

Budhu, M., Giese, R. F., Campbell, G., and Baumgrass, L. (1991). The permeability of soils with organic fluids. *Can. Geotech. J.*, 28:140–147.

Casagrande, A., and Wilson, S. D. (1951). Effect of rate of loading on the strength of clays and shales at constant water content. *Geotechnique*, 2:251–263.

Chamberlain, E. J., Iskandar, I. and Hunsicker, S. E. (1990). Effect of freeze-thaw cycles on the permeability and macrostructure of soils. *Proc. Conf. on Frozen Soil Impacts on Agriculture, Range and Forest Lands*, Spokane, WA, pp. 145–155.

Daniel, D. E., and Koerner, R. M. (1993). Cover systems. In: *Geotechnical Practice for Waste Disposal*, D. E., Daniel, ed. Chapman and Hall, New York, pp. 455–496.

D'Appolonia, D. J. (1982). Slurry trench cut-off walls of hazardous waste isolation. *Proc. 13th Annual Geotechnical Lecture Series, Geotechnical Group*, ASCE, Philadelphia.

Evans, J. C., Fang, H. Y., and Kugelman, I. J. (1985). Organic fluid effects on the permeability of soil-bentonite slurry walls. *Proc. Natl. Conf. on Hazardous Wastes and Environmental Emergencies*, Cincinnati, OH.

Evans, J. C., Sambasivam, Y., and Zarlinski, S. (1990). Attenuating materials in composite liners. Proc. Symp. on Waste Containment Systems Held in Conjunction with the National Convention of the American Society of Civil Engineers, San Francisco, CA, pp. 247–263.

Fang, H. Y. (1994). Cracking and fracture characteristics of contaminated fine-grained soil. Material developed for presentation at a Session on Application of Fracture Mechanics to Geotechnical Engineering, Annual Convention of the American Society of Civil Engineers, Atlanta, GA.

Fernandez, F., and Quigley, R. M. (1985). Hydraulic conductivity of natural clays permeated with simple liquid hydrocarbons. *Can. Geotech. J.*, 22:205–214.

Fernandez, F., and Quigley, R. M. (1991). Controlling the destructive effects of clay-organic liquid interactions by application of effective stresses. *Can. Geotech. J.*, 28:388–398.

Fleming, L. N., and Inyang, H. I. (1995). Permeability of clay-modified fly ash under thermal gradients. *J. Mater. in Civil Eng., ASCE*, 7(3):178–182.

Grubbs, D. M., Haynes, C. D., Hughes, T. H., and Stow, S. H. (1972). Compatibility of subsurface reservoirs with injected liquid wastes, Report no. 721. Natural Resources Center, University of Alabama, Huntsville, Alabama.

Haxo, H. E., Nelson, N. A., and Miedema, J. A. (1985). Solubility parameters for predicting membrane-waste liquid compatibility. Proc. 11th Annual Research Symp. EPA/600/9-85/013. U.S. Environmental Protection Agency, Washington, DC, pp. 198–212.

Inyang, H. I. (1992). Aspects of landfill design for stability in seismic zones. *J. Environ. Syst.*, 21(3):223–235.

Inyang, H. I. (1994). A Weibull-based reliability analysis of waste containment systems. Proc. First Int. Congress on Environmental Geotechnics, Alberta, Canada, pp. 273–278.

Inyang, H. I., Betsill, J. D., Breeden, R., Chamberlain, G. H., Dutta, S., Everett, L., Fuentes, R., Hendrickson, J., Koutsandreas, J., Lesmes, D., Loomis, G., Mangion, S. M., Morgan, D., Pfeifer, C., Puls, R. W., Stamnes, R. L., Vandel, T. D., and Williams, C. (1996). Performance Monitoring and Evaluation. Chapter 12 of Assessment of Bar-

rier Containment Technologies (Editors: Rumer, R. R. and Mitchell, J. K.), National Technical Information Service, Springfield, VA, pp. 355–400.
Kargbo, D. M., Fanning, D. S., Inyang, H. I., and Duell, R. W. (1993). The environmental significance of acid sulfate clays as waste covers. *Environ. Geol.*, 22:218–226.
Khera, R. P. (1995). Calcium bentonite, cement, slag and fly ash as slurry wall materials. *Proc. Specialty Conf. (Geoenvironment 2000) of the American Society of Civil Engineers*, New Orleans, LA, pp. 1237–1249.
Koerner, R. M., Halse, Y. H., and Lord, A. E. (1990). Long-term durability and aging of geomembranes. *Proc. ASCE Geotech. Eng. Div. Specialty Conf. on Waste Containment Systems*, San Francisco, CA, pp. 106–134.
Koerner, R. M., and Richardson, G. N. (1987). Design of geosynthetic systems for waste disposal. *Proc. Geotech. Eng. Div. Speciality Conf. on Geotechnical Practice for Waste Disposal*, Ann Arbor, MI, pp. 65–86.
Mitchell, J. K. and Madsen, F. T. (1987). Chemical effects on clay hydraulic conductivity. *Proc. ASCE Geotechnical Engineering Division Specialty Conf.*, Ann Arbor, MI, pp. 87–116.
Moo-Young, H. K., and Zimmie, T. F. (1995). Design of landfill covers using paper mill sludges. *Proc. Research Transformed into Practice: Implementation of NSF Research*, J. Colville and A. M., Amde, eds., pp. 16–28.
Mulkey, L. A., Salhotra, A., and Brown, L. (1989). Risk-based approach to evaluation of groundwater contamination from land-based waste disposal. In a Draft Monograph Prepared by the Groundwater Risk Assessment Task Committee for the American Society of Civil Engineers, pp. 1–12.
Murray, R. S., and Quirk, J. P. (1982). The physical swelling of clay in solvents. *Soil Sci. Soc. Am. J.*, 46:865–868.
Pask, J. A., and Davis, B. (1945). Thermal analysis of clays and acid extraction of alumina from clays, Technical Paper 644. U.S. Bureau of Mines, U.S. Department of the Interior, Denver, CO, pp. 56–58.
Pavilonsky, V. M. (1985). Varying permeability of clayey soils. *Proc. 11th Int. Conf. on Soil Mechanics and Foundation Engineering*, San Francisco, CA, pp. 1213–1216.
Peirce, J. J., Sollfors, G., and Peel, T. A. (1987). Effects of selected inorganic leachates on clay permeability. *J. Geotech. Eng.*, ASCE, 13(8):915–919.
Rowe, R. K., and Booker, J. R. (1991). Modeling of two-dimensional contaminant migration in a layered and fractured zone beneath landfills. *Can. Geotech. J.*, 28:338–352.
Sarsby, R. W., and Finch, S. (1995). The use of industrial byproducts to form landfill caps. *Proc. Conf. (Green '93) on Geotechnics Related to the Environment*, Bolton, U.K., pp. 267–273.
Selig, E. T., and Mansukhani, S. (1975). Relationship of soil moisture to the dielectric property. *J. Geotech. Eng. Div., ASCE*, 101(GT8):755–771.
Smith, J. A., and Li, J. (1995). Organobentonites as components of earthen landfill liners to minimize contaminant transport. *Proc. Conf. (Geoenvironment 2000) of the American Society of Civil Engineers*, New Orleans, LA, pp. 806–814.
Tchobanoglous, G., Theisen, H., and Vigil, S. (1993). *Integrated Solid Waste Management*. McGraw-Hill, New York.

Uppot, J. O., and Stephenson, R. W. (1988). Permeability of clays under organic permeants. *J. Geotech. Eng., ASCE*, 115(1):115–131.

U.S. EPA. (1980). Lining of waste impoundment and disposal facilities, Technical Guidance Document SW-870. Office of Water and Waste Management, U.S. Environmental Protection Agency, Washington, DC.

U.S. EPA. (1991). Groundwater monitoring, SW-86, Office of Solid Waste, U.S. Environmental Protection Agency, Washington, DC, chap. 11.

U.S. EPA. (1993a). Subsurface characterization and monitoring techniques: a desk reference guide, Vol. 11: Vadose Zone, field Screening and Analytical Methods, EPA/625/R93/003b. Office of Research and Development, U.S. Environmental Protection Agency, Washington, DC.

U.S. EPA. (1993b). Use of airborne, surface and bore hole geophysical techniques at contaminated sites: A reference guide, EPA/625/R/007. Office of Research and Development, U.S. Environmental Protection Agency, Washington, DC.

Walther, E. G., Pitchford, A. M., and Olhoeft, G. R. (1986). A strategy for detecting subsurface organic contaminants. *Proc. NWWA/API Conf. on Petroleum Hydrocarbons and Organic Chemicals in Groundwater*, Houston, Texas, pp. 357–381.

Whang, J. M., and Committee. (1995). Chemical-based barrier materials. In: *Assessment of Barrier Containment Technologies*, R. R. Rumer and J. K. Mitchell, eds. National Technical Information Service, pp. 211–246.

Index